Advances in Intelligent Systems and Computing

Volume 455

Series editor

Janusz Kacprzyk, Polish Academy of Sciences, Warsaw, Poland
e-mail: kacprzyk@ibspan.waw.pl

About this Series

The series “Advances in Intelligent Systems and Computing” contains publications on theory, applications, and design methods of Intelligent Systems and Intelligent Computing. Virtually all disciplines such as engineering, natural sciences, computer and information science, ICT, economics, business, e-commerce, environment, healthcare, life science are covered. The list of topics spans all the areas of modern intelligent systems and computing.

The publications within “Advances in Intelligent Systems and Computing” are primarily textbooks and proceedings of important conferences, symposia and congresses. They cover significant recent developments in the field, both of a foundational and applicable character. An important characteristic feature of the series is the short publication time and world-wide distribution. This permits a rapid and broad dissemination of research results.

Advisory Board

Valentina Emilia Balas · Lakhmi C. Jain
Xiangmo Zhao
Editors

Information Technology and Intelligent Transportation Systems

Volume 2, Proceedings of the 2015 International Conference on Information Technology and Intelligent Transportation Systems ITITS 2015, held December 12–13, 2015, Xi'an China

Editors
Valentina Emilia Balas
Department of Automation and Applied Informatics, Faculty of Engineering
Aurel Vlaicu University of Arad
Arad
Romania

Xiangmo Zhao
School of Information Engineering
Chang'an University
Xi'an
China

Lakhmi C. Jain
Bournemouth University
Poole
UK

ISSN 2194-5357 ISSN 2194-5365 (electronic)
Advances in Intelligent Systems and Computing
ISBN 978-3-319-38769-7 ISBN 978-3-319-38771-0 (eBook)
DOI 10.1007/978-3-319-38771-0

Library of Congress Control Number: 2016945142

Printed on acid-free paper

This Springer imprint is published by Springer Nature
The registered company is Springer International Publishing AG Switzerland

Preface

These volumes constitute the Proceedings of the 2015 International Conference on Information Technology and Intelligent Transportation Systems (ITITS 2015) held in Xi'an, China during December 12–13, 2015. The Conference ITITS 2015 was sponsored by Shaanxi Computer Society and co-sponsored by Chang'an University, Xi'an University of Technology, Northwestern Poly-technical University, CAS, Shaanxi Sirui Industries Co., Ltd.

The book covers a broad spectrum of intelligent techniques, theoretical and practical applications employing knowledge and intelligence to find solutions for intelligent transportation systems and other applications.

The conference papers included in these proceedings, published post conference, were grouped into the following parts:

Volume I—Part II: Proceedings Papers: Theory Research in Intelligent Transportation Systems
Volume I—Part III: Proceedings Papers: Application and Technologies in Intelligent Transportation
Volume II—Part I: Proceedings Papers: Management Issues on Intelligent Transportation
Volume II—Part II: Proceedings Papers: Information Technology, Electronic and Control System

In ITITS 2015, we had 12 eminent keynote speakers: Prof. Asad J. Khattak (USA), Prof. Robert L. Bertini (USA), Prof. Heng Wei (USA), Prof. Ping Yi (USA), Prof. Haizhong Wang (USA), Prof. Jonathan Corey (USA), Prof. Zhixia (Richard) Li (USA), Prof. Guohui Zhang (USA), Prof. Luke Liu (USA), Prof. Yu Zhang (USA), Prof. Valentina E. Balas (Romania), and Prof. Lakhmi C. Jain (UK). Their summaries talks are included in this book.

Intelligent transport systems vary in technologies applied, from basic management systems to more application systems, and information technology also plays tightly with intelligent transportation systems including wireless communication, computational technologies, floating car data/floating cellular data, sensing technologies, and video vehicle detection; technologies of intelligent transportation systems also include topics from theoretic and application such as emergency vehicle notification systems, automatic road enforcement, collision avoidance systems, and some cooperative systems. The conference also fostered cooperation among organizations and researchers involved in the merging fields, invite worldwide well-known professors to further explore these topics, and discuss in-depth technical presentations with the presenters, 12 invited speakers, and over 200 participants. The conference absorbed overwhelm 330 submissions from 5 countries and regions, and each paper was under double peer reviewed by at least 3 reviewers, and more than 120 papers were accepted finally.

We would like to thank the authors of the submitted papers for keeping the quality of the ITITS 2015 conference at high levels. The editors of this book would like to acknowledge all the authors for their contributions and also the reviewers. We have received an invaluable help from the members of the International Program Committee and the chairs responsible for different aspects of the workshop.

Special thanks go to Janusz Kacprzyk (editor in chief, Springer, Advances in Intelligent Systems and Computing Series) for the opportunity to organize this guest-edited volumes.

We are grateful to Springer, especially to Dr. Thomas Ditzinger (senior editor, Applied Sciences & Engineering Springer-Verlag) for the excellent collaboration, patience, and help during the evolvement of this volume.

We hope that the volumes will provide useful information to professors, researchers, and graduated students in the area of intelligent transportation.

Arad, Romania — Valentina Emilia Balas
Poole, UK — Lakhmi C. Jain
Xi'an, China — Xiangmo Zhao

Contents

Part II Information Technology, Electronic and Control System

Part I
Management Issues on Intelligent Transportation

Design of Comprehensive Quality Management System Based on SSI

Zhang Hui-li, Zhou Ming-xia, Zhang Shu, Yuan Shuai and Cheng Xin

Abstract In this paper, we studied the SSI frame, and proposed a new plan of the comprehensive quality Management System by means of the advantage of SSI frame, which needs less codes. In order to solve the problem which the system cannot be compatible with the different version of the browser, we imported the JQuery technology. In order to make the user interface of the system more beautiful and concise, we designed the user interface by means of the EsayUI technology.

Keywords JQuery · EasyUI · SSI · Ajax · mySQL

1 Introduction

The comprehensive quality Management System has been used successfully in our jobs of the university college students' comprehensive quality management, and has been widely welcomed by the teachers and students of our school. The application of the software is significant for the plan of promoting college students quality. The software promotes the college students to take an active part in the activity. The software provides an effective way for the credit quantitative management in the quality education.

Z. Hui-li (✉) · Z. Ming-xia
Zhengzhou Electric Power College, Zhengzhou 450000, China
e-mail: huiliwelcome@yeah.net

Z. Ming-xia
e-mail: nokia6108@126.com

Z. Shu · Y. Shuai · C. Xin
Henan Shangqiu Power Supply Company, Shangqiu 476000, China
e-mail: 6363512@qq.com

Y. Shuai
e-mail: 510582939@qq.com

C. Xin
e-mail: 12303828@qq.com

V.E. Balas et al. (eds.), *Information Technology and Intelligent Transportation Systems*, Advances in Intelligent Systems and Computing 455,
DOI 10.1007/978-3-319-38771-0_1

The research content of the system is based on the object oriented software engineering method.

2 Related Technology

2.1 JQuery

JQuery is a fast, small, and feature-rich JavaScript library. It makes things like HTML document traversal and manipulation, event handling, animation, and Ajax much simpler with an easy-to-use API that works across a multitude of browsers [1]. With a combination of versatility and extensibility, jQuery has changed the way that millions of people write JavaScript. One important thing to know is that jQuery is just a JavaScript library. All the power of jQuery is accessed via JavaScript, so having a strong grasp of JavaScript is essential for understanding, structuring, and debugging the code [2].

2.2 EasyUI

Easyui is a collection of user interface plugin based on jQuery. Easyui provides the necessary functions for creating the modern interactive JavaScript application [3].

Using easyui you don't need to write a lot of code, you only need by writing some simple HTML tags, you can define the user interface. Easyui is a complete framework supporting HTML5 pages perfectly. Easyui saves your time and scale of the web development. Easyui is very simple but powerful.

2.3 SSI Framework

The SSI framework is the three layers architecture including the struts2, spring and ibatis framework. They are responsible for the interaction and collaboration between each layer, so as to realize the whole web implementation and integration of the function [4].

The Struts is mainly responsible for data transfer and control at present, the spring is relying on its strong dependency injection technology can realize the function such as bean managed and integration. Of course this is just the tip of the iceberg of the spring function, as a kind of lightweight and ibatis OR Mapping framework, provides the realization of automation object relational Mapping, higher degree of freedom relative to hibernate.

2.4 *Ajax Technology*

The Ajax is Asynchronous JavaScript and XML, and is the web development technology in order to create interactive web application. Ajax technology is the set of all the running JavaScript technology in the browser [5].

3 Requirement Analysis of System

In the early stage of the software development developers need to communicate with the users in order to determine the software function modules, and the software requirements specification.

The software requirements specification guide the jobs of design, detailed design and coding, so the software requirements specification is very important in the process of software development.

The system administrator can set small modules in each module, and administrators at all levels can release all kinds of activities, submit the activities for students to choose from. The credit summary must tally with the credit of standards at the end of the semester, and the student can achieve the credit of the quality education, as well as other subjects in this semester grades. If the student achievement is unqualified, the student can choose all sorts of activities in the next semester.

4 Analysis and Design of System

4.1 *General Design*

The administrator role has the highest authority in the system, and can achieve the function as shown in Fig. 1.

4.2 *Detail Design*

The main work of the detailed design is to build the development framework building. The system has realized the detailed design by means of the software development ideas of high cohesion and low coupling. The development of each package adopts MVC architecture implementation. The MVC includes presentation layer, control layer, business logic layer, and achieves the high cohesion and low coupling.

The package files function of the system includes is shown in Fig. 2.

The structure description of the main packages is as follows:

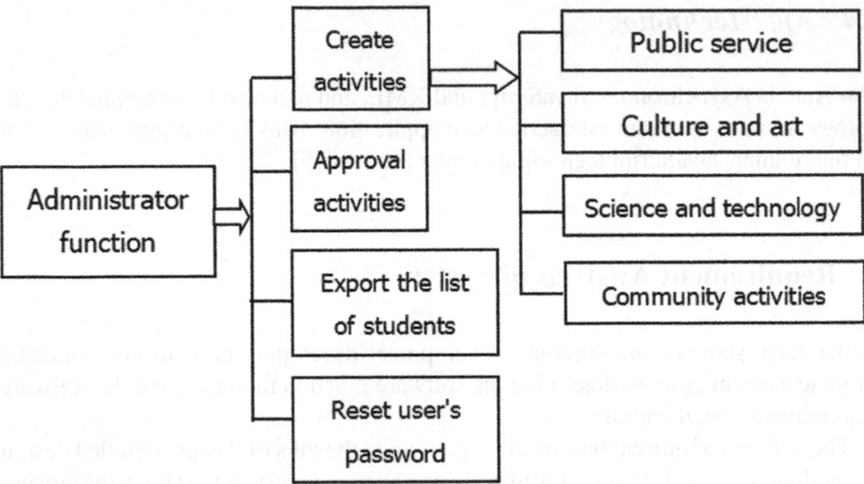

Fig. 1 The authority of the administrator

Fig. 2 The authority of the administrator

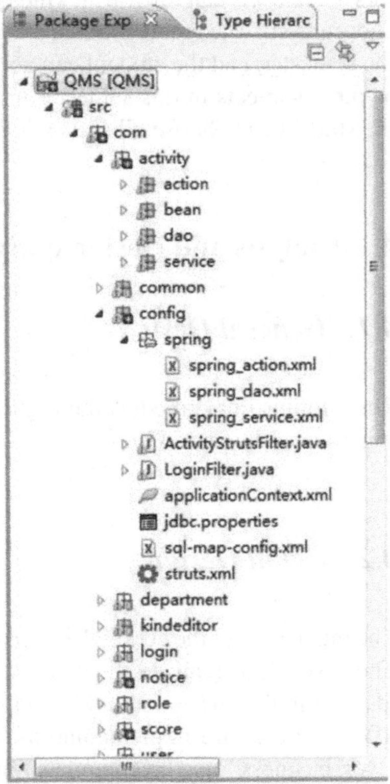

Activity package: the package completes the function of creating activities, and the function of examination and approval activities. Common package: it is the public package, the package completes the function of viewing activities. The student roles, the administrator role, head of characters all use this package.

Config package: it is the system setup package, this package completes the role management, user management functions. Department package: this package completes the functions of adding and managing departments. Login package: this package completes the login system. Score package: it is the credit management package, this package completes the credit allocation and the credit authentication management.

5 Database Design

The Database of the system is based on the mySQL database management platform. The mySQL database management system is the very popular cross-platform database management tools, widely used in the B/S structure program design. The database of the system in MySQL is shown in Fig. 3.

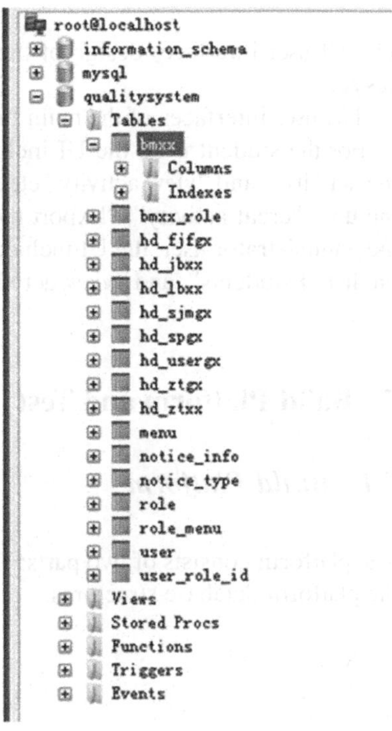

Fig. 3 The authority of the administrator

Table 1 The table bmxx field information

Field	Comment
bmxx-role-id	Role number
bmxx-id	Department number

Table 2 The table hdfjf field information

Field	Comment
bmxx-role-id	Role number
bmxx-id	Department number
fjScore	Extra point
bz	Note
cjr	Creation id
cjsj	Creation time

The field information of the department information table bmxx is in the Table 1. The field information of the bonus points relationship table hdfjf is in the Table 2.

6 UI Design

The UI(user interface) design of the system is based on the technology of Jquery, easyui.

The user interfaces of the main function module are as follows.

For the student user, the UI includes “Individual management”, “Participate in the activity”, and “view activity” etc. For the user which is a department head, the UI includes “creat activity”, “Export the list of students”, and “view activity” etc. For the administrator user, the UI includes “approval activity”, “creat activity”, “Export the list of students”, and “view activity” etc.

7 Build Platform and Test

7.1 Build Platform

The platform consists of two parts: Tomcat, the Web services platform. MySQL5.1, the platform database structures.

7.2 Software Testing

Before the testing system we need to configure test environment.

Hardware environment includes a server and several clients. Software environment includes network operating system, SQL Server database management system, and TOMCAT Server software. Unit test settles five aspects of the problem: local data interfaces module structure, boundary conditions, independent path and error handling [6].

Function test checks whether program function according to the provisions of the requirement, test each functional whether there are omissions, and the detection performance characteristics whether meet the requirements. Function test detects whether interface is wrong, and detectes whether data structure or external database access is wrong.

The task of the pressure test includes that this system normal operation can support 300 terminal parallel work, and the pressure test 1000 terminals of parallel work.

7.3 Choice of Development Tools and Frameworks

The development tools and frameworks of this system are as follows:

The front desk technology includes Jquery, javascript, ajax, and easyui technology. The database technology includes mysql. The background technology includes JSP, struts 2 + Spring + ibatis framework. The hosting tool is SVN or Subversion.

8 Conclusion

At present, the new technologies of the Web application develop more and more quickly.

This system make use of the advanced software development framework ssi. This system designs the evaluation index system of the college students comprehensive quality.

The scientific evaluation method is based on the credit system. Students can choose activities online.

The system can be compatible with the version of the browser by means of the technology of JQuery and EsayUI. Also, JQuery and EsayUI make the interface display more beautiful and concise.

This paper solved some problems of Web application development, but the level have some limitations and need to have been the perfect.

References

1. Qianjin TAN, Weimin GOU, Peili SUN (2015) The research on the construction of monitoring and evaluation system for the operation of marine economy in liaoning province. Asian Agric Res 39–43
2. Hong M (2013) Tech services on the web: jquery mobile. Techn Serv Q 302:231–232. www.jquerymobile.com
3. Parida BK, Panda S, Misra N, Panda PK, Mishra BK (2014) BBProF: an asynchronous application server for rapid identification of proteins associated with bacterial bioleaching. Geomicrobiol J 314:299–314
4. Shahzad F (2013) Satellite monitoring of wireless sensor networks. Procedia Comput Sci 21:479–484
5. Song XM, Pan W, Chen LY, Song X, Li XD (2014) A web application for poloidal field analysis on HL-2M. Fusion Eng Des 89(5):572–575
6. Kulkarni AT, Mohanty J, Eldho TI, Rao EP, Mohan BK (2014) A web GIS based integrated flood assessment modeling tool for coastal urban watersheds. Comput Geosci 64:7–14

Design of Active Safety Warning System for Hazardous Chemical Transportation Vehicle

Guiping Wang, Lili Zhao, Yi Hao and Jinyu Zhu

Abstract As the hazardous chemical transportation traffic accident is getting more and more serious, this paper designs a kind of active safety warning system for hazardous chemical transportation vehicle. The content of this paper mainly includes the composition and principle, establishment of theoretical model of safety distance, design of avoidance control strategy and algorithm and so on. This paper firstly introduces the composition and function of the active safety warning system for hazardous chemical transportation vehicle, and claries its working principle. And then establishes a safety distance model according to the theory of vehicle dynamics, and another four models of key vehicle distance which include safety critical, target vehicle, dangerous critical and limit critical. According to the different dynamic models, the paper designs four control strategies including safety control, alarm control, deceleration control and brake control and the algorithm. In this paper, the active safety warning system for hazardous chemical transportation vehicle can help the driver avoid traffic accident in the course of transportation, detect the safety state of the tank, improve the safety performance of the vehicle at the greatest degree and realize the intelligent monitoring of the hazardous chemical transportation vehicle.

Keywords Active safety · Safety distance · Acoustooptic alarm · Hazardous chemical transportation

G. Wang · L. Zhao (✉) · Y. Hao · J. Zhu
School of Electronic and Control Engineering, Chang'an University,
Xi'an 710064, China
e-mail: 1249163704@qq.com

G. Wang
e-mail: gpwang@chd.edu.cn

Y. Hao
e-mail: 294433691@qq.com

J. Zhu
e-mail: 474812097@qq.com

V.E. Balas et al. (eds.), *Information Technology and Intelligent Transportation Systems*, Advances in Intelligent Systems and Computing 455,
DOI 10.1007/978-3-319-38771-0_2

1 Introduction

With the acceleration of the industrialization process, the demand for the hazardous chemical is increasing, the production and transportation of hazardous chemical are increasing year by year. The hazardous chemical is flammable, explosive, corrosive, radioactive or toxic, and so on. In the course of transportation, a traffic accident, high pressure, or high temperature, may lead to leak, burning out, corrosive, poison or other malignant accidents. According to incomplete statistics, hazardous chemical transportation accidents accounted for 30–40 % of the total number of hazardous chemical. A series of hazardous chemical transportation accidents seriously threaten the safety of road transportation and the life and property of the people, resulting in huge economic loss and bad social influence.

Through the analysis of the causes of a large number of hazardous chemical transportation accidents, the main reasons are summarized [1]:

(1) The driver's negligence, fatigue driving or failure of judgment and other reasons will lead to traffic accidents.

(2) The vehicle is at a high speed or doesn't maintain a safety distance on the highway, so when the danger comes the driver can't timely take a brake.

(3) High temperature, large pressure, high liquid level or the discharge valve's abnormal opening may lead to spontaneous combustion, explosion of the hazardous chemical.

In view of the above reasons, this paper designs a kind of active safety warning system for hazardous chemical transportation vehicle based on sensor technology, communication technology and data processing technology, which can realize the intelligent monitoring of road transportation of hazardous chemical and improve the safety level of road transportation of hazardous chemical in China.

2 System Composition and Working Principle

The system is composed of information collection module, data processing module and sound and light alarm module.

2.1 Information Collection Module

Information collection module is composed of speed sensor, laser ranging sensor, temperature sensor, pressure sensor, liquid level sensor, and discharge valve state monitoring switch, used to obtain the speed of the vehicle, relative distance, temperature, pressure, liquid level in the tank, and the discharge valve state and other information.

2.2 *Data Processing Module*

Data processing module receives the signals of the laser sensor and the speed sensor, determines the vehicle travel state according to the safety state of the logic algorithm; receives the singles of temperature sensor, pressure sensor, liquid level sensor and discharge valve state monitoring switch, determines whether it beyond the safety range. At last, the analysis results transmit to the sound and light alarm module.

2.3 *Sound and Light Alarm Module*

The sound and light alarm module is installed in the driving room of the vehicle, with a display and a buzzer. The relative distance and relative speed between the vehicle and the target vehicle and the temperature, pressure, liquid level and the state of the discharge valve in the tank are shown on the display.

2.4 *Working Principle*

When the system starts to work, the pre laser sensor measures the distance between the vehicle and the target, and the vehicle speed sensor which is installed on the vehicle measures the speed of itself [2, 3]. Data processing module receives the above singles and calculates the real time vehicle distance using the vehicle information and target vehicle information according to the safety distance models. Compared to the target distance obtained by the laser sensor, determines whether the driving state is safe and determines the risk level according to analysis results [4]. In the end, the alarm signal is send to sound and light alarm module.

Temperature sensor, pressure sensor, liquid level sensor, and the discharge valve state monitoring switch can be used to receive the singles in the tank. The data processing module receives the singles and judges whether it beyond the safety range and sends alarm signal to sound and light alarm module.

3 Establishment of Theoretical Model of Safety Distance

3.1 *Automobile Braking Process*

From the point when the driver realizes the danger, takes the brake pedal to the stop of the vehicle, the process is composed of five stages [5, 6].

In the Fig. 1, F_p is the maximum braking force, a_{max} is the maximum deceleration.

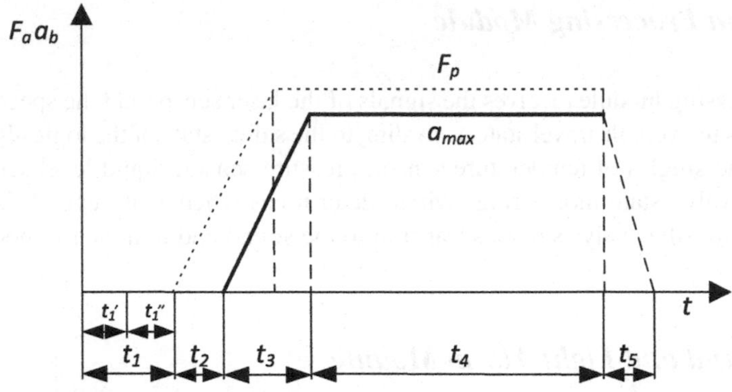

Fig. 1 Automobile brake process

(1) The driver's reaction time t_1: when danger occurs, the driver realizes it after t_1', then moves the foot from the accelerator pedal to the brake pedal after t_1''.

(2) Elimination of braking clearance time t_2: vehicle brake reserve brake clearance to prevent sensitivity, it spends t_2 to produce ground braking force after stepping the pedal.

(3) Build-up time of braking force t_3: when the brake pedal is stepped down by the driver and the brake clearance is overcome, the ground brake force increases linearly, and the braking deceleration increases linearly, too.

(4) Sustained braking phase t_4: in this period, the deceleration doesn't increase and stains the maximum, which is the key stage of the braking process.

(5) Release time t_5: after the vehicle stops, the driver releases the brake pedal, and the braking force gradually reduces to 0 after t_5.

3.2 *Derivation of Braking Distance*

The braking distance is the distance from the danger signal is found by driver to the stop of the vehicle. According to the analysis of Fig. 1, the first four stages of vehicle braking should be calculated [7, 8].

(1) During the period when the driver finds the single and reacts that lasts t_1, the vehicle is at a constant speed state, the distance is:

$$l_1 = v_0 t_1 \tag{1}$$

(2) During the period when the brake clearance is overcome that lasts t_2, the vehicle is at a constant speed state, the distance is:

$$l_2 = v_0 t_2 \tag{2}$$

(3) During the build-up time of braking force period that lasts t_3, the vehicle's braking deceleration increases linearly and the distance is:

$$l_3 = v_0 t_3 - \frac{a_{max} t_3^2}{6} \tag{3}$$

(4) During the sustained braking period that lasts t_4, the vehicle is at a constant deceleration state, the distance is:

$$l_4 = \frac{v_0^2}{2a_{max}} - \frac{v_0 t_3^2}{2} + \frac{a_{max} t_3^2}{8} \tag{4}$$

Ignore other secondary factors, the distance from where the driver receives an emergency stop signal to the point the vehicle stops completely:

$$L = l_1 + l_2 + l_3 + l_4 = v_0 \left(t_1 + t_2 + \frac{t_3}{2} \right) + \frac{v_0^2}{2a_{max}} - \frac{a_{max} t_3^2}{24} \tag{5}$$

As t_3 is too small, $\frac{a_{max} t_3^2}{24}$ is ignored in the theoretical calculation, so the theoretical braking distance is:

$$L = v_0 \left(t_1 + t_2 + \frac{t_3}{2} \right) + \frac{v_0^2}{2a_{max}} \tag{6}$$

3.3 Braking Safety Distance

As shown in Fig. 2, when vehicle A detects two vehicles may be a collision, it will warn the driver to slow down through the sound and light alarm module. Now A is at initial velocity v_1, acceleration a_1; B is at initial velocity v_2, acceleration a_2; the distance is S_1. When they come into another safety state, A is at speed v_1', acceleration a_1'; B is at speed v_2', acceleration a_2'; the distance is S_2. L_0 is the safety distance. Over the whole process, A travels L_1, B travels L_2. Set theory safety distance S', so $S' = L_1 - L_2 + L_0$. Imagine the two vehicles both can reach the maximum deceleration a_{max}.

(1) B is at static state

When B is static, $v_2 = a_2 = L_2 = 0$, $L_1 = v_1 \left(t_1 + t_2 + \frac{t_3}{2} \right) + \frac{v_1^2}{2a_{max}}$. So:

$$S' = v_1 \left(t_1 + t_2 + \frac{t_3}{2} \right) + \frac{v_1^2}{2a_{max}} + L_0 \tag{7}$$

(2) B is at constantly driving state

When B is constantly driving, if A is at a higher speed and doesn't brake, they will make a crash. When A is at the same speed of B, they will be safe. So:

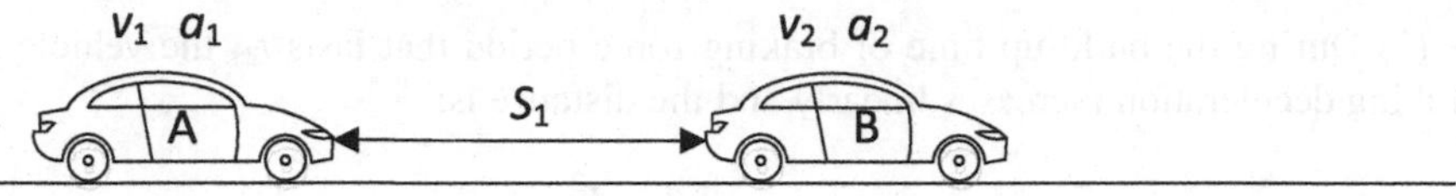

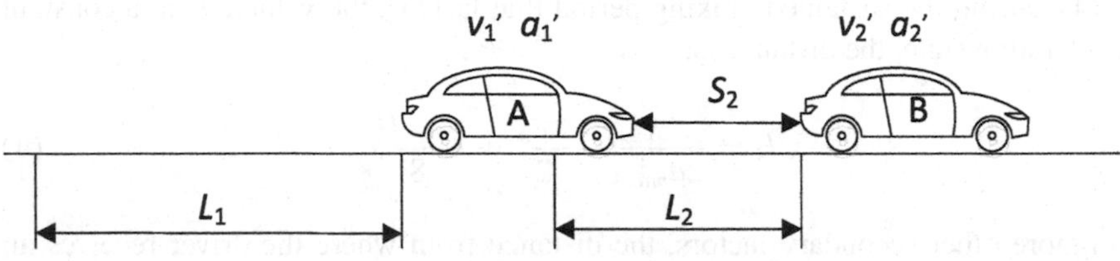

Fig. 2 Braking safety distance model

$$L_1 = v_1\left(t_1 + t_2 + \frac{t_3}{2}\right) + \frac{v_1^2 - v_2^2}{2a_{max}} \tag{8}$$

$$L_2 = v_2\left(t_1 + t_2 + \frac{t_3}{2} + \frac{v_1 - v_2}{a_{max}}\right) \tag{9}$$

$$S' = (v_1 - v_2)\left(t_1 + t_2 + \frac{t_3}{2}\right) + \frac{(v_1 - v_2)^2}{2a_{max}} + L_0 \tag{10}$$

(3) B is at uniform deceleration brake state

B is uniform deceleration brake, if A is at a higher or the same speed and doesn't brake, they will make a crash. So:

$$L_1 = v_1\left(t_1 + t_2 + \frac{t_3}{2}\right) + \frac{v_1^2}{2a_{max}} \tag{11}$$

$$L_2 = v_2\frac{t_3}{2} + \frac{v_2^2}{a_{max}} \tag{12}$$

$$S' = v_1(t_1 + t_2) + (v_1 - v_2)\frac{t_3}{2} + \frac{(v_1^2 - v_2^2)}{2a_{max}} + L_0 \tag{13}$$

If $v_1 = v_2$, then:

$$S' = v_1(t_1 + t_2) + \frac{(v_1^2 - v_2^2)}{2a_{max}} + L_0 \tag{14}$$

Compare the formula (7), (10) and (13), we can find the braking safety distance in formula (13) is the biggest. Considering the hazardous chemical transportation vehicle has high standard for safety, we choose formula (13) as the braking safety distance.

3.4 Safety Distance Model

(1) Safety critical distance

Vehicles keep a certain distance to the vehicle ahead when driving, this distance can ensure that in most cases the vehicle is safe, even sudden obstacles or sudden deceleration of the vehicle ahead will not threaten the safety. We call it safety critical distance d_s. From the above derivation:

$$d_s = v_1(t_1 + t_2) + (v_1 - v_2)\frac{t_3}{2} + \frac{(v_1^2 - v_2^2)}{2a_{max}} + L_0 \tag{15}$$

(2) Target vehicle distance

The safety critical distance d_s can ensure the safety of the vehicle in most cases, but it can lead to a slow or operational failure due to the slow or poor psychological quality of some drivers. The target vehicle distance is composed by safety critical distance and another distance that the vehicle drives during t_0.

$$d_t = v_1(t_0 + t_1 + t_2) + (v_1 - v_2)\frac{t_3}{2} + \frac{(v_1^2 - v_2^2)}{2a_{max}} + L_0 \tag{16}$$

(3) Dangerous critical distance

Safety critical distance d_s is the index that determines whether the vehicle's driving state is safe, when above this value it means that the vehicle is at a safe state, and another one is dangerous critical distance d_d, it can determine whether the driving state of the vehicle is dangerous, when below the value it means that the vehicle is at a dangerous state, and it may have a traffic accident. On the base of safety critical distance, the calculation of the dangerous critical distance is that distance minuses the distance of the vehicle in the driver's reaction stage.

$$d_d = v_1 t_2 + (v_1 - v_2)\frac{t_3}{2} + \frac{(v_1^2 - v_2^2)}{2a_{max}} + L_0 \tag{17}$$

(4) Limit critical distance

Safety critical distance d_s is used to determine whether the driving state is safe, dangerous critical distance d_d is used to determine whether the vehicle's driving state is dangerous. To this correspondence, limit critical distance d_L is the smallest distance from rear-end collisions.

If the vehicle B slows down at deceleration a_2, vehicle A brakes at deceleration a_{max}. If a_2 is small, vehicle A and B will be both out of danger when $v_1 = v_2$ if the distance between them is at least L_0. If a_2 is large, vehicle A is still driving when B has stopped, that is $v_1 > v_2$, vehicle A and B will be both safe if the distance between them is at least L_0 after A has stopped.

Assume that at a certain time two vehicles are at the same speed, that is $v_1 = v_2 = v_3$, then $v_3 = v_1 - a_{max}t = v_2 - a_2 t$, $t = \frac{v_1 - v_2}{a_{max} - a_2}$, $v_3 = \frac{v_2 a_{max} - v_1 a_2}{a_{max} - a_2}$.

When $a_2 < \frac{v_2 a_{max}}{v_1}$, $v_3 > 0$, it means two vehicles are at the same speed when they get out of danger, It's v_3, thus $L_1 = v_1(t_2 + \frac{t_3}{2}) + \frac{(v_1^2 - v_3^2)}{2a_{max}}$, $L_2 = \frac{(v_2^2 - v_3^2)}{2a_2}$. So

$$d_L = v_1\left(t_2 + \frac{t_3}{2}\right) + \frac{(v_1 - v_2)^2}{2(a_{max} - a_2)} + L_0 \tag{18}$$

When $a_2 < \frac{v_2 a_{max}}{v_1}$, $v_3 < 0$, it means v_3 is negative when two vehicles are at the same speed, it's not reasonable, so vehicle A and B have stopped when they get out of danger. Thus $L_1 = v_1(t_2 + \frac{t_3}{2}) + \frac{v_1^2}{2a_{max}}$, $L_2 = \frac{v_2^2}{2a_2}$. So:

$$d_L = v_1(t_2 + t_3) + \frac{v_1^2}{2a_{max}} - \frac{v_2^2}{2a_2} + L_0. \tag{19}$$

4 Avoidance Control Strategy and Algorithm

4.1 *Avoidance Control Strategy*

Avoidance control unit is the core part of the active safety warning system, it calculates real-time critical distance between vehicles after receiving input information. According to avoidance control algorithm, the unit analysis current driving state of the vehicle, and gets the best control scheme to circumvent, conveys the singles to the output corresponding element.

Avoidance control strategy means measures the vehicles take when they avoid the traffic accident in a certain state. The working mode of the active safety warning system is determined by the avoidance control strategies.

(1) Safety control strategy

In general, it's safe when the actual relative distance is greater than safety critical distance. If a period of time (such as 0.5s) will be locked in front of a danger target, it can not only ensure the safety of driving, but also extend the reaction time of the driver, and it's conducive to improve the safety of driving. So in fact, when the actual distance is greater than target vehicle distance, the vehicle is safe and the green indicator light on the display works.

(2) Alarm control strategy

When the actual vehicle distance is less than the target vehicle distance, the system considers that the driving state can not be fully guaranteed safe at this time, but as long as the actual vehicle distance is not less than the safety critical distance, there will be no danger. Therefore, when the actual vehicle distance is between the safety critical distance and target vehicle distance, the system will remind the driver to drive carefully, and at this time there is a low frequency "drops" alarm sound, and the yellow-green indicator light on the display flashes.

(3) Deceleration control strategy

When the alarm is not timely responded or responded error, and the actual vehicle distance is less than the safety critical distance, the system determines the current driving state is dangerous. But as long as the distance is not less than the dangerous critical distance, the dangerous state can be improved by vehicle deceleration. So when the relative distance is between the dangerous critical distance and safety critical distance, the system will remind the driver to decelerate, and at this time there is a higher frequency "drops" alarm sound, and the orange indicator light on the display flashes.

(4) Braking control strategy

After the deceleration control strategy, if the driver doesn't take measures or the target vehicle slows down at a high deceleration or other danger targets appear suddenly, and the relative distance is less than dangerous critical distance, the system will determine that the current driving state is very dangerous, the possibility of accident is very large. So the system will remind the driver emergency brake, and at this time there is a more high frequency "drops" alarm sound, and the red indicator light on the display flashes.

4.2 Avoidance Control Algorithm

Set a safety variable: $SF = \frac{R_{ref}-d_L}{d_s-d_L}$, where, R_{ref} means the relative distance to the target vehicle at the moment, d_L means limit critical distance, d_s means safety critical distance, SF is called safety variable.

(1) Safety state

If $SF > 1$, that is $R_{ref} > d_s$, it means the vehicle is at a safe driving state, and is safer when SF increases, and should take safety control strategy.

(2) Alarm state

Defining variable $m = \frac{d_d-d_L}{d_s-d_L}$, obviously $0 < m < 1$.

If $m < SF \leq 1$, that is $d_d < R_{ref} \leq d_s$, it means the vehicle is at a unsafe driving state, and is more unsafe when SF decreases, and should take alarm control strategy.

(3) Deceleration state

If $0 < SF \leq m$, that is $d_L < R_{ref} \leq d_d$, it means the vehicle is at a dangerous driving state, and is more unsafe when SF decreases, and should take deceleration control strategy.

(4) Braking state

If $SF \leq 0$, that is $R_{ref} \leq d_L$, it means the vehicle is at an extremely dangerous driving state, and is more unsafe when SF decreases, and should take braking control strategy.

4.3 Tank Safety Monitoring Algorithm

After receiving the sensor singles from the tank, the data processing module analysis and processes them.

In order to avoid the vehicle body temperature or pressure measurement's error is too large, the system uses two temperature sensors and two pressure sensors, which are installed in the front and rear of the hazardous chemical transportation vehicle. After receiving the sensor datas, the data processing module calculates the average data of temperature and pressure, then compares the average data with the preset safety range. If the range of safety is exceeded, the sound and light alarm device will be driven.

The singles that the liquid level sensor and discharge valve state monitoring switch send to the data processing module are switch variables. When the liquid level exceeds the preset value or the discharge valve state monitoring switch opens abnormally, the sensor sends a high level signal. The data process module receives it and sends it to the sound and light alarm module.

5 Conclusion

The active safety warning system for hazardous chemical transportation vehicle designed by the paper real-time detects the relative distance, relative speed and the speed of the vehicle. According to the control algorithm, the system determines whether the current vehicle driving state is safe, and timely warns the driver. At the same time, the system can monitor the vehicle tank body temperature, pressure, liquid level and the state of the discharge valve state monitoring switch, if the temperature, pressure or liquid level beyond the preset safety range or the discharge valve state monitoring switch opens abnormally, the system will timely warn to remind the driver.

The system can judge whether the vehicle's driving state is safe correctly and timely and monitor the state of the tank and assist the driver to complete the transportation task and realize the intelligent monitoring of the road transportation of hazardous chemical.

Acknowledgments This work has been supported by Fundamental Research Funds for the Central Universities (0009-2014G1321037).

References

1. Liang T, Beibei Z (2010) Study on experiment platform of integrated steering and braking control for vehicle active safety system. In: Information engineering and computer science (ICIECS). IEEE Press, Wuhan, pp 1–4
2. Herrmann S, Utschick W, Botsch M (2015) Supervised learning via optimal control labeling for criticality classification in vehicle active safety. In: Intelligent transportation systems. IEEE Press, Gran Canaria, Spain, pp 2024–2031
3. Hongzhao D, Weifeng C, Mingfei G (2014) A study on key technologies of vehicle active collision avoidance system based on VII technology. J. Automot Eng 8:28–43
4. Sepulcre M, Gozalvez J, Botsch M (2013) Cooperative vehicle-to-vehicle active safety testing under challenging conditions. J Transp Res Part C Emerg Technol 26:233–255
5. Wang Y, Weifeng Q (2014) Design of an active automotive safety system. J Eng Sci Technol Rev 6:28–43
6. Vandanjon P-O, Coiret A, Lorino T (2014) Application of viability theory for road vehicle active safety during cornering manoeuvres. J Vehicle Syst Dyn 52:244–260
7. Litkouhi BB, Owechko Y, Kukshya V (2012) Imaging sensor technology for intelligent vehicle active safety and driver assistant systems. J Int J Vehicle Autonom Syst 10:198–228
8. Xiaoqin Y, Mingxia W (2014) Safety distance mathematical model of pro-active head restraint based on fuzzy theory. J Appl Mech Mater 1:710–714

Influencing Factors for Carbon Information Disclosure of Chinese Listed Companies Based on Network Media Data

Li Li, He Xi and Liu Dongjun

Abstract This research utilizes company carbon information disclosure results obtained by web crawler from network media as data sources, discusses the factors which influenced carbon disclosure for the SSE social responsibility index. 100 stocks of Shanghai social responsibility index are selected as research samples. The multivariate regression analysis model is applied to study influence factors of carbon disclosure of listed companies. The research results show that the carbon disclosure of the listed companies in China has positive correlation with enterprise size, involved industry, growth ability, enterprise property and whether listed abroad or not. The first three factors have obvious significance, while the carbon disclosure has negative correlation with corporate profitability and liabilities level.

Keywords Carbon information disclosure · Influence factor · Web crawler · Chinese listed companies

1 Introduction

With the global citizens awareness of environmental protection, public disclosure of information and carbon emissions information control become important. The carbon information disclosure has become an important tool for the US Department to

Research works in this paper are financially supported by Research Planning Foundation in Humanities and Social Sciences of the Ministry of Education of China (Grant No. 13YJAZH044) and National Science Foundation of China (Grant No. 61173052).

L. Li (✉) · H. Xi · L. Dongjun
Shenzhen Graduate School, Harbin Institute of Technology,
Shenzhen, People's Republic of China
e-mail: ximlli@126.com

H. Xi
e-mail: 916680813@qq.com

L. Dongjun
e-mail: laueastking3168@163.com

V.E. Balas et al. (eds.), *Information Technology and Intelligent Transportation Systems*, Advances in Intelligent Systems and Computing 455,
DOI 10.1007/978-3-319-38771-0_3

regulate the greenhouse gas emissions for major companies [1]. In the global situation of the low-carbon, many international companies have begun to voluntarily disclose information on carbon emissions, and actively participate in mitigation actions.

This phenomenon has drawn wide attention of scholars, and they have studied the global climate change and corporate carbon disclosure. However, studies on carbon disclosure of enterprises in China were still rare. Most of the study samples were based on CDP project, annual reports, and CSR reports, by which indicators were extracted to measure the corporate carbon disclosure levels. So these researches may have limitations and their values also were restricted. With the rapid development of the Internet, online media has become a new media for corporate carbon disclosure. Its role should not be overlooked. Williams [2] believed that the disclosure on the way of online media had several advantages compared to other ways for disclosures. The online media was timely, fast and cheaper, and is not restricted by the space. There-fore, this study used the web crawler to extract carbon information disclosure in enterprises on the Internet, and discussed the factors which influenced carbon disclosure for the SSE Social Responsibility Index based on empirical researches.

2 Research Questions and Hypotheses

This study summarized the results of previous researches to make the hypothesis. The influencing factors were analyzed to affect the carbon information disclosure levels in the following seven aspects.

(1) Industry characteristics

The companies that belong to heavy pollution industry and are regulated tend to disclose more information on carbon emissions. The enterprises are defined to be heavy polluting companies according to "Guidelines of listed companies industry classification" [3].

Shamima studied the Australian sample enterprises, and also found that the high pollution industry enterprises are more inclined to disclose carbon information [4]. However, Audrey and other scholars found that companies with low environmental performance from polluting industries are not significantly positive response to climate change phenomenon [5]. In order to verify the correlation, we propose the following hypothesis 1:

H1: Industry characteristics and carbon disclosure levels were positively correlated.

(2) Company size

The larger the company, the ownership structure is more complex, the company is more powerful, and also it has the more strong influence on society which leads to the greater agency costs accordingly.

Warsame and other scholars and Chen [6] researchers found that a sufficient condition of the disclosure of information at high level of the company is that the company is big enough. However, the common views of these studies are all based on the samples of foreign countries. For Chinese listed companies, the company size may affect the levels of their carbon disclosure. Thus we propose the following hypothesis 2:

H2: Company size and the carbon disclosure levels were positively correlated.

(3) Corporate profitability

Choi found correlations between the enterprise profitability and carbon disclosure levels are not significant [7]. Shamima found that the profitability was not affecting factors carbon information disclosure levels based on the sample enterprises in Australia [4]. To further validate the correlation, we propose the following hypothesis 3:

H3: The profitability and the carbon disclosure levels were positively correlated.

(4) Debt level

The greater the extent of the company's debt, the corresponding financial risk is al-so higher. According to the contract theory, the listed companies with higher proportion of debt not only need to disclose the carbon disclosure to obtain the trust and support of the shareholders, but also need to prove its legitimacy to the creditors, and strictly comply with the requirements of debt and contractual obligations to meet some requirements of the creditors [8].

Cormier [9] proved there were negative correlation between the extent of corporate debt and its environment information disclosure levels. However, Gu argued [10] that there were significant positive correlation between carbon disclosure and the level of corporate debt. In order to verify the affecting factors between the level of corporate debt and the level of disclosure, we propose the following hypothesis 4:

H4: Corporate debt levels and carbon disclosure levels were positively correlated.

(5) Enterprise nature

The enterprises with the nature of state-owned were directly under the control of state and government, and they often faced more political pressure, especially the large-scale state-owned enterprises, which bared greater public expectations.

Choi [7] held the opposing views, that is, the operating condition, ownership structure and governance effectiveness of a listed company might be of obvious differences because of the different ownership of the company's largest shareholder. They believed that if the company's largest shareholder was non-state ownership, the corporate management and governance would be more flexible and more efficient. Based on the research results, we propose the following hypothesis 5:

H5: the companys carbon information disclosure has a positive correlation with enterprise nature.

(6) Growth opportunity.

The capital demand of the enterprises with strong growth opportunities was relatively large, so the financing demand was more intense. Once the relevant negative news occurred, the enterprises may affect the financing cost and corporate values. Therefore, the enterprises at this stage tended to attach great importance to maintain the good enterprise image, and to obtain more investors. Nowadays, the problem of environmental pollution is increasingly serious, and the contradiction among the stakeholders also gradually highlights. The related business information provided by financial statements may gradually be unable to meet the needs of investors. So the carbon disclosure could get the trust of shareholders and other stakeholders, reduce financing cost, and obtain more capital support for the healthy development. However, some scholars found Growth opportunities and carbon company information disclosure levels were with inversely proportional, and the enterprises with strong growth opportunities had low levels of carbon disclosure [11]. Due to the uncertainty of previous studies, we propose the following hypothesis 6:

H6: the companys carbon information disclosure has a positive correlation with the growth opportunities.

(7) Whether listed overseas.

Enterprises in the overseas listing of the situation, increased their financing channels, and obtained more regulation. For the enterprises, whose financing opportunity increased, would also face greater competition pressure at the same time. According to the theory of legitimacy, the enterprise through the carbon disclosure can enhance the good social image, and obtain more stakeholders trust. Carbon emissions trading markets abroad were more sound than domestic, so there were more stakeholders to understand enterprises' carbon emissions and other related information. This prompted the enterprises to take the initiative to disclose carbon information, so as to raise more capital to promote the rapid development itself [11]. However, Choi [7] found that the correlation between whether the enterprise belonged to overseas listing corporation and enterprise carbon disclosure was not obvious. Therefore, we propose the following hypothesis 7:

H7: the companys carbon information disclosure has a positive correlation with the company listed overseas.

3 Empirical Results and Analysis

3.1 Data and Descriptive Statistic Analysis

The 100 companies in Shanghai in 2012 were selected as the research samples in this study. The information that enterprises disclosed through the network media was gotten with the web crawler. The carbon disclosure levels of the enterprises were measured according to the search results. This study utilized the python programming language to implement the web crawler and the functions of the system. The system structure was shown in Fig. 1.

Carbon disclosure levels were set as dependent variable. The independent variable included: industry characteristics, company size, corporate profitability, debt level, enterprise nature, growth opportunities and whether listed overseas. The influencing factor model for carbon information disclosure was shown in Fig. 2.

All the variables were defined as follows, which were shown in Table 1, Descriptive statistics results were shown in Table 2.

As was shown in Table 2, the average value of CDI equaled to 0.0491, the median was equals to 0.05, the standard deviation was 0.03872, the maximum was 0.15, and the minimum was 0. That even for the sample companies with the Shanghai social

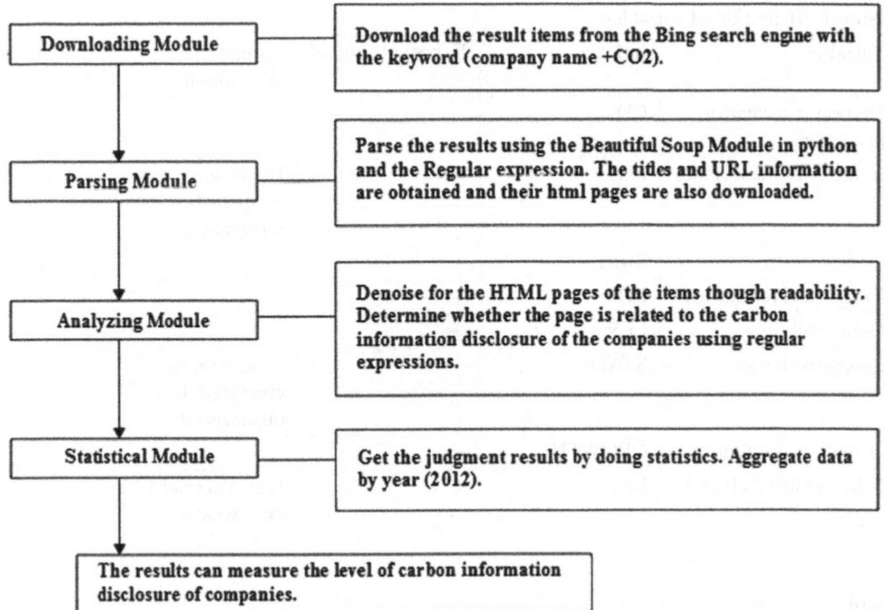

Fig. 1 System structure of data processing

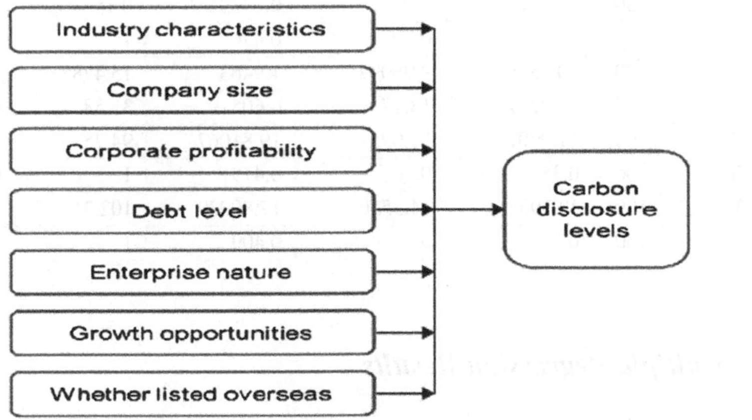

Fig. 2 Influencing factor model for carbon information disclosure

responsibility index, the carbon disclosure situation was still not very optimistic. The carbon information disclosure levels of the listed companies in China were not high. There are 35 sample enterprises in China, accounting for 35 % of all the valid samples; 21 companies were listed on the Shanghai stock exchange, and also listed in Hong Kong or other stock exchanges overseas; 34 % of the enterprise belonged to the metallurgical, chemical, petrochemical or other heavily polluting industries.

Table 1 Definition of variables

Variable	Indicator	Expected symbol	Variable description
Carbon in formation disclosure	CDI		
Industry characteristics	IND	+	Heavy pollution industry 1, otherwise 0
Company size	SIZE	+	
Corporate profitability	ROE	+	
Debt level	LEV	+	
Enterprise nature	STATE	+	State-owned enterprise 1, otherwise 0
Growth opportunities	GROWTH	+	
Whether listed abroad or not	LIST	+	Listed abroad 1, otherwise 0

Table 2 Descriptive statistics

Variable	N	Mean	Median	Standard deviation	Maximum	Minimum
CDI	100	0.0491	0.05	0.03872	0.15	0
IND	100	0.34	0.00	0.476	1	0
SIZE	100	10.45168	9.96097	1.9463	15.4782	7.5191
ROE	100	13.9227	13.17	6.60519	33.58	0.25
LEV	100	61.0907	61.095	19.81067	94.75	22.22
STATE	100	0.35	0.00	0.479	1	0
GROWTH	100	17.5935	14.3550	17.80029	102.31	–8.85
LIST	100	0.21	0	0.409	1	0

3.2 *Multiple Regression Results*

Carbon disclosure levels were set as the dependent variable. The independent variable included: industry characteristics, company size, corporate profitability, debt level, enterprise nature, growth opportunities and whether listed overseas. A logistic model was developed to examine the factors that influenced disclosure of carbon information:

$$CDI_i = \beta_0 + \beta_1 IND_i + \beta_2 SIZE_i + \beta_3 STATE_i + \beta_4 LEV_i + \beta_5 GROWTH_i + \beta_6 ROE_i + \beta_7 LIST_i + \varepsilon$$

Table 3 Regression results

Variable		IND	SIZE	ROE	LEV	STATE	GROWTH	LIST
Expected symbol		+	+	+	+	+	+	+
Explanatory	Coefficient	0.251	0.645	–0.085	–0.233	0.061	0.189	0.007
Variable	P value	0.006	0	0.347	0.053	0.542	0.034	0.94

The results of multiple regression results were shown in Table 3. In addition, $R^2 = 0.379$ Adjusted-$R^2 = 0.332$, F-statistic= 8.017 ($P = 0.0000$), D.W stat=1.960.

3.3 Analysis and Discussion

From Table 3, F value equaled to 8.017, $P = 0.000 < 0.01$, this suggested that the regression analysis was significant correlation at the 0.01 level. In addition, all the independent variables of the tolerance values were within the range of 0.1–1. This indicated that there was no multicollinearity between variables in regression models. The DW value was close to 2, which illustrated there was no autocorrelation among variables. The data comply with the requirements of multiple linear regressions.

Based on the multiple regression results, the company size had a significant correlation with the carbon information disclosure at 1 % ($P < 0.01$) level. Industry characteristic had a significant correlation with carbon information disclosure at 1 % level. The carbon information disclosure of the company between heavy pollution and the light pollution industry had a significant difference. The company with strong growth ability had a high carbon disclosure level. The company at the stage of fast development would have a large demand for money. The carbon disclosure could raise their company's image and competitiveness, and attract stakeholders to gain more financing opportunities. Corporate debt level had a non-significant negative correlation with the carbon information disclosure levels. Both of them were significant correlation at 5 % level. Company profitability had a negative correlation with the carbon disclosure, and its effect was not significant. The enterprises with poor profitability may disclose more carbon information to deflect attention of the stakeholders. The nature of the company has a non-significant correlation with carbon information disclosure. Whether companies listed overseas also had a non-significant correlation with the carbon information disclosure. The abroad market has more strict environmental regulation. In the current background of rapid development of information technology, listed enterprises has no significant effect on the carbon disclosure levels.

4 Conclusion

In the framework of voluntary disclosure, the actual situation of carbon information disclosure in enterprises through the network media was measured. Empirical researches were made to study the influencing factors of carbon information disclosure of the listed companies by using the multiple linear regression method. 100 companies with social responsibility index in the Shanghai Stock Exchange were selected as the research samples. Some important factors were comprehensive considered. It was found that the company size, industry characteristic and growth opportunity had a significant correlation with the carbon information disclosure. The corporate profitability and debt level had a negative correlation with carbon disclosure, and the correlation of the former was not significant. In addition, whether companies listed overseas also had a non-significant correlation with the carbon information disclosure.

References

1. Li L, Yang YH, Niu GH, Sun L (2014) A research review on carbon information disclosure. Sci Technol Manag Res 7:234–240
2. William SM, Pei HW (1999) Corporate social disclosures by listed companies on the web sites: an international comparison. Int J Acc 34(3):389–414
3. Kuo L, Vivan Chen YJ (2013) Is environmental disclosure an effective strategy on establishment of environmental legitimacy for organization? Management 7(51):1463–1487
4. Shamima H, Craig D, Robert I (2013) Disclosure of climate change-related corporate governance practices. Qld Univ Technol 1:34
5. Audrey WH, Wang T (2013) Does the market value corporate response to climate change. Omega 41(2):195–206
6. Chen JP, Charles P, Jaggi BL (2000) The association between independent nonexecutive directors of family control and disclosure. J Acc Public Policy 13(1):16–37
7. Choi B, Lee D, Jim P (2013) An analysis of Australian company carbon emission disclosures. Pac Acc Rev 1(25):58–76
8. Freedman M, Jaggi B (2005) Global warming, commitment to the kyoto protocol, and accounting disclosures by the largest global public firms from polluting industries. Int J Acc 40(3):215–232
9. Cormier D, Gordon IM (2001) An examination of social and environmental reporting strategies. Acc Audit Account J 5(14):587–616
10. Gu ZJ, Guo XM, Jian LX (2012) An empirical study on the factors affecting carbon ac-counting information disclosure of Chinese listed companies. In: Proceedings of international conference on low-carbon transportation and logistics, and green buildings, 995–1000
11. Junrui Zhang, Huiting Guo (2008) Research on influential factors on disclosure of environmental accounting information -practical data from public listed companies in the chemical products industry in China. Stat Inf Forum 23(5):32–39

A Novel Object Localization Method in Image Based on Saliency Map

Ying Wang, Jinfu Yang, Fei Yang and Gaoming Zhang

Abstract Saliency has been widely applied to detecting the most attractive object in an image. This paper presents a novel method based on saliency map for object localization in an image. We regard saliency map computation as a preprocessing step, which is obtained using a discriminative regional feature integration approach. Then, the saliency map is processed to highlight more saliency contour and remove some subordinately salient points using region-grow segmentation algorithm. Finally, saliency object is located with efficient sub-window search algorithm (ESS) in the binarized saliency map. This approach is able to identify the exact location and roughly the size of saliency object in an image. The performance evaluation using MSRA-B dataset demonstrates our approach performs well in object localization in images.

Keywords Saliency map · Efficient subwindow search · Region growing · Object localization

This work is partly supported by the National Natural Science Foundation of China under Grant no. 61201362, 61273282 and 81471770, Graduate students of science and technology fund under no. ykj-2004-11205.

Y. Wang (✉) · J. Yang · F. Yang · G. Zhang
Department of Control and Engineering, Beijing University of Technology, No. 100 Chaoyang District, Beijing 100124, People's Republic of China
e-mail: 15650750952@163.com

J. Yang
e-mail: 494761664@qq.com

F. Yang
e-mail: 354003021@qq.com

G. Zhang
e-mail: 604581234@qq.com

V.E. Balas et al. (eds.), *Information Technology and Intelligent Transportation Systems*, Advances in Intelligent Systems and Computing 455,
DOI 10.1007/978-3-319-38771-0_4

1 Introduction

Recent years object recognition for natural images has gained great progress. But many state-of-art object recognition methods cannot decide the exact location and size of an object in an image, which only deal with a binary classification problem. Object localization is a problem deserving to discuss in the field of image processing and computer vision, which plays an important role in object recognition, image understanding and analysis. Traditional object localization approaches apply sliding window search technique to natural images [1]. However, an image of $320 * 240$ with a low resolution may include one billion and even more rectangular sub-windows. In this paper, we propose a novel object localization method which is based on saliency map and adopts efficient sub-window search algorithm (ESS). ESS algorithm finds the global optimum of the quality function over all possible sub-windows relying on branch-and-bound approach, which can perform fast globally optimal object localization for arbitrary objects in images and the results are equivalent to sliding window search.

Visual saliency detection has become increasingly popular in neuroscience, psychology and computer vision [2], the goal of which is to find out the most important part of an image. Methods of saliency detection are generally classified as bottom-up and top-down approaches. Top-down methods [3] are task-driven, which require supervised learning with manually labeled ground truth. Bottom-up methods [4] usually exploit low-level cues to gain saliency maps, such as color features, intensity features, orientation features, texture features and so on. Saliency detection can be used in object detection and recognition [5], image cropping [6], visual tracking [7], image segmentation [8] and so on.

In this paper, we focus on the case of a single salient object in an image. We combine saliency object detection with ESS algorithm implementing localization of salient object in an image. Firstly, conduct saliency maps which consists of three steps. The first step is to obtain multi-level segmentations exploiting graphed-based approach. The second step is region saliency computation. Extract three types of features forming a vector and learn a random forest regressor which maps the feature vector to a score. Last, the final saliency map is obtained by fusing all level saliency maps. Secondly, saliency maps achieved are processed to highlight more saliency contour and remove some subordinately salient points using region-grow segmentation algorithm. Last, the size and location of saliency object in the image is obtained by effective sub-window search algorithm in the binarized saliency map. Our algorithm finally outputs a rec-tangle. We organize the rest of this paper as follows. We introduce related work and compare our method with other methods in Sect. 2. In Sect. 3 we describe the saliency map computation process and approach. Section 4 introduces region growing algorithm and efficient sub-window search algorithm. Section 5 presents the localization results of our method and compares our method with other methods. Finally, Sect. 6 summarizes the paper.

2 Related Work

2.1 Saliency Detection

A majority of saliency detection algorithms are based on the feature integration theory [9], which is divided into two stages, divided attention stage and focused attention stage. In divided attention stage, assume that only independent features can be detected including color, size, direction, contrast, skewed, curvature, line endpoint and so on. In attention stage, the perceptual system combines divided features into representation of an object. Later, a computational attention model is presented [10], which represents the given image from color, orientation and intensity channels forming three saliency maps using center-surround differences. The three saliency maps are fused to yield the final saliency map.

Recently, a variety of saliency features are designed to represent sal ient objects or regions. Most methods adopt the center-surround difference framework, which is often used to compute region-based saliency representation. Color histograms were computed to evaluate the saliency value between a region and its neighboring regions in [11]. In [12], the saliency score was evaluated by the regional contrast, spatial distribution and element color uniqueness. Other methods are also presented for saliency computation. Object prior such as background prior [13] was introduced for saliency computation. Center-bias is studied in [14].

Existing algorithms combine saliency maps of different kinds of features to yield the final saliency map. The discriminative regional feature integration approach [15] uses a contrast vector to directly compute the saliency map unlike the existing algorithms using a contrast value.

2.2 Saliency for Object Localization

The objects are usually more conspicuous than background. Saliency detection has been applied to object recognition in images. The images were classified directly based on saliency maps in [16]. A hierarchical graph structure was built to represent objects based on saliency map in [17]. Moosmannet et al. [18] exploited the features of saliency to boost the classifier of randomized decision trees. However, object recognition cannot decide the location and the size of an object in an image. Applying saliency detection to object localization is important. Laurent Itti et al. [10] proposed to use a dynamically neural network to select attended locations on the saliency maps. In [11], A CRF model based saliency features was used to object localization. Y.-F. Ma et al. [19] presented a method based on contrast and fuzzy growing for object localization.

The methods directly apply saliency map to object localization. In this paper, we firstly compute saliency maps. Then, the saliency maps are processed to remove subordinately salient points adopting region growing. Finally, the binarized saliency map will be applied to object localization using efficient subwindow search algorithm.

3 Visual Saliency Map

The saliency map is obtained based on the discriminative regional feature integration approach [15], which regards the saliency map computation as a regression problem. This approach includes three main steps. The first step is multi-level segmentation which uses graph-based segmentation to segment an image into many regions. The second step is region saliency computation which maps the features extracted from each region to a saliency score. The third step is multi-level saliency fusion which combines all level saliency maps to yield the final saliency map.

3.1 Multi-level Segmentation

An image is represented by M-level segmentations $S = \{S_1, S_2, \ldots, S_M\}$, each segmentation S_m is consisted of K_M regions. S_1 is obtained by applying graph-based segmentation algorithm [20]. It is finest segmentation, which is consisted of the most regions. Other segmentations $\{S_2, \ldots, S_M\}$ are computed based on S_1. Each segmentation is computed based on its front level segmentation e.g. S_m is obtained by combination of regions in S_{m-1}. S_m is the coarsest segmentation, which is consisted of the smallest number of regions.

3.2 Region Saliency Computation

The saliency score for each region is computed. Each region is represented by three types of features: regional contrast, regional property, and regional back roundness [15]. The feature is denoted using a vector x, which is mapped to a saliency score by using a random forest regressor f. The random forest regressor f is trained by using the regions from the training images. Each region is represented using a feature vector v, including texture and color features. The region contrast descriptor is computed by the differences between features of a region and its neighborhood features. The region property descriptor is computed by extracting appearance features and geometric features. The region back roundness descriptor is computed with a pseudo-background region B. B is defined as the 15-pixel wide narrow border region of image. The region back roundness descriptor is computed by the differences between features of a region and features of the presudo-backgroundness region.

3.3 Multi-level Saliency Fusion

After computing region saliency in each level, every region gets a saliency value which is assigned to its included pixels generating M saliency maps $\{P_1, P_2, ..., P_M\}$. Then, all saliency maps are combined using a combinatory function to get the final saliency map $P = h(P_1, P_2, ..., P_M)$.

4 Saliency Object Localization

4.1 Region Growing Algorithm

Region growing algorithm is a region-based image segmentation approach, which is also classified as a pixel-based image segmentation approach because of the selection of initial seed points. Region growing is a process clustering pixels or sub regions into a larger area based on predefined criteria. Starting with a group of seed points, adjacent pixels or regions which are similar to seed points merge with seed points to form a new seed point. Repeat this process until seed points cannot grow.

We use 8-connected neighborhood for our pixels adjacent relationship. $f(x, y)$ indicates an input image array. $S(x, y)$ indicates a seed array. The value of seed points is 1 in array and other value is 0. Q indicates the property of location (x, y). A region growing algorithm based on 8-connected is as follows.

(1) The first step is looking for all connected components in $S(x, y)$ and each connected component is corrosion by one pixel. All pixels found are labeled as 1 and the rest in S are labeled as 0.

(2) Image f_0 is formed at coordinate (x, y): if the input image satisfies the given properties Q in the coordinate place, $f_0(X, Y)$ equals 1, otherwise $f_0(X, Y)$ equals 0.

(3) Image is formed: 8-connected seed points in mage f_0 which equal 1 are added to each seed point in S.

(4) Each connected component is labeled in image g using different region labels.

The seed points of region segmentation are selected from the salient points on object edges. Some subordinately salient points are removed by using Region Growing algorithm. Figure 1 shows the result.

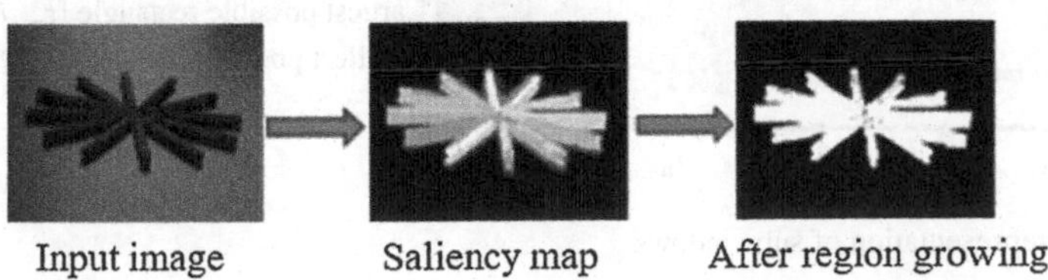

Fig. 1 The result after region growing

4.2 Efficient Sub-Window Search

In the process of search, a large number of regions are used in judging the existence of an object, but only a few numbers of them contain objects. The cost of evaluating the quality function of all candidate regions is expensive. The regions with the highest score should be directly identified and the rest regions in search space are ignored.

Efficient sub-window search is an efficient algorithm for object localization in an image, which adopts the branch-and-bound method as the core of searching for optimal solution finding the rectangular sub-window which is the most closed to the size of object. Due to the number of the rectangular sub-windows in an image is large, such as, an image of 320 × 240 resolution may contain a billion rectangular sub-windows, [21] adopts the branch-and-bound method and adds a function to decide whether an object is present in the sub-window or not. Computation complexity is between $O(n^2)$ and $O(n^4)$. Effectiveness of the ESS is to discard a mass of invalid solution space using the branch-and-bound method.

The solution space of ESS algorithm is a set which contains all possible rectangular boxes in an image. Subsets are obtained by imposing restrictions on the rectangular box coordinate values. Rectangles (sub-windows) are represented by four tuples (*TBLR*). *T* and *B* indicate two vertical intervals between the largest possible rectangle and the smallest possible rectangle and indicate two horizontal intervals between the smallest rectangle and the largest rectangle. ESS terminates when all values (*TBLR*) are equal to zero i.e. the largest rectangle is unique in sub-window set. Sub-windows are split as shown in Fig. 2, each decomposition is to calculate the sum of weights of all the pixels in each sub-window.

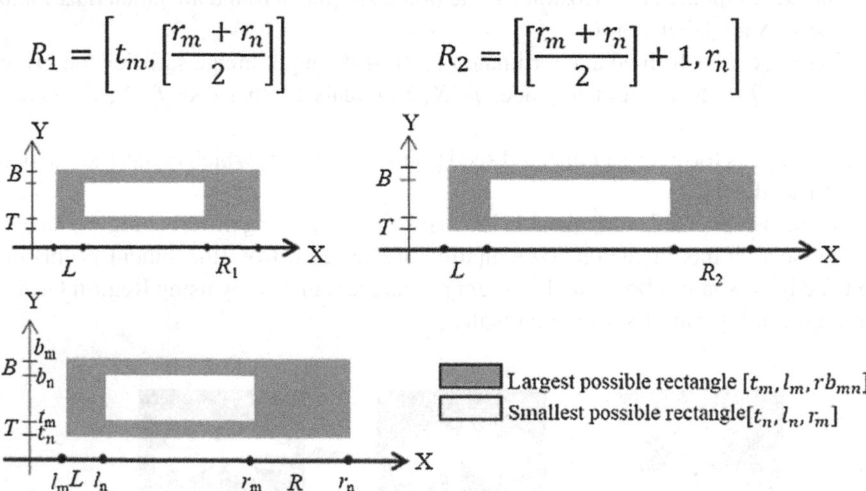

Fig. 2 The representation of subwindows

5 Experimental Results

We evaluate the performance of our algorithm and compare with previous algorithms over MSRA-B dataset [11], which contains 5000 images. Each image in MSRA-B is labeled by nine different users using a rectangle, which contains the most conspicuous object in an image. Most images contain a single salient object. We randomly select 1000 images from the MASR-B to train random forest regressor and 500 images from the rest images to test.

We compare our algorithm with three leading methods: One is the CRF model presented in [11]. The second method is the contrast and fuzzy growing [19], called FG. The last method is the saliency model proposed in [10] called SM. Figure 3 shows results of each step of our approach.

The evaluation results of the four algorithms on MSRA-B dataset are shown in Table 1. Our approach improves 25 and 16 % precision compared with SM and FG and reduced 1 % compared with CRF. The reason may be that CRF model use multi-scale contrast feature. Figure 4 shows some examples with ground truth rectangles for

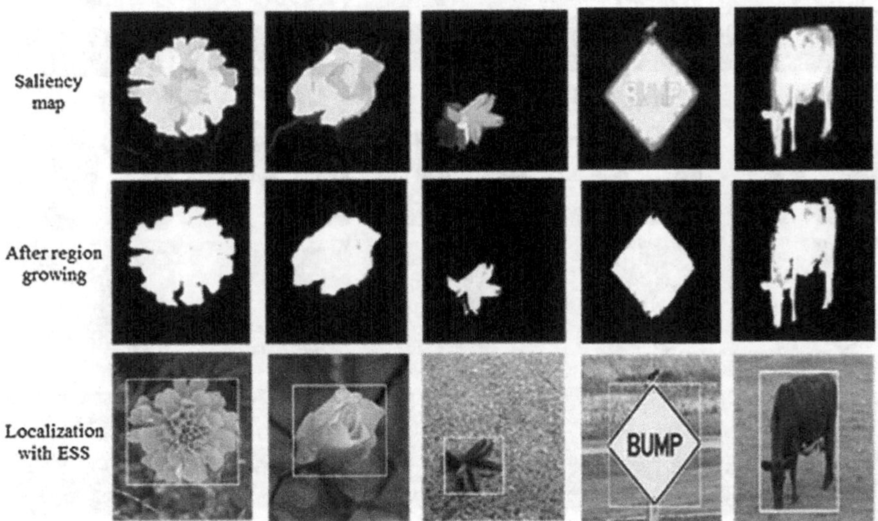

Fig. 3 The results of our approach

Table 1 Precision comparison of saliency object localization

Saliency object localization approaches	Precision of saliency object localization (%)
FG	56
SM	65
CRF	82
OURS	81

qualitative evaluation. SM and FG approach produce a larger rectangle. Our approach is similar with CRF, which is closer to the size of saliency objects. Figure 5 shows our localization results on MSRA-B dataset. Figure 6 shows two failure examples, which indicate that our approach is challenging for objects which have hierarchical parts.

Fig. 4 Comparison of different approaches

Fig. 5 Our localization result on some images

6 Conclusion and Future Work

In order to locate an object in an image more actually, we present a new method which combines the binarized saliency map with ESS algorithm implementing object localization. Firstly, the saliency map is obtained based on the discriminative regional feature integration approach. Secondly, Saliency map achieved is processed to highlight more saliency contour and remove some subordinately salient points using region growing segmentation algorithm. Last, saliency object location is obtained by ESS algorithm in the binarized saliency map. This approach is able to fast identify the location of saliency object and roughly the size.

Additionally, our approach is able to be further improved. Our approach focus on a single object in images. We can implement multi-objects localization in an image and the proposed object localization method is applied to object recognition and detection as a preprocessing step.

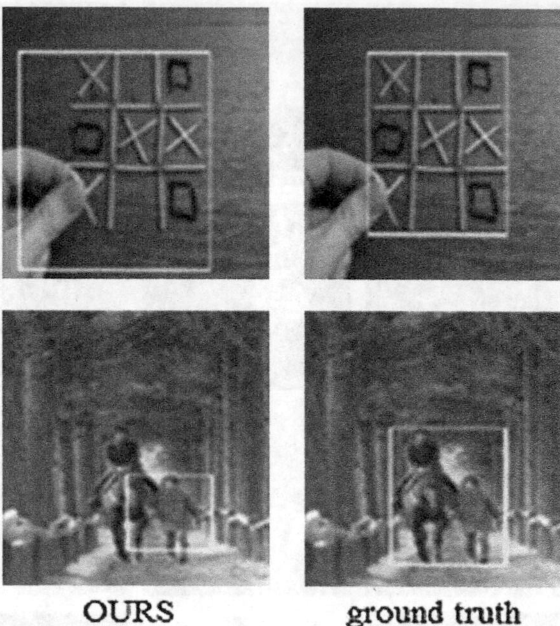

Fig. 6 Failure results

References

1. Chum O, Zisserman A (2007) An exainplar model for learning object classes. In: CVP
2. Achanta R, Hemami S, Estrada F, Susstrunk S (2009) Frequency-tuned salient region detection. In: IEEE conference on computer vision and pattern recognition, 2009. CVPR 2009. IEEE, pp 1597–1604
3. Yang J, Yang M-H (2012) Top-down visual saliency via joint crf and dictionary learning. In: 2012 IEEE conference on computer vision and pattern recognition (CVPR). IEEE, pp 2296–2303
4. Yan Q, Xu L, Shi J, Jia, J (2013) Hierarchical saliency detection. In: 2013 IEEE conference on computer vision and pattern recognition (CVPR). IEEE, pp 1155–1162
5. Kanan C, Cottrell GW (2010) Robust classification of objects, faces, and flowers using natural image statistics. In: CVPR, pp 2472–2479
6. Marchesotti L, Cifarelli C, Csurka G (2009) A framework for visual saliency detection with applications to image thumbnailing. In: ICCV, pp 2232–2239
7. Mahadevan V, Vasconcelos N (2009) Saliency-based discriminant tracking. In: IEEE conference on computer vision and pattern recognition, 2009. CVPR 2009. IEEE, pp 1007–1013
8. Rother C, Kolmogorov V, Blake A (2004) Grabcut: interactive foreground extraction using iterated graph cuts. In: ACM transactions on graphics (TOG), vol. 23. ACM, pp 309–314
9. Treisman A, Gelad G (1980) A feature-integration theory of attention. Cogn Psychol 12(1):97–136
10. Itti L, Koch C, Niebur E (1998) A model of saliency based visual attention for rapid scene analysis. IEEE Trans PAMI 20(11):1254–1259
11. Liu T, Yuan Z, Sun J, Wang J, Zheng N, Tang X, Shum HY (2011) Learning to detect a salient object. IEEE Trans Pattern Anal Mach Intell 33(2):353–367
12. Jiang H, Wang J, Yuan Z, Liu T, Zheng N, Li S (2011) Automatic salient object segmentation based on context and shape prior. In: British machine vision conference (BMVC)

13. Wei Y, Wen F, Zhu W, Sun J (2012) Geodesic saliency using background priors. In: ECCV (3), pp 29–42
14. Wang P, Wang J, Zeng G, Feng J, Zha H, Li S (2012) Salient object detection for searched web images via global saliency. In: CVPR, pp 3194–3201
15. Jiang H, Wang J, Yuan Z, Wu Y, Zheng N, Li S (2013) Salient object detection: a discriminative regional feature integration approach. In: CVPR, pp 2083–2090
16. Shokoufandeh A, Marsic I, Dickinson SJ (1999) View-based object recognition using saliency maps. Image Vis Comput 17:445–460
17. Ren Z, Gao S, Chia L-T, Tsang IW-H (2014) Region-based saliency detection and its application in object recognition. IEEE Trans Circuits Syst Video Technol 24(5):769–779
18. Moosmann F, Larlus D, Jurie F (2006) Learning saliency maps for object categorization. In: Proceedings of european conference on computer vision workshop represent. Use Prior Knowl. Vision
19. Ma Y-F, Zhang H-J (2003) Contrast-based image attention analysis by using fuzzy growing. In: Proceedings of ICMM, pp 374–381
20. Felzenszwalb PF, Huttenlocher DP (2004) Efficient graph-based image segmentation. Int J Comput Vis 59(2):167–181
21. Walther D, Koch C (2006) Modeling attention to salient proto-objects. Neural Netw 19(9):1395–1407

An Improved Dynamic Programming for Power Allocation in Cognitive Radio

Qian Wang, Qian Chen, Hang Zhang and Menglan Fan

Abstract In this paper, a novel power allocation algorithm to maximize the throughput under the bit error rate (BER) constraint and the total power constraint in cognitive radio is proposed. Although water filling (WF) algorithm is the optimal power allocation in theory, it ignores the fact that the allocated power in use is discrete, and the algorithm doesnt take the waste power into consideration. In the improved algorithm, the total power is asymmetrically quantized to apply to the practice and reduce the computation complexity before adopting the dynamic programming which is commonly used to solve the knapsack problem, so this improved algorithm is called as asymmetrically quantized dynamic programming (AQDP). Moreover, AQDP reused the residual power to maximize the throughput further. The simulation results show that AQDP algorithm has improved the transmit throughput of all CR users compared with the classical power allocation algorithms referred as WF algorithm in this paper.

Keywords Cognitive radio · Power allocation · Bit error rate constraints · Water filling algorithm · Dynamic programming

Q. Wang (✉) · Q. Chen · H. Zhang · M. Fan
PLA University of Science and Technology, Nanjing 210007, China
e-mail: wangqian_lgdx@163.com

Q. Chen
e-mail: qqianchen@126.com

H. Zhang
e-mail: hangzh_2002@163.com

M. Fan
e-mail: fanmenglan@126.com

V.E. Balas et al. (eds.), *Information Technology and Intelligent Transportation Systems*, Advances in Intelligent Systems and Computing 455,
DOI 10.1007/978-3-319-38771-0_5

1 Introduction

Cognitive radio (CR) is a new communication paradigm to solve the spectrum shortage problem by allowing secondary users (SU) to access the spectrum holes which refer to the frequency bands not occupied by primary users (PU) temporarily. Once the information about the spare frequency bands is detected via spectrum sensing, power allocation becomes an important step because it decides the utilization ratio of the spectrum directly.

Resource allocation has always been a significant point in all fields concerning energy conservation problem [1]. Knopp R [2] has proved that iterative water-filling algorithm is the optimal power allocation algorithm in theory, but the algorithm ignores the fact that the power needed in practice is actually discrete and doesnt take the waste power into consideration. In [3–5], the power allocation problem is formulated as a knapsack problem (KP) which regards symmetrically discretized total power as objects to be put into the knapsack, and some heuristic algorithms such as the greedy man-min algorithm [3] minimal revenue efficiency removal algorithm [4] and competitive decision algorithm [5] are chosen to solve the knapsack problem. Even though the complexity of heuristic algorithms is low enough, heuristic algorithms can only have a suboptimal solution. Dynamic programming [6–8] is another algorithm widely used for solving knapsack problem which can not only get the global optimal result [9], but also is polynomial complexity [10]. However, these papers fix the modulate pattern when transmitting information, thus limit the efficiency of spectrum and power resource. In fact, there exit multi-modulation combinations that can further improve the total transmit throughput.

This paper formulates the power resource allocation problem as maximizing the transmit throughput of CR users subject to the constraints of bit error rate (BER) and a total transmit power constant for CR users. Considering that total power is the sum of a discrete set of power in practical use and the modulate pattern is MQAM in which M is variable, knapsack problem is chosen as the model and dynamic programming is chosen as the algorithm. Based on the above, we propose the asymmetrically quantized dynamic programming (AQDP) algorithm. Simulation results show that the AQDP algorithm has improved the transmit throughput of all CR users compared with the classical power allocation algorithms.

The rest of this paper is organized as follows. In Sect. 2, the system model is presented. The proposed power allocation algorithm is described in Sect. 3. In Sect. 4, the evaluation analysis and the simulation results involving comparisons with iterative WF algorithm are given. Finally, the conclusion is presented in Sect. 5.

2 System Model

Consider a cognitive radio system with one PU and several SUs, in which the total bandwidth can be divided to K available spectrum bands with corresponding bandwidth array $\mathbf{B}[j] = \{B_1, B_2, \ldots, B_K\}$. We suppose that the modulation is MQAM of which the order M is changeable, so the symbol rate of the cognitive terminal is adjustable. To utilize the spectrum holes efficiently, we set the bit rate of the band $j(j = 1, 2, \ldots, K)$ to be $B_j/2$ bits/s which doesnt account for protective interval between bands. Each secondary user chooses the channel and the order value $M_n(M_n = 2^n \in \mathbf{M}, n = 1, 2, 3\ldots)$ of the modulation MQAM according to BER constraint as well as allocated power, then different information rates are obtained accordingly.

The optimization problem is how to allocate the total power to idle channels within the BER restriction in order to maximize the transmit throughput. The minimum power allocated to certain channel is calculated by BER function in regard to the order M_n, the noise power density spectrum (PDS) n_{0j} of band j, bit error rate condition P_b and the bandwidth B_j. The power allocation can be described as follows under the condition that the total power is C.

$$\begin{aligned} &\max\left\{\sum_{j=1}^{K}\frac{B_j}{2}\log_2 M_j\right\} \\ &\text{s.t.} \sum_{j=1}^{K} P_j \le C \end{aligned} \tag{1}$$

The BER bound of MQAM signal can be described as [11]

$$p_b \le 0.2e^{-1.5\gamma_s/(M-1)} \tag{2}$$

In the inequality, M refers to the order of MQAM and γ_s means signal to noise ratio (SNR)

$$\gamma_s = \frac{P}{B \cdot n_0} \tag{3}$$

Traditional power allocation such as WF algorithm aims at maximizing channel throughput function and the optimal solution can be obtained by matrix theory and convex optimization [7]. WF algorithm can achieve the theoretical optimal solution, however, when considering the decimal power in practical use, the WF algorithm cannot effectively utilize all the allocated power to get more transmit throughput. That is to say, some power is wasted by WF algorithm. For example, one of the available spectrums needs 2.2606 W to realize 2QAM modulation and needs 6.7819 W to realize 4QAM modulation, which means that the power in practical use is added discontinuously. If the power allocated to the channel is 3.5 W, then the power can only afford 2QAM modulation and leaves 1.2394 W wasted.

Since the power allocation problem is similar to knapsack problem and dynamic programming (DP) is a solution to KP with only polynomial complexity. To solve the decimal problem described above, we improve DP as AQDP by two steps. More detail is shown in.

3 The Proposed Power Allocation Algorithm

Dynamic programming is one of the general solution for optimal problem without acknowledged function or procedure. It is designed differently according to diverse applications. The principal usage of dynamic programming is to solve multiphase decision problem. If one kind of problem can be divided into several related sub-problems and each decision of sub-problem has an effect on the next decision, then we regard this problem as dynamic programming and obtain the optimal result by solving the multiphase step by step.

We cannot directly cite the dynamic programming to solve general knapsack problem case because of two special characteristics in power allocation. On the one hand, the power to be allocated is decimal while the quantity to be allocated in general knapsack problem is integral and has equal difference. On the other hand, general dynamic programming deals with integral problem, so the waste will appear after allocating. However, since the allocated power in practical cases is decimal, the wasted power in each frequency band may add up to some amount that can change the order of one channel for increasing the spectrum efficiency. That is to say, waste power needs to be reallocated to maximize the transmit rate.

To solve above two problems, we improve the recursion algorithm as asymmetrically quantized dynamic programming (AQDP) in two steps to apply to the power allocation problem. Firstly, the total power is asymmetrically quantized into a discrete sequence as described in detail in Sect. 3.1. Then, we accumulate the residual power in each frequency band to see if the sum can add one order of some channel as described in Sect. 3.2.

3.1 Asymmetrically Quantize the Total Power

Firstly, we asymmetrically discrete the total power into a set of power and we use these quantities to avoid infinite recursion. The procedure is as follows:

(a) Calculate the minimum power needs of each frequency band according to the specifications of error bit rate p_b, bandwidth B_j and the noise power density spectrum n_{0j} according to inequality (2). Table 1 lists the minimum power required

Table 1 The minimum power $\mathbf{P}[i, j]$ required by each frequency band B_j, $(j = 1, 2, \ldots, K)$ with the BER threshold $P_b = 10^{-5}$

j	B_j (kHz)	n_{0j} (mW/Hz)	$\mathbf{P}[i, j]$ (W) $M_1 = 2$	$M_2 = 4$	$M_3 = 8$	$M_4 = 16$	$M_5 = 32$	$M_6 = 64$
1	1	0.4624	3.0529	9.1587	21.3704	45.7937	94.6404	192.3337
2	4	0.0856	2.2606	6.7819	15.8245	33.9095	70.0797	142.4201
3	3	0.2958	5.8589	17.5767	41.0123	87.8835	181.6260	369.1109

by randomly chosen 3 frequency bands. Suppose that there are K idle channels and $S = \max\{M_n\}, M_n = 2, 4, 8...$, then we can obtain the array $\mathbf{P}[i, j], 1 \leq i \leq S, 1 \leq j \leq K$ of which the entry is the required minimum power. If the entry of $\mathbf{P}[i, j]$ extends the total power threshold, 0 is used to replace it. According to (1), the corresponding transmit throughput array is $\mathbf{R}[i, j], 1 \leq i \leq S, 1 \leq j \leq K$.

(b) Set the total power as p_N and rank the discrete power values in $\mathbf{P}[i, j]$ that are suitable for BER constraint and the total transmit power constraint from small to large. Then add 0 and the total power p_N respectively as the first element and the last element to compose the one-dimensional array of discrete power $\mathbf{P}_d[w] = \{p_1, p_2, ..., p_N\}$, $w = 1, 2, ..., N$ in which N denotes the number of elements in $\mathbf{P}_d[w]$.

3.2 Reallocate the Residual Power

Residual power is reused by two steps. Firstly, after each allocation step during the recursion procedure, residual power is recorded and added to the total residual power for allocation. Secondly, when all the channels are allocated with proposal power, AQDP checks that if the last residual power can improve the order of certain frequency band. These two steps is the key improvement that makes our algorithm better than WF algorithm.

Moreover, when several allocation schemes achieve the same throughput, we choose the least consumption one which offers more improvement opportunities to other frequency bands. Based on above analysis, maybe the residual power during allocation has a great contribution to improve the total transmit rate, so it is reasonable to record them during the recursion procedure.

3.3 The Proposed Power Allocation Algorithm

It is necessary to describe the definitions of parameters before introducing the specific algorithm is the number of available idle channels (frequency bands) and $\mathbf{P}_d[w] = \{p_1, p_2, \ldots, p_N\}$ indicates the set of discrete power. $\lfloor p \rfloor$ means the closest power of the discrete set that is smaller than or equal to p. That is to say, when $p_l \in \mathbf{P}_d$ and $p_l \leq p < p_{l+1}$, $\lfloor p \rfloor = p_l$. The symbols $\lfloor \cdot \rfloor$ and $\lceil \cdot \rceil$ denote the flooring operation and ceiling operation [7] respectively.

$\mathbf{P}[i, j], 1 \leq i \leq S, 1 \leq j \leq K$ denotes the array of the chosen minimum power of K idle channels in the assumption that each channel has S minimum power values to choose from with corresponding transmit throughput array $\mathbf{R}[i, j], 1 \leq i \leq S, 1 \leq j \leq K$.

The Recursion Function After step in Sect. 3.1 and step in Sect. 3.2, the recursion function of improved power allocation algorithm is as follows. $\mathbf{R}_{sys}[w, j], 1 \le w \le N, 1 \le j \le K$ denotes the maximum system transmit throughput we can get after $\mathbf{P}_d[w]$ is allocated to front j frequency bands, and the entry of $\mathbf{R}_{sys}[w, j]$ is the sum of certain elements of array $\mathbf{R}[i, j]$.

$$j = 1, i = 1, 2, ..., S, \mathbf{R}_{sys}[w, j] = \begin{cases} 0 & , \mathbf{P}_d[w] < \mathbf{P}[i, j] \\ \mathbf{R}[i, j] , & \mathbf{P}_d[w] \ge \mathbf{P}[i, j] \end{cases}$$
$$j \ne 1, \ \mathbf{R}_{sys}[w, j] = \begin{cases} \mathbf{R}_{sys}[w, j-1] & , \mathbf{P}_d[w] < \mathbf{P}[i, j], \ \ i = 1, 2, ..., S \\ \max\limits_{i=1}^{S} \left\{ \mathbf{R}_{sys}[w, j-1] , \mathbf{R}_{sys}[w-l, j-1] + \mathbf{R}[i, j] \right\} \\ , \mathbf{P}_d[w] \ge \mathbf{P}[i, j] \end{cases} \tag{4}$$

Among the above formula, l is the subscript of p_l, $p_l \in \mathbf{P}_d$, and $\mathbf{P}_l = \lfloor \mathbf{P}_d[w] - \mathbf{P}[i, j] \rfloor$. The residual power $\mathbf{\Delta} \boldsymbol{P}[w, j] = \mathbf{P}_d[w] - \mathbf{P}[i, j] - p_l$ is calculated to be rearranged if it can increase the order M of one channel during the recurring procedure.

The Backtracking Procedure $\mathbf{P}_{opt}[w, j], 1 \le w \le N, 1 \le j \le K$ denotes the power allocated to the jth band when the maximum system transmit throughput is $\mathbf{R}_{sys}[w, j]$ and this term is a array used to backtrack the optimal scheme; $\mathbf{P}_{sch}[j] = \{p_1, p_2, \ldots, p_K\}$ denotes the power allocated to the j th band when the system transmit throughput is maximal.

According to the recursion algorithm, the elements in array $\mathbf{R}_{sys}[w, j]$ can be calculated from top to bottom and from left to right, and the maximum $\mathbf{R}_{sys}[N, K]$ is just at right bottom corner. To obtain the detailed allocation scheme, we need to backtrack from the last frequency band. The backtracking algorithm is as follows.

Firstly, the sum of allocated power is defined as $p_{all} = p_N - \mathbf{\Delta} \boldsymbol{P}[N, K]$, then backtracking recursion starts with $w = N$ and $j = K$. Secondly, the power allocated to the jth ($j = 1, 2, \ldots, K$) bands is calculated based on array $\mathbf{P}_{opt}[w, j]$, which is $p_j = \mathbf{P}_{opt}[w, j]$, $p_j \in \mathbf{P}_{sch}$, $j = 1, 2 \ldots K$. Then by updating the allocated power the location subscript of allocated power and the serial number of frequency band until $j = 1$ as follows,

$$\begin{aligned} & p_{all} = p_{all} - p_j \\ & p_n = \lceil p_{all} \rceil \ , \quad p_n \in \mathbf{P}_d \\ & j = j - 1 \end{aligned} \tag{5}$$

the whole set of $\mathbf{P}_{sch}[j] = \{p_1, p_2, ..., p_K\}$ can be finally obtained.

4 Simulation Results

According to the power allocation model described above, we choose MQAM as modulation pattern with the order M_n changeable within 2^n ($n = 1, 2, 3, 4, 5, 6, 7$). The BER threshold is $P_b = 10^{-5}$ while the total power constraint is 100 W. Assume that there are 5 available frequency bands and the bandwidth B between 1 and 5 kHz can be achieve by random function as well as the noise power density spectrum n_0 ranging from 0 mW/Hz to 0.05 mW/Hz. To demonstrate that the algorithm we introduced is close to practical use and consumption saving, we evaluate the WF algorithm at the same time. When the total power is assumed to be 100 W, we can obtain the allocation scheme as Table 2. According to the simulation results, dynamic programming is better in practice because the power allocated is fully used.

By keeping 5 available frequency bands constant and changing total power from 50 to 500 W with a step of 10 W, the allocation results of dynamic programming and WF both in theory and practice are shown in Fig. 1. Figure 1 shows that WF algorithm only takes the lead in theory, which indicates the improved dynamic programming we proposed outperforms WF algorithm and is more suitable for practical use.

By keeping the total power as 300 W and changing available frequency bands from 5 to 15 with a step of 10 W, the allocation results of dynamic programming and WF both in theory and in practice are shown in Fig. 2. The four curves indicate that

Table 2 The comparison of two algorithms with the total power $p_N = 100$ W

The ordinal of frequency bands		1	2	3	4	5	Residual power
Dynamic Programming algorithm	Alloted power (W)	21.2549	31.0771	23.0850	13.9164	10.4819	0.1847
	Theoretical throughput	4.3790	6.6444	5.5612	4.3790	2.9264	–
	Total theoretical throughput	23.8900					
	Actual throughput	1.5000	4.5000	4.5000	2.000	4.000	–
	$\mathbf{R}_{sys}[N, K]$	17.5000					
Water Filling algorithm	**Alloted power(W)**	19.7622	20.5215	20.3358	20.1327	19.2477	0
	Theoretical throughput	4.2792	6.0530	5.3824	4.8902	3.7141	–
	Total theoretical throughput	24.3189					
	Actual throughput	1.5000	4.5000	3.0000	2.0000	2.0000	–
	$\mathbf{R}_{WF}[N, K]$	13.0000					

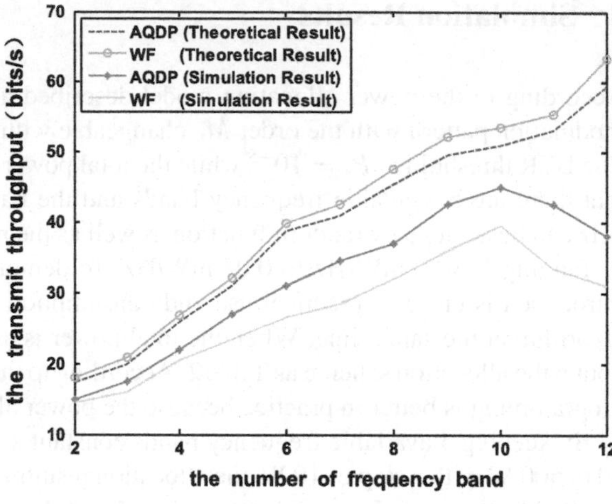

Fig. 1 The results of the total power ranging from 50 to 500 W with 5 available frequency bands

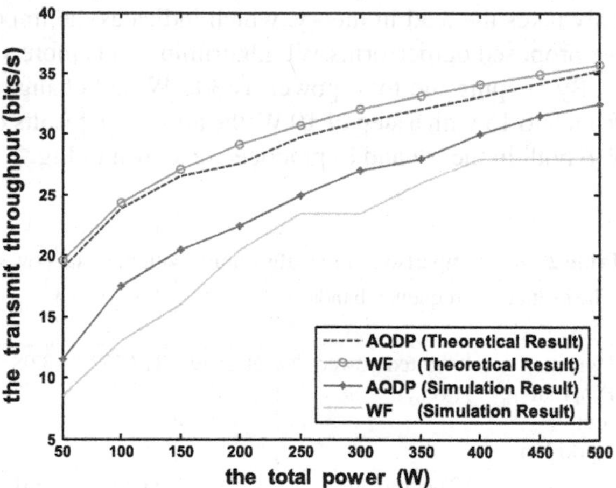

Fig. 2 The results of the total power as 300 W with available frequency bands changing from 2 to 12

the total transmit throughput of two algorithms increase with increasing frequency in both theory and in practice. The phenomenon corresponds to the communication theory that dividing the total power to several bands produces more throughput than just one single band. It should be noted that the improved dynamic programming performs better than WF algorithm regardless of the amount of frequency bands. Additionally, the gain value becomes larger with frequency increasing. Since in WF algorithm there is wasted power in almost each band, the amount of wasted power increases with frequency bands growing, which leads to throughput declining. At the meantime, the simulation line of AQDP suffers declining because of the same reason. However, the AQDP approach still performs better than WF algorithm owing to the reusing of waste power.

5 Conclusion

A novel power allocation algorithm named as AQDP is introduced to maximize the throughput under certain BER and total power constraint in cognitive radio. Based on classical dynamic programming that solves the knapsack problem, we improve the dynamic programming to apply to practical power allocation. Simulation results show that the proposed algorithm solves the problem better than widely used WF algorithm. In future work, we will take outage probability into consideration, and make the algorithm closer to practical applications.

References

1. Lee JW, Mazumdar RR, Shroff NB (2006) Joint resource allocation and base-station assignment for the downlink in CDMA networks [J]. IEEE/ACM Trans Netw 14(1):1 C 14
2. Knopp R, Humblet PA (1995) Information capacity and power control in single cell multiuser communications [C]. In: IEEE International conference on communications (ICC95), vol 1. Seattle, USA, p 335
3. Zhang Y, Leung C (2009) Resource allocation in an OFDM-based cognitive radio system [J]. IEEE Trans Commun 57(7):1928–1931
4. Xiang J, Zhang Y, Admission Skeie T (2010) Power control for cognitive radio cellular networks: a multidimensional knapsack solution [C]. In: (2010) 3rd international symposium on applied sciences in biomedical and communication technologies (ISABEL). IEEE, pp 1–5
5. Xiong XH, Ning AB, Liang M, Wang AB (2009) Competitive decision algorithm for multidimensional knapsack problem [C]. In: 2009 international conference on management science and engineering. ICMSE. IEEE, pp 161-167
6. Ejaz W, Hasan NU, Kim HS (2012) KENSS: knapsack-based energy-efficient node selection scheme for cooperative spectrum sensing in cognitive radio sensor networks [J]. Iet Commun 6(17):2998–3005
7. Gao L, Cui S (2009) Power and rate control for delay-constrained cognitive radios via dynamic programming [J]. IEEE Trans Veh Technol 58(9):4819 C 4827
8. Xiang X, Wan J, Lin C et al (2013) A dynamic programming approximation for downlink channel allocation in cognitive femtocell networks [J]. Comput Netw 57(15):2976 C2991
9. Sinha A, Chaporkar P (2012) Optimal power allocation for a renewable energy source [C]. In: 2012 national conference on communications (NCC). IEEE, pp 1–5
10. Kellerer H (1999) A polynomial time approximation scheme for the multiple knapsack problem [J]. Lect Notes Comput Sci 1671:51–62
11. Goldsmith AJ, Chua SG (1997) Variable-rate variable-power MQAM for fading channels [C]. IEEE Trans Commun 45(10):1218 C 1230

Analysis and Forecast of Regional Freight Characteristics in the Silk Road Economic Belt

Jianqiang Wang and Zhipeng Huang

Abstract The proposal of building the Silk Road Economic Belt aiming to strengthen cooperation among Asian and European nations will benefit all the people along this Economic Belt. To establish the Silk Road Economic Belt, comprehensive transport systems are a principal premise. Freight is an important part of comprehensive transport systems. Thus, in this study, a curve regression model is established for statistics analysis and forecast freight turnover quantity and freight volume of nine provinces in China, in the Silk Road Economic Belt. Detailed parameters of the model could be calculated by SPSS, including a significant and fitting test. The calculation results by SPSS show that this curve regression model has a good fitness to actual freight data and could be significant. Finally, the development trend of freight turnover quantity and freight volume of nine provinces in future from 2015 to 2020 is predicted using the model, according to the supposed GDP growth rate (7 %). The results of the actual case analysis prove the feasibility and efficiency of the curve regression model. All these research results can be used to guide the construction of the comprehensive transportation system in the Silk Road Economic Belt.

Keywords Freight volume · Freight turnover quantity · Forecast · Nonlinear regression analysis · Silk road economic belt

This research is partially supported by the National Social Science Foundation of China (Grant no. 14XGL011), the Humanity and Social Science Youth Foundation of Ministry of Education in China (Grant no. 12YJC630200), and the Natural Science Foundation of Gansu Province in China (Grant no. 145RJZA190).

J. Wang (✉) · Z. Huang
School of Traffic and Transportation, Lanzhou Jiaotong University, Lanzhou, China
e-mail: xinxiwjq@126.com

Z. Huang
e-mail: huangzp@mail.lzjtu.cn

V.E. Balas et al. (eds.), *Information Technology and Intelligent Transportation Systems*, Advances in Intelligent Systems and Computing 455,
DOI 10.1007/978-3-319-38771-0_6

1 Introduction

The Silk Road Economic Belt is a significant national strategy of China to implement new regional coordination and westward opening in future, which is not only a vast economic belt throughout Europe and Asia, but also the transnational and cross-regional cooperation platform, known as the longest mainland economic "artery" in the world. In the construction process of the Economic Belt, to establish comprehensive transport corridors is a principal premise [1–3]. Freight will be a major component of future transport corridors, in the Silk Road Economic Belt.

In the comprehensive transport corridors, freight has been the focus of transportation researches. Shi et al. discussed the basic conditions for the logistics development and space disequilibrium of transportation infrastructure in the Silk Road Economic Belt [4]. Alexey et al. proposed the development of international transport corridors in Russian and China [5]. Zhang analyzed the regional difference of transport infrastructure and used panel data to investigate the relationship between the development level of transport infrastructure and the regional economic growth [6]. Tang et al. pointed out the high nonlinearity, coupling and time changing character of indexes data which reflects regional freight turnover quantity. He confirmed the interrelated forecast variable and established regional road freight turnover quantity forecast model based on support vector regression [7]. Fu et al. put forward a gray model algorithm to forecast the railway freight category volume [8]. The research shows that the freight turnover quantity and added value of the logistics industry can pull economic growth highly, but the contribution degree is different [9]. It is a superior way to promote economic growth though boosting freight turnover quantity, promoting cooperation between different provinces in the New Silk Road Economic Zone [10]. Thus, this study carried out data analysis and forecast for freight turnover quantity, freight volume and GDP value in the western region of China. It is aimed to promote the infrastructure and three-dimensional traffic system construction to pull the regional economic development well.

2 Nonlinear Regression Analysis and Forecast Model

The statistical analysis process mainly includes correlation analysis and causal regression analysis. Correlation analysis is accomplished by two or more variables with correlation, so as to measure the correlation between the two variables.

In the process of establishing the regression model, it is a very important issue to choose the fitness variables. Its better to choose the less but high precision variables so as to complete the regression model and the prediction process.

Thus, a nonlinear regression analysis model is established, which is based on the function expression of the causal relationship between the independent variable and the dependent variable. If there is only one independent variable of the nonlinear regression analysis, the regression equation is a curve equation. Using the curve estimation module in SPSS, we can get the dependency relation. Through the regression equation, we can carry out the relevant variables.

After comparison, we choose the GDP variable and fixed asset investment variable as the main factors that affect the freight turnover rate and freight volume. After the correlation analysis of the choice variables, we found that the economic indicators of GDP and fixed asset investment present high co-linearity. So, we decided to take the GDP variable that is higher as independent variables to establish a regression model.

After testing actual data, the trend extrapolation forecast method (exponential curve model) is adopted to reveal the development trend, as shown in Eq. (1):

$$y_i = y_0 \cdot e^{k \cdot t_i} \tag{1}$$

where y_i represents the estimated value of the ith item among dependent variable sequences; y_0 is an initialization constant; k is a coefficient; t_i represents the sample value of the ith item among independent variable sequences. After taking the logarithmic transformation, we can get Eq. (2):

$$\ln y_i = ln y_0 + k \cdot t_i \tag{2}$$

Supposed that $y_i^{'} = \ln y_i, a = ln y_0, b = k$, then there is $y_i^{'} = a + b \cdot t_i$.

Then, we get the coefficients though the application of single linear regression method as shown in Eqs. (3) and (4):

$$a = \frac{\sum {t_i}^2 \sum y_i^{'} - \sum t_i \sum t_i y_i^{'}}{n \sum {t_i}^2 - (\sum t_i)^2} \tag{3}$$

$$b = \frac{n \sum t_i y_i^{'} - \sum t_i y_i^{'}}{n \sum {t_i}^2 - (\sum t_i)^2} \tag{4}$$

where n is the number of data set.

In addition, the correlation coefficient R is an important index to evaluate the linear correlation between the two variables, as shown in Eq. (5):

$$R = \frac{n \sum t_i y_i^{'} - \sum t_i y_i^{'}}{\sqrt{n \sum {t_i}^2 - (\sum t_i)^2} \sqrt{n \sum {y_i^{'}}^2 - (\sum y_i^{'})^2}} \tag{5}$$

3 Case Analysis

In this case analysis, IBM SPSS Statistics 20(Statistical Package for the Social Sciences) software is chosen for data statistical analysis and forecast. Data are analyzed using variance and multiple regression analysis. After establishing the relationship between a variable (dependent variable) and the others or multiple variables (independent variables), we can get the relational model between them so as to predict the change of the dependent variable with the independent variable.

3.1 Data Selection

The data sources used in the analysis and forecast of regional freight characteristics are selected in "China Statistical Yearbook" from 2008 to 2013, regarding nine provinces including Sichuan, Chongqing, Yunnan, Guangxi, Shanxi, Gansu, Qinghai, Ningxia and Xinjiang, in the Silk Road Economic Belt. We select three indexes, including GDP, passenger turnover quantity and freight volume, as analysis indexes.

3.2 Freight Turnover Quantity Forecast

Freight Turnover Quantity Curve Regression Model In order to establish a single curve regression model, the freight turnover quantity and GDP regression analysis are carried out by analysis—regression and curve estimation module in SPSS. In the curve regression result as shown in Fig. 1, the optimal fitting curve is obtained by testing the value of "Sig." and "R cube" as shown in Table 1. The results show that the value of "Sig." in this case is 0.000, which should be less than 0.05. The

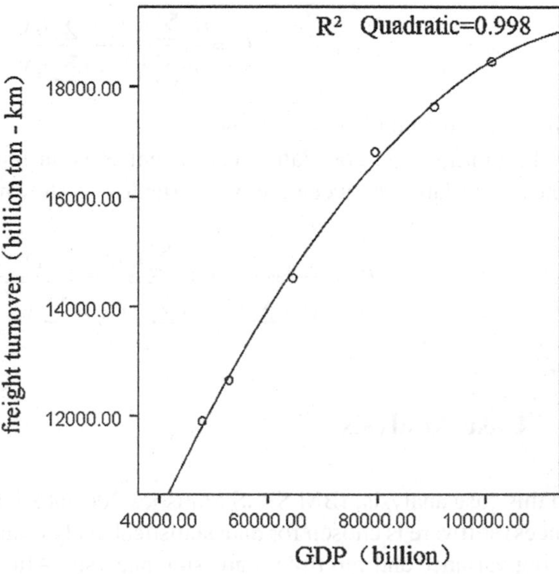

Fig. 1 Scatter diagram of freight turnover quantity-GDP

Table 1 Model coefficients

Coefficients					
	Non-Standard B	Standard errors	Standard Beta	t	Sig.
1/GDP (billions)	–41188.568	1164.266	–0.998	–35.377	0.000
(Constant)	10.233	0.018	573.606	0.000	
Variable is ln (Freight turnover (billion ton–km))					

Table 2 Variance analysis

ANOVA					
	Quadratic sum	df	Mean square	F	Sig.
Regression	0.164	1	0.164	1251.552	0.000
Residual	0.001	4	0.000		
Total	0.165	5			
Independent Variable is GDP (Billion)					

"R cube" indicates that the fitting degree between model and the curve. The closer to 1 the value of "R", the better the model fitting curve.

Thus, we achieved the curve regression model as shown in Eq. (1)

$$\ln(v) = \exp\left(100.233{-}411880.568/x\right) \tag{6}$$

where v represents the provincial freight turnover quantity vector x represents the GDP values vector. In order to identify the validity of the regression model, we should carry out variance analysis, including significant test of regression equation and fitting test.

In Table 1, the value of "Sig." is less than 0.05, which represents that the reciprocal of and constant term can pass the tests. The "sig." value of regression equation is 0.000, which means not only the equation is very significant, but also the model meets the requirements (Table 2). F-measure is a significant degree of the "R", where F-measure is significant that means the regression method is correct. From Table 3, "R" represents the complex correlation coefficient of the regression model; "R power" represents the fitting degree of the curve; "Adjusted R square" is 0.996 which is very close to 1, which shows that the nonlinear relationship between freight turnover quantity and GDP is significant. The regression model to describe the actual data had a high approximate degree that fits the sample data well. The results of regression equation fitting test are shown in Table 4. The mean absolute percentage error rate of the regression equation is 0.80 %, which indicates the model error is small. Accordingly, we can draw predict freight turnover quantity from 2008 to 2013 as

Table 3 Model checking

Model summary			
R	R power	Adjust R power	Standard error of estimate
0.998	0.997	0.996	0.011
Independent variable is GDP (Billion)			

Table 4 Regression equation fitting test

Year	GDP (billion)	Freight turnover quantity (billion ton km)	Freight turnover quantity forecast (billion ton km)	Residual	MAPE (%)
2008	47995.16	11903.44	11786.66	116.78	0.98
2009	52906.19	12654.30	12764.01	–109.71	0.87
2010	64626.87	14523.19	14699.62	–176.43	1.21
2011	79567.41	16803.99	16568.35	235.64	1.40
2012	90470.99	17627.05	17634.94	–7.89	0.04
2013	101137.15	18450.11	18502.32	–52.21	0.28

shown in Fig. 2. It can be seen that the predicted value is basically in agreement with the actual value.

Freight Turnover Quantity Forecast In order to predict the freight turnover quantity in future, it is supposed that the average growth rate of GDP in China from 2015 to 2020 is 7 %. Then, we can adopt the curve regression model to predict the freight turnover quantity in the next six years. The results of freight turnover quantity forecast are shown in Table 5.

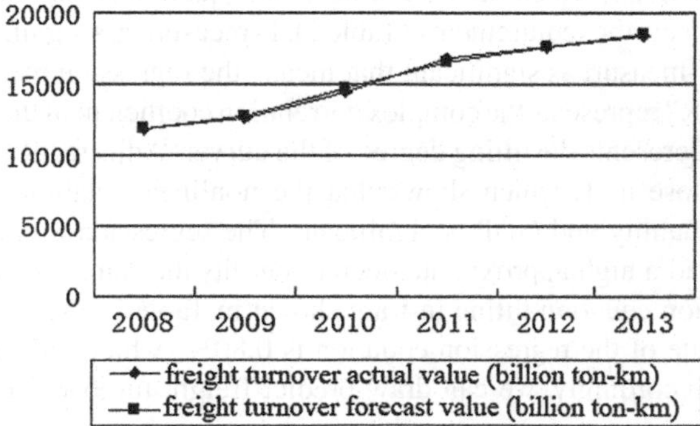

Fig. 2 Freight turnover quantity comparison of between forecast and actual value

Table 5 Results of freight turnover quantity forecast from 2015 to 2020

Year	2015	2016	2017	2018	2019	2020
Freight turnover quantity forecast (Billion Ton–km)	19480.98	19939.63	20378.04	20796.48	21195.30	21574.95

3.3 Freight Volume Forecast

Freight Volume Curve Regression Model Regional freight volume refers to the goods circulation of transportation enterprises in a certain period of time and certain area. The forecast of freight volume is closely related to the GDP of the region. So, we can get the change curve of compound type between the freight volume and GDP. Next, we take Gansu province as a sample to make a freight volume forecast. Firstly, we make the scatter diagram based on freight volume of Gansu province and GDP as shown in Fig. 3. In the Figure, we can affirm that freight volume increase with the growth of GDP. Then, we make the correlation analysis as shown in Table 6. It can be seen that the correlation between freight volume and GDP is 0.982, which has a very high correlation. Furthermore, we can get the single curve regression analysis results as shown in Table 7.

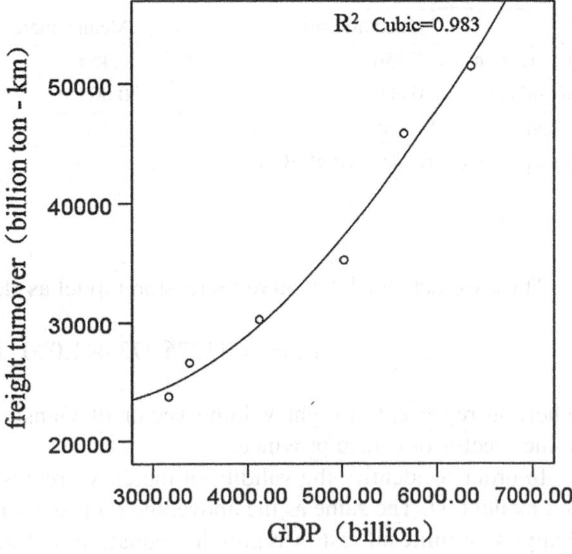

Fig. 3 Scatter diagram of freight volume-GDP of Gansu province

Table 6 Correlation analysis

Correlation			
		GDP (billion)	Freight volume (million ton)
GDP (billion)	Pearson correlation	1	0.982**
	Significant (both sides)		0.000
	N	6	6
Freight (million ton)	Pearson correlation	0.982**	1
	Significant (both sides)	0.000	
	N	6	6

**Significant correlation at 0.01 level (bilateral)

Table 7 Model coefficients

Coefficients					
	Non-Standard B	Standard errors	Standard Beta	t	Sig.
GDP (billion)	1.000	0.000	2.693	61174.867	0.000
(Constant)	11376.273	884.357		12.864	0.000

Variable is ln (Freight volume (billion ton–km))

Table 8 Variance analysis

ANOVA					
	Quadratic sum	df	Mean square	F	Sig.
Regression	0.456	1	0.456	212.701	0.000
Residual	0.009	4	0.002		
Total	0.465	5			

Independent variable is GDP (Billion)

Thus, we achieved the curve regression model as shown in Eq. (2)

$$\ln(w) = 11376.273 + 1.000 \cdot \ln(y) \tag{7}$$

where w represents freight volume vector of Gansu province; y represents GDP values vector of Gansu province.

In order to identify the validity of the curve regression model, we should carry out model test. The same as the above, the regression model is checked by variance analysis, significant test of regression equation and fitting test. Firstly, we make a standard deviation analysis as shown in Table 8. In Table 8, it can be seen obviously

Table 9 Model checking

Model summary			
R	R power	Adjust R power	Standard error of estimate
0.991	0.982	0.977	0.046
Independent variable is GDP (Billion)			

Table 10 Regression equation fitting test

Year	Freight volume	Forecast	Residual	MAPE (%)
2008	23741	24203.78	–462.78	1.95
2009	26605	25511.62	1093.38	4.11
2010	30270	30384.37	–114.37	0.38
2011	35269	37652.54	–2383.54	6.76
2012	45832	43752.72	2079.28	4.54
2013	51463	51458.96	4.04	0.01

that the regression equation is very significant and F-measure indicate the model is correct. Secondly, we make a significant test of regression equation as shown in Table 9. The significance test shows that the "R power" is 0.982, and the "adjusted R" is 0.977, which indicates that the nonlinear relationship between the two is significant well. Meanwhile, the curve has a high fitting degree. Thus, the test results of regression equation fitting are shown in Table 10.

In Table 10, we notice that the mean absolute percentage error rate of the regression equation is small, 2.96 %. Then, we can adopt the curve regression model to predict the freight volume from 2008 to 2013 as shown in Fig. 4. It can be seen that the predicted value is basically in agreement with the actual value.

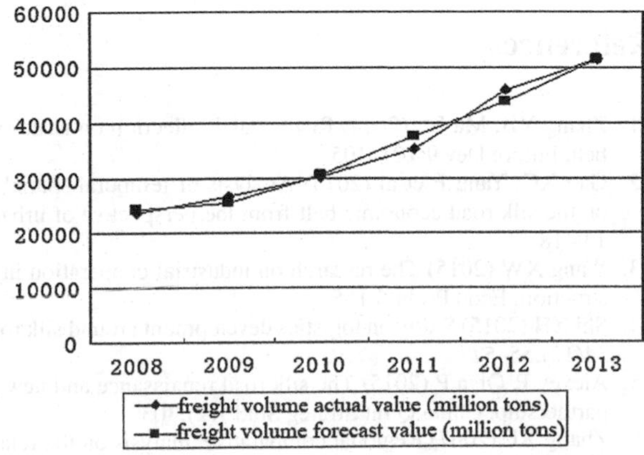

Fig. 4 Freight volume comparison of between forecast and actual value in Gansu

Table 11 Freight volume forecast of Gansu from 2015 to 2020

Year	2015	2016	2017	2018	2019	2020
Freight volume forecast (million)	64037.93	72271.63	82257.52	94475.22	109564.58	128388.78

Freight Volume Forecast In order to predict the freight volume in future, it is still supposed that the average growth rate of GDP in China from 2015 to 2020 is 7 %. Then, we can adopt the curve regression model to predict the freight volume in the next six years. The results of freight turnover quantity forecast are shown in Table 11.

4 Conclusion

To analyze and predict reliably regional freight characteristics in the Silk Road Economic Belt is a preliminary research work. In this study, we put forward to a simple and effective forecast approach for analysis of regional freight characteristics in the Silk Road Economic Belt. A curve regression model is adopted to analyze and forecast the freight turnover quantity and freight volume based on the GDP in recent year. The detailed parameters of the curve regression model could be obtained by SPSS. The results of calculation by SPSS show that this model could be preferably applied to analysis and forecast of freight characteristics. Furthermore, we adopt the model to predict the freight turnover quantity and freight volume of nine provinces in the Silk Road Economic Belt. All these research results can be used to guide the construction of the comprehensive transportation system, in the Silk Road Economic Belt.

References

1. Zhang YB, Ma LL (2014) Review and reflection on issues related to the silk road economic belt. Future Dev 9:101–105
2. Gao XC, Yang F et al (2015) Analysis of temporal-spatial changes of urban economic ties on the silk road economic belt-from the perspective of urban flow intensity. J Lanzhou Univ 1:9–18
3. Wang XW (2015) The research on industrial cooperation in the silk road economic belt construction. Econ Probl 5:1–5
4. Shi XH (2015) Study on logistics development around silk road economic belt. Logist Technol 34(12):55–57
5. Alexey P, Olga P (2015) The silk road renaissance and new potential of the Russian-Chinese partnership. China Q Int Strateg Stud 1(2):305
6. Zhang XL (2007) Regional comparative analysis on the relationship between transport infrastructure and economic growth in China. Study Financ Econ 33(8):51–63

7. Tang J, Yang LP (2011) Regional road freight turnover forecast based on support vector re-gression model. Logist Eng Manag 3:67–69
8. Fu JF, An ZW, Song XM et al (2014) Traffic volume forecast of railway different freight categories based on grey model. J Transp Eng Inf 12(3):38–42
9. Wei XJ, Chen H (2014) Research on the driven factor of logistics development that contribute to economic growth which based on the regional diversity—based on the panel data model analysis of the silk road economic belt in northwest region. Shanghai Econ Rev 6:14–22
10. Li ZM, Yu QY (2014) Study of the spatial heterogeneity of logistics promoting for economic growth: taking the new silk road economic zone as an example. Econ Probl 6:121–125

Controller-Based Management of Connected Vehicles in the Urban Expressway Merging Zone

Chaolei Zhu, Yangzhou Chen and Guiping Dai

Abstract The merging zone in urban expressway is easy to turn into a traffic bottleneck just because of the complex interaction among vehicles there. But with the development of connected vehicle technology, it seems that this problem can be alleviated. This paper designs a system that can harmonize the interaction among connected vehicles in the merging zone, so that they can cross this region in a smaller cost—less average travel time and average fuel consumption. The controller of the system receives the real-time data from vehicles periodically, and sends instructions of the computed solution back to them. At last, Netlogo is utilized to act as a simulation platform to evaluate the proposed system, and the results show that the coordination system of connected vehicles can reduce cost significantly in the merging zone.

Keywords Connected vehicles · Urban expressway merging zone · Cost function

This work was supported by the National Natural Science Foundation of China (61273006, 61573030), and Open Project of Beijing Key Laboratory of Urban Road Intelligent Traffic Control (XN070).

C. Zhu · Y. Chen (✉) · G. Dai
Beijing Key Laboratory of Transportation Engineering, College of Metropolitan Transportation, Beijing University of Technology, Beijing 100 Pingleyuan, Chaoyang District, Beijing 100124, China
e-mail: yzchen@bjut.edu.cn

C. Zhu
e-mail: zhuchaolei101@emails.bjut.edu.cn

G. Dai
e-mail: daigping@bjut.edu.cn

V.E. Balas et al. (eds.), *Information Technology and Intelligent Transportation Systems*, Advances in Intelligent Systems and Computing 455,
DOI 10.1007/978-3-319-38771-0_7

1 Introduction

The urban expressway system enables vehicles to move in a higher and safer speed, so it has been getting a particular development recently. However, due to the rapid increase of vehicle amount, the congestions occurred in expressway are deteriorating, which manifest in larger jam regions and a longer jam duration. In consequence, numerous cities have suffered huge losses, such as the longer traffic delay time and the worse air condition. One of the main reasons is that the expressway merging zones tend to be bottlenecks once disturbed, just because two different traffic flows merge here and the interactions among these vehicles are too complex to predict and control [1]. However, with the development of Intelligent Transportation System (ITS), especially the connected vehicle technology, this problem will be alleviated in the near future [2].

Based on vehicle-to-infrastructure (V2I) and vehicle-to-vehicle (V2V) interaction technologies, connected vehicles can communicate with other vehicles and infrastructures around them through dedicated short-range communications (DSRC) and other wireless transmission media [3]. Therefore, connected vehicle technology has the potential to provide transportation agencies with improved real-time traffic data, making it easier to manage transportation systems for maximum efficiency and minimum cost [4]. In addition, the connected vehicle system can generate and capture relevant real-time traffic data and utilize the data to create actionable information to support and facilitate green and safe transportation choices, as a result, that technology will be able to reduce the environmental impacts and improve the driving safety [5].

Lots of research efforts have been expended on the cooperation mechanism of connected vehicles passing through expressway merging zones and intersections. Jackline [6] proposed a framework and a closed-form solution that optimized the acceleration profile of wireless connected vehicles in terms of fuel economy while avoiding collision at the merging zone. Qiu Jin [7] addressed a platoon-based multi-agent intersection management system where vehicle agents could strategically form platoons through V2V communication. Md Tanveer Hayat [8] developed a connected vehicle enabled freeway merge assistance system, with the aim of reducing conflicts among merging vehicles in freeway ramp area. And they had conducted a field test of this system. Milanes [9] used fuzzy logic to design a controller that allowed a fully automated vehicle to yield to an incoming vehicle in the conflicting road, if collision risk was not present. Zohdy [10] developed a cooperative algorithm based on game theory for the connected vehicles crossing the intersections.

In this paper, the objective is to design a system that can harmonize the interaction among connected vehicles in the urban expressway merging zones, so that they can pass through these regions safer, greener and faster. It is noted that this paper is different from previous research such as Jackline [6] and Qiu Jin [7], which considered two one-lane roads merging. It also differs with Milanes's [8] and Zohdy's [10] research, which focused on the cooperation of the connected vehicles in downtown intersections. The most notable differences are given as follows: (1) a realistic situation considering the multi-lane expressway merging with an on-ramp and an acceleration lane; (2) various management means maintaining connected vehicles crossing the merging zone in a low cost; (3) a novel cost function considering both travel time and fuel consumption; (4) simulation experiments evaluating the performance of the proposed algorithm under varying traffic conditions.

The remainder of this paper is organized as follows: Sect. 2 introduces the principle of this system, including connected vehicle kinematic model, problem formulation, and control algorithm. Section 3 describes the simulation of this system, including the settings, results and analyses. Finally, conclusions are shown in the last section.

2 Problem and Approach

This section proposes the dynamics model of the vehicle, its operational assumptions and limits. In general, two kinds of motions are included in the vehicle's whole moving process: longitudinal motion and lateral motion. For a certain vehicle, its motions are strictly limited by the shape of the lane. For example, the longitudinal motions are always occurred in its acceleration and deceleration process along the current lane, the lateral motions are probably executed by the drivers when they would like to change lanes or overtake the leading vehicle. So two-dimensional state vectors can be used to represent the vehicle's characteristics, and each vehicle's dynamics model can be described as follows:

$$\begin{cases} \dot{x}_i(t) = v_i(t) \\ \dot{v}_i(t) = u_i(t) \end{cases} \tag{1}$$

where $x_i(t)$ is the position of the vehicle on its path, $v_i(t)$ is the speed, $u_i(t)$ is the control input (acceleration).

In addition, we should not only consider the vehicle's dynamic limits related to acceleration shown as Eq. (1), but also we need take into account the limits by its own properties and the surroundings:

$$v^i_{\min} \le v_i(t) \le \min\{v^i_{\max}, v_{\lim}\} \tag{2}$$

where $v^i_{\min} = 0$, $v^i_{\max}$ is the top speed of the vehicle, $v_{\lim}$ is the speed limit of the certain lane. In general, vehicles usually travel in a medium speed in the lane, which is less than its max speed and the lane limit speed.

2.1 Problem Description

This paper considers a common situation, in which an on-ramp merges into a main expressway (Fig. 1). Currently, drivers from on-ramp must be as careful as possible

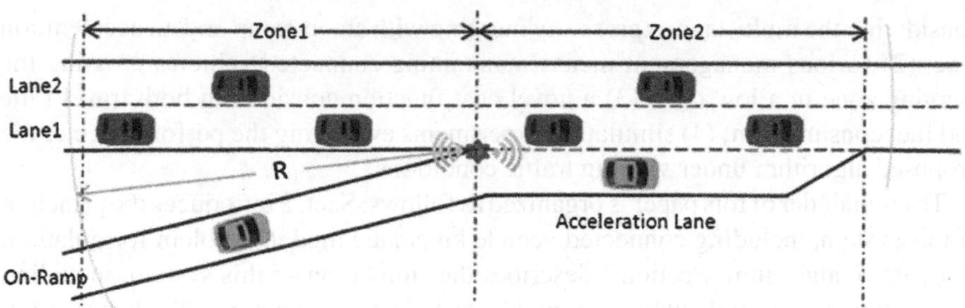

Fig. 1 An on-ramp merging into the main expressway

when they want to merge into the main road safely and quickly. So when they move to the end of the on-ramp, they will always slow down to ensure safety. What's more, if the traffic of the main road is too busy, the drivers will have to stop to wait for a safe merging chance, thus the waiting queue forms a bottleneck. Now this problem may be alleviated by connected vehicle technology. We can set a controller in the merging zone to harmonize the interaction among the multiple traffic flows. During its working hours, it can communicate with all of the local connected vehicles and lead the vehicles to cross the merging zone smoothly without stopping and waiting.

The controller locates around the merging zone, as shown in Fig. 1. Its communication range is a circular region whose radius is R. It assumes that every vehicle within the communication range connects with the controller automatically, then sends its own real-time information (position, speed, acceleration, etc.) to the controller and receives the relevant feedback (instructions) from it periodically. After that, the vehicles will follow the instructions precisely. Thus, the controller achieves the intelligent management of the merging zone.

2.2 *Process Analysis*

To describe the detailed management process and the mechanism, the control zone is divided into two parts: Zone 1 (Pre-merging Zone) and Zone 2 (Merging Zone), as shown in Fig. 1, and whose lengths are L_1 and L_2. Obviously, both of them are less than the communication range radius R. When connected vehicles enter Zone 1, they will create link with the controller automatically, and send their own messages to the controller periodically.

It is assumed that the free flow speed of the mainline is v_f, and the management objective in Zone 1 is that all vehicles can travel in a safe and steady speed around v_f just before they enter Zone 2. If the minimum time headway is $T_{\min}$, the minimum distance headway $L_{\min} = v_f \cdot T_{\min}(11)$. In order to ensure that the vehicles from the on-ramp can merge into the expressway safely, the headways between vehicles in the expressway must be large enough. Let L_{safe} be the safe distance headway threshold in Zone 1, and $L_{safe} = L_{\min}$, in another way, $T_{safe} = T_{\min}$, T_{safe} is the safe time headway threshold. This is the safety condition of Zone 1, or called the first safety condition. Next, let t_i^{arr} be the ith vehicle's arrival time to the separating line between Zone 1 and Zone 2, and let $\Delta t_{i(i+1)}^{arr}$ be the arrival time gap of ith and $(i+1)$th vehicle. Based on the given parameters, we define the maximum number of vehicles in Zone 1: $N_1 = L_1 / L_{safe}$. Therefore, if the real-time vehicle count $n_1(t)$ in Zone 1 is greater than the capacity, the controller will also order some vehicles to change lanes. In a word, the safety conditions in Zone 1 are $n_1(t) \le N$ and $\Delta t_{i(i+1)}^{arr} \le T_{safe}$.

Although the vehicles probably enter Zone 2 in a similar speed, their relative positions may not be fit for merging. If so, some vehicles' positions should be adjusted before the on-ramp vehicles enter the Lane 1. In this part, we will describe the safety merging condition. As shown in Fig. 2, in which l_{ij} is the actual distance between ith and jth vehicle, l is the length of a vehicle. It is easy to get the safety condition of Zone 2 from Fig. 2:

$$\begin{cases} l_{i(i+1)} \ge L_{\min} - l \\ l_{i(i-1)} \ge L_{\min} - l \end{cases} \tag{3}$$

When the ith vehicle merges into the traffic flow of the mainline, it should adjust its speed to satisfy the safety condition of Zone 2.

Based on the received real-time messages, the controller can compute the safety conditions and test if they have been met. As to Zone 1, if the safety condition is not met, the controller will send a lane-changing instruction to some vehicles in lane 1, and a speed-adjusting instruction to some other vehicles in lane 1 and lane 2 to create the safe headway gap. At the same time, the controller will also send a speed-adjusting order to some vehicles in the on-ramp. By those means, all connected vehicles can keep safe headways and have similar speeds just before they enter the merging zone. However, even all of the vehicles cruise in the similar speed before they enter, it also could be dangerous. Because we do not know the exact merging point—or called the latent conflict point, vehicles from the on-ramp can merge into the expressway anywhere they want. Before the vehicles enter the traffic flow of the mainline, the controller will also test if safety conditions are met. If not, the controller will send orders to adjust the vehicle's speed in the acceleration lane.

The entire flow chat is shown in the Fig. 3:

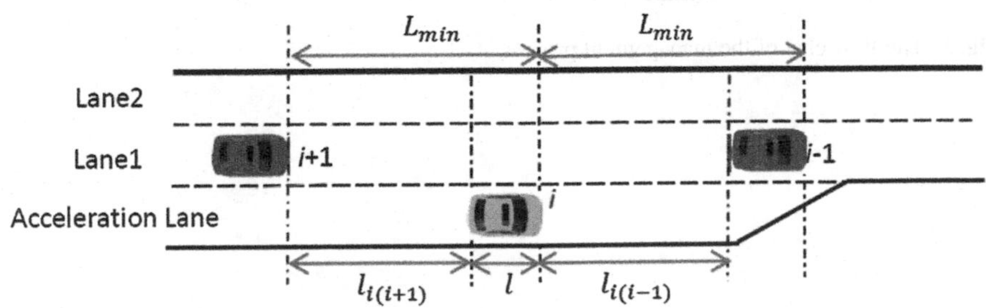

Fig. 2 The safety merging condition in Zone 2

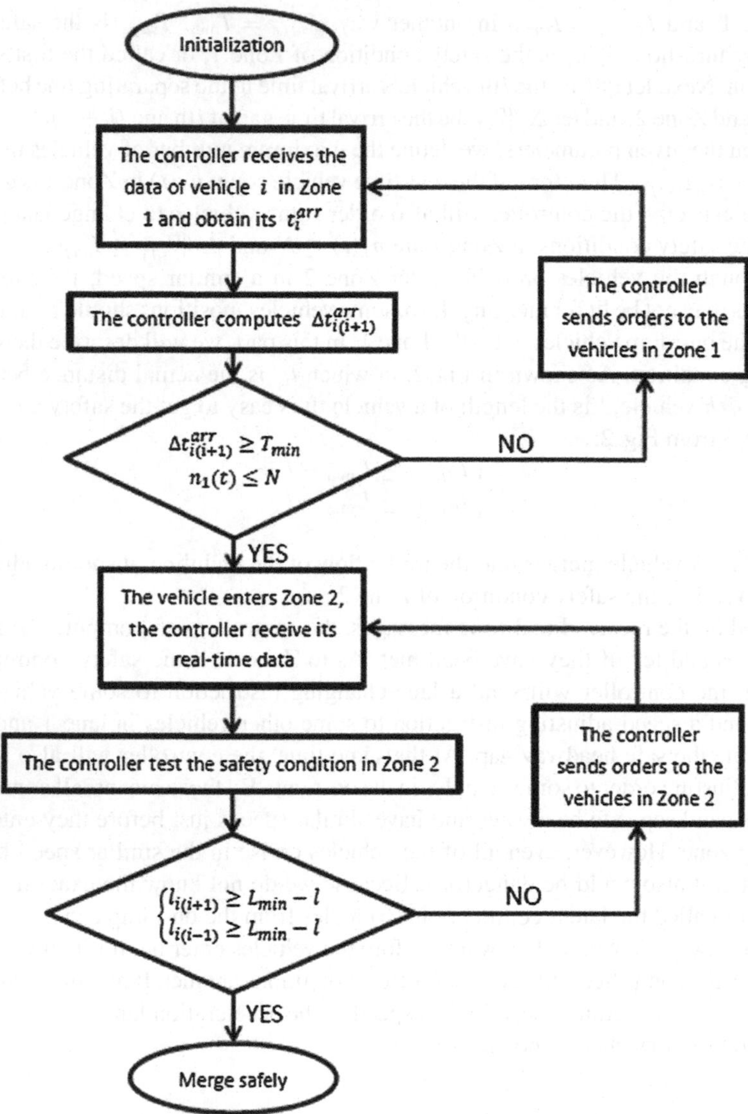

Fig. 3 The flow chat of the management process

2.3 Performance Evaluation

As described in the above section, the management objective is that all of the connected vehicles can cross the merging zone in a smaller average cost. In this paper, the cost consists of two parts: the average travel time and the average fuel consumption. One thing which should be noted is that the management system works all the time, but in order to evaluate the performance of it, we choose a certain period of time as the studying period when we compute the cost function.

The travel time is defined as a specified period of time in which the vehicle spends in the merging zone (Zone 1 and Zone 2). And one thing to be noted is that it includes the time spent by vehicles while waiting in the on-ramp lane. The average travel time is the average value of all the vehicles' real travel time which has been defined before. Under ideal condition, all vehicles could pass through the merging zone in the free flow speed. So the ideal travel time is:

$$t_s = \frac{L_1 + L_2}{v_f} \tag{4}$$

In reality, however, the vehicles cannot always move in the free flow speed because of the interactions among these vehicles and the surroundings. So the real travel time will be greater than the ideal time. Let t_i^r be the real travel time of ith vehicle, it should satisfy the equation:

$$L_1 + L_2 = \int_0^{t_i^r} v_i(t)dt \tag{5}$$

During the studying period, if N vehicles have crossed the merging zone, the average travel time $\bar{t}_r$ can be computed by

$$\bar{t}_r = \frac{\sum_{i=1}^{N} t_i^r}{N} \tag{6}$$

In this part, we define a new parameter k_t as a performance index to measure the traffic crossing the merging zone, and $k_t = \frac{\bar{t}_r}{t_s}$.

The fuel consumption is the total fuel that the vehicle consumed while moving in the merging zone. For a typical vehicle [12], the total fuel consumption consists of two parts: one is the fuel consumed when the vehicle moves in a constant speed, the other part is consumed while the vehicle's speed is not steady, such as the condition of acceleration and deceleration. Like the definition of the average travel time, the average fuel consumption is the average value of all the vehicles' real fuel consumption which has been defined before. The typical vehicle's fuel consumption per period f can be estimated as follows: can be estimated as follows:

$$f = f_1 + f_2 \tag{7}$$

where f_1 is the fuel consumed per period at a steady speed, and

$$f_1 = w_0 + w_1 v + w_2 v^2 + w_3 v^3 \tag{8}$$

f_2 is additional consumption due to presence of acceleration, and

$$f_2 = a(r_0 + r_1 v + r_2 v^2) \tag{9}$$

v is the speed of the vehicle, a is the acceleration, w_i and r_j are parameters. In the fuel consumption experiment, the parameters were identified as follow: $w_0 = 0.1569$, $w_1 = 2.45 \times 10^{-2}$, $w_2 = -7.415 \times 10^{-4}$, $w_3 = 5.975 \times 10^{-5}$, $r_0 = 0.07224$, $r_1 = 9.681 \times 10^{-2}$, $r_2 = 1.075 \times 10^{-3}$. Under the ideal condition, the vehicles could pass through the merging zone in a constant velocity, so the individual vehicle's fuel consumption is a constant as well, and $f_s = f_1 = w_0 + w_1 v + w_2 v^2 + w_3 v^3$, where $v = v_f$. In reality, the vehicles' speeds are always changing, so the real fuel consumption of a vehicle is:

$$f_r = (w_0 + w_1 v + w_2 v^2 + w_3 v^3) + a(r_0 + r_1 v + r_2 v^2) \tag{10}$$

In the period, if N vehicles passed through the region, the average fuel consumption can be computed by: $\overline{f_r} = \frac{\sum_{i=1}^{N} f_r^i}{N}$. This paper defines a new variable k_f as a performance index to measure the fuel consumption in the region, and $k_f = \frac{\overline{f_r}}{f_s}$.

Given the earlier mentioned variables k_t and k_f, we introduce a new measuring function as the performance:

$$J = \alpha \cdot k_t + (1 - \alpha) \cdot k_f \tag{11}$$

where α is the weight of travel time, and $\alpha \in [0, 1]$. $\alpha = 0$, that means only fuel consumption is considered when we measure the cost. $\alpha = 1$, that stands for the cost function just takes the travel time into account. In addition, under the ideal condition, J would always equal to 1, whatever α is.

3 Simulations and Analysis

In order to evaluate the performance of the proposed system, the multi-agent programming and modeling environment–NetLogo is used to act as the simulation platform. At first, a model with the merging zone is built, whose background is shown in Fig. 4. In this model, we select the roadway dimensions strictly based on the rules of Standard of P.R.C. *Specification for Design of Urban Expressway*, so as to ensure the simulation results as accurate as possible. The length of Zone 1 is 120 m, the length of Zone 2 is 165 m, the free flow speed is set as 60 Km/h, the flow rate of the mainline freeway is 1800 vehicles per hour per lane, the flow rate of the on-ramp

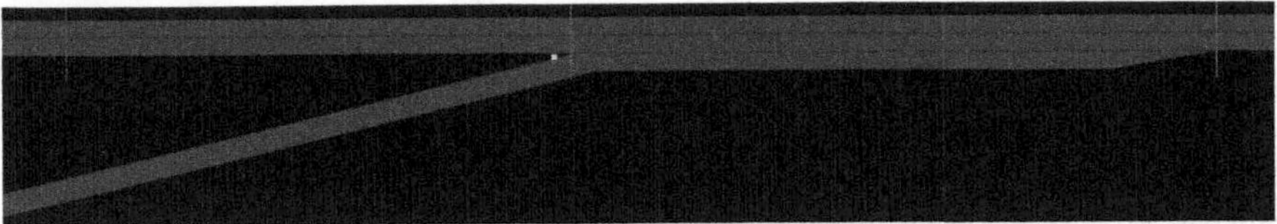

Fig. 4 The background of the simulation

is 1200 vehicles per hour per lane. The vehicle generation model uses the Poisson distribution with the variable of time headway [13].

In the simulation, results will be compared in the same merging zone and same parameters with and without management. In the simulation without any management, all vehicles are assumed to be fully operated by their own drivers independently. In the second simulation, it is assumed that all connected vehicles have already been equipped with DSRC device and they can perfectly follow the instructions [14].

Each group of simulations will share the same initial conditions to ensure the accuracy. The simulation results are shown in Figs. 5 and 6. The first group of figures shows that how the average speed changes in different management strategies. As presented in the figures, the speed seriously fluctuates under the free flow speed without any management measures conducted on the merging zone. As a comparison, after conducting the management system, the average speed increases and fluctuates in a smaller scale. The trend also occurs in the average cost figures.

For evaluating the efficiency of the system quantitatively, this paper has conducted various different groups of simulation. Each group has two situations as presented

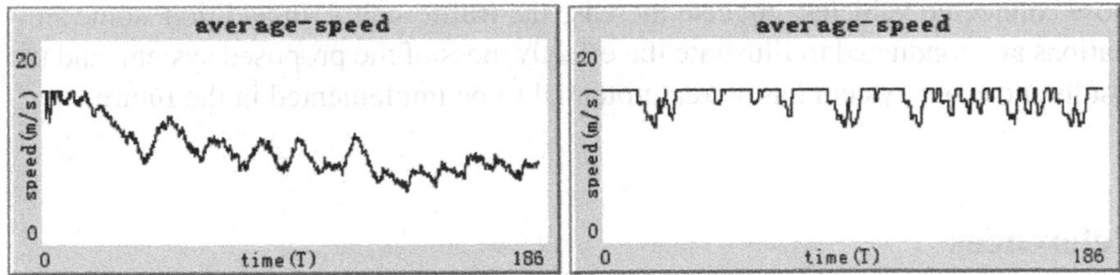

Fig. 5 The average speed without/with management

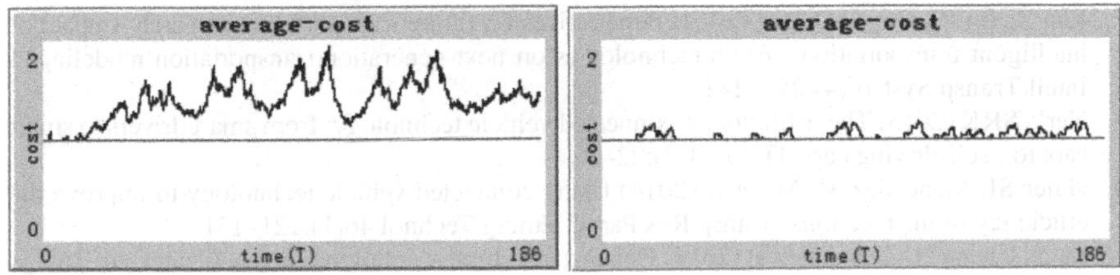

Fig. 6 The cost function without/with management

Table 1 Relative improvement of the performance

	Group1			Group2			Group3		
	$\overline{t_r}$	$\overline{f_r}$	J	$\overline{t_r}$	$\overline{f_r}$	J	$\overline{t_r}$	$\overline{f_r}$	J
Simulation1	27.3	17.9	1.55	30.6	19.6	1.7	32.1	21.4	1.82
Simulation2	21.1	14.4	1.22	22.5	15.8	1.31	23.8	16.5	1.38
Improvement (%)	22.7	19.5	21.2	26.4	19.2	22.9	25.8	22.8	24.1

in the above, different groups have different parameters, such as vehicle dispatch model, the initial speed and acceleration, the duration of simulation and so on. Table 1 shows the three groups of simulation results. The line of Simulation1 is the statistics results of simulation without any management, and the line of Simulation 2 is the statistics results of simulation in which the proposed management has been carried out. Obviously, the average travel time, average fuel consumption and cost of the system all gain significant improvements.

4 Conclusion

In this paper, we develop a system that can harmonize the interaction among connected vehicles within the urban expressway merging zone. Based on the received real-time traffic data, the controller sends the corresponding instructions to the individual vehicles, so that they can pass through that region steadily. As a result, the system could not only decrease the average travel time and fuel consumption for those connected vehicles, but also increase the traffic safety. In addition, some simulations are conducted to illustrate the effectiveness of the proposed system, and the results show our system has a great potential to be implemented in the future.

References

1. Sarvi M, Kuwahara M, Ceder A (2007) Observing freeway ramp merging phenomena in congested traffic. J Adv Transp 41(2):145–170
2. Ran B, Jin PJ, Boyce D et al (2012) Perspectives on future transportation research: Impact of intelligent transportation system technologies on next-generation transportation modeling. J Intell Transp Syst 16(4):226–242
3. Narla SRK (2013) The evolution of connected vehicle technology: from smart drivers to smart cars to.. self-driving cars. ITE J 83(7):22–26
4. Guler SI, Menendez M, Meier L (2014) Using connected vehicle technology to improve the efficiency of intersections. Transp Res Part C Emerg Technol 46(1):121–131
5. Lu N, Cheng N, Zhang N et al (2014) Connected vehicles: solutions and challenges. Int Things J 1(4):289–299

6. Rios-Torres J, Malikopoulos A, Pisu P (2015) Online Optimal control of connected vehicles for efficient traffic flow at merging roads. Oak Ridge National Laboratory, National Transportation Research Center
7. Jin Q, Wu G, Boriboonsomsin K, et al. (2013) Platoon-based multi-agent intersection management for connected vehicle. In: Intelligent transportation systems conference, 1462–1467
8. Hayat MT, Park H, Smith BL (2014) Connected vehicle enabled freeway merge assistance system-field test: preliminary results of driver compliance to advisory. In: Intelligent Vehicles Symposium, pp 1017–1022
9. Milanés V, Pérez J, Onieva E et al (2010) Controller for urban intersections based on wireless communications and fuzzy logic. Intell Transp Syst 11(1):243–248
10. Zohdy IH, Rakha H (2012) Game theory algorithm for intersection-based cooperative adaptive cruise control (CACC) systems. In: Intelligent transportation systems conference, pp 1097–1102
11. Risto M, Martens MH (2013) Time and space: the difference between following time headway and distance headway instructions. Transp Res Part F Traffic Psychol Behav 17(1):45–51
12. Kamal M, Samad A, Mukai M et al (2013) Model predictive control of vehicles on urban roads for improved fuel economy. Control Syst Technol 21(3):831–841
13. Yue MZYJS (2003) Poisson distribution based mathematic model of producing vehicles in microscopic traffic simulator. J Wuhan Univ Technol 1:21–26
14. Lee J, Park B (2012) Development and evaluation of a cooperative vehicle intersection control algorithm under the connected vehicles environment. Intell Transp Syst 13(1):81–90

Part II
Information Technology, Electronic and Control System

Max-Pooling Convolutional Neural Network for Chinese Digital Gesture Recognition

Zhao Qian, Li Yawei, Zhu Mengyu, Yang Yuliang, Xiao Ling, Xu Chunyu and Li Lin

Abstract A pattern recognition approach is proposed for the Chinese digital gesture. We shot a group of digital gesture videos by a monocular camera. Then, the video was converted into frame format and turned into the gray image. We selected the gray image as our own dataset. The dataset was divided into six gesture classes and other meaningless gestures. We use the neural network (NN) combining convolution and Max-Pooling (MPCNN) for classification of digital gestures. The MPCNN presents some differences on the data preprocessing, the activation function and the network structure. The accuracy and the robustness have been verified by the simulation experiments with the dataset. The result shows that the MPCNN classifies six gesture

---

Zhu Mengyu—This work is supported by the national High Technology Research and Development Program of China (2015AA042300).

Z. Qian
College of Computer Science and Technology, Jilin University, Changchun, China
e-mail: supernext007@163.com

L. Yawei · Y. Yuliang
School of Computer and Communication Engineering,
University of Science and Technology Beijing, Beijing, China
e-mail: ustbweiwei@163.com

Y. Yuliang
e-mail: teacheryangustb@126.com

Z. Mengyu (✉)
School of Life Science and Technology, Beijing Institute of Technology, Beijing, China
e-mail: zmy@bit.edu.cn

Z. Qian · X. Ling · X. Chunyu
Luoyang Hydraulic Engineering Technical Institute, Luoyang, China
e-mail: xiaoling1234@eyou.com

X. Chunyu
e-mail: xcy860915@sina.com

L. Lin
Beijing Canbao Architectural Design and Research Institute, Beijing, China
e-mail: bdsttt@163.com

V.E. Balas et al. (eds.), *Information Technology and Intelligent Transportation Systems*, Advances in Intelligent Systems and Computing 455,
DOI 10.1007/978-3-319-38771-0_8

classes with 99.98 % accuracy using the Max-Pooling, the Relu activation function, and the binarization processing.

Keywords Convolutional neural network · Chinese digital gesture recognition · Data preprocessing · Activation function

1 Introduction

Digital gesture recognition [1, 2] has tremendous potential in the field of user interaction and human communication. Since China has such a large population, Chinese digital gesture recognition is an extremely significant issue. Therefore, this article presents an exclusive solution for the Chinese digital gesture recognition to achieve a natural and intelligent human-computer interaction [3, 4]. Nowadays, the Chinese digital gesture recognition is in the initial stage and needs further research.

Digital gesture still exist limitations to its real environment application. As an example, R.Y. Wang [5] used Hausdorff-like distances and the color glove as metric to recognition static gestures from camera. Even if this method is robust in indoor scenes, it sacrifices flexibility and accuracy. Another method is based on the hidden Markov model (HMM) gesture recognition [6, 7]. The specialty of general topological structure of HMM is to describe the gesture signal. However, large number of state probability density should be calculated and large number of parameters should be estimated. The training and recognition speed is relatively slow.

In order to solve the issues mentioned above for digital gesture recognition, We propose a method with the Max-Pooling convolutional neural network [8] to simplify the complicated questions. Convolutional neural network [9, 10] is an effective recognition approach in the field of pattern recognition and image processing. It has the advantages of simpler structure, fewer training parameters and stronger adaptability. Unlike the neural network, the MPCNN presents some differences on the data preprocessing, the activation function [11] and the network structure.

The rest of this paper is organized as follows: The convolutional neural network is briefly reviewed in Sect. 2, our proposed approach is described in Sect. 3 and the experiment results for digital gesture recognition are given in Sect. 4. We present our conclusions in Sect. 5.

2 Convolutional Neural Network

Convolutional neural network combines three methods to achieve the recognition of displacement, the scaling, and the other forms of distortion in-variance. They are local receptive field, weight sharing, and subsampling. Local receptive field means that each layer neurons connect with a small neighborhood of nerve cells in the upper layer. Weight sharing technology significantly reduces the training parame-

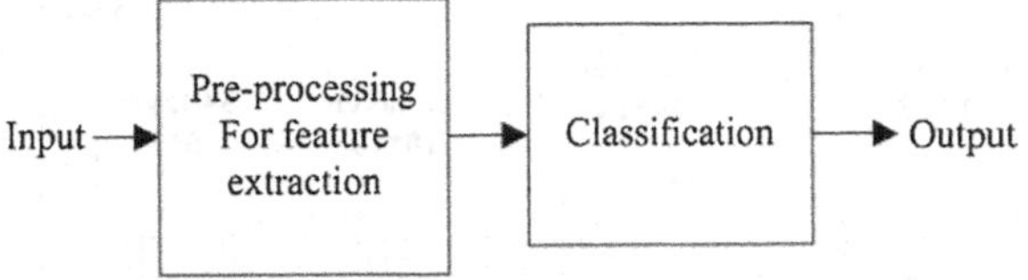

Fig. 1 Traditional classification method

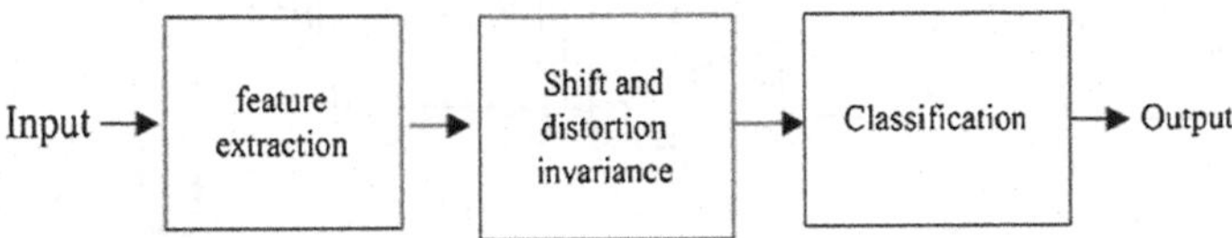

Fig. 2 CNN classification method

ters of the network. Therefore, the structure of the neural network becomes more simple and adaptable. Subsampling features can reduce the resolution to achieve the displacement, the scaling, and the other forms of distortion in-variance.

The traditional classification model is shown in Fig. 1. The image features are extracted [12] by a series of complex preprocessing, eventually the results are given in the output of the classification. The differences of the CNN is that it can directly input a two-dimensional image and the results are given in the output of the classification, among them image features are extracted through convolutional layers and subsampling layers. As is shown in Fig. 2.

2.1 Network Structure

The Fig. 3 is a convolutional neural network structure, including input and output layer, convolution layer, subsampling layer and full connection layer. The layer inputs the images in a size of 3232 pixels, by alternating convolutional layer, the subsampling layer and finally fully connected layer. The result is given in the output layer of the network.

2.2 Convolutional Layer

Convolutional layer is also called as feature extraction layer. Each input neuron connects with the former layer of local receptive fields and extracts the local feature. Once the local feature is extracted, the positional relationship between it and other characteristics will be determined. An important feature of the convolution operation

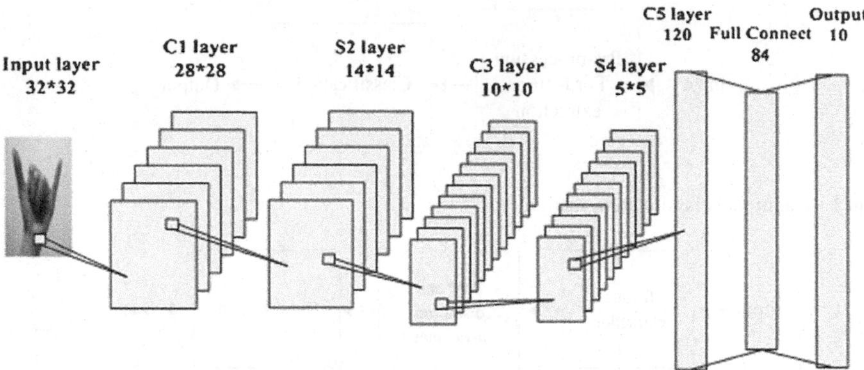

Fig. 3 Convolution neural network structure

is that the original signal is enhanced and the noise is reduced. In general, the form of the convolutional layer is shown in the form of a formula (1) [13]:

$$x_j^l = f(\sum_{i \in M_j} x_i^{l-1} * k_{ij}^l + b_j^l) \quad (1)$$

The above function shows: x_j^l represents the jth feature map of the L layer in the convolution layer, $f(\cdot)$ is the representation of the activation function, and the K is the convolution kernel, M_j represents the collection of the input graph, the convolution operation $*$, and the bias B.

The $C1$ layer is a convolutional layer, which is composed of six feature maps. Each neuron in the feature map is connected to the neighborhood of the input the size of 5×5. The feature map size is 28×28, which can prevent the connection input fell outside of the bounds. The C3 layer is a convolutional layer. The feature maps are only 10×10 neurons. It has sixteen different convolution kernels, so there are sixteen feature maps. The $C5$ layer is also a convolutional layer, and there are 120 feature maps. Each unit is connected with the 55 neighborhood of the layer of the $S4$ layer. The size of the $C5$ feature map is 1×1, which constitutes the full connection between the $C5$ layer and the $S4$ layer.

2.3 Subsampling Layer

Subsampling layer takes the sample operation on the input image. If the number of input feature map is n, then the output of the feature map number is also n, but the output of the feature map is smaller than the input graph. Using the principle of local correlation, the image can be sampled and the data processing capacity can be reduced. The structure uses the Sigmoid function as the activation function of the convolutional network. The feature map has the displacement in-variance. The general form of the subsampling layer is shown in formula (2) [13]:

$$x_j^l = f(\beta_j^l down(x_j^{l-1}) + b_j^l) \tag{2}$$

The above function shows: $down(\cdot)$ represents subsampling functions. The subsampling function is the weighted sum of the size of $n \times n$ of the input image, therefore, output image size is $\frac{1}{n}$ times of the size of the input image. β as the weighted coefficients, and b for the bias.

The $S2$ layer is a subsampling layer with six feature maps of the size of 14×14. Each element in the feature map is connected to the 2×2 neighborhood of the $C1$ layer in the corresponding feature map. The $S4$ layer is a subsampling layer with 16 feature maps.

2.4 Fully Connected Layer and Output Layer

The $F6$ layer has 84 units (the reason for choosing this number comes from the output layer), which is connected with the $C5$ layer. Dropout is used for learning of the fully connected layer. The output layer has 10 neurons, which are composed of the radial basis function unit (RBF), and each neuron of the output layer corresponds to a gesture class. The calculation method for the RBF unit is shown in the formula (3) [4]:

$$y_i = \sum_j (x_j - w_{ij})^2 \tag{3}$$

3 Proposed Method

In view of the difference between Chinese digital gesture recognition and handwritten digital recognition [14], the variants of Max-Pooling convolutional neural network has been list out:

(1) The output layer of the convolutional neural network is changed from ten to six neurons. The dataset is divided in six gesture classes and other meaningless gestures. Therefore, the number of output neurons should select six.

(2) The convolutional neural network requires simple preprocessing, and we choose the most appropriate method by comparing different results.

(3) Two activation functions are commonly used in the neural network. Two kernel functions are shown in Fig. 4.

The Sigmoid function [15] is the choice of the general neural network, but for the depth of the network, many people choose the Relu kernel function [15], the reason lies in:

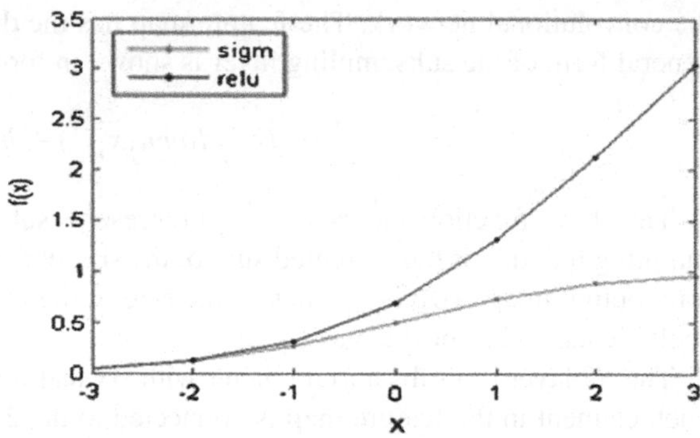

Fig. 4 Comparison of two activation functions

- to simplify the calculation of Back Propagation
- to make learning faster
- to avoid saturation issues

The Relu function has three major changes in the Relu function:

- unilateral suppression
- relatively wide excitement border
- sparse activation

(4) The subsampling layer is also called as the pooling layer, the common pool operation is the average pool and the maximum pool [16]. The average pool: the average value of the image area is calculated as the value of the pool. The maximum pool: the maximum value of the selected image area is calculated as the value of the pool after the region.

Max-Pooling leads to faster convergence rate by selecting superior invariant features which improves generalization performance. Each sub region (rectangular area) output is the maximum value. Its typical value is 2×2. The maximum pool is used to reduce the dimension of the intermediate representation.

4 The Experiment Results

There is no official available gesture dataset, so we shoot a group of digital gesture videos for the experiment. The gesture dataset is divided into six gesture classes, from number 6 to number 10 and other meaningless gestures. The digital gesture videos are made by a monocular camera, each gesture class for five minutes. The video will be converted into frame format and be turned into the gray image. During the data acquisition process, hand and camera position remains unchanged, gestures can be placed at random and can be rotated, up, down, left and right arbitrary. Part of the gestures are shown in Fig. 5.

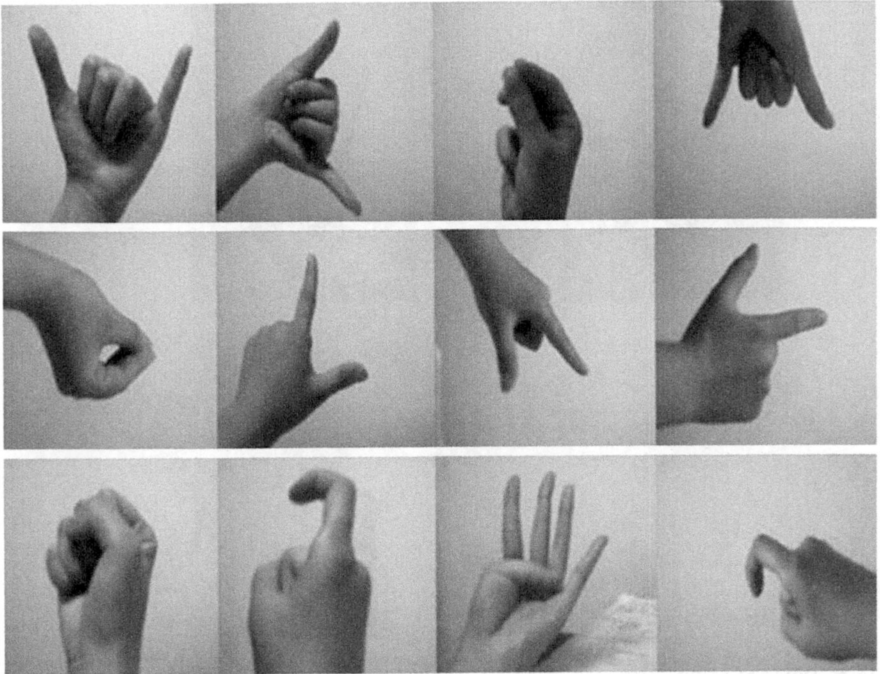

Fig. 5 Gesture images

The acquired 18000 images are split in ratios of 80 % and 20 % for the training set and test set respectively. The dataset contains 3000 images for each class. Each gesture class is randomly selected from 2500 samples, and 500 samples are randomly selected as the test set, and the other 2000 samples are used in training set.

4.1 Data Preprocessing

Simple preprocessing can improve the efficiency and accuracy of the whole recognition system, including the processing of gesture area detection and gesture processing.

The gesture region needs to be detected. All images are stored in an 8-bit unsigned Portable Network Graphics (PNG) format. All images taken by the frame are converted into gray scale images. The gray value is used to select a threshold which is processed by binarization. The gesture part is Zero, and the background part is One. All the gestures are scaled to 32×32 pixels in the region as experimental data input. As is shown in Fig. 6.

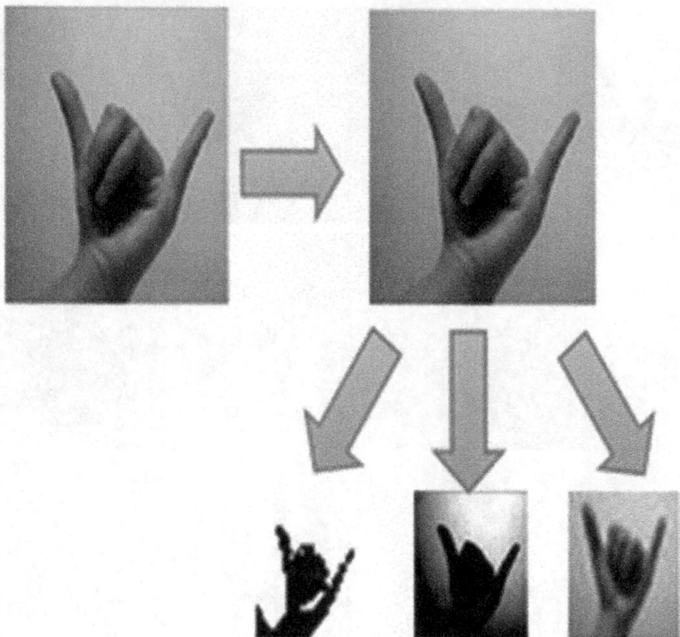

Fig. 6 Different gesture data

Table 1 Error rate of different preprocessing

Numepochs pre-processing	2 times (%)	5 times (%)	10 times (%)	15 times (%)	30 times (%)
Gray processing	21.96	4.2	2.56	1.44	0.36
Whitening processing	19.24	4.1	1.76	1.08	0.32
Binarization processing	15.52	4.04	1.72	0.84	0.32

In this paper, different datasets are obtained by different processing methods. Then through the MPCNN model, we get the following error rate (the data in the Table 1 is the error rate of the test set).

By comparing three datasets, the accuracy rate of the dataset obtained by binarization processing is the highest. It also displays the number of iterations influence on the accuracy rate. Table 1 indicates that the binarization processing of the input images, making it the best choice for the Chinese digital gesture recognition.

Table 2 Error rate of different activation functions

Whitening processing	2 times	5 times	10 times	15 times
Sigmoid	19.24 %	4.1 %	1.76 %	1.08 %
Relu	1.92 %	0.48 %	0.28 %	0.20 %
Binarization processing	2 times	5 times	10 times	15 times
Sigmoid	15.52 %	4.04 %	1.72 %	70.84 %
Relu	1.40 %	0.28 %	0.08 %	0.08 %

4.2 Different Activation Functions

Relu is the popular activation function, because of its faster convergence. The MPCNN model uses the Relu activation function as the substitute of the Sigmoid activation function. Table 2 indicates that the error rate of the test set is reduced to four percent after two iterations. The use of Relu function makes the learning cycle greatly shortened. Considering accuracy rate and efficiency rate, most of the activation functions in the convolutional neural network should select the Relu function. Table 2.

The MPCNN model displays the mean square error curve, which is shown in Fig. 7. The horizontal coordinate represents the sample number of the test dataset, and the vertical coordinate is the mean square error of the test dataset. The final learning rate is 100 %, the iteration number is 30 times, the activation function is Relu, the Max-Pooling value is 2 × 2, and the time required for the iteration is about 50 s.

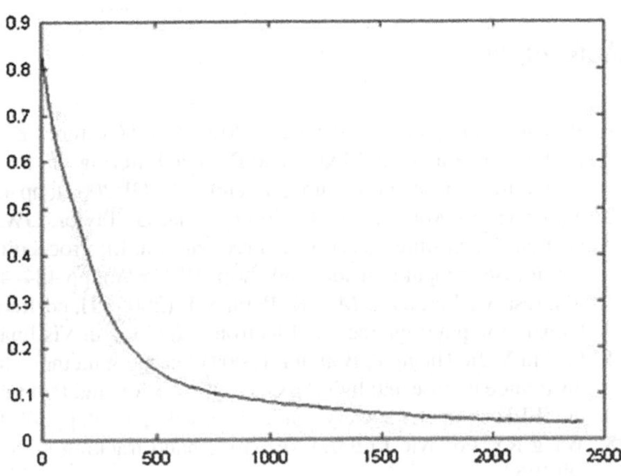

Fig. 7 Mean square error curve

Fig. 8 Error gesture samples

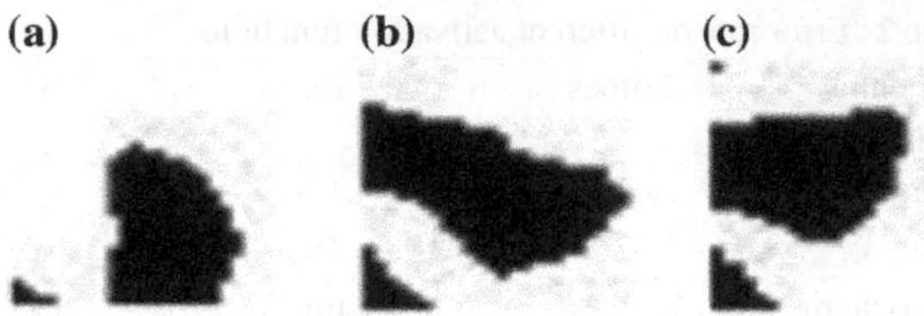

4.3 Error Sample Analysis

The error samples are caused by the effect of pretreatment (character segmentation effect is not ideal) and the effect of noise (critical region on the gesture has prodigious noise). The digital gesture number 7 is mistaken for number 10. It becomes visually challenging for humans to distinguish the correct finger number by inspecting the images. The error rate can be reduced by an effective preprocessing procedure. In addition, it can improve the accuracy rate of the Chinese digital gesture recognition by expanding the dataset. The error samples are shown in Fig. 8.

5 Conclusion

Chinese digital gesture recognition is presented based on the Max-Pooling convolutional neural network. We studied different processing on the overall performance. A result of 99.92 % accuracy of six gesture classes is classified using the Max-Pooling, the Relu activation function, and the binarization processing. The MPCNN is an effective recognition model in the field of the image recognition.

References

1. Bretzner L, Laptev I, Lindeberg T (2002) Hand Gesture Recognition using Multi-Scale Colour Features, Hierarchical Models and Particle Filtering. In: Proceedings of the IEEE confer-ence on automatic face and gesture recognition (FGR2002), pp 423–428
2. Neverova N, Wolf C, Paci G, Sommavilla, G, Taylor, GW, Nebout, F (2013) A Multi-scale approach to gesture detection and recognition. In: Proceedings of the IEEE international conference on computer vision workshops (ICCVW), pp 484–491
3. Manresa C, Varona J, Mas R, Perales F (2005) Hand tracking and gesture recognition for human-computer interaction. Electron Lett Comput Vis Image Anal 5(3):96–104
4. LeCun Y, Jie Huang F, Bottou L (2004) Learning methods for generic object recognition with invariance to pose and lighting Computer Vision and Pattern Recognition". In: Proceedings of the IEEE computer society conference, vol 2, No 2, pp 97–104
5. Wang RY, Popovic J (2009) Real-time hand-tracking with a color glove. ACM Trans Graph 28(3):63

6. Liu N, Lovell BC, Kootsookos PJ (2004) Model structure selection & training algorithms for an HMM gesture recognition system, Ninth international workshop on frontiers in handwriting recognition, IEEE conference publications, IWFHR-9. pp 100–105
7. Bluche T, Ney H, Kermorvant C (2013) Tandem HMM with convolutional neural network for handwritten word recognition, In: Proceedings of the IEEE international conference on acoustics, speech and signal processing (ICASSP), pp 2390–2394
8. Nagi J, Ducatelle F, Di Caro GA, Ciresan D, Meier U, Giusti A, Nagi F, Schmidhuber J, Gambardella LM (2011) Max-pooling convolutional neural networks for vision-based hand gesture recognition. In: Proceedings of the IEEE international conference on signal and image processing applications (ICSIPA), pp 342–347
9. Duffner S, Berlemont S, Lefebvre G, Garcia C (2014) 3D gesture classification with convolutional neural networks. In: Proceedings of the IEEE international conference on acoustics, speech and signal processing (ICASSP), pp 5432–5436
10. Yamashita T, Watasue T (2014) Hand posture recognition based on bottom-up structured deep convolutional neural network with curriculum learning. In: Proceedings of the IEEE international conference on image processing (ICIP), pp 853–857
11. Kim, H-J, Lee JS, Park J-H (2010) Dynamic hand gesture recognition using a CNN model with 3D receptive fields. In: Proceedings of the international conference on neural networks and signal processing, 2008, pp 14–19
12. Simonyan K, Zisserman A (2014) Very Deep Convolutional Networks for Large-Scale Image Recognition, arXiv preprint arXiv, pp 1409–1556
13. LeCun Y, Bottou L, Bengio Y, Haffner P (1998) Gradient-based learning applied to document recognition. In: Proceedings of the IEEE, vol 86, No 11, pp 2278–2324
14. Lauer F, Suen CY, Bloch G (2007) A trainable feature extractor for handwritten digit recognition, Pattern Recognit (S0031-3203), 40(6):1816–1824
15. Francke H, Ruiz-del-Solar J, Verschae R (2007) Real time hand gesture detection and recognition using boosted classifiers and active learning, In: Proceedings of the 2nd Pacific Rim conference on advances in image and video technology (PSIVT07), pp 533–547
16. Nagi J, Giusti A, Nagi F, Gambardella LM, Di Caro GA (2014) Online feature extraction for the incremental learning of gestures in human-swarm interaction. In: Proceedings of the IEEE international conference on robotics and automation (ICRA), pp 3331–3338
17. LeCun Y, Kavukcuoglu K, Farabet C (2010) Convolutional networks and appli-cations in vision. In: Proceedings of the IEEE international symposium on circuits and systems, pp 253–226

An Evaluation of Auto Fault Diagnosis Complexity Based on Information Entropy

Lei Wang and Jianyou Zhao

Abstract The auto fault diagnosis complexity coefficient should be applied for repairing and maintaining price model according to modern auto complex structure. Based on the information entropy theory, this paper proposed the fault diagnosis complexity evaluation for the auto engine fault as a example, five hierarchies structure and evaluate index system were built, then worked out the uniform complexity information entropy value of the auto engine fault diagnosis, which was consistent with the auto repair and maintenance industrial actual situation. The information entropy evaluation provided theoretical basis for the existence and generation of the fault diagnosis complexity coefficient.

Keywords Auto fault diagnosis · Information entropy · Complexity coefficient · Hierarchy structure

1 Introduction

Now the equipment fault diagnosis complexity coefficient is widely used to formulate the work time quota [1], the complex coefficient is an assume unit which the figure mainly depends on the equipment maintainability. The equipment is easy to repair then complex coefficient figure is small and vice versa [6]. Modern equipment is not only composed of traditional mechanical structure, but also a large number of new cross disciplinary technology were applied in this area, which causes large errors if we simply makes use of repair work to formulate the complex coefficient. And

---

L. Wang (✉)
Aviation and Automotive Department, Tianjin Sino-German University of Applied Sciences, Tianjin, China
e-mail: lei.wang6@aliyun.com

J. Zhao
School of Automobile, Chang'an University, Xi'an, China
e-mail: jyzhao@chd.edu.cn

V.E. Balas et al. (eds.), *Information Technology and Intelligent Transportation Systems*, Advances in Intelligent Systems and Computing 455,
DOI 10.1007/978-3-319-38771-0_9

the design of modern equipment is biased towards module structure, although the internal small unit design is complex, the repair procedure is often performed in the assembly replacement, so the corresponding repair complexity is not high.

At present, the general application of repair working hours pricing formula is: Working Hours = Work Time Quota × Unit Price, with the continuous application of the automotive new electronic technology, the performance and comfort ability of the auto is improved, but the whole electrical and electronic control system is becoming more and more complex.

The structure complexity of modern auto determines the auto repair and maintenance industry having great technical complexity. Obviously, the traditional sense of complex coefficient can not accurately and comprehensively reflect the complexity of modern auto repair and maintenance, and can not reflect the difference of repair and maintenance work. And modern car repair quota should consider the complexity of the auto fault diagnosis, that needs to establish a reasonable and scientific model of the fault diagnosis working hours. In China, there is no research on the complexity evaluation of auto fault diagnosis. Based on the information entropy, this paper evaluates the auto fault diagnosis complexity, which provides theoretical basis for the existence and generation of fault diagnosis complexity coefficient.

In recent years, the information entropy fault diagnosis method in other fields of equipment fault diagnosis is mentioned widely. Based on the information entropy, Taylor and Francis proposed new method of supply chain complexity measurement and analysis; S.M. Pncius studied the rationality of the approximate entropy complexity, Liu put forward multi-angle evaluation of the equipment management complexity; based on the basic concept of information fusion, Chen Fei et al. used the singular spectrum entropy, power spectrum entropy, and wavelet characteristic entropy spectrum, analyzed the characteristics of time domain, frequency domain and time-frequency domain, and proposed the method of vibration fault monitoring and diagnosis for rotating machinery based on information entropy distance [1], and used an example to support the feasibility of information entropy distance calculating the fuzzy property. Huang Chengdong, used the information entropy method to analyze the fault diagnosis of scroll compressor; according to the information entropy principle, Kong Fansen et al. evaluated complexity of electrical equipment fault diagnosis, provided theoretical support for the complex degree of electrical equipment maintenance. The entropy principle can play an important role in the fault diagnosis and maintenance strategy formulation. In this paper, the evaluation method of auto fault diagnosis complexity is studied based on information entropy, and the evaluation method validity is proved by using the fault tree example. The method mentioned above can be applied to determine auto repair complex coefficient of the specific fault, to the improve the rationality and scientificity of the auto repair quota.

2 Feature of Auto Fault Diagnosis

The auto fault diagnosis complexity is suitable for the evaluation by information entropy, the characteristics of the modern auto fault is as following [8]. Working on Auto fault diagnosis, it needs to classify the fault phenomenon and causes, for auto continuous fault it should find out and replace the problem parts, for the auto accidental fault if it returns to normal after repair or maintenance, it needs to find out the fault causes and take corresponding measures.

(1) Auto fault diagnosis has the hierarchical characteristics

In the process of auto fault diagnosis, auto maintenance engineers analyze the auto fault with view of hierarchical characteristics, for the purpose of auto repair or maintenance is to resume work soon, so its enough that the auto maintenance engineers check out the fault parts and replace it, no needs to test each tiny electrical components of the fault part assembly any more.

(2) Modern autos are equipped with much complex electronic devices, it determines the auto fault diagnosis is more difficult than fault parts replacement.

In the process of practical repair and maintenance, engineers need to observe and ask the auto owner about the fault phenomenon, to test and find the point of fault. The possibility information of all kinds of fault factors can be gathered in the process, the process of the auto fault diagnosis is the process of selecting, analyzing, judging and choosing the information of the fault elements, and the information entropy is a good point and analyze tool to describe the complexity of the system.

3 Complexity Entropy Theory for Vehicle Fault Diagnosis

3.1 Auto Fault Diagnosis Complexity

Fault diagnosis system is a multi-hierarchy, multi-factor complex structure system. The complexity of the composite structures reflected by the hierarchies structure relationship, factors correlation relationship and information transfer relationship, etc. In this paper, the complexity of auto fault diagnosis system is evaluated from three aspects, such as hierarchies structure relationship, factors correlation relationship and information transfer relationship [3], as shown in Fig. 1.

The complexity of fault factors conversion Auto fault diagnosis system is mainly composed of diagnostic equipment, diagnostic objects (auto assembly and parts) and the diagnostician [1, 3–5, 9].

(1) The fault information feedback loop (Xx)

The important feature of the complexity is being with information feedback loop, using the feedback network diagram to evaluate the fault conversion rate. The complexity of fault diagnosis increased, which is caused by fault diagnosis without enough accurate and effective information transformation. The fault conversion rate

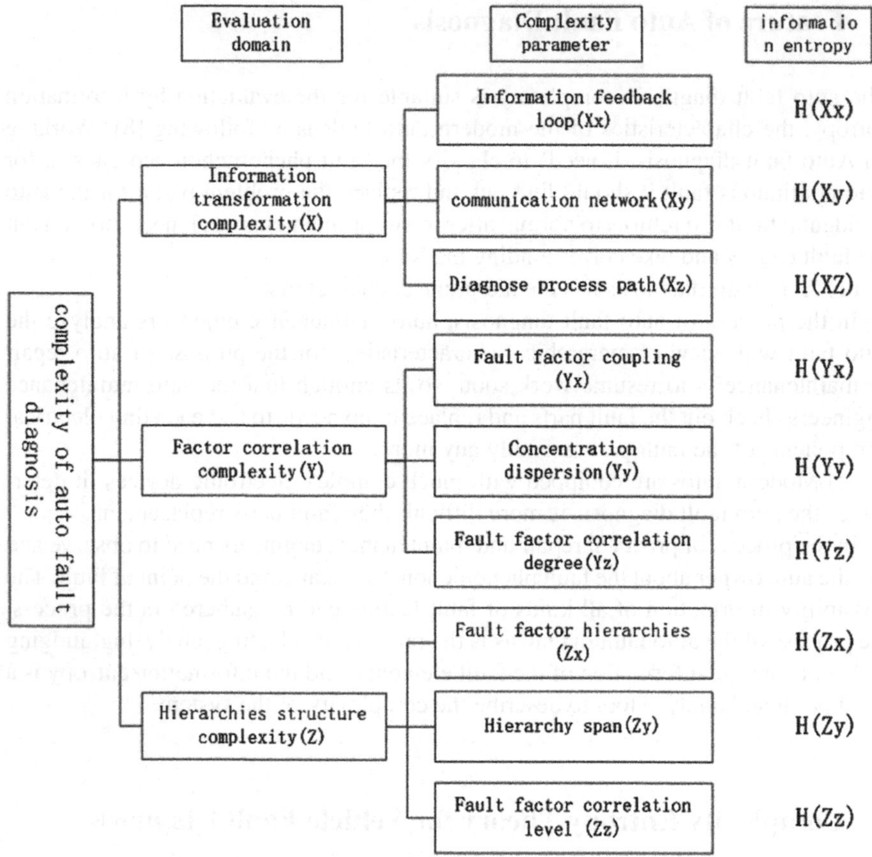

Fig. 1 Structure of auto fault diagnosis system complexity

is an important indicator for measuring fault diagnosis. The greater the feedback loop or the feedback network is, the less information the diagnosis system got, and the diagnosis complexity is increased.

(2) Fault information communication network (Xy)

The complexity of the communication network affects the fault information conversion in the longitudinal direction. The fault elements are composed of the level and the link to form a straight line, Y type, etc. Too many links or levels cause the loss of the information of the fault, making it to be more complex.

(3) Judge process path (Xz)

The complex degree of fault judge path is a critical factor to evaluate the conversion rate and fault complexity. As for auto fault diagnosis, the more the fault judge factors, the more complex the judge hierarchies are.

Fault factors correlation complexity Analysis fault factors correlation complexity from the following 3 aspects:

(1) Fault factor coupling (Yx)

The coupling system is the factors combination relationship linked by causality among the factors. The functional coupling of fault factors reflects the logic relationship among the faults, and the factor relationship chain length is composed of each fault factors hierarchy, which reflects the difficulty and complexity of the fault information judge.

(2) The cohesion degree (concentrated-dispersion degree) of fault factors (Yy)

The cohesion degree of fault factors is determined by the number of faults in the next hierarchy, the more fault possibilities the next hierarchy is, the more complex the auto fault diagnosis is.

(3) Fault factors correlation degree (Yz)

For the auto fault diagnosis, there are 4 kinds of correlations among the various fault factors: inclusion, feedback, progressive, mesh. These four correlations reflect the degree of the fault factors. the longer the correlation hierarchies between the fault factors, the greater the fault diagnosis complexity is.

Fault hierarchies structure complexity Auto fault diagnosis system is a multi-hierarchy and multi-factor composite structure system, which is an organic combination of the flexibility system, artificial systems and hardware systems. The correlations among the systems, the structures and information transmission factors reflect the complexity of the system hierarchies structure.

(1) Fault factor hierarchies (Zx)

Fault factor hierarchies is the number of fault hierarchies, which reflects the hierarchies of top events to bottom events, the fault diagnosis complexity is larger if the more hierarchies are judged.

(2) Hierarchy span (Zy)

Hierarchy span is based on the diagnosis of a certain fault hierarchy n, then needs to further determine the number of the next fault hierarchy (n − 1). If the span is very large, it needs to judge more hierarchies, then the fault diagnosis complexity is larger and Vice versa.

(3) Fault factor correlation level (Zz)

It is an important factor to evaluate the fault diagnosis complexity. The larger the next hierarchy of the fault span, the correlation among the number of various faults points needed to be diagnosed are growing faster; the more number of the correlation, the more complex of the fault diagnosis system is. The correlation level among the fault factors includes 4 types: inclusion, subordinate, coordinate, and compound.

3.2 Complexity Measurement of Auto Fault Diagnosis

System complexity factors The fault diagnosis process is a system composed of causal links, the basic evaluating elements are the nodes causal links and linking arcs built by system based on certain correlation. Given directed graph $G(t) = \{V(t), X(t)\}$, G is a non empty finite collection, in which the V is a vertices set of the directed graph, X is a set of subsets of V.

In the fault diagnosis system, the fault factor variant $V_i(t)$ of upper hierarchy is changed with the variation of the fault factors $V_j(t)$ in the next hierarchy, then there is a causal link $V_i(t) \to V_j(t), t \in T$, t is the number of different fault factors.

Causal chain is the basic factor to calculate the fault complexity, assumed the number of causal chain for fault judge system is n, $\phi(a) = n$, then $n = \sum_{r=1}^{T} n_r$. The relative frequency is $f_r = \frac{n_r}{n}$, f_r is the distribution function of the entropy function.

According to the relationship between f_r and its probability p_r, the entropy function can be defined as

$$H = -\sum_{r=1}^{T} \frac{n_r}{n} \lg \frac{n_r}{n} = -\sum_{i=1}^{T} p_r \lg p_r \tag{1}$$

System complexity evaluation entropy metric space According to the complexity core theory, the complexity always can be expressed by the correlations among three factors, the three-elements structure is the basic structure of the system. Its a good method of the mathematical techniques and means to dealing with the problems in three-dimensional (3D) space, by reducing the dimension method, the general problems can be solved by using advanced and mature mathematical methods in 3D space, so the 3D is the basic dimension of expressing complex system [2, 7]. Here take 3×3 space as an example, to establish a 3D complexity measurement model.

Definition 1: a complex vector space E which contains X, Y and Z three dimensions. The elements of E are defined as the complex vector $E_{mi} = (e_{m,xi}, e_{m,yi}, e_{m,zi})$, $e_{m,xi}$, $e_{m,yi}$ and $e_{m,zi}$ is the information entropy in the subordinate 3 dimensions of E_{mi}, thereinto m represents the fault diagnosis hierarchy; i represents the judging index of fault factors complexity. $E_{mi} \in H^3$, here, the H^3 is the 3D space in H, is a set of information entropy. $\|E_{mi}\|$ represents the information of fault factors or the length of complexity vector. The distance between two vectors in H is defined as

$$\left\|E_{mi} - E_{m-1,i}\right\| = \left\|e_{m,xi} - e_{m-1,xi}, e_{m,yi} - e_{m-1,yi}, e_{m,zi} - e_{m-1,zi}\right\| \tag{2}$$

The matrix form of the vector is

$$B_m = \begin{bmatrix} E_{m1} \\ E_{m2} \\ E_{m3} \end{bmatrix} = \begin{bmatrix} e_{m,x1} & e_{m,y1} & e_{m,z1} \\ e_{m,x2} & e_{m,y2} & e_{m,z2} \\ e_{m,x3} & e_{m,y3} & e_{m,z3} \end{bmatrix}$$

B_m is the complexity information entropy matrix for the fault diagnosis of the m hierarchy.

Definition 2: E_x, E_y and E_z is information transformation vector space, function space vector and structured vector space in H^3, in order to get a scale of judging fault diagnosis system complexity, establish a unified complexity space, combining three space complexity E_x, E_y and E_z to be a new complexity scale, $E_x \times E_y \times E_z$ then is called 3D entropy space. Assume Φ is a reflection, defined by $\Phi : E_x \times E_y \times E_z \rightarrow H$. Φ is three-dimensional linear form or three-dimensional covariance tensor in H, defined the tensor as a vector space $T + 3(H)$ in H. If a 3×3 matrix B is constituted by the of the three-dimensional complexity vector of the fault Bi on E_x, E_y and E_z, then $\|\varphi_i\| = \left\|\varphi(E_{xi}, E_{yi}, E_{zi})\right\| = |\det(B_{mi})|$ is the form of the tensor, Bi represents the unified entropy of a certain hierarchy fault, $\|\varphi_i\|$ is the information content value (tensor complexity value). The set $(T_3(H), \|\Phi_i\|)$ is the complexity of multi linear space. Therefore, the distance from E_{i-1} to E_i is

$$\left\|\Phi(E_{x_i}, E_{y_i}, E_{z_i}) - \Phi(E_{x_{i-1}}, E_{y_{i-1}}, E_{z_{i-1}})\right\| = \left|\det(B_{m_i} - B_{m_{i-1}}\right| \tag{3}$$

This is the total amount of information required to be generated by after the fault hierarchy. So,

$$\|B\|_T = \sum_{i=1}^{m} \|\Phi_i\| = \sum_{i=1}^{m} \left\|\Phi(E_{x_i}, E_{y_i}, E_{z_i})\right\| = \sum_{i=1}^{m} |\det(B_{mi})| \tag{4}$$

is all the tensor entropy, and

$$\|B\|_E = \sum_{i=1}^{m} \left\|\Phi(E_{x_i}, E_{y_i}, E_{z_i}) - \Phi(E_{x_{i-1}}, E_{y_{i-1}}, E_{z_{i-1}})\right\| = \sum_{i=1}^{m} \left|\det(B_{m_i} - B_{m_{i-1}})\right| \tag{5}$$

is the result of all information entropy for auto fault diagnosis system evaluation B.

4 Complexity Evaluation of Auto Fault Diagnosis Based on Information Entropy

In practical application, it needs to classify the fault phenomenon and cause [6], for auto continuous fault it should find out and replace the problem parts, for the auto accidental fault if it returns to normal after repair or maintenance, it needs to find out the fault causes and take corresponding measures. In this paper, we select several common auto fault diagnosis process, and carry out the empirical study of the complexity evaluation. In the case of the fault diagnosis flow chart and relevant experience, the engine cannot start as an example, the structure of the fault diagnosis is set up as shown in Fig. 2.

The information feedback loop is summarized by decision of Boolean function, then feedback to next fault hierarchy. The complexity calculating process of the fault information conversion is shown in Table 1.

Note: $X_{\text{di}} = H(X_X)_i + X_{d(i-1)}$ means fault feedback is a cumulative process.

Table 1 The complexity calculation of fault factors information conversion

Fault hierarchy	Fault factors	Fault feedback (X_x)	Communication type	Link(X_y)	Diagnose path(X_z)	$H(X_x)_i$	$e\ X_{di}$	$e\ HX_y$	$e\ H(X_z)$
1st hierarchy	Engine can not start	3	Straight line	1	1				
	Total of M1	3		1	1	0.1236	0.1236	0.0630	0.0350
2nd hierarchy	Fuel supply system fault	3	Straight line	1	1				
	Cylinder pressure insufficient	3	Straight line	1	1				
	No spark	3	Straight line	1	1				
	Total of M2	9		3	3	0.1561	0.2796	0.1207	0.0753
3rd hierarchy	Lack of fuel	0	Straight line	1	2				
	No fuel inject	2	Straight line	1	2				
	Fuel pipe fault	0	Straight line	1	2				
	Seals leakage	0	Straight line	1	2				
	Piston can not move	4	Straight line	1	2				
	Piston rings leakage	0	Straight line	1	2				
	Spark plug fault	0	Straight line	1	2				
	Wire fault	0	Straight line	1	2				
	Broken electrical fault	0	Straight line	1	2				
	Total of M3	6		9	18	0.1569	0.4365	0.1577	0.1597
4th hierarchy	Sensors fault	0	Straight line	1	3				
	Electrical injector fault	0	Straight line	1	3				
	Piston deformation	0	Straight line	1	3				
	Rotational energy shortage	2	Straight line	1	3				
	Connecting rod and bearing fault	0	Straight line	1	3				
	vehicular network system fault	0	Straight line	1	3				
	Total of M4	2		6	18	0.1000	0.5365	0.1554	0.1597
5th hierarchy	Start motor fault	0	Straight line	1	4				
	Sensors fault	0	Straight line	1	4				
	Total of M5	0		2	8	0.0000	0.5365	0.0973	0.1297
	Total	20		21	48				

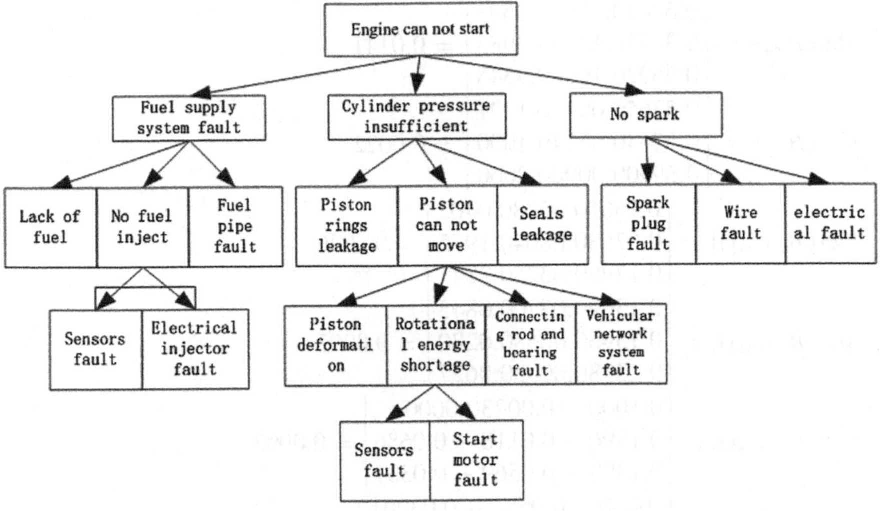

Fig. 2 Structure of the engine fault diagnosis

The calculation principle of other complexity evaluation field is same as above, the calculation process is omitted, the result is shown in Table 2. Based on the correlation calculation results, the three-dimensional vector space determinant of each fault structure hierarchy can be obtained by using aforementioned complexity scale formula: $|\det(B_{m1})| = \begin{vmatrix} 0.1236 & 0.0630 & 0.0350 \\ 0.0000 & 0.0724 & 0.2415 \\ 0.1398 & 0.1236 & 0.1207 \end{vmatrix} = 0.0008$

$$|\det(B_{m2})| = \begin{vmatrix} 0.2796 & 0.1207 & 0.0753 \\ 0.0763 & 0.1528 & 0.4390 \\ 0.2796 & 0.1561 & 0.4049 \end{vmatrix} = 0.0069$$

$$|\det(B_{m3})| = \begin{vmatrix} 0.4365 & 0.1577 & 0.1597 \\ 0.2359 & 0.1596 & 0.4670 \\ 0.4194 & 0.1569 & 0.4676 \end{vmatrix} = 0.0093$$

Table 2 Calculation results of information entropy

Structure hierarchy	e_{x1}	e_{y1}	e_{z1}	e_{x2}	e_{y2}	e_{z2}	e_{x3}	e_{y3}	e_{z3}
1	0.1236	0.0630	0.0350	0.0000	0.0724	0.2415	0.1398	0.1236	0.1207
2	0.2796	0.1207	0.0753	0.0763	0.1528	0.4390	0.2796	0.1561	0.4049
3	0.4365	0.1577	0.1597	0.2359	0.1596	0.4670	0.4194	0.1569	0.4676
4	0.5365	0.1554	0.1597	0.3955	0.1479	0.3983	0.5592	0.1000	0.4345
5	0.5365	0.0973	0.1297	0.5264	0.0724	0.1000	0.6990	0.0000	0.2000

$$|\det(B_{m4})| = \begin{vmatrix} 0.53650.15540.1597 \\ 0.39550.14790.3983 \\ 0.55920.10000.4345 \end{vmatrix} = 0.0141$$

$$|\det(B_{m5})| = \begin{vmatrix} 0.53650.09730.1297 \\ 0.52640.07240.1000 \\ 0.69900.00000.2000 \end{vmatrix} = 0.0022$$

$$|\det(B_{m2-m1})| = \begin{vmatrix} 0.15610.05780.0402 \\ 0.07630.08040.1975 \\ 0.13980.03250.2841 \end{vmatrix} = 0.0026$$

$$|\det(B_{m3-m2})| = \begin{vmatrix} 0.15690.03700.0845 \\ 0.15960.00680.0280 \\ 0.13980.00080.0627 \end{vmatrix} = 0.0002$$

$$|\det(B_{m4-m3})| = \begin{vmatrix} 0.1000 - 0.00230.0000 \\ 0.1596 - 0.0118 - 0.0686 \\ 0.1398 - 0.0569 - 0.0331 \end{vmatrix} = 0.0003$$

$$|\det(B_{m5-m4}| = \begin{vmatrix} 0.0000 - 0.0582 - 0.0300 \\ 0.1309 - 0.0755 - 0.2983 \\ 0.1398 - 0.1000 - 0.2345 \end{vmatrix} = 0.0007$$

Thus the complexity of the fault diagnosis process is shown in Table 3. Note: $\|B\|_T = 0.0333$, $\|B\|_E = 0.0038$, is the uniform complexity information entropy value of the auto engine fault diagnosis. From Table 3, $\|B\|_T = 0.0333$, $\|B\|_E = 0.0038$, this is a unified complexity entropy value of fault diagnosis for engine cannot start. The complexity of engine fault diagnosis can be compared and evaluated according to the size of the value, and auto repair and maintain complexity increases with the increase of the value, that needs to devote more manpower and material resources to diagnosis a specific fault. Therefore, in order to utilize the application of new electronic technology in automobile, it is necessary to set up a fault diagnosis complex coefficient according to information entropy valuation.

Table 3 The uniform complexity of engine cannot start

Structure hierarchy	$\|det(B_{mi})\|$	$\|det(B_{mi-m(i-1)})\|$
m1	0.0008	–
m2	0.0069	0.0026
m3	0.0093	0.0002
m4	0.0141	0.0003
m5	0.0022	0.0007
Total	0.0333	0.0038

5 Conclusion

The modern auto applied the module design concept, the equipment repair and maintenance has not been carried out at the hierarchy of small components, but at the higher hierarchy of the larger unit (composite parts, module), even replace the entire assembly to complete the repair tasks, such as electrical equipment, hydraulic components fault repair. Therefore, the auto repair complex coefficient can not reflect fully the repair work, and the modern car repair quota have to consider the fault diagnosis complexity.

In this paper, we utilized the method of information entropy to build the evaluation system of the auto fault diagnosis complexity. The practice shown that the complexity scale value calculated by the formula given in this paper is consistent with the actual situation. The application will be conducive to the reasonable pricing and healthy development of the auto repair and maintenance industry, but also to protect the interests of consumers. However, the application of this method requires the government and industry management organizations to develop a unified model classification and auto fault diagnosis complexity coefficient, which is the basic prerequisite for the promotion and application of new pricing model.

References

1. Fan-sen K, Ya-fu W, Cong L (2011) Assessment of fault diagnosis complexity about electrical fault diagnosis of equipment based on information entrop. J Jilin Univ (Eng Technol Ed) 41(3):698–701
2. Hong Y (2014) Application of information entropy in transformer fault dianosis and condition based planned maintenance, North China Electric Power University
3. Hualin S (2010) Researches of complexity measurement theory, method and case study for enterprise complex systems based on the check space and entropy metrics. Economic management press
4. Jianqiao T, Ciguang W (2013) Adaptability between supply structure and demand structure of regional freight age based on information entropy theory. J Highw Transp Res Dev 30(4):149–153
5. Qi-zhi L, Guo-ping H (2012) Optimization algorithm for fault diagnosis strategy based on failure feature information entropy. J Comput Appl 32(4):1064–1066
6. Shiting L, Qi C, Bo Z (2010) Research on the quantification of emergency maintenance task complexity for marine power facility. J Wuhan Univ Technol (Transp Sci Eng) 34(4):813–817
7. Shouq C, Yanling H (2008) Research on decision-making model and evaluation method for intelligent fault diagnosis. Mach Tool Hydraul 36(8):150–155
8. Wei H (2007) Application of information entropy in coordination complexity supply chain, Hefei University of Technology
9. Wei W, Chang-jian C (2011) The complexity evaluation of chemical emergency response plans based on graphical entropy. 7(2):67–72

A Dynamic Extraction of Feature Points in Lip Regions Based on Color Statistic

Yingjie Meng, Huasong Du, Xin Liu, Koujin Wang and Chen Chen

Abstract This paper deals with the problem of insensitivity and low-effect in the detecting feature points in lip regions. We propose a new feature points extraction model based on Harris algorithm, combining ideas of color statistic and graph connectivity. To detect feature points, firstly, we build a proper possibility statistical module of color component, with a specifying color model in lip regions, then we realize feature points filter module under the thought of graph connectivity. Experiments show that our algorithm has a better accuracy and stability, especially a competitive advantage in lip regions feature extraction.

Keywords Lip regions · Extraction · Color statistic · Graph theory

1 Introduction

Feature points extraction is an important stage in facial recognition research, which is stuck by manual marking. Almost each facial matching algorithm needs to begin with

---

This research is supported by the Fundamental Research Funds for the Central Universities (lzujbky-2015-107).

Y. Meng · H. Du (✉) · X. Liu · K. Wang · C. Chen
School of Information Science and Engineering,
Lanzhou University, Lanzhou 730000, China
e-mail: duhs13@lzu.edu.cn

Y. Meng
e-mail: mengyj@lzu.edu.cn

X. Liu
e-mail: 591462038@qq.com

K. Wang
e-mail: 863480670@qq.com

C. Chen
e-mail: 514981717@qq.com

V.E. Balas et al. (eds.), *Information Technology and Intelligent Transportation Systems*, Advances in Intelligent Systems and Computing 455,
DOI 10.1007/978-3-319-38771-0_10

manually feature points marking and modeling. Therefore, an accurate algorithm of feature points selfdetecting and locating can increase automation and practicability. A brief summary of some of the relevant research work can be separated as follows:

Corner detection algorithms based on grayscale are mainly Harris algorithm and SUSAN algorithm.

Harris algorithm judges grayscales in two directions, which defines a point as a feature point if its changes in two directions are both respectively higher than thresholds, based on the assumption that points with sharp changes can be used as descriptors of an image. Thresholds are key elements for Harris corner detection because oversize threshold causes false detection and undersize threshold misses feature points. This algorithm relies on high accurate of threshold, especially in lip regions, changes of which are complex. Furthermore, its performance is barely satisfactory when it comes to different scales of brightness and shape.

Corner detection algorithms based on binary image, such as [1], are not widely concerned, for the reason that binary image is a midtransform from grayscale image to edge image.

Corner detection algorithms based on edge feature [2] are hardly used for their complicate processing, which constrains their applicability in reality.

All of the algorithms above have one thing in common: a manual-set pre-value, which takes most responsibility on low performance of these algorithms in cases such as complicate grayscale edge, inconspicuous regionalism and variable brightness and outline. In this paper, we design an algorithm about feature detection and feature extraction for lip regions to recognize the speakers identity in RGB color space.

The experiments are all done in the image set S whose contents have been scale normalized to simplify description.

2 Model of Detection for Lip Regions and Extraction for Eigenvalue

The main process of detecting lip regions and extraction feature points can be described as follows: (1) Detect the edge of F in RGB spaces to obtain X; (2) Process X with binaryzation depending on the threshold W to get binary image H; (3) Extract feature points in H to create T; (4) Map T and original S to acquire TS; (5) Filter TS based on red cluster and spatial connectivity, fusing the results to generate eigenvalue, where S refers to pending image set; F refers to each frame of image; X refers to gradient image; W refers to binary threshold; H refers to binaryzation image; T refers to middle feature points set; TS refers to pixel set for respective feature point; TR refers to feature points set after Filtering based on red cluster; TE refers to feature points set after Filtering based on spatial connectivity. The Fig. 1 demonstrates the work flow of whole process.

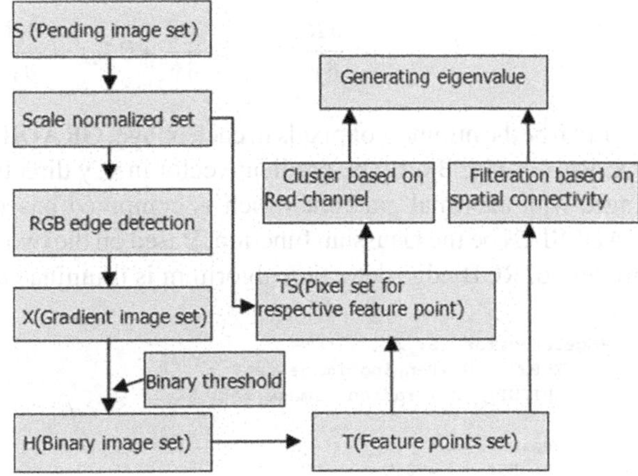

Fig. 1 Model of detection for lip regions and extraction for eigenvalue

More details of RGB edge detection in Fig. 1 can be found in [3]. The research has shown the basic method of gradient binaryzation in Fig. 1 [4, 5], and the algorithm of Feature points extraction is based on the previous study [5–8].

3 Algorithm Design of Detection and Extraction of Eigenvalue

Our method is mainly consist of RGB edge detection, gradient binaryzation, feature points extraction, cluster based on Red-channel and filtration based on spatial connectivity. Section 2 gives the specific algorithm for each core module.

3.1 RGB Edge Detection

The principal task of RGB edge detection is to preprocess the original image with analyzing its change of gradient in RGB color space. This step produces spatial gradient image by acquiring spatial gradient and extracting edge.

Experiment results show that red channel weighs most in all 3 channels, green channel ranking second and barely influence from blue channel. Feature points focus in internal lip regions, if only red channel is used to be transform unit vector. Therefore, we adopt the idea in [9] to add different weights to each channel, where 2 for red channel, 0.2 for green and blue channel. Hence, the direction vector u and v can be deduced as:

$$u = \frac{\partial R}{\partial x} * 2r + \frac{\partial G}{\partial x} * 0.2g + \frac{\partial B}{\partial x} * 0.2b \tag{1}$$

$$v = \frac{\partial R}{\partial y} * 2r + \frac{\partial G}{\partial y} * 0.2g + \frac{\partial B}{\partial y} * 0.2b \qquad (2)$$

Let n be the numbers of pixels in each image, GRADIENT be the gradient function, Rx, Ry, Gx, Gy, Bx, By be gradient vector in x, y direction, THETA be the direction angle with maximal gradient which is computed based on the previous study [9], GAUSSIAN be the Gaussian function. Based on the two equations above, the specific process of RGB edge detection algorithm is illuminated as:

```
EdgeDetection (S, X)
    INPUT:  S (Pending image set)
    OUTPUT:  X (Gradient image set)

    Begin
        For i := 0 To n  DO
            [R := GET (Each pixel); G := GET (Each pixel);
             B := GET (Each pixel);
             [Rx,Ry] := GRADIENT(R);[Gx,Gy] := GRADIENT(G);
             [Bx,By] := GRADIENT (B);
             gxx := Rx*Rx*4+Gx*Gx*0.04+Bx*Bx*0.04;
             gyy :=  Ry*Rx*4+Gy*Gy*0.04+By*By*0.04;
             gxy := Rx*Ry*4+Gx*Gy*0.04+Bx.*By*0.04;
             THETA := 0.5*ARCTAN (2*gxy/(gxx - gyy));
             //ARCTAN is the answer of anctan function
             X := SQRT(0.5*(gxx+gyy)+(gxx-gyy)*cos(2THETA)+2*gxy*sin(2THETA)) ];
        X := GAUSSIAN(X);
        //Conduct x with Gaussian smoothing
    End
```

3.2 The Binaryzation of Gradient

The main task of this step is to generate binaryzation image set H with a binary threshold W, gradient image set X, number of rows of each gradient image R and number of line of each gradient image L.

We filter pixels in X with W, setting 255 to which are higher than this threshold and 0 to the lowers. Let (x, y) be the coordinate for each pixel, the process of pre-set value is computed as:

```
GradientDetection (X, H)
    INPUT:  X (Gradient image set)
    OUTPUT:  H (Binary image set)

    Begin
        For x := 0 To R Do
            For y := 0 To L Do
                If  X(x,y)< W
                Then X(x,y)  := 0;
                Else X(x,y)  := 255;
            H := X;
    End
```

3.3 Extraction of Feature Points Set

Feature points set can be generated from binaryzation image set H, with the help of Harris algorithm. Parameters in the experiments are set as: the grayscale of pixels range from 0 to 255 and constant coefficient K = 0.05, T = 0.2 which are the most optimal value concluded by a series of experiments.

3.4 Aggregation Filtering Based on Red Channel

In this part, we produce a new feature points set TR from original set T, relying on pixel value set TS. The color analysis is conducted to get rid of feature points whose red channel component surpass the average value, as a result that pixels far away from the lip region get a large red channel component.

On account of thoughts mentioned above, we assume that MEANS is the particle of T and SUM is the sum of it. Ti represents each feature point with a pixel value named F (Ti). The algorithm is realized as:

```
RedAggregation (T, TR)
    INPUT:   T (Middle feature points set)
    OUTPUT:   TR (Feature points set after Filtering based on red cluster)

    Begin
        SUM := 0;
        For   i:=0 To n Do
            SUM := SUM+T i;
        MEANS := SUM/n;
        //MEANS is the mass point of T
        For i := 1 To n Do
            If   F (Ti) - MEANS>0
            Then  DELETE  Ti;
        TR := T;
    End
```

3.5 Filtering Based on Spatial Connectivity

In this paper, we screen feature points to create a new feature points TE from T. We take the spatial connectivity and density into account to eliminate extreme isolation points and those deviated from outline.

As a viewpoint from graph theory, internal lip region is indicated as a circle consist of several points no matter what the lip region is. Hence, we can make some definitions: edge is the Euclidean distance between two points.

Joint point is a pair of points whose edge is lower that connectivity threshold SW; continuous path is the path connecting joint points without circles. The number of edge in a continuous path represents its length.

We affirm a random feature point A is in the outline of lie region ruled by the assumption that the length of one of the continuous paths containing point A is over edge threshold M.

In this way, points away from lip regions can be bolted by the theory proposed above.

Let DISTANCE be the Euclidean distance of two feature points, TABLE be the distance list of all feature points, SW be the connectivity threshold which is set 20 in this paper, R be status parameter including true and false, SET be the set, ADD () be write operation of set, Tx be the xth number of set T.

COPERATE () is recurrence iteration function with a maximum iteration time M. COPERATE () calculates as:

```
COPERATE(x, COUNT)
    INPUT:  x (Iteration parameter)
    OUTPUT:    COUNT (Count parameter)

    Begin
        If (COUNT>M)
        Then [R := TRUE; RETURN]
        //Return COPERATE function
        For  y := 1 To n Do
            [SET := ADD(x);
             If (TABLE(x, y) < SW) &&(x = SET)
             Then COUNT COUNT+1;
             COPERATE(y, COUNT)];
        If R=FALSE Then DELETE Tx;
        // Delete the x-th number of set T
    End
```

The whole spatial connectivity filtering algorithm SpaceDetection can be divided into two parts establishment of Euclidean distance table and analysis of connectivity.

Specific steps are as follows:

```
SpaceDetection (T, TE)
    INPUT: T (Middle feature points set)
    OUTPUT: TE (Feature points set after Filtering based on spatial connectivity)

    Begin
        FOR Ti := T0 To Tn Do
          FOR Tj := T0 To Tn Do
            [ DISTANCE := |F(Ti)-F(Tj)|;
            TABLE (i, j) := DISTANCE];
        R := FALSE;
        SET := 0;
        COUNT := 0;
        M := 4;
        FOR x := 1 To n
           COPERATE(x, COUNT);
        TE := T;
    End
```

We set M as a value of 4 in the condition of hardware situation. The ultimate feature points are those selected by red-channel cluster filtration and spatial connectivity filtration.

4 Experiment and Results

We will provide in this section experiments and results of our algorithm a comparison with Harris algorithm. Experiments were done by MATLAB in the stage of Windows. Two images containing S0 and S3, the results of lip images preprocessed with the different pronunciations, are the input data. We process S0 with Harris algorithm and our algorithm to create S1 and S2 and process S3 with Harris algorithm and our algorithm to create S4 and S5. The constant coefficient parameters K and T in Harris algorithm are set as 0.05 and 0.2.

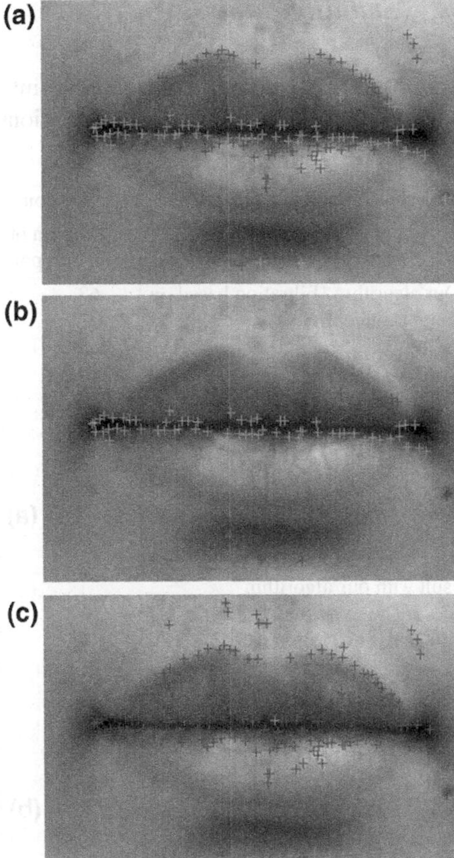

Fig. 2 The feature points images with the S0 (input data). **a** The experiment result of filteration based on Red-channel. **b** The experiment result with our algorithm (S1). **c** The experiment result with Harris algorithm (S2)

4.1 Accuracy Analysis

Accuracy is a fatal factor of judging feature points extraction algorithm. We calculate accuracy by hit rate of feature points.

Figure 2 shows the comparison of S1 and S2. Points in (b) are generated by the means of spatial connectivity based on (a).

We create Table 1 according to the result of statistical analysis based on Fig. 2.

As shown in Table 1, the hit rate of our model is 74.6 %, superior than 36.8 % of Harris algorithm.

4.2 Stability Analysis

Stability is one of criteria of feature points extraction algorithm. In feature points detection algorithm, especially in lip regions, results are strongly affected by bright-

Table 1 Comparison with hit rate of feature points based on S0

	The sum of feature points	The sum of feature points which are on the internal lip	Hit rate (%)
Our algorithm (Filtration based on spatial connectivity)	68	44	64.7
Our algorithm (Filtration based on Red-channel) (S1)	59	44	74.6
Harris algorithm (S2)	141	52	36.8

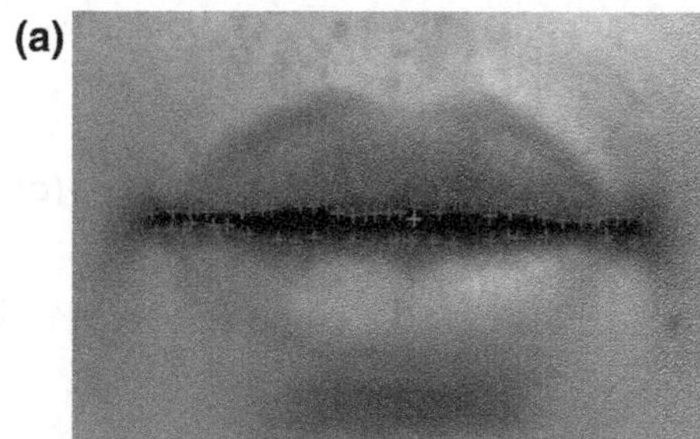

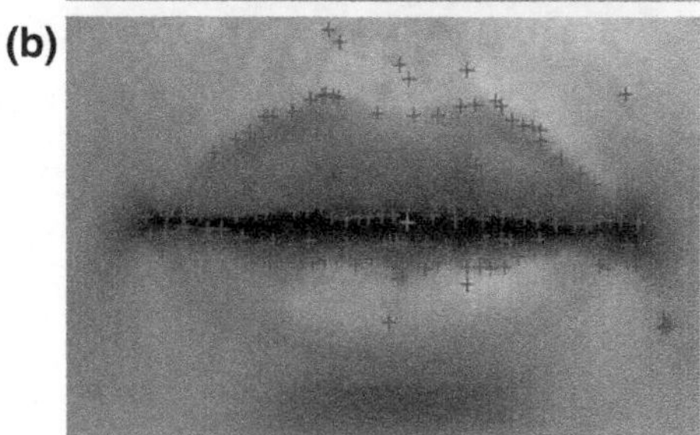

Fig. 3 The feature points images with the S3 (input data). **a** The experiment result with our algorithm (S4). **b** The experiment result with Harris algorithm (S5)

Table 2 Comparison with hit rate of feature points based on S3

	The sum of feature points	The sum of feature points which are on the internal lip	Hit rate (%)
Our algorithm (S4)	71	56	78.9
Harris algorithm (S5)	161	73	45.34

Table 3 Comparison of change degree of hit rate

	Change degree of hit rate (%)
Harris algorithm	18.68
Our algorithm	5.44

ness, contrast, shape of lips and some other factors. We adopt the change degree of feature points hit rate as representation of stability.

Figure 3 is the result image of S4 and S5. The feature points hit rate is shown in Table 2.

The comparison of change degree of hit rate is represented in Table 3.

The result of Harris algorithm is computed as $45.34\,\% - 36.87\,\%/45.34\,\% = 18.68\,\%$ while our algorithm changes 78.9 to 74.6 %, whose result is 5.44 %, much lower than Harris. Therefore, our algorithm owns a higher stability countering to lip shape changing.

5 Conclusions

The basic questions are discussed and we have drawn some conclusions. We present a new model for feature points extraction in lip regions with the support of color gradient image in RGB color space. This model is based on color component statistics model which can detect feature points precisely. Experiments show a promising result on locating feature points and a strong robustness in cases of changing lip regions. With a precise location of feature points, the problem of manual-marking feature points can be perfect solved, enforcing the practicability of the future work. In the next stage, we will embed our algorithm into AAM (Active Appearance Model) and ASM (Active Shape Model) algorithm template to test the effect of detecting lip regions.

References

1. Cui J, Xie J, Liu T, Guo X, Chen Z (2014) Corners detection on finger vein images using the improved Harris algorithm. Optik 125:4668–4671
2. Cooper J, Sveth A, Kitechen L (1993) Early jump-out corner detectors [J]. IEEE Trans PAMI 15:823–828

3. Hsu RL, Jain AK (2002) Face detection in color images [J]. IEEE Trans. Pattern Anal Mach Intell, 696–706
4. Wu R, Huang J (2009) Method of text image binarization processing using histogram and spectral clustering [J]. J Electron Inf Technol 31(10):2461–2464
5. Ostu N (1979) A threshold selection method from gray-level histograms [J]. IEEE Trans Syst, Man Cybern 9(1):62–66
6. Harris C, Stephens MJ (1988) A combined corner and edge detector. In: Alvey vision conference, pp 147–152
7. Shi J, Tomasi C (1994) Good features to track. International conference on computer vision and pattern recognition, pp 593–600
8. Mikolajczyk K, Schmid C (2004) Scale and affine invariant interest point detectors. Int J Comput Vis 60(1):63–86
9. Rosten E, Drummond T (2006) Machine learning for high-speed corner detection. In: European conference on computer vision, pp 430–443
10. Feng X, Wang W, Xuzhen W, Pan J (2009) Adaptive mouth region detection algorithm based on multi-color spaces. J Comput Appl 29(7):1001–9081

Electronic Countermeasures Jamming Resource Optimal Distribution

Yang Gao and Dong-sheng Li

Abstract In the electronic warfare, the jamming resource optimal distribution is always an important research topic. Through the existing interference effect assessment criteria, jamming effectiveness evaluation index, general jamming resource allocation model is established. Through the study of the integration of interference target, the distribution model is divided into two categories, one-to-one and much-to-little. The one-to-one model is dealt by Hungarian algorithm and the much-to-little model is dealt with the combination of Hungarian algorithm and dynamic programming algorithm, which is called secondary optimal distribution method. Finally the radar jamming resource assignment example shows the method is effective.

Keywords Electronic countermeasures · Resource distribution · Secondary distribution model · Hungarian algorithm · Dynamic programming algorithm

1 Introduction

In the electronic countermeasures, the game of both sides, how to effectively utilize the limited jamming resources cause the maximum interference on enemy, which is called jamming resources optimization allocation problem, has always been a research hot spot. Whether it is radar confrontation, communication countermeasure or identification friend or foe confrontation, etc., establishing the jamming effect evaluation index structure for enemy targets is important step. Also it is essential to evaluate the different enemy targets threat degree. The jamming resource should be assigned to the higher threat target generally. Then the integration of jamming targets is also useful to maximize the jamming distribution efficiency.

Jamming resources optimal allocation methods have two kinds, the simple combinatorial optimization and heuristic optimization algorithm. Combinatorial optimization algorithm have the basis of close degree [1], the Hungary algorithm [2], the 0-1

Y. Gao (✉) · D.-s. Li
Electronic Engineering Institute, Hefei 230037, China
e-mail: gao_yang_mail@163.com

V.E. Balas et al. (eds.), *Information Technology and Intelligent Transportation Systems*, Advances in Intelligent Systems and Computing 455,
DOI 10.1007/978-3-319-38771-0_11

programming [3], etc., which have specialties in solving the small scale assignment problem. Heuristic optimization algorithm are mainly genetic algorithm [4, 5], evolutionary algorithm, ant colony algorithm [6] and so on, which have an advantage in solving large-scale resource allocation problems.

Making a research on radar jamming resources assignment with the combinational optimization. Due to the Hungary algorithm can effectively solve the problem of allocation under a certain size, but the algorithm using rules are limited; Dynamic programming algorithm are efficient and practical on small-scale resource allocation problem, but as the scale of the increase in the number, the algorithm cost too much time. According to the above problem, this paper considers the combination of the two algorithm. The one-to-one model is dealt with Hungary algorithm; the much-to-little model is dealt with the Hungary algorithm and the dynamic programming algorithm [7], which is called secondary distribution model, to make the jamming benefits maximize.

2 Jamming Effect Evaluation and Jamming Targets Integration

2.1 *Jamming Effect Evaluation Rules and Indexes*

The existing jamming effect assessment criteria, such as information rule, power principles and the tactics standards (probability criterion), suppression coefficient criterion, the EMJ principle, etc. These guidelines are all the basic principles to assess the jamming effect. The commonly used rules are power, the information standard, tactical use criterion and time standard. To comprehensively consider the factors of both sides, enemy and ourselves, jamming power, jamming pattern, time and space frequency factor index are chosen to evaluate the jamming effect.

(1) The membership functions of jamming power

$$E_{Jj} = \begin{cases} 1 & P_j/P_s \geq 2K_j \\ \frac{3P_j/P_s}{4K_j} - 0.5 & else \\ 0 & P_j/P_s \leq 0.5K_j \end{cases} \tag{1}$$

In which P_j is jamming power, P_s is signal power, K_j is the minimum signal to noise ratio when system in normal work.

(2) The membership functions of jamming pattern Defining the membership function of jamming pattern: the ratio of a jammers pattern number to the total jammers pattern numbers

$$E_{Js} = \frac{N_{one}}{N_{total}} \tag{2}$$

(3) Time-space-frequency evaluation factor membership function

$$E_{Jk} = K_t K_s K_f \tag{3}$$

In which K_t is time factor, K_s is space factor, K_f is frequency factor.

(1) Describe K_t as a suppress-time concept:

$$K_t = T\left(T_j, T_s, T_{js}\right) \tag{4}$$

In which T_j, T_s, T_{js} is jamming signal action time, enemy signal action time, jamming signal and signal mutual action time; (2)

$$K_s = \frac{\left(\Omega_j + \theta_j\right)}{T} \tag{5}$$

In which Ω_j is the largest beam pointing range of jamming resources transmitting antenna; θ_j is the interference beam coverage at any time; T is the rotation period of the jamming antenna. (3)

$$K_f = F\left(F_j, F_s\right) \tag{6}$$

In which F_j, F_s is the jammer frequency coverage and the signal frequency range. The factors are decided by actual need. The factors expressions are not the key of this paper, so don not give a deeper description.

2.2 Jamming Targets Integration

After getting the membership function of every index and setting the weight of each index, the system jamming benefit decision matrix can be gotten through fuzzy synthetic method. The jamming resource distribution is based on system decision matrix, aims to make the jamming profits maximum in multiple system to multiple targets resource assignment. The jamming targets integration rules:

(1) when the number of jamming resource m is equal to the number of jamming targets n, the distribution model is typical one-to-one model;

(2) when $m \succ n$, the jamming resources are enough, which is much-to-little model. Firstly ensuring each targets has a jammer to jam, then making a reasonable assignment to the rest resources;

(3) when $m \prec n$, the jamming resources are lacking, these targets which meet the time, space, frequency consistent principle could integrated to the same series and be jammed by one jammer. In this situation the little-to-much model could turn into one-to-one or much-to-little model;

(4) when $m \prec n$, but all of the targets could not be integrated one series, the little-to-much model could not be turned. In this situation the n jammer resource only can be assigned to the n targets whose threat degree are rounding out the top n, which also means a one-to-one model.

3 Jamming Resource Distribution Model and Solution

According the above analysis, the assignment models are summed up in two kinds, one-to-one and much-to-little. (1) When $m = n$:

$$Z = \max \sum_{i=1}^{m} \sum_{j=1}^{m} x_{ij} t_j e_{ij} \; s.t. \begin{cases} \sum_{j=1}^{n} x_{ij} = 1 & i = 1, 2, \ldots m \\ \sum_{i=1}^{m} x_{ij} = 1 & j = 1, 2, \ldots m \\ x_{ij} = 1 \;\; or \;\; \mathrm{x_{ij}} = 0 \end{cases} \tag{7}$$

In which t_j is the enemy target threat degree; x_{ij} is decision variable, when $x_{ij} = 1$, the jammer i jam the target j, when $x_{ij} = 0$, the jammer i does not jam the target j; e_{ij} is the jamming profit; Z is the total jamming profits.

By using the Hungarian method [2] to solve the above model, the Hungarian algorithm is put forward to solve assignment problem by Kuhn in 1955, he cited the Hungarian mathematician Connie's independent zero element theorem. The algorithm requires that the objective function is linear, the objective function is minimum value and benefit matrix is phalanx. So the jamming distribution model should turn to as the expression (8) shows.

$$Z = -\min \sum_{i=1}^{m} \sum_{j=1}^{m} x_{ij} u_{ij} \; s.t. \begin{cases} \sum_{j=1}^{n} x_{ij} = 1 & i = 1, 2, \ldots m \\ \sum_{i=1}^{m} x_{ij} = 1 & j = 1, 2, \ldots m \\ x_{ij} = 1 \;\; or \;\; \mathrm{x_{ij}} = 0 \\ u_{ij} = -t_j e_{ij} \end{cases} \tag{8}$$

(2) When $m > n$:

$$Z = \max \sum_{i=1}^{m} \sum_{j=1}^{n} x_{ij} t_j e_{ij} \; s.t. \begin{cases} \sum_{j=1}^{n} x_{ij} = 1 & i = 1, 2, \ldots m \\ \sum_{i=1}^{m} x_{ij} \succ 1 & j = 1, 2, \ldots n \\ x_{ij} = 1 \;\; or \;\; \mathrm{x_{ij}} = 0 \end{cases} \tag{9}$$

Adding $m - n$ virtual targets, making the model turn to one-to-one model, the jamming profit to the virtue targets is 0:

$$u_{ij} = \begin{cases} -t_j e_{ij}, & i = 1, 2, \ldots m; \; j = 1, 2, \ldots n \\ 0, & i = 1, 2, \ldots m; \; j = n + 1, \ldots m \end{cases} \tag{10}$$

The model can be solved by Hungarian algorithm. But the resource allocation for the virtual system is invalid, so after Hungary algorithm distribution, the remaining

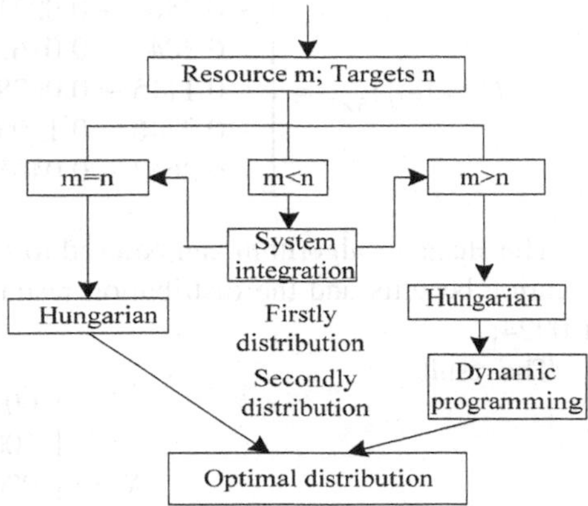

Fig. 1 Distribution process diagram

jamming resources could be allocated by the dynamic programming approach, called secondary distribution, to ensure the optimization of overall benefit value as much as possible.

The jamming resource distribution process can be described in Fig. 1.

4 Radar Jamming Resource Distribution Example

4.1 One-to-One Model

Assuming that, in a battle, we detect enemy five radar (serial number 1 5) and we have five jamming resources (serial number 1 5), which is $m = n = 5$. According to this, the jamming resource distribution is requested to achieve the maximum jamming effect.

First, the five radar threat degree (after standard) are given

$$T = (0.404, 0.136, 0.205, 0.284, 0.325)$$

The jamming effect decision matrix:

$$E = (e_{ij})_{5\times 5} = \begin{bmatrix} 0.70940.16260.58530.69910.1493 \\ 0.75470.11900.22380.89090.2575 \\ 0.27600.49840.75130.95930.8407 \\ 0.67970.95970.25510.54720.2543 \\ 0.65510.34040.50600.13860.8143 \end{bmatrix}$$

After weighted

$$U = \left(u_{ij}\right)_{5\times5} = \begin{bmatrix} -0.2866 - 0.0221 - 0.1200 - 0.1985 - 0.0485 \\ -0.3049 - 0.0162 - 0.0459 - 0.2530 - 0.0837 \\ -0.1115 - 0.0678 - 0.1540 - 0.2724 - 0.2732 \\ -0.2746 - 0.1305 - 0.0523 - 0.1554 - 0.0826 \\ -0.2647 - 0.0463 - 0.1037 - 0.0394 - 0.2646 \end{bmatrix}$$

The Hungary algorithm can be used to solve the problem. The time and the allocation benefits and the distribution result matrix are: Time = 0.005589 s; Z = 1.0924;

The result

$$X = \begin{bmatrix} 00100 \\ 10000 \\ 00010 \\ 01000 \\ 00001 \end{bmatrix}$$

The result is: resource 2 to target 1; resource 4 to target 2; resource 1 to target 3; resource 3 to target 4; resource 5 to target 5. Thus it can be seen that the Hungarian algorithm can better solve the jamming resources distribution problem.

4.2 Much-to-Little Model

when $m = 7$, $n = 4$, namely seven jamming resource to four enemy targets. Firstly adding three virtual, then directly giving the benefit matrix U, after threat degree weighted and it is suitable for Hungarian algorithm.

$$U = \left(u_{ij}\right)_{7\times7} = \begin{bmatrix} -0.8147 - 0.5469 - 0.8003 - 0.0357000 \\ -0.9058 - 0.9575 - 0.1419 - 0.8491000 \\ -0.1270 - 0.9649 - 0.4218 - 0.9340000 \\ -0.9134 - 0.1576 - 0.9157 - 0.6787000 \\ -0.6324 - 0.9706 - 0.7922 - 0.7577000 \\ -0.0975 - 0.9572 - 0.9595 - 0.7431000 \\ -0.2785 - 0.4854 - 0.6557 - 0.3922000 \end{bmatrix}$$

Using the Hungarian algorithm to get the distribution matrix and benefit value, which is the first distribution:

$$X = \left(x_{ij}\right)_{7\times7} = \begin{bmatrix} 0000100 \\ 1000000 \\ 0001000 \\ 0010000 \\ 0100000 \\ 0000001 \\ 0000010 \end{bmatrix},$$

$$Z_1 = 3.7261.$$

The distribution result is efficient to target 1−4, is inefficient to target 5−7. So the jamming resource 1, 6, 7 need to be allocated once more. The second distribution to resource 1, 6, 7 is allocated by dynamic programming method. Giving the new profit distribution matrix U':

$$U' = \left(u_{ij}\right)_{3\times4} = \begin{bmatrix} 0.81470.54690.80030.0357 \\ 0.09750.95720.95950.7431 \\ 0.27850.48540.65570.3922 \end{bmatrix}$$

Normally, dynamic programming algorithm is divided different step by jamming targets, but the number of jamming resources is less than the targets here. So we assign the target to the jamming resource, which means the step division is by jamming resource not jamming targets. State transition equation is expressed as the specific ways of distribution. The secondary distribution are behaved by diagram in three steps:

(1) first step: for the resource 1, there are four choices, target 1–4.
(2) second step: for the resource 6 and 1, there are twelve choices
(3) third step: for the resource 7, 6 and 1, there are twelve choices at the basis of second step (Figs. 2, 3, 4).

After the second distribution, the finally distribution is

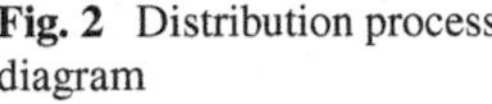

Fig. 2 Distribution process diagram

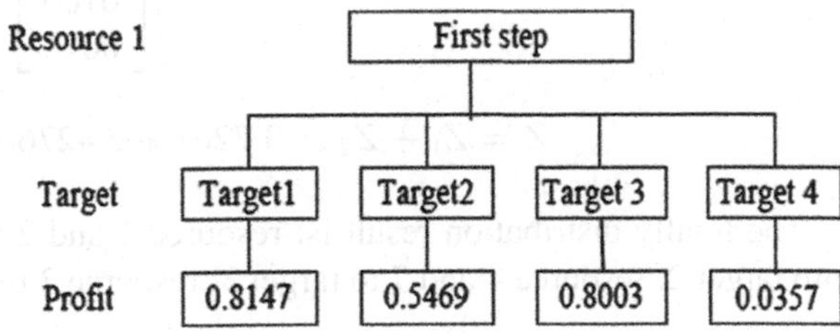

Fig. 3 Distribution process diagram

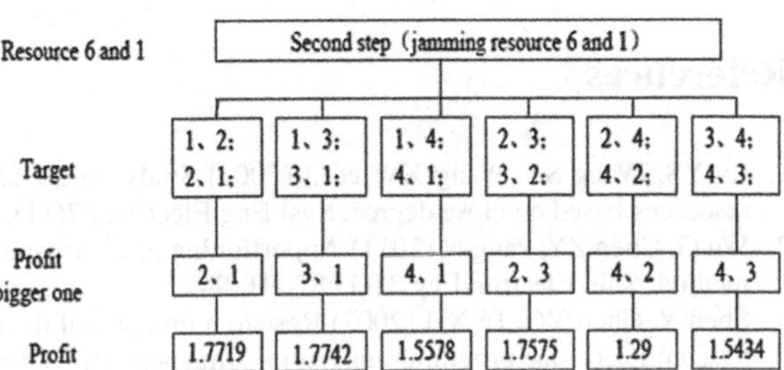

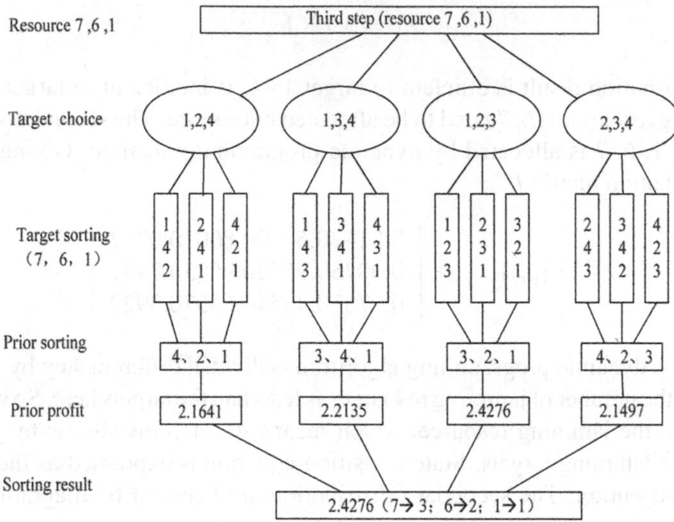

Fig. 4 Distribution process diagram

$$X = (x_{ij})_{7\times 4} = \begin{bmatrix} 1000 \\ 1000 \\ 0001 \\ 0010 \\ 0100 \\ 0100 \\ 0010 \end{bmatrix},$$

$$Z = Z_1 + Z_2 = 3.7261 + 2.4276 = 6.1537$$

The finally distribution result is: resource 1 and 2 to target 1; resource 5 and 6 into target 2; resource 4 and 7 to target 3; resource 3 to target 4.

References

1. Lv YS, Wang SZ, Wang XW et al (2005) Study on the allocation tactics for radar jamming resources based on close degree. Syst Eng Electron 27(11):1893–1894, 1974
2. Wu G, Chen ZY, Pang N (2011) Apportioning of radar jamming resources based on Hungarian method. Ship Electron Eng 31(1):87–91, 94
3. Shen Y, Chen YG, Li XH (2007) Research on optimal distribution of radar jamming resource based on zero-one programming. Acta Armamentarii 28(5):528–532
4. Xue Y, Zhuang Y, Ni TQ et al (2010) One improved genetic algorithm applied in the problem of dynamic jam resource scheduling with multi-objective and multi-constraint. In: Proceedings of the IEEE 5th international conference on bio-inspired computing: theories and applications, pp 708–712

5. Xue Y, Zhuang Y, Ni TQ et al (2014) Self-adaptive learning based discrete differ-ential evolution algorithm for solving CJWTA problem. J Syst Eng Electron 25(1):59–68
6. Wang YX, Qian LJ (2008) Weapon target assignment problem satisfying expected damage probabilities based ant colony algorithm. J Syst Eng Electron 19(5):939–944
7. Ren S, Si CZ, Lei J (2008) Fuzzy multi-attribute dynamic programming model for radars distributing machine allocation. Syst Eng Electron 30(10):1909–1913

Analysis on the Test Result of the Novel Power Wireless Broadband System for Smart Power Distribution and Utilization Networks

Jinping Cao, Wei Li, Suxiang Zhang, Dequan Gao and Tao Xu

Abstract Communication network plays an important part in the construction of the smart power distribution and utilization system, and it covers wide area and has a great number of scattered signals. The paper summed up the advantages and disadvantages of several common power wireless communication technologies, and proposed a novel power broadband wireless system. In order to verify its performance, we tested the coverage, transmission rate etc. Meanwhile, the paper also analyzed the communication requirements of the smart distribution and utilization system in detail. Compared with the requirements, the test results show that the novel power broadband wireless system has the characteristics of wide coverage, high transmission rate and short transmission delay, which provides an effective wireless communication solution for the power distribution and utilization services.

Keywords Smart power distribution and utilization network · Power wireless communication · TD-LTE

J. Cao (✉) · S. Zhang · D. Gao · T. Xu
State Grid Information & Telecommunication Branch, Beijing 100761, China
e-mail: lemon_cjp@163.com

S. Zhang
e-mail: suxiang-zhang@sgcc.com.cn

D. Gao
e-mail: gao_dequan@sgcc.com.cn

T. Xu
e-mail: xutao@sgcc.com.cn

W. Li
North Branch of State Grid Corporation of China, Beijing 100053, China
e-mail: llww1226@sina.com

V.E. Balas et al. (eds.), *Information Technology and Intelligent Transportation Systems*, Advances in Intelligent Systems and Computing 455,
DOI 10.1007/978-3-319-38771-0_12

1 Introduction

According to the "distribution network construction and transformation action plan (2015–2020)" issued by the National Energy Administration, China will upgrade smart distribution network, improve the core competitiveness of distribution network. As an important part of the smart power distribution and utilization system, the adaptability and innovation of communication network also need to be improved.

With the construction and development of the smart grid, Ethernet passive optical network (EPON), power line carrier communication (PLC), data transceiver, wireless sensor network (WSN) and other communication technologies are widely applied in power distribution and utilization networks. Among them, wireless communication gets more and more attention for its low cost, strong adaptability, easy expansion and convenient maintenance.

Through several wireless communication technologies have been used in the power distribution and utilization network, their performances of transmission rate, costs, security can't meet the requirements of power system, we need a new wireless communication technology to fit the requirements.

This paper introduces a new power wireless broadband communication system, shows the test results of its major parameters. Compared with the communication requirements of power services in coverage, transmission rate, latency, the test results show that the new power wireless broadband system satisfies the developing demand of services in the power distribution and utilization communication system.

The rest of this paper is organized as follows: Sect. 2 presents the related work. Section 3 introduces several common wireless communication technologies and proposes an electric power wireless private network based on TD-LTE. Section 4 shows the test results of the major parameters. Finally in Sect. 5, we conclude our paper.

2 Related Work

Some literatures are researched in [1–8], these papers designed some communication schemes for distribution and utilization networks.

In order to build power distribution networks more standardized, state grid corporation of China published some construction specifications for the smart power distribution communication system in [1, 2] according to its construction experiences. In [3–5], authors respectively presented the state of smart power distribution and utilization communication system, and research on characters and applications of some communication technologies. In [6–8], authors focused on the application of wireless communication in smart power grid.

Through these researches have done a lot of works on the communication schemes, they are based on the mature communication technologies, it's difficult for them to achieve breakthrough progress.

3 Background

Currently many kinds of wireless communication have been used in power distribution and utilization networks, such as general packet radio service(GPRS), code division multiple access (CDMA), 230Mdata transceiver, worldwide interoperability for microwave access (WiMAX), bidirectional power frequency automatic communication (TWACS).

In North America, most power consumption information collection systems are built based on GPRS, WiMax or TWACS, only a few systems are built by leased optical fiber. In Europe, the power consumption information collection systems are not large-scaled except Italy, only has a small number of applications in France, Spain, the UK and etc., they often use radio communication. Italy mainly uses GPRS in remote channel. Korea electric power corporation established intelligent community in Seoul, it provides the intelligent services through the broadband power line communication and wireless communication technology.

3.1 GPRS/ CDMA

In China, the major wireless communication technologies have been used in power distribution and utilization systems are GPRS, CDMA and 230M data transceiver.

GPRS and CDMA are wireless public networks. Using public spectrum resources to carry power services, they do not need to build dedicated networks. As mature technologies, they are widely used in the power system.

However, the network on GPRS or CDMA system is built for public mobile communication, not for power services, so it is hard to cover remote area; the mobile operators usually put public services in the first place, so the transmission quality of electric power services can't be guaranteed when the network is busy, and it can't meet the requirements of real-time, security and reliability too. Furthermore, it needs to pay mobile operators expensive channel lease fees.

3.2 230M Data Transceiver

Currently, the available dedicated spectrum for electric power system is consisted of 40 sub-bands at the frequency band of 230 MHz, each width of these sub-bands is 25 kHz.

The mainly deployed system at this band is 230M data transceiver. There is only one device may send data at a time on one frequency. Furthermore, since the 230M data transceiver often adopts point-to-multipoint communication allocation and polling query mode, when a great number of monitoring points connected to the

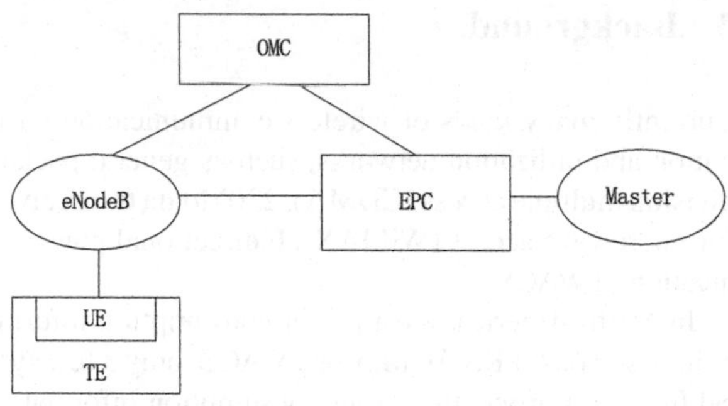

Fig. 1 Architecture of the TD-LTE230 system

master station, it will lead to a very long delay time. If a terminal could not access to the master station because equipment failure or frequency interference, the system will wait until it reaches timeout threshold, the sampling will spend more time.

Therefore, the transmission rate of 230M data transceiver is not high, which is often between 300 bps and 19.2 kbps [9].

3.3 *TD-LTE230*

TD-LTE230 system is a novel power wireless private network at the frequency band of 230 MHz. This new power wireless communication system is developed at the dedicated 230 MHz spectrum. It applies the technologies of orthogonal frequency division multiplexing (OFDM), cognitive radio and carrier aggregation (CA) to carry the services in smart distribution and utilization networks. The system keeps the advantages of TD-LTE and has the security of dedicated network.

TD-LTE230 network consists of master station (Mater), evolved Node B (eNodeB),evolved packet core (EPC), operation maintenance center (OMC), terminal equipment (TE), user equipment (UE) and so on. The architecture of the system is showed in Fig. 1.

The system can support these power services: distribution automation, power consumption information collection, load management and emergency communication, power system security monitoring and so on.

4 Tests and Results

In order to verify the performance of TD-LTE230 network, some tests have been taken on the coverage, transmission rate, latency, system capacity and others.

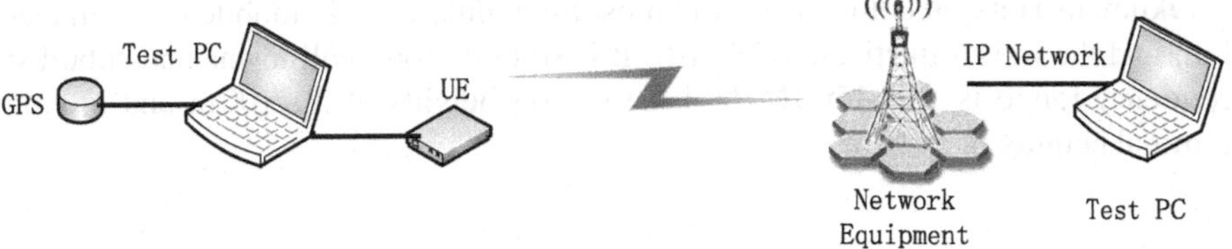

Fig. 2 Configuration scheme of test

Table 1 Test equipments

Type	Number	Application
Road test software	3	Monitoring terminal data
GPS	5	Recording position information
Dumeter	1	Statistical throughput information
Ring back tool	1	Send and receive test data packets
simulator	1	Simulating base station

4.1 *Tests Configuration*

In view of the actual test environment, the test included two parts: field test and laboratory test. The field test is taken at Yangzhou, Jiangsu province.

Tests including coverage at urban and suburban, maximum uplink and downlink transmission rate on multi-band, maximum uplink and downlink transmission rate on single band, the maximum number user terminals can be connected to a single base station and the transmission delay etc.

(1) Configuration scheme (Fig. 2)
(2) Test equipments (Table 1)

4.2 *Tests Results*

4.2.1 Coverage

There is a great number of terminal equipments covered a large geographical territory extending from urban centers to rural areas, and inside buildings and homes in the distribution and utilization system. Improving the coverage can reduce the number of base station, save funds on structure and maintenance, so it is an important criterion for the wireless communication.

The TD-LTE230 system works in the power dedicated 230 MHz spectrum. Compared with 1800 MHz, the 230 MHz frequency band has obvious advantages in the coverage area.

Okumura-Hata model is one of the most frequently used in mobile communication, and the most effective model [10]. It is suitable for small towns and suburbs; frequency range is 150–1500 MHz, base station heights of 30–100 m and mobile terminal heights of 1–10 m.

$$P_{L,urban}(dB) = 69.55 + 26.16\lg f_c - 13.82\lg h_b - \alpha(h_m) + (44.9 - 6.55\lg h_b)\lg d - K \quad (1)$$

where f_c is the carrier frequency, d is the distance between sender and receiver, L is the path loss, h_b is the base station antenna height gain factor, h_m is the mobile antenna height gain factor, $a(h_m)$ is a correction factor for the mobile antenna height based on the size of the coverage area, for small to medium sized cities, this factor is given by

$$\alpha(h_m) = [1.11gf_c - 0.7]h_m - [1.56\lg f_c - 0.8] \quad (2)$$

The European science and Technology Association (EURO-COST) proposed that the Hata model is extended to the 2 GHz band model [10], and the formula is:

$$P_{L,urban}(dB) = 46.3 + 33.9\lg f_c - 13.82\lg h_b - \alpha(h_m) + (44.9 - 6.55\lg h_b)\lg d + CM \quad (3)$$

where $a(h_m)$ is correction factor, *CM*is 0dB for medium-sized cities and suburbs, carrier frequency range to 1500–2000 MHz.

Assuming transmitting power of base station is 25 W at urban and 5 W at rural, receiving sensitivity of user equipment is −105 dB, the base station height is 30 m, the terminal height is 1.5 m.

From the above parameters, we can get that the available coverage at 1800 MHz in theory is 1.46 km at urban, 3.19 km at suburban and 18.58 km at rural areas.

In the field test, the system has been tested in two different areas of urban and suburban. The test results are: the coverage is 4.2 km at urban, 9.2 km at suburb (Table 2).

As can be seen from the test results, the new power wireless broadband system has the characteristics of low transmission loss and wide coverage.

Table 2 Coverage

Area	Theory (1800 MHz)	Test (230 MHz)
Urban (km)	1.46	4.2
suburban (km)	3.19	9.2

4.2.2 Transmission Rate

There are some major services in the power distribution and utilization network, they are power consumption information collection, distribution automation, emergency repair, overhaul and mobile assets visualization management system etc., these services have different requirements for the transmission rate. Refer paper [11], at conventional model, the transmission rate of these systems are as follows: the power information acquisition system can reach 8.33 kbps, the distribution automation system is 64 kbps ~ 1.2 kbps, the emergency repairing system is 512 kbps–1.2 Mbps (Table 3).

TD-LTE230 system uses 40 sub bands. The transmission rates at single band and multi-band were tested, and the test results are showed as follows:

By using orthogonal frequency division multiplexing, carrier aggregation and other advanced technologies, the spectrum efficiency of TD-LTE230 is greatly improved. From the test results, we can find the spectrum efficiency of TD-LTE 230 system is higher than 2 bps/Hz, which is about 3 times higher than 0.768 bps/Hz of 230M data transceiver.

4.2.3 Latency

Distribution automation is a real-time monitoring system, but the latency of wireless communication is uncertain. Therefore, it is important to test whether the transmission delay of TD-LTE230 system can meet the requirements of the power system.

Table 3 The maximum average transmission rate

Type	Single band Uplink	Single band downlink	Multi-band Uplink	Multi-band downlink
Transmission rate	41.2 kbps	14.2 kbps	1.5 mbps	551 kbps

Table 4 Data delivery time required

Information types internal to substation external to substation	Information types internal to substation external to substation	Information types internal to substation external to substation
Protection Information, high-speed	1/4 cycle	8–12 ms
Monitoring and Control Information, medium speed	16 ms	1 s
Operations and Maintenance Information, low-speed	1 s	10 s
Text Strings	2 s	10 s
Processed Data Files	10 s	30 s
Program Files	60 s	10 min
Image Files	10 s	60 s
Audio and Video Data Streams	1 s	1 s

According to IEEE Std1646–2004, we can get the maximum message delivery time required for selected types of information exchanging among applications inside the substation and outside the substation as shown in Table 4 [9].

In the test, three test points are selected by the transmission distance, and the transmission length of each test point is 32, 256 and 1024 bytes respectively. From the test, we get that: the average transmission delay is 125 ms and the maximum average transmission delay is 234 ms.

5 Conclusion

Several wireless communication technologies have been used in the power distribution and utilization network, but they do not meet the performance requirements more or less. In the paper, we propose a novel wireless broadband system (TD-LTE230)for those power services on the dedicated power spectrum and test its major performances. The tests include coverage, transmission rate, latency, system capacity and others, test results have proved that the novel wireless system is suitable for the smart power distribution and utilization network.

References

1. Q/GDW382-2009 Technical Guide for Distribution Automation. China Electric Power Press, Beijing (2010)
2. State Grid Corporation of China (2011) Q/GDW625-2011Technical rule for distribution automation and retrofit of standardized design. China Electric Power Press, Beijing
3. Yuqing Lei, Jianqi Li, Baosu Hou (2011) Power distribution and utilization communication network for smart grid. Power Syst Technol 35(12):14–18
4. Wang Wenye, Yi Xu, Khanna Mohit (2011) A survey on the communication architectures in smart grid. Comput Netw 55:3604–3629
5. Lan Zhang, Peng Gao, Deng Wang (2010) Construction of distribution automation communication system of china southern power grid. Telecommun electric power Syst 31(217):20–24
6. Daoud M, Fernando X (2011) On the Communication Requirements for the Smart Grid . Energy and Power Eng 3(1):53-60
7. Department of Energy (2010). A Survey of Wireless Communication for the Electric Power System. pp 62–66
8. Clark A, Pavlovski, CJ (2010) Wireless networks for the smart energy grid: application aware networks. In: Proceedings of the international multi conference of engineers and computer scientists, IMECS, vol II, March 17–19
9. China State bureau of Quality and Technical Supervision. GB/T 16611-1996. Generic specification for data transceiver (1996)
10. Goldsmith Andrea (2005) Wirel Commun 37–38:360–370
11. Jinping Cao, Jianming Liu, Xiangzhen Li (2013) A power wireless broadband technology scheme for smart power distribution and utilization networks. Autom Electric Power Syst 37(11):76–80

Improvement in Text-Dependent Mispronunciation Detection for English Learners

Guimin Huang, Changxiu Qin, Yan Shen and Ya Zhou

Abstract This paper put forth two novel approaches to effectively improve the performance of mispronunciations detection in English learners speech. On one hand, a distance measure called Kullback–Leibler Divergence (KLD) between Hidden Markov Models (HMMs) is introduced to optimize the probability space of a posteriori probability; On the other hand, back end processing of normalization based on the variants of speakers is introduced to improve the performance of the system. Experiments on a database of 6360 syllables pronounced by 50 speakers with varied pronunciation proficiency indicate the promising effects of these methods by decreasing the FRR from 58 to 44 % at 20 % FAR.

Keywords Mispronunciation detection · Text-dependent · Kullback–Leibler divergence · Normalization

1 Introduction

Computer assisted language learning (CALL) system potentially can offer some advantages over human assessment by overcoming time consuming and labor intensive. There have been laudable researches on the requirement of obtaining pronunciation score for sentences with grading consistency similar to that of an expert teacher [1, 2]. Although the score is an intuitional indicator to reflect proficiency of pronunciation, English learners need more specific feedback in producing a wrong sound.

G. Huang (✉) · C. Qin (✉) · Y. Shen · Y. Zhou
Guilin University of Electronic Technology, Guilin 541004, China
e-mail: sendhuang@126.com

C. Qin
e-mail: chason1990@126.com

Y. Shen
e-mail: rosagllbt888@aliyun.com.cn

Y. Zhou
e-mail: ccyzhou@guet.edu.cn

V.E. Balas et al. (eds.), *Information Technology and Intelligent Transportation Systems*, Advances in Intelligent Systems and Computing 455,
DOI 10.1007/978-3-319-38771-0_13

Therefore, in CALL, information should be provided at not only sentence level but also phone level, and then learners can pay their attention on the problematic sounds [3]. Hence, our current efforts focus on the mispronunciation detection and helping English learners to correct or improve their pronunciations.

Existing work on automatic pronunciation detection can be grouped into two categories: One is based on prior knowledge and discriminative features. The other is based on the frame of statistical speech recognition. In the first category, Truong used duration, formant and pitch period as the discriminative features to define several mispronunciation pairs, then he introduced decision trees and Linear Discriminant Analysis (LDA) to distinguish these mispronunciation pairs [4]. Dong et al. proposed an approach to distinguish two kinds of Chinese phones: there is a discriminative feature between flat tongue phones and rising tongue phones that energy distribution of the former focused on the higher frequency than that of the latter [5]. All these researches share some common limitations that they only focus on several kinds of typical mispronunciations and neglect others. In other words, if a mispronunciation is not within any categories which already defined, then the system would ignore the mispronunciation no matter how serious it is. Hence, this paper introduces KLD to build a confusion error set with confusable phones as much as possible. The accuracy of mispronunciation detection can improve by calculating a posteriori probability based on this set. As to the second category, some laudable works focus on the processing on both front end and back end. For the front end, An et al. proposed a method called Error-Detecting Network of Pronunciation (EDNP), which inserted mispronunciation detection routes into task-specific Finite State Grammar (FSG) network [6]. Ge et al. exploited a new adaptive feature called adaptive frequency scale which was obtained through an adaptive warping of the frequency scale prior to computing the cepstral coefficients to improve mispronunciation detection [7]. For the back end, the normalization basing on speaker variants, phone and context had been proved to effectively enhance the performance [8], and we will adopt this method in this paper.

The remainder of this paper is organized as follows. Section 2 presents the basic system for mispronunciation detection. Section 3 describes the KLD and its application in optimizing the probability space. Section 4 introduces how to use normalization for mispronunciation detection. Experiments and analysis are showed in Sect. 5. Conclusions are given in Sect. 6.

2 Baseline System

The baseline system for mispronunciation detection is show in Fig. 1. Firstly, the input speech is preprocessed and segmented into frames. Then the feature parameters are extracted from these speech frames. The function of force alignment is to provide boundary information of each phoneme in input speech with given text. Based on the boundary information, the score of a posteriori probability can be calculated with the corresponding feature parameters. At last, by comparing the score with the threshold, we can determine whether the pronunciation is correct.

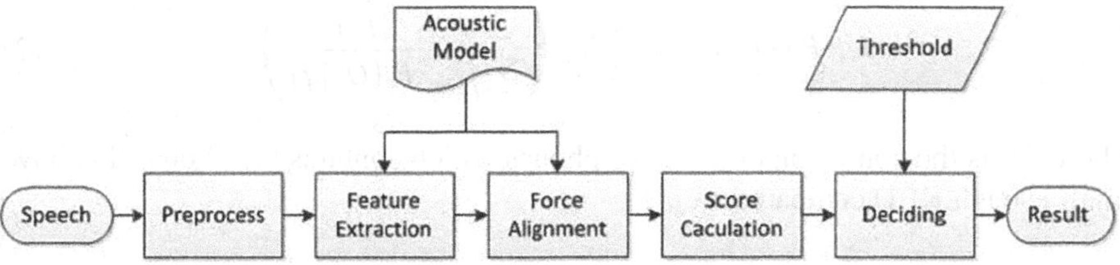

Fig. 1 The structure of baseline system for mispronunciation detection

3 Optimizing Probability Space

3.1 Traditional Method to Calculate Posteriori Probability

In the procedure of score calculation, the score of a posteriori probability is calculated with the corresponding feature parameters. Traditionally, for each O_i corresponding to the phone p_i, the logarithmic posteriori probability (LPP) can be defined as the following formula (1):

$$\begin{aligned}
&\mathrm{LPP}\,(O_i, p_i) \\
&= \frac{1}{d_i} \log\left(P(p_i|O_i)\right) \\
&= \frac{1}{d_i} \log\left(\frac{P(O_i|p_i)P(p_i)}{\sum_{q\in M} P(O_j|q)}\right) \\
&\approx \frac{1}{d_i} \log\left(\frac{P(O_i|p_i)}{\sum_{q\in M} P(O_j|q)}\right)
\end{aligned} \tag{1}$$

where M is the full set of phones in standard acoustic model, d_i is the duration of p_i, $P(p_i)$ and $P(q)$ are prior probability, $P(O_i|p_i)$ is the likelihood of p_i.

3.2 Improved Method to Calculate Posteriori Probability

In formula (1), the candidate set of phones M is the full set. Whereas the frequency of mispronunciation varies from kinds, some occur frequently, some hardly occur. Hence, this paper introduces Kullback–Leibler divergence (KLD) to optimize probability space by building a confusion error set of phones. Then the denominator of formula (1) is calculated based on the confusion error set of phones rather than the full set. Thus the improved logarithmic posteriori probability (LPP) can be defined as the following formula (2)

$$LPP\left(O_i, p_i\right) = \frac{1}{d_i}\log\left(\frac{P\left(O_i|p_i\right)}{\sum_{q\in M_e} P\left(O_j|q\right)}\right) \tag{2}$$

where M_e is the confusion error set of phones which contains the phones that have small enough KLD compared to p_i.

3.3 Calculating KLD Between HMMs

The primary task of establishing error set of phones based on the difference of distance between HMMs is to determine a dissimilarity measure. In [9], a distance measure between GMMs called KLD which based on the unscented transform is opposed to address the issue. According to [9], when two HMMs have the same number of states which are described by GMM, the KLD between two HMMs is defined as follow:

$$D[H|\tilde{H}] \approx \sum_{i=1}^{N} D[S_i|\widetilde{S_i}] \tag{3}$$

where H and $\tilde{H}$ are two HMMs, S_i and $\widetilde{S_i}$ are the i_{th} state of H and $\tilde{H}$ respectively, N is the state number of HMM, $D[S_i|\widetilde{S_i}]$ is defined as follow:

$$D[S_i|\widetilde{S_i}] \approx \frac{1}{2d}\sum_{m=1}^{M} w_m \sum_{k=1}^{2d} log\frac{b\left(O_{m,k}\right)}{\widetilde{b}\left(O_{m,k}\right)} \tag{4}$$

such that:

$$O_{m,k} = \begin{cases} \mu_m + \left(\sqrt{d\Sigma_{m,k}}\right) u_{m,k} & 1 \le k \le d \\ \mu_m - \left(\sqrt{d\Sigma_{m,k}}\right) u_{m,k-d} & d < k \le 2d \end{cases} \tag{5}$$

where M is the number of mixture GMMs, w_m is the m_{th} weight of GMM, d is the dimension number of feature vector, b( ) represents output probability, μ_m is the m_{th} mean vector of Gaussian function, $\Sigma_{m,k}$ is the k_{th} eigenvalue of variance matrix of m_{th} GMM, $u_{m,k}$ is the k_{th} eigenvector of variance matrix of m_{th} GMM.

After above calculation, we can get the KLD between a specific phone and the other phones, then the phones with smaller enough KLD are considered as the confusable phones and added into confusion set.

4 Back End Processing of Normalization

As described above, basing on each O_i corresponding to the phone p_i, we can get the LPP of p_i. The hypothesized phone would be classified as correct only if its LPP is no less than the threshold, or wrong otherwise, which can be defined as follows:

$$\text{LPP} > \text{Threshold}\begin{cases} Y\ correct \\ N\ wrong \end{cases} \tag{6}$$

Nevertheless, human annotators would take into consideration the pronunciation proficiency at not only phone level but also sentence level when they detect mispronunciations. For example, for a deficient phone, if the speaker has high pronunciation proficiency, then the annotators tend to classify the hypothesized phone as wrong, or correct otherwise. That is to say, the annotators tend to be strict with the speaker with high pronunciation proficiency, and lower their expectations to the speaker with low pronunciation proficiency. Hence, for the speaker with high pronunciation proficiency, the threshold of mispronunciation should be increased, or decreased otherwise. Then the formula will be modified as follows:

$$\text{LPP} + \alpha \times P_{\text{spk}} > \text{Threshold}\begin{cases} Y\ correct \\ N\ wrong \end{cases} \tag{7}$$

where α is the parameter need to be trained, P_{spk} is speaker score at sentence level, which is defined as:

$$P_{\text{spk}} = \frac{\sum_{i=1}^{N} LPP_{p_i}}{N} \tag{8}$$

where LPP_{p_i} is the logarithmic posteriori probability corresponding to phone p_i, N is the total number of phones in sentences. The experiment will be show in following section.

5 Experiments and Analysis

5.1 *Test Database*

Speeches were recorded from 50 Chinese students with varied pronunciation proficiency. Each student was asked to read 4 phonetically rich sentences from Arctic corpus. Moreover, the pronunciation of each phone in database is labelled as correct or not by three experienced human annotators. Then a phone is decided as mispronunciation if either of the evaluation labels is incorrect. There are 1,080 mispronounced phones and 5,280 correctly pronounced phones in the test set.

5.2 *Performance Measure*

This paper uses False Alarm Rate (FAR) and False Rejection Rate (FRR) as the performance measure for mispronunciation detection. The detection result can be divided into 4 categories as below Table 1.

Table 1 Categories of mispronunciation detection

		Machine detection	
		Wrong	Right
Human detection	Wrong	N_{ww}	N_{wr}
	Right	N_{rw}	N_{rr}

Where N_{ww} is the number of pronunciations that both human evaluator and machine mark as wrong, N_{wr} is the number of pronunciations that human evaluator marks as wrong but machine marks as right, N_{rw} is the number of pronunciations that human evaluator marks as right but machine marks as wrong, N_{rr} is the number of pronunciations that both human evaluator and machine mark as right. Then the FAR and FRR are defined as follows:

$$\text{FAR} = \frac{N_{rw}}{N_{rw} + N_{rr}} \times 100\,\% \tag{9}$$

$$\text{FRR} = \frac{N_{wr}}{N_{wr} + N_{ww}} \times 100\,\% \tag{10}$$

By changing the threshold of posteriori probability, we can get several pairs of values for FAR and FRR which can form a FAR–FRR curve.

5.3 Experimental Results and Analysis

This section introduces the performances of the mispronunciation detection systems, including the baseline system, and the two improved methods using KLD and normalization, respectively.

The mispronunciation detection results on the test set are shown with FAR?CFRR curves. KLD and Normalization in the Fig. 2 mean the KLD method and the normalization method, while baseline means the performance of baseline system.

From Fig. 2 we can see that the overall FRR rate decreases from 72, 58 and 45 % to 60, 48 and 35 % in the KLD method and to 67, 50 and 41 % in the normalization method at three different FAR levels of 10, 20 and 30 %, respectively.

Considering the structure between the methods of KLD and normalization is different, we can combine these two methods and the performance is show in Fig. 3.

The overall FRR rate decreases from 72, 58 and 45 % to 59, 44 and 33 % in the KLD + Normalization method at three different FAR levels of 10, 20 and 30 %.

Some conclusions can be drawn from the previous figures. First, we can see from the Fig. 2 that the KLD method outperforms the normalization method, and achieves better performance. Second, even adopting combined method, the FRR rate only decreases 1 % from 60 % in the KLD method to 59 % in the KLD + Normalization

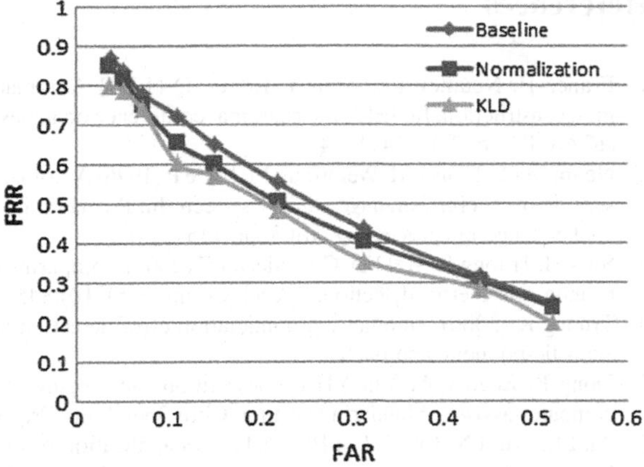

Fig. 2 Performances of the three mispronunciation detection methods

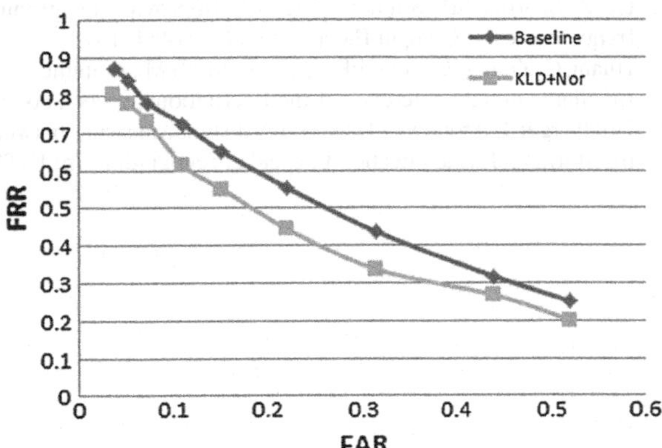

Fig. 3 Performances of baseline and KLD + Normalization methods

method at FAR level of 10 %. We can make a conjecture that KLD method limits the patterns of mispronunciation, so the system lost the ability to detect other patterns of mispronunciation.

6 Conclusions

This paper has shown the feasibility of the proposed two methods of KLD and normalization, and the combined method achieves the best performance. Experiments show the promising effects of the combined method by decreasing the FRR from 58 to 44 % at 20 % FAR.

Acknowledgments This work is supported by the Foundation of Key Laboratory of Cognitive Radio and Information Processing, Ministry of Education (Guilin University of Electronic Technology, No. CRKL150105). This work is also supported by the Innovation Project of GUET Graduate Education, No. YJCXS201543.

References

1. Franco H, Neumeyer L, Kim Y, Ronen O (1997) Automatic pronunciation scoring for language instruction. In: IEEE international conference on acoustics, speech, and signal processing (ICASSP), vol 2, pp 1471–1474
2. Neumeyer L, Franco H, Weintraub M, Price P (1996) Automatic text-independent pronunci-ation scoring of foreign language student speech. In: Proceedings of the international conference on spoken language processing, vol 3, pp 1457–1460
3. Strik H, Truong K, Wet FD, Cucchiarini C (2009) Comparing different approaches for automatic pronunciation error detection. Speech Commun 51(10):845–852
4. Truong K (2006) Automatic pronunciation error detection in Dutch as a second language: an acoustic-phonetic approach
5. Dong B, Zhao QW, Yan YH (2006) Automatic scoring of flat tongue and raised tongue in computer-assisted Mandarin learning. Chin Spok Lang Process 580–591
6. An LL, Wu YN, Liu Z, Liu RS (2012) An application of mispronunciation detecting network for computer assisted language learning system. J Electron Inf Technol 34(9):2085–2090
7. Ge Z, Sharma SR, Smith, MJ (2013) Improving mispronunciation detection using adaptive frequency scale. Comput Electr Eng 39(5):1464–1472
8. Huang C, Zhang F, Soong FK, Chu M (2008) Mispronunciation detection for mandarin Chinese. In: Ninth annual conference of the international speech communication association
9. Goldberger J, Aronowitz H (2005) A distance measure between GMMs based on the un-scented transform and its application to speaker recognition. In: INTERSPEECH, pp 1985–1988

Research on Fast Extraction Algorithm of Cross Line Laser Image Center

Mingmin Yang, Xuxiao Hu and Ji Li

Abstract When measuring three dimensional information of workpiece and object by laser triangulation method, one of the key issues about precision detection is how to quickly and accurately extract the center coordinate position of projected laser stripe. In the detection process of medium and heavy plate flatness, we use the cross line laser as the light source, attain image information by CMOS camera and store them in the computer. With digital image processing technology, we extract the center coordinate position, and utilize space projection matrix to convert image coordinate into a three-dimensional coordinate, thus obtaining the flatness of medium and heavy plate. Therefore, it will help improve the detection accuracy of the medium and heavy plate flatness to accurately extract the center coordinate position of the cross line laser stripe. Based on the characteristic that the pixels sum in the eight neighboring field of cross line laser center are four, this paper determine the coordinate of the center. Compared with the least square method, it verifies the reliability of the algorithm presented in this paper. The results can reach sub-pixel level and the system has high speed with processing time about 0.1252 s,improved by 92.22 %.

Keywords Cross line laser · Laser triangulation method · Flatness · Sub-pixel

M. Yang (✉) · X. Hu · J. Li
Department of Mechanical Engineering, Zhejiang Sci-Tech University,
Hangzhou 310018, China
e-mail: ymingmin@126.com

X. Hu
e-mail: huxuxiao@zju.edu.cn

J. Li
e-mail: blueliji@163.com

V.E. Balas et al. (eds.), *Information Technology and Intelligent Transportation Systems*, Advances in Intelligent Systems and Computing 455,
DOI 10.1007/978-3-319-38771-0_14

1 Introduction

With the rapid development of modern industry, intelligent detection technology has been widely used in vehicle, defense, aerospace and so on. And machine vision is gradually becoming one of the most important parts in detection technology [1–3]. Especially for laser triangulation measurement technology, because of its non-contact, active controllability, high precision and good real-time, it has been widely applied in modern industrial inspection, medical diagnosis, computer vision and reverse engineering. In laser detecting system, we collect light stripe projected on the surface of the measured object by CCD or CMOS digital camera, extract characteristic information of the light stripe by image processing technique, and determine the object-image relationships by camera calibration technique. Then we calculate the three-dimensional information of measured object. Therefore, the processing of laser stripe image is a key link in the whole measurement [4, 5]. And the key to keeping the stability, real-time and accuracy of the system is the stable, fast and accurate extraction algorithm of laser stripe image center.

2 Factors Affecting the Extraction of Laser Stripe Center

The laser triangulation measurement system is composed of laser emitter, camera and computer. The working principle is: the laser emitted by laser emitter is projected on the surface of measured object, and the two-dimensional image of light stripe is captured by camera; the velocity sensor detects the transmission speed of steel plates and the information above are transferred to the computer; then we extract the characteristic information of light stripe by image processing technique and calibrate transformation matrix between three-dimensional space and two-dimensional image in the camera; after these, we can determine the three-dimensional space coordinate of laser stripe center and realize three-dimensional measurement of object. The flow chart is shown in Fig. 1.

In laser measurement system, it mainly uses monochromatic laser emitter whose light intensity on its cross-sectional surface can be considered to follow Gaussian distribution, as is shown in Fig. 2. However, in reality light intensity is interfered and no longer strictly obeys Gaussian distribution due to various factors, thus making the extraction of light stripe center difficult [6]. These interference factors mainly include three types: (1) Problem of light source. It will form a spot when point laser is projected on the surface of measured object, and the spot size will affect its edge extraction; when line laser is used, the width changes of light stripe will also affect extraction of the center. Besides, its best for the brightness of light stripe to clearly show characteristics of measured objects surface in camera. If the brightness of light stripe is too high, several largest gray points will appear at the same time, which will cause error to the judgement of laser stripe center; if the light brightness is too low, the image of laser stripe captured by camera will appear breakpoints

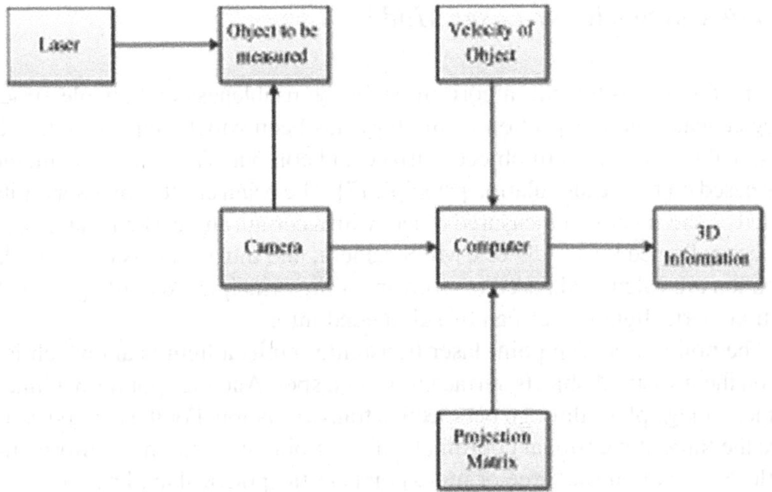

Fig. 1 The flow chart of laser triangulation measurement system

Fig. 2 Light intensity distribution

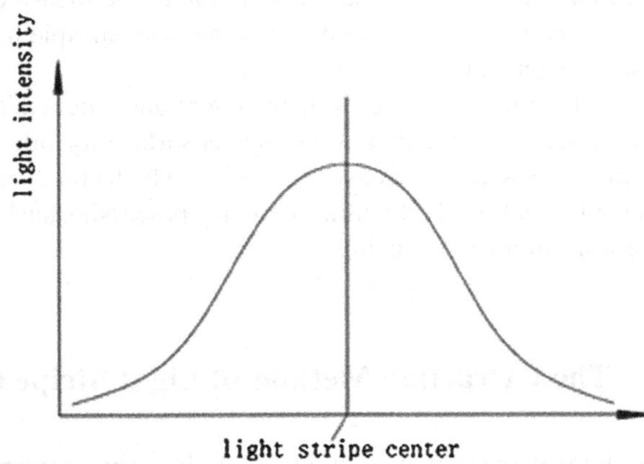

due to roughness of the surface, thus increasing the difficulty of light stripe center extraction. (2) Problem of measured objects surface. Due to the difference in shape, color, roughness, grain and texture of the measured objects surface, the ability of diffuse reflection and light intensity reflection may perform differently. This will cause the images collected by camera to include a lot of noise, which increase the difficulty of image preprocessing. (3) Problem of environmental noise. The outside environment noise, electrical thermal noise caused by image acquisition card and camera internal structure are difficult to eliminate with its randomness, complexity and non repeatability.

2.1 The Categories of Laser Mode

Due to its fast measurement, algorithm stability, usableness and simple structure, the mechanical vision inspection technology has been widely applied in the three-dimensional measurement of objects surface and contour. The measuring method is mainly based on laser triangulation principle [7]. The beam emitted by laser emitter is projected on the surface of measured object with a certain angle.Then images of light stripes are collected by a CCD or CMOS camera, and three-dimensional coordinate information are calculated based on corresponding principle. According to the laser projection mode, light source can be subdivided into:

(1) The point laser. The point laser transmitter emits a light beam which is projected on the measured objects surface to form a spot. And the spot form a image on the camera image plane through perspective transformation. For the method can only acquire the three-dimensional coordinate of one point, the amount of information is too little. So we often use three or more points in the practical application.

(2) The line laser. The line laser transmitter emits a plane which is projected on the measured objects surface to form a bright stripe with a certain width. We can get more three-dimensional information of the measured surface by spatial coordinate transformation. And we commonly use multiple line lasers for measurement in practical application.

(3) The multi-line laser. The line laser transmitter emits multiple light beams which are projected on the measured objects surface to form bright stripes. These stripes with certain widths will cover each other, which greatly increases the measuring range as well as technical difficulties in image processing such as light stripe segmentation, identification and matching.

3 The Extraction Method of Light Stripe Center

Traditional extraction methods of light stripe center mainly include extremum method, threshold method, gravity method, curve fitting method, edge method, geometric center method and so on. Assume that the gray value of an image is $f(x, y), 0 < x < i, 0 < y < j$, and (x_c, y_c) is the center of laser light stripe.

The ideal light stripe intensity follows Gaussian distribution, and the peak of curve is the center of light stripe with largest intensity. Based on this principle, extremum method [8] realizes the extraction of light stripe center by gradually searching and comparing pixels in each column to find the maximum light intensity. This method is very sensitive to noise and will produces a false center when the intensity of noise is greater than that of real center. Combining with the gravity method of rigid body in mechanics, we consider the gray level distribution centroid of pixels on each cross section as the light stripe center on this section. The extraction formula under traditional gravity method [9] is:

$$x_c = \frac{(\sum_{ij} x \times f(x,y))}{(\sum_{ij} f(x,y))}, y_c = \frac{(\sum_{ij} y \times f(x,yT))}{(\sum_{ij} f(x,y))} \tag{1}$$

Before the extraction of center, we set a threshold valueT and subtract it from the gray level of original image. The extraction formula under threshold method [10] is:

$$x_c = \frac{\sum_{ij} x \times (f(x,y) - T)}{\sum_{ij}(f(x,y) - T)}, y_c = \frac{\sum_{ij} y \times (f(x,y) - T)}{\sum_{ij}(f(x,y) - T)} \tag{2}$$

Considering large error and low precision in traditional extraction method, scholars both at home and abroad have done lots of researches and proposed many improved and novel methods on this basis. To overcome the shortcomings of extremum method, Yang Xuejiao etc. [11] calculated the normals direction of all light stripes by fitting the centers obtained through extremum method. Then they optimize and get the point with greatest gray level by using genetic algorithm on normals section. Cui Xijun etc. [12] put forward adaptive gravity iterative algorithm which makes the gravity of each line in image tend to stable and smooth finally with accuracy of 0.01 mm. Kong Panqing etc. [13] proposed to conduct a series of transformations for Gaussian function before solving equation and obtaining laser stripe center by Householder method. Gao Shiyi etc. [14] put forward a variable-margin Gaussian fitting method which has good extraction effect for laser stripes with different widths. Besides, this method has the accuracy of sub pixel and high robustness. Jeong-Hun Jang etc. [15] detected image edge by Canny operator and built a Euclidean distance mapping graph to the distance from each nearest edge then they refined the extracted curve into one-pixel width and removed redundant points to get the light stripe center.

4 The Processing of Cross-Line Laser Stripe

In the flatness detection experiment of medium plate, the laser emitted by a single cross-line laser emitter is projected on the surface of plate, and images are collected by a CMOS camera, as shown in Fig. 3. In this paper, we conduct research on the extraction algorithm of cross-line laser center in the collected images.

In the plate flatness information system based on cross-line laser, the extraction of light stripe center is a key link in the data calculation and will determine the measurement accuracy of plate flatness. The extraction process of light stripe center is generally divided into three steps: (1) image segmentation, this step mainly intends to separate the image of light stripe from background; (2) the extraction of light stripe, this step mainly intends to identify light stripe and realize the processing of speckle; (3) the extraction algorithm of light stripe center, this step mainly intends to extract the center of light stripe.

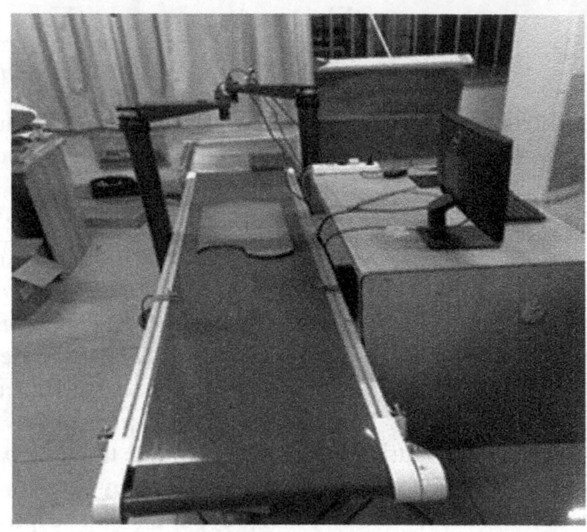

Fig. 3 The test platform of medium plate

4.1 Image Segmentation

Image segmentation is the process of segmenting an image or scene into several parts or subsets according to certain rules. The main purpose of image segmentation is to extract the useful or needed characteristics of an image from the background. As the first step of image processing and analyzing, image segmentation technique is widely used in sub scientific researches or engineering technology at present.

Image segmentation methods vary with the difference in image types and characteristics needed to extract. Gray level threshold method is mainly applied in this paper. This method is mainly suitable for images with gray level difference between target and background. After setting a threshold, this method determine each pixel as target or background based on the threshold, thus dividing the image into two parts with different gray levels. The expression of image threshold segmentation principle is:

$$g(x, y) = \begin{cases} Z_B, f(x, y) \in Z \\ Z_E, else \end{cases} \tag{3}$$

In this expression, Z represents the threshold value, Z_B and Z_E represents the gray level of object and background selected randomly. The main idea of Ostu method is to select the threshold value that maximizes the variance between background and target. The difference between target and background becomes greater with the increase of their variance; the decrease of variance may indicates the segmentation error. Therefore, the selection of maximum variance in this method means the lowest probability of segmentation error. The realization process of the maximum between-cluster variance is as follows:

Assume that the gray value of an image is m, pixel that the gray value is i is n_i, and the total pixels are:

$$N = \sum_{i=1}^{m} n_i \tag{4}$$

The probability of the each grey value is:

$$P_i = \frac{n_i}{N} \tag{5}$$

Using k value to divide it into two groups: $C_0 = [1 \ldots k]$ and $C_1 = [k+1 \ldots m]$, and the probability of the each group is ϖ_0, ϖ_1:

$$\varpi_0 = \frac{\sum_{i=1}^{k} n_i}{N} = \sum_{i=1} k P_i \tag{6}$$

$$\varpi_0 = \frac{\sum_{i=k+1}^{m} n_i}{N} = \sum_{i=k+1} m P_i = 1 - \varpi_0 \tag{7}$$

The average grey value of each group is:

$$u_0 = \frac{\sum_{i=1}^{k} n_i * i}{\sum_{i=1}^{k} n_i} = \frac{\sum_{i=1}^{m} P_i * i}{\varpi_0} \tag{8}$$

$$u_1 = \frac{\sum_{i=k+1}^{m} n_i * i}{\sum_{i=k+1}^{m} n_i} = \frac{\sum_{i=k+1}^{m} P_i * i}{\varpi_1} \tag{9}$$

The whole average grey value of is:

$$u = \sum_{i=1}^{m} p_i * i \tag{10}$$

When the threshold is k, the average grey value is:

$$u(k) = \sum_{i=1} k P_i * i \tag{11}$$

Sampling of the gray scale average is $\mu = \varpi_0 u_0 + \varpi_1 u_1$, and the formula for the variance between the two groups is:

$$d(k) = \varpi_0 (u_0 - u)^2 + \varpi_1 (u - u_1)^2 \tag{12}$$

In the range of [1, m], we can calculate k^* by altering the value of k to make $d(k^*)$ equal max $(d(k))$, and then acquire the best segmentation effect by using the k^* value as the threshold.

4.2 The Extraction of Laser Stripe Center

After the image segmentation of target and background, the next step is to extract laser stripe. For the laser stripe is white and the background is gray, we conduct binarization processing directly and the gray levels of laser stripe and background are 1 and 0 separately. After getting the binary image, we conduct morphological operation for the light stripe outlines and thin the stripe into line. The target without holes will shrink into minimal connected edges, and target with holes will shrink into connected rings, thus getting the final laser stripe. Then the refined image can be used for the extraction of laser stripe center. After enlarging the pixels of refined image, we can find the difference in coordinates of center and internal line. As is shown in the following table, we analyze the pixel value of a particular pixel and its surrounding pixels (Tables 1 and 2).

The tables above show these information: in theory, cross center wont appear breakpoints when a cross line laser is projected on a plate; gray level of the center must be 1 within the 8 area of laser stripe center; gray level of the areas left and right, up and down is also 1. Through these features, we can know that extracting cross line laser center can be turned into searching pixel coordinates meeting the characteristics in the two-dimensional image. We can conduct algorithm processing around each point by circulation function, and judge each pixel in turn. If all the areas at left, right, up and down side of the pixel have a gray level of 1, the pixel is considered as the center of the laser image. Then we can extract its coordinates, namely:

Table 1 The distribution of pixel point and the surrounding

The horizontal coordinates of pixel point (j)			
The vertical coordinates of pixel point (i)	$f(i-1, j-1)$	$f(i-1, j)$	$f(i-1, j+1)$
	$f(i, j-1)$	$f(i, j)$	$f(i, j+1)$
	$f(i+1, j-1)$	$f(i+1, j)$	$f(i+1, j+1)$

Table 2 The gray value of standard cross center and the surrounding

The horizontal coordinates of pixel point (j)			
The vertical coordinates of pixel point (i)	0	1	0
	1	1	1
	0	1	0

$$f(i, j) = f(i, j-1) = f(i, j+1) = f(i-1, j) = f(i+1, j) = 1 \tag{13}$$

In this expression, $f(i, j)$ represents the pixel value in i row and in j column.

For the uneven surface of plate may lead to dislocation phenomenon when the laser stripe get close to center, the center may appear to be one or more dislocated pixels instead of a standard cross. Table 3 shows the distribution of horizontally dislocated pixels due to the uneven surface of plate. Table 4 shows the distribution of vertically dislocated pixels.

Considering the situations above, we can hold the view that if the sum of gray level is 4 within 8 areas of laser stripe center, namely:

$$\begin{matrix} (\mathrm{i}-1,\mathrm{j}-1)+\mathrm{f}(\mathrm{i}-1,\mathrm{j})+\mathrm{f}(\mathrm{i}-1,\mathrm{j}+1)+\mathrm{f}(\mathrm{i},\mathrm{j}-1) \\ +\mathrm{f}(\mathrm{i},\mathrm{j}+1)+\mathrm{f}(\mathrm{i}+1,\mathrm{j}-1)+\mathrm{f}(\mathrm{i}+1,\mathrm{j})+\mathrm{f}(\mathrm{i}+1,\mathrm{j}+1) \end{matrix} == 4 \tag{14}$$

In this expression, $f(i, j)$ represents the pixel value in i row and in j column. We consider the point as center. Then we find out all points satisfying the condition by circulation, and average these points to get the center coordinates. We take many plate images during the experiment and select six of them as research objects, as shown in Fig. 4.

Among these nine images, we pick out five images as mainly research objects. And the preprocessing and extracting processes of them are presented below. When preprocessing, we filter the image first and get different results by different filtering templates, as shown in Fig. 5 (Figs. 6 and 7).

From above results we can see that the processing effect of 3×3 median filtering method is better. After filtering the image noise, we conduct Ostu segmentation to the image, take $1/((T + 10))$ of optimum threshold T as the threshold and get the binary image, as shown in Fig. 8.

Table 3 Pixels of lateral displacement of the cross center and the surrounding

The horizontal coordinates of pixel point (j)				
The vertical coordinates of pixel point (i)	0	1	0	0
	1	1	1	1
	0	0	1	0

Table 4 Pixels of the longitudinal displacement of the center and the surrounding

The horizontal coordinates of pixel point (j)			
The vertical coordinates of pixel point (i)	0	1	0
	1	1	0
	0	1	1
	0	1	0

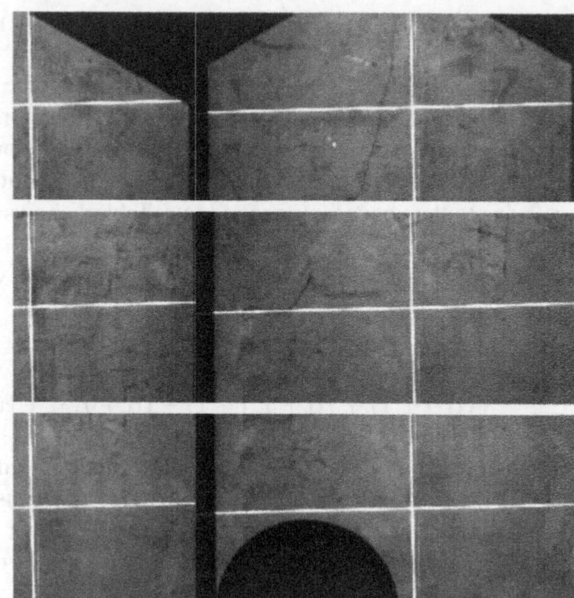

Fig. 4 Plate images

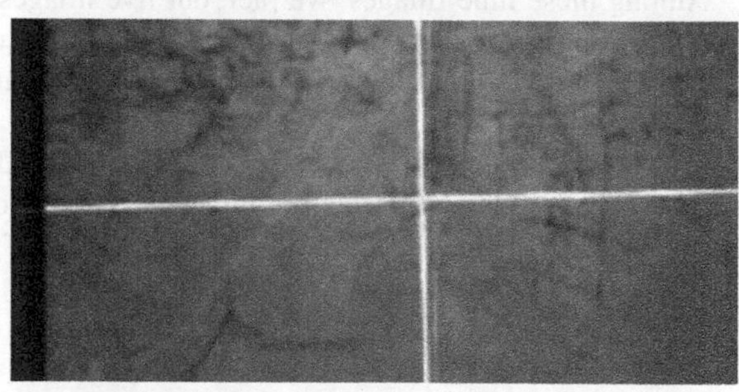

Fig. 5 3×3 median filtering method

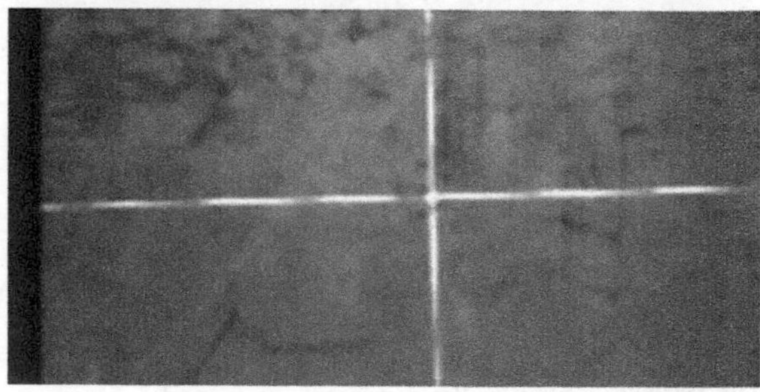

Fig. 6 5×5 median filtering method

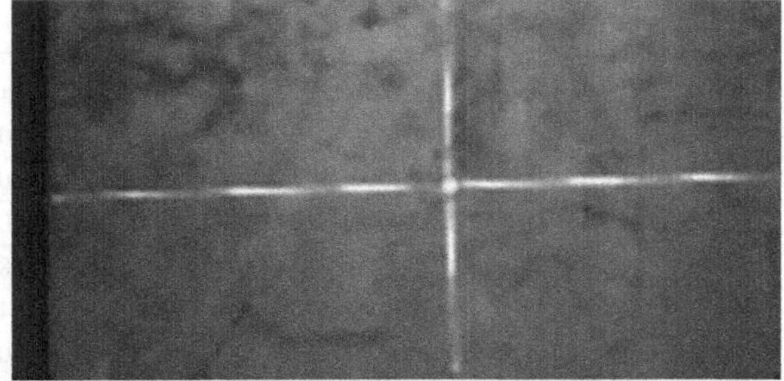

Fig. 7 7 × 7 median filtering method

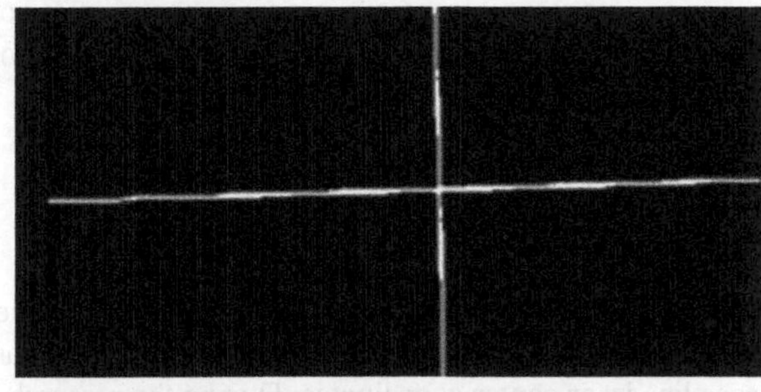

Fig. 8 Binary image

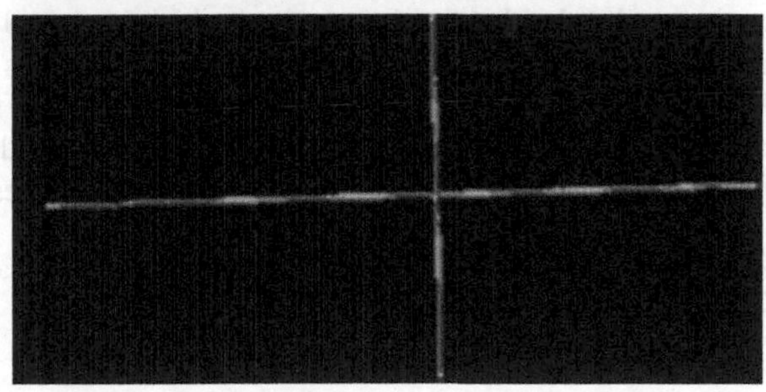

Fig. 9 Least squares fitting method

Then we conduct morphology operations to the binary image, refine it to obtain the center line of laser stripe and get center coordinates of cross line laser stripe by formula (1). To verify the reliability of this method, we compare its data with the laser stripe center under least squares fitting method, as shown in Table 5. The laser stripe center under least squares fitting method is shown in Fig. 9, in which blue lines represent the center line.

From the experimental data in Table 5, we can see that in this paper, the extraction method of cross line laser center coordinates has error within 1 compared to the results of least squares method. And the precision is even within 0.1 pixel for some image processing. The extraction time taken under the method in this paper and least squares method is 0.1252 s and 1.6090 s respectively. Namely, the extraction speed has been improved by 92.22 % in this paper and the precision of the method also meets requirements.

Table 5 Experimental data

Group	Methods x	This paper y	Least x	Square method y	Error of center Δx	Coordinates Δy	The square root error d
1	113.0000	259.0000	112.9647	258.9026	0.0353	0.0974	0.1026
2	113.0000	259.0000	113.0010	258.8731	−0.0010	0.1269	0.1269
3	113.0000	258.2500	113.0769	258.7738	−0.0769	−0.5238	0.5294
4	114.0000	258.2500	113.1711	258.6805	0.8289	−0.4305	0.9340
5	113.5000	258.5000	113.0663	258.6307	0.4337	−0.1307	0.4530
6	113.0000	258.0000	113.0526	258.7373	−0.0526	−0.7373	0.7392
7	114.0000	258.2500	113.0522	258.7375	0.9478	−0.4875	1.0682
8	113.0000	258.0000	113.0220	258.7967	−0.0220	−0.7967	0.7970
9	113.0000	258.5000	112.9775	258.8076	0.0225	−0.3076	0.3084

5 Conclusion

Based on the characteristic that the pixels sum in the eight neighboring field of cross line laser center are four, this paper puts forward the extraction method of cross line laser center coordinates. During the research, this paper also explores the optimum threshold algorithm method which has good image segmentation effects for experiments with noise on the surface of plate. This helps the extraction of cross line laser center and reduces the influences of noise on final results. The experiment results show that compared with least squares method, the method applied in this paper has a basically sub-pixel precision level and takes short time.

Acknowledgments This work is supported by Science Foundation of Zhejiang Province under Grant No.LZ14E050003.

References

1. Zhao J, YangGe Li Y (2010) The cold rolling strip surface defect on-line inspection system based on machine vision. In: Proceedings of the second Pacific-Asia conference on circuits, communications and system (PACCS), pp 402–405
2. Molleda J, Usamentiaga R, Garca DF (2011) Shape measurement of steel strips using a laser-based three-dimensional reconstruction technique. Trans Ind Appl 47(4):1536–1544
3. Xiaojie D, Fajie D, Fangfang H (2011) Study on surface defect vision detection system for steel plate based on virtual instrument technology. In: Proceedings of the international conference on control, p 1–4
4. Zhang S, Huang PS (2006) High-resolution real-time three-dimensional shape measurement. Opt Eng 12(45):1–8
5. Naesset E, Nelson R (2007) Using airborne laser scanning to monitor tree migration in the boreal-alpine transition zone. Remote Sens Environ 110(3):357–369
6. Ren FY, Xu ZP, Demin, Wang YQ.Several factors that affect laser strip center and meyhods accordingly in the three-dimensional measurement using structured laser light. Microcomputer Information, 2-6,22(11-2):259-261

7. Sun CK, He M, Peng, W (2001) Laser detection technology. Tianjin University Press, Tianjin
8. Junji H, Guangjun Z (2003) Study on method for processing image of strip in structured-light 3D vision measuring technique. J Beijing Univ Aeronaut Astronaut 29(7):594–595
9. Zhongwei L, Congjun W, Yusheng, S (2008) An algorithm for detecting center of structured light stripe combining gradient sharpening with barycenter method. J Image Graph, 13(1):65
10. Jianbo Wu, Zhen Cui, Hong Zhao, Yushan Tan (2001) An adaptive threshold method for light-knife center acquisition. Semicond Optoelectron 27(2):189–191
11. Yang X, Haihong C (2009) An improved method of extracting structured light strip center. Appl Sci Technol 36(12):41–44
12. Xijun C, Chuan Y, Baohua Liu et al (2007) Self-adaptive iterative method of extracting center of linear-structured light stripe. J Xian Jiao tong Univ 41(1):73-76
13. Sun P, Yang Y, Liangliang H (2012) An improved Gaussian fitting method used in light-trap center acquiring. Electron Des Eng 20(13):179–185
14. Shiyi G, Kaizhen Y, (2011) Research on central position extraction of laser based on varied-boundary Gaussian fitting. Chin J Sci Instrum 32(5):1132-1137
15. Jang J-H, Hong K-S (2002) Detection of curvilinear structures and reconstruction of their regions imgray-scale images. Patt Recog 35(4):807–824
16. Wang A, Shen W, Wu C, Zhang L (2010) Research on a new extraction method of laser lights stripe center based on soles information scanned. J Mech Electr Eng 27(2):21–23

Adhesion Ore Image Separation Method Based on Concave Points Matching

Ning Zhigang, Shen Wenbin and Cheng Xiong

Abstract In order to segment adhesion ore image well, the adhesion ore image separation method based on concave points matching has been proposed. Firstly, adhesion ore images of conveyor belt were obtained by high-speed digital camera, and those images were preprocessed. Secondly, improved Harris corner detection operator was used to detect the corner points, and match concave point pair was determined by corner detection operator coupled with the circular template. Thirdly, sector searching was carried out in simply connected region to match concave point pair and find out the best match concave point pair by rectangular limiting. Finally, division line was determined between concave point pair, and adhesion ore image was completely segmented. Experimental results show the method has high segmentation accuracy and good segmentation effect.

Keywords Concave points matching · Corner detection · Circular template · Adhesion ore image

1 Introduction

Image segmentation technology is always difficult point and hot topic of digital image processing. Domestic and foreign scholars have proposed lots of algorithms in different application fields. Xiangzhi Bai proposed a new touching cells splitting algorithm based on concave points and ellipse fitting [1]. The concave points were used to divide contour into different segments. Ellipse processing was used to process

N. Zhigang (✉) · S. Wenbin · C. Xiong
College of Electrical Engineering, University of South China,
Hunan, Hengyang 421001, China
e-mail: 373062228@qq.com

S. Wenbin
e-mail: 781555135@qq.com

C. Xiong
e-mail: 568635523@qq.com

V.E. Balas et al. (eds.), *Information Technology and Intelligent Transportation Systems*, Advances in Intelligent Systems and Computing 455,
DOI 10.1007/978-3-319-38771-0_15

different segments of contour into possible single cells by using the properties of fitted ellipses. Rong Fu proposed automatically separating overlap cell image algorithm based on searching concave spot and concave point pair [2]. Ke Dong proposed ore image segmentation algorithm based on concave point detecting and matching [3]. Concave points was detected at first, then concave point pair was matched, and division line was found out. Adhesion ore image was finally segmented. Ore image which had a few particles could preferably be segmented. When ore image which had many particles was segmented, there existed over-segmentation and under-segmentation phenomenon. Weihua Liu proposed automatic separation algorithm of overlapping power particle based on searching concave point [4]. Different particle overlapping types were analyzed, and different concave point matching criteria was used to separate adhesion ore image. Dongdong Wei proposed image segmentation algorithm for overlapped particles based on concave point matching [5]. Edge contour was used to detect concave point, and overlapped particles were segmented. Digital image separation algorithm based on concave points matching is used to adhesion ore image in the paper.

2 Flow Diagram of Adhesion Ore Image Separation Algorithm

Ores are broken through crusher at first, then transferred through transmission band. Because gray difference of adjacent ore images is very little, adjoining ore are often touched and overlapped when ore images are collected. If adhesion ore images aren't separated, the accuracy of ore particle size measurement is influenced. Concaves exist in touched ore field, and concave points can be used to segment adjoined ore images. The flow diagram of adhesion ore image separation algorithm is showed in Fig. 1. Improved Harris corner detection algorithm is used to determine corner points of ore image [6, 7], which are judged concave points by circular template. Concave points are matched by sector searching and rectangular limiting method in order to find the best match concave point pair and determine division line. Improved digital image separation algorithm [8] is used to segment ore image.

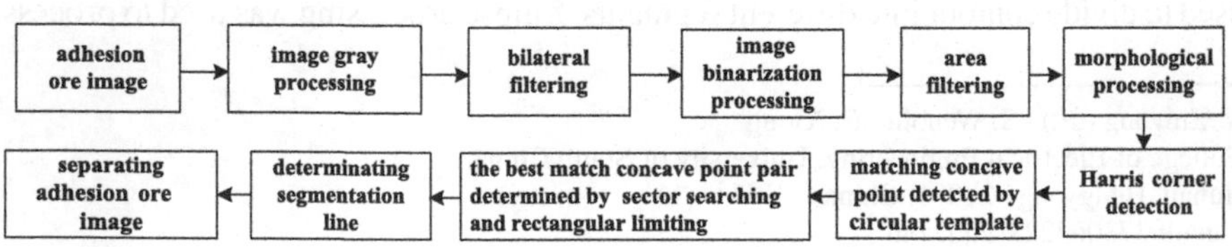

Fig. 1 Flow diagram of adhesion ore image separation algorithm

3 Image Gray Processing, Image Filtering and Image Segmentation

Collected ore image is RGB color image. In order to decrease amount of calculation, RGB color image is carried out gray processing. Because green is most sensitive to human eye, G component is accounted for the largest proportion, and B component is accounted for the smallest proportion in the conversion process. In order to obtain the best gray scale image, color image is converted into gray scale by forming a weighted sum of the R, G and B components: 0.2989 R + 0.5870 G + 0.1140B. Bilateral filter is a kind of nonlinear filter, whose kernel is composed of two functions. One is pixel space distance function, the other is pixel difference function. Both functions determine filter coefficients. Bilateral filter filters noises and keeps intact image edge, which connects spatial relationship of pixels with gray relation. Corresponding defined expressions are as follows:

$$G_{\sigma_d} = \exp\left(-\frac{(i-k)^2 + (j-l)^2}{2{\sigma_d}^2}\right) \tag{1}$$

$$G_{\sigma_r} = \exp\left(-\frac{\|f(i,j) - f(k,l)\|^2}{2{\sigma_r}^2}\right) \tag{2}$$

$$W_{k,l} = \mathrm{G}_{\sigma_\mathrm{d}} G_{\sigma_r} \tag{3}$$

$$BF\left[f(i,j)\right] = \frac{\sum_{k,l} f(k,l) W_{k,l}}{\sum_{k,l} W_{k,l}} \tag{4}$$

where, weight coefficient $W_{k,l}$ is the product of function G_{σ_d} and function G_{σ_r}. Expression $BF\,[\bullet]$ is used to apply a bilateral filtering operation to original image [9]. Original color ore image, gray level image, median filtering image, bilateral filtering image and segmentation image are shown in Fig. 2. Compared median filtering image with bilateral filtering image, it can conclude that bilateral filtering can filter out ore image noise more effectively and maintain better edge information. So bilateral filtering is selected in the paper.

Maximum inter class variance algorithm (abbr OTSU) is common image segmentation method, which segments image into foreground and background according to gray-level feature. OTSU segmentation method [10] and pulse coupled neural network (abbr PCNN) segmentation method [11] are used to segment ore image respectively in the paper. OTSU segmentation image and PCNN segmentation image are shown in Fig. 2e, f respectively. OTSU segmentation method can't remove shadow image of ore and can't segment adhesion ore image. PCNN segmentation method can remove shadow image of ore, but the method can't segment adhesion ore image. In order to segment adhesion ore image well, the adhesion ore image separation method based on concave points matching is adopted in the paper. The hollowness between

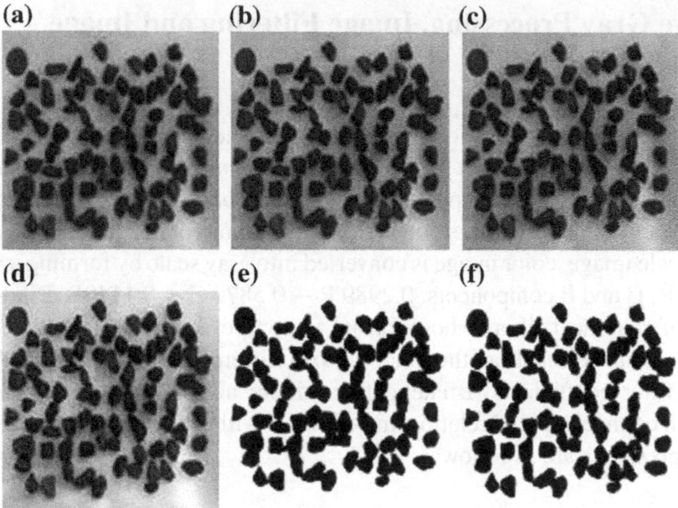

Fig. 2 Image gray processing, image filtering and image segmentation. **a** Original image. **b** Gray scale image. **c** Medium filtering image. **d** Bilateral filtering. **e** OTSU segmentation. **f** PCNN segmentation

adjacent ore is used to find the best match concave point pair, then touching ores are separated according to the best match concave point pair. PCNN binarization segmentation algorithm may form hollows in the foreground region, which can be filled up by area filtering method. Image burr is removed by morphological expansion and corrosion method. Smooth edges are obtained. Thus angular points can be extract by Harris corner detection operator.

4 Adhesion Ore Image Separation Method Based on Concave Points Matching

Concave points are found out by Harris corner detection algorithm and circular template detection method. Harris corner detection algorithm is used to extract angular points in ore binary image, as shown in Fig. 3. Circular template detection method is used to detect concave point at every angular point. The radius of circular template is adaptively determined by the length of division line. Central angle which is formed by angular point and intersection points of circular template and inner edge of adhesion ore image is detected at every angular point. If the central angle is larger than the threshold θ, then the angular point is judged concave point. The number L of concave point adds one, and the coordinates of concave point is saved in the array P. The point 1, the point 2, the point 3 and the point 4 are supposed to concave points, which are to be matched according to matching criterion. The center of gravity g may

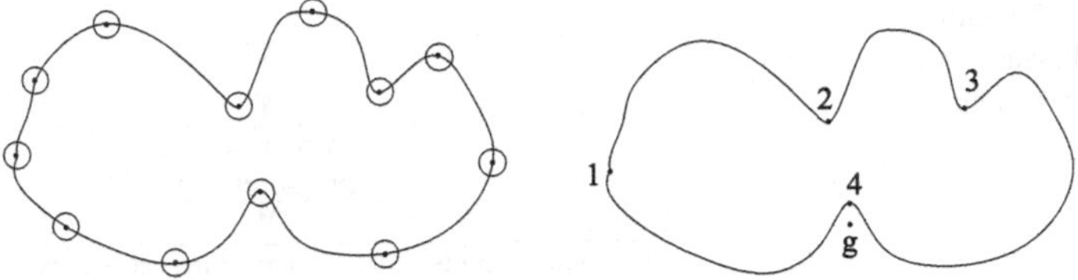

Fig. 3 Concave point detection by Harris operator and circular template

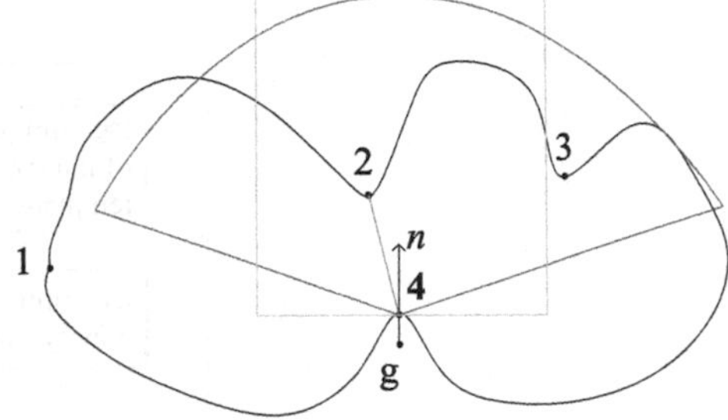

Fig. 4 The best match concave point pair by sector searching and *rectangular* limiting

be determined by pixel coordinates of inner edge of adhesion particles in the circular template. The principal vector n can be determined at each concave point, as shown in Fig. 4. The flow diagram of maximum distance and maximum offset angle algorithm is shown in Fig. 5. The offset angle A_{ij} is the angle between the connection line of concave point pair and the principal vector. The distance D_{ij} between concave point pair and the offset angle A_{ij} which deviates from the principal vector are calculated at each match concave point of simply connected region. The maximum distance (maxD) and maximum offset angle (maxA) are determined at each match concave point, which are saved in the array D_L and the array A_L respectively. The maximum distance, maximum offset angle and the principal vector are used to determine sector searching region and rectangular limiting region. The maximum offset angle is used to determine the central angle of sector.

Matching concave point pair is determined by sector searching and rectangular limiting method. The division line of adhesion ore is determined according to match concave point pair, which is used to segment adhesion ore. The flow diagram of adhesion ore image separation algorithm is shown in Fig. 6. The key point of digital image separation method is to determine the region of searching sector, match concave point pair and the best division line. The center point of searching sector is concave point i, and the radius of searching sector is the maximum distance (maxD). The central angle of searching sector is two times as big as the maximum offset angle (maxA). The principal vector is the angular bisector of central angle of searching sector. The point i is the middle point of width of limiting rectangle, and the length of width is c times as long as the maximum width of sector (empirical value 0.4). The length of limiting rectangle is the radius of searching sector. If concave point j is locate in the limiting rectangle and the sector (not including edge), the point i

Fig. 5 Flow diagram of maximum distance and maximum offset angle algorithm

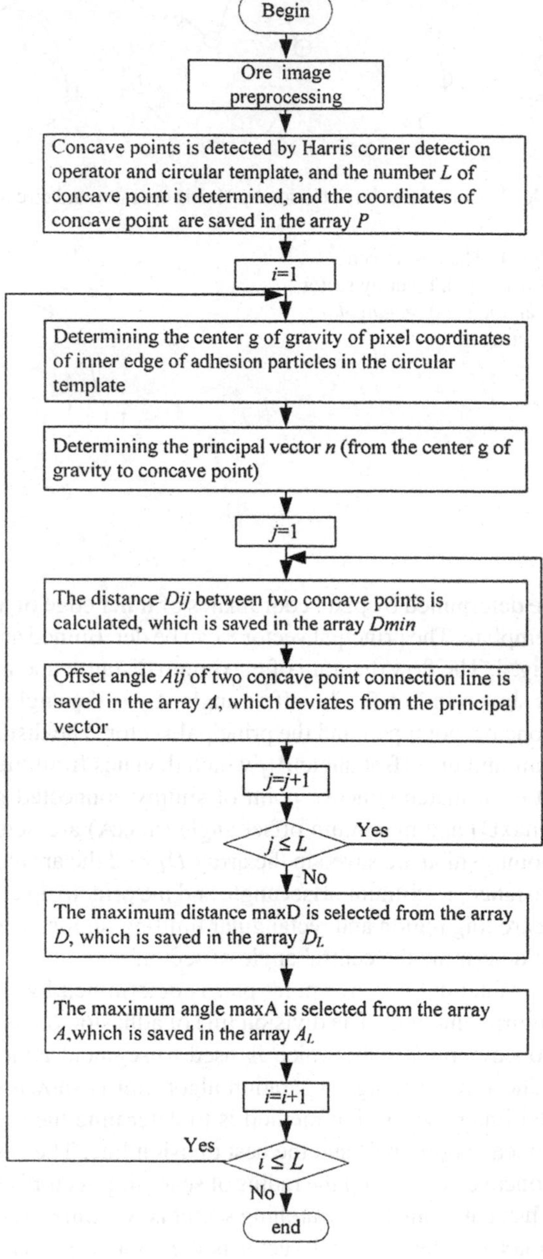

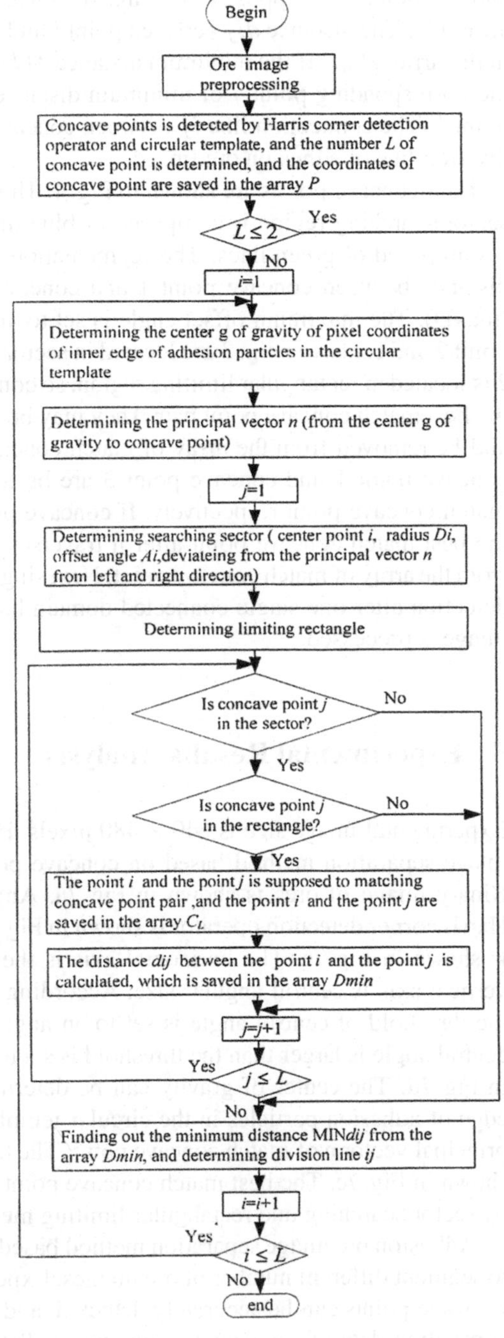

Fig. 6 Flow diagram of adhesion ore image separation algorithm

and the point j is supposed to be match concave point pair which are saved in the array C_L. The distance dij between point i and point j is calculated, which is saved in the array D_{min}. If the minimum distance $MINdij$ is found from the array $Dmin$, the corresponding point j of minimum distance is judged to be the match point of point i. The connection line ij between concave point i and concave point j is the division line of adhesion ore image.

Four concave points are shown in Fig. 4. The point 4 is concretely analyzed. The sector searching region is composed of blue lines, and rectangular limiting region is composed of green lines. The segmentation line is denoted by the red line. The distance between concave point 1 and concave point 4 is the maximum distance (maxD). The maximum offset angle is set to an angle of 60 in the paper. Concave point 2 and concave point 3 are located in sector searching region, and concave point 2 is located in rectangular limiting region, so concave point 2 and concave point 4 are the best match concave point pair. They may be used to determine segmentation line and be removed from the array of match concave point. According to the method, concave point 1 and concave point 3 are be to match in order to search the best match concave point respectively. If concave point 1 or concave point 3 can't find the best match concave point after it traverses the entire array, it will be removed from the array of match concave point. Next single connected domain carries through detection after one single connected domain has finished detection until the whole image is processed.

5 Experimental Results Analysis

Experimental image size is 640 × 480 pixels. Experimental results of adhesion ore image separation method based on concave points matching are shown in Fig. 7. Binary image of ores is shown in Fig. 7a. Angular points which are detected by Harris corner detection operator is shown in Fig. 7b. The distribution of central angle is shown in Fig. 7c. The horizontal axis is the ordinate of angular point, and the vertical axis is central angular size. According to the distribution of central angle, the threshold of central angle is set to an angle of 225. The concave point whose central angle is larger than the threshold is set to the match concave point, as shown in Fig. 7d. The center of gravity can be determined by pixel coordinates of inner edge of adhesion particles in the circular template, which is used to determine the principal vector n of match concave point. The blue points is the center of gravity, as shown in Fig. 7e. The best match concave point pair and segmentation line is found by sector searching and rectangular limiting method, as shown in Fig. 7.

Adhesion ore image separation method based on concave points matching is used to segment different number of ore images. Experimental results are shown in Fig. 8. Concave points can be accurately detected, and adhesion ore can be separated. The correlation data of segmentation image are listed in Table 1. The accuracy rate of segmentation is the ratio of accurate segmentation lines to practical segmentation lines. The complexity of segmentation is the ratio of particle number counted before

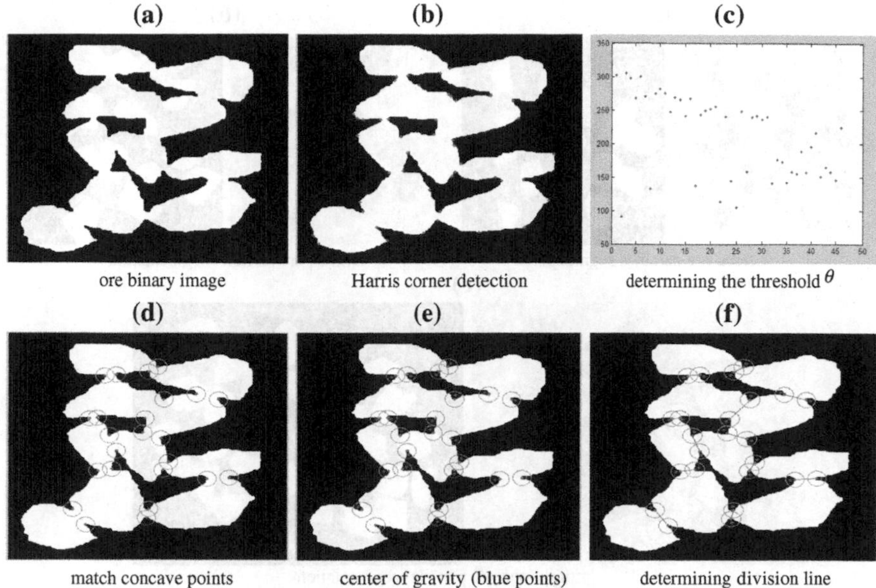

Fig. 7 Experimental results of adhesion ore image separation method. **a** Ore binary image. **b** Harris corner detection. **c** Determining the threshold θ. **d** Match concave points. **e** Center of gravity (*blue points*). **f** Determining division line

and after segmentation. The higher is the accuracy rate of segmentation, the more accurate is segmentation of ore image. The less is the complexity of segmentation, the more complex is segmentation situation. There isnt under-segmentation and over-segmentation phenomenon in higher or lower complexity of segmentation image under the circumstance of small quantities of ore particles, as shown in Fig. 8a, c. The effect of ore segmentation is good, and the accuracy rate of segmentation is 100 %. There is under-segmentation and over-segmentation phenomenon in higher complexity of segmentation image under the circumstance of large quantities of ore particles, as shown in Fig. 8d, e. The accuracy rate of segmentation decreases to approximately 90 %. The practical number of particle is approximately equal to particle number counted after segmentation, as shown in Table 1. The data show that the algorithm has a high accuracy rate of segmentation for higher or lower complexity of adhesion ore image. Adhesion digital image separation algorithm based on concave points matching is satisfied with practical demands, which can improve measure precision of granularity parameters of ore.

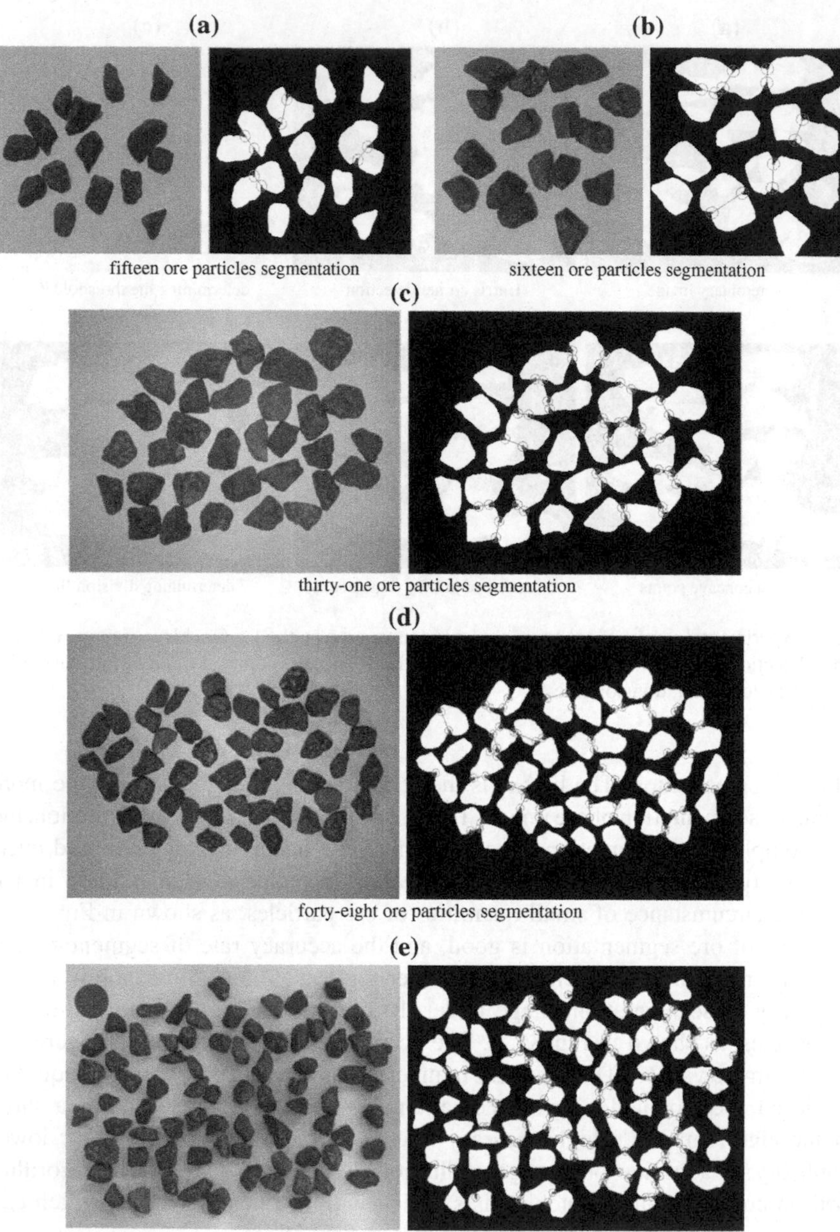

Fig. 8 Different number of ore images segmentation. **a** Fifteen ore particles segmentation. **b** Sixteen ore particles segmentation. **c** Thirty-one ore particles segmentation. **d** Forty-eight ore particles segmentation **e** Seventy-seven ore particles segmentation

Table 1 Elated data of ore image segmentation

	Practical particle number	Particle number before segmentation	Particle number after segmentation	Segmentation line number	Under-segmentation line number	Over-segmentation line number	Right segmentation line number	Accuracy rate of segmentation (%)	Complexity of segmentation
Fig. 7f	10	1	10	12	0	0	12	100	0.100
Fig. 8a	15	11	15	4	0	0	4	100	0.733
Fig. 8b	16	10	16	6	0	0	6	100	0.625
Fig. 8c	31	9	31	22	0	0	22	100	0.290
Fig. 8d	48	21	49	28	1	2	25	89.3	0.429
Fig. 8e	77	34	80	43	0	3	40	93.0	0.442

References

1. Bai X, Sun C, Zhou F (2009) Splitting touching cells based on concave points and ellipse fitting. Pattern Recognit 42(11):2434–2446
2. Rong F, Hong S, Hao C (2007) Research of automatically separating algorithm for overlap cell based on concave spot. Comput Eng Appl 43(17):21–23, 28 (In Chinese)
3. Ke D (2013) Research on ore granularity detection technology based on machine vision. Beijing University of Technology, Beijing (In Chinese)
4. Weihua L, Qingmei S (2010) Automatic segmentation of overlapping power particle based on searching concavity points. J Electron Meas Instrum 24(12):1095–1100 (In Chinese)
5. Dongdong W, Yuhong Z (2010) An image segment algorithm for overlapped particles based on concave points matching. Comput Appl Chem 27(1):99–102 (In Chinese)
6. Ping G, Xiangbin L, Peng Z (2010) Improved Harris-based corner algorithm. Comput Eng Appl 46(11):173–175 (In Chinese)
7. Wanjin Z, Shengrong G, Chunping L, Xiangjun S (2008) Adaptive Harris corner detection algorithm. Comput Eng 34(10):212–214 (In Chinese)
8. Van den Berg EH, Meesters A, Kenter JAM (2002) Automated separation of touching grains in digital images of thin sections. Comput Geosci 28(2):179–190
9. Yuling W (2010) Study of algorithm in image processing based on bilateral filter [D]. Xidian University, Xian (In Chinese)
10. Zhigang N, Hui C, Heng Y, Xiong C (2015) Dynamic recognition of oscillating orange for harvesting robot. J Jiangshu Univ (Nat Sci Ed) 36(1):53–58 (In Chinese)
11. Xin G, Zou B, Li J, Chen Z, Cai M (2011) Image segmentation with PCNN model and maximum of variance ratio. J Image Graph 16(7):1310–1316 (In Chinese)

Design and Implementation of Hybrid Formation System

Yibo Li, Cangku Wang, Yi Wang and Senyue Zhang

Abstract In the paper the three-dimensional visualization system based on FlightGear, MFC and Matlab (Simulink) was designed. For future war mode that man-aircraft mixed with UAV cooperative engagement and enhance the demand of virtual realization effect of 3D visualization, the three-dimensional visualization system by functional modules was constructed and integrated each module to visual simulation; The system unions time-sequence actuation mechanism, virtual reality technology, UAV model rendering software, FlightGear visual simulation flight software and UAV dynamics Simulink and makes the flight data and flight visual to be Time-Succession, visible, and brings high participant operation to operator, enables each simulation module to be renewable and exchangeable. During numbers of simulation experimentation, this system implements a series of 3D visual formation flight simulation, air combat demonstration, weather conditions and geographical environment, which has three-dimensional visualization effect.

1 Introduction

The hybrid formation flight of Manned/unmanned aircraft makes the systems comprehensive effectiveness and the redundancy performance of the task to be improved significantly, and makes up for the lack of the formation of the UAV to identify and judge the complex environment, which is becoming new hot of the hybrid formation

Y. Li · C. Wang (✉) · S. Zhang
School of Automation, Shenyang University, Shenyang 110136, China
e-mail: 13354247128@163.com

Y. Li
e-mail: liyibo_sau@163.com

S. Zhang
e-mail: 3310848@qq.com

Y. Wang
AVIC Chengdu Aircraft Industrial (group) CO.LTD, Sichuan Chengdu 610073, China
e-mail: scjjwy@21cn.com

V.E. Balas et al. (eds.), *Information Technology and Intelligent Transportation Systems*, Advances in Intelligent Systems and Computing 455,
DOI 10.1007/978-3-319-38771-0_16

of Manned/unmanned aircraft and its application technology. Currently in formation control, information exchange, and specific implementation verification has carried out extensive research. Along with the research, the differences of the setting environment and the formation model have brought about the difficulty to the method and the means of the verification research. Due to the entity authentication requires a comprehensive engineering design, takes a lot of work to verify, and expensive experimental equipment and material consumption, therefore the virtual simulation verification method as a flexible, low cost is suitable for most of the research and theoretical study [1–5].

At present, the research on the design and implement of the virtual simulation experiment platform for the flight of the manned aircraft and UAV formation flight has been carried out abroad [6–8]. A visual simulation system of multi action device is established in the literature [6], which is able to disguise the behavior of a number of mobile devices, but the Control instruction cannot be applied during simulation and the effect of 3D visualization is also not good. The complex aerospace system is established in the literature [7], which can enhance information flow and its harmonization, but the virtual environment can be real-time provided and can add control instructions in the experiment. To sum up the shortcomings of the above existing literature in the design of the system, thus, in the paper a visual flight simulation system based on MFC, Simulink and FlightGear is designed and there is no literature on the joint simulation of Matlab, FlightGear and MFC in the existing literature. Thus a visual joint simulation system based on MFC, Simulink and FlightGear is designed. The interface terminal is developed by MFC; the Simulink is used as data solution of manned/unmanned aircraft models; the FlightGear is used as visual simulation of manned/unmanned aircraft.

2 System Structure Design

Multi aircrafts hybrid formation visualization system consists of a ground station, several manned aircraft units, several unmanned aircraft units, several enemy aircraft units, a visual display unit, a high speed Ethernet switch and a server, which a unit stands for a computer, as shown in Fig. 1. The ground station is responsible for the simulation system initialization and the whole formation flight of a series of process control and other tasks. The manned aircraft units consist of the driving terminal and task terminal. The driving terminal shows cockpit and the manned aircraft can also be steered by an operator control air-craft operating lever. The task terminal is partitioned into multiple task regions that respectively illustrate situation including battlefield situation chart, formation, threaten and so on, and attitude and status information of each aircraft; the unmanned aircraft units consist of interface terminal and Simulink kinetic model. The interface terminal is used for showing attitude and position information of each aircraft; Simulink kinetic model is capable of solving flight data to drive UAV model; The enemy aircraft units consist of the aircraft operating lever and FlightGear terminal. The flight data is transport to FlightGear

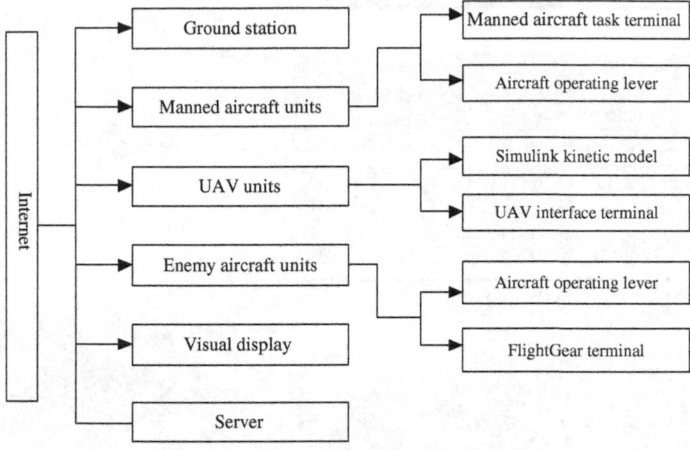

Fig. 1 The overall architecture of the system

terminal to drive the enemy aircraft model. The enemy aircraft can also be steered by an operator control aircraft operating lever; the sever unit makes manned aircrafts, UAV and enemy aircrafts display in the same 3D scene. The high speed Ethernet switch is used for data transmission between manned aircraft and UAV and enemy aircraft.

3 The Realization of the System

The system can be divided into the UAV module, the UAV dynamics system simulation module and the FlightGear 3D visualization module.

3.1 FlightGear Virtual Scene Implements

Because the huge amount of code of the OpenGL realize scene and the model of the aircraft is not conducive to the modification of the different UAV size model, the AC3D software is used to construct the UAV's external model. The flight data is transmitted to the FlightGear simulation module by the communication module, and drives UAV module simulation. The aircraft module is controlled by environmental system and dynamic system and by way of 3D visual simulation system for rendering, the system log record flight attitude and position information, the noise of the engine and the voice/sound system simulation, the control results Greatly simplify the development of complexity and UAV model can be modified accordingly for different models, which make the establishment of the model and the visual simulation

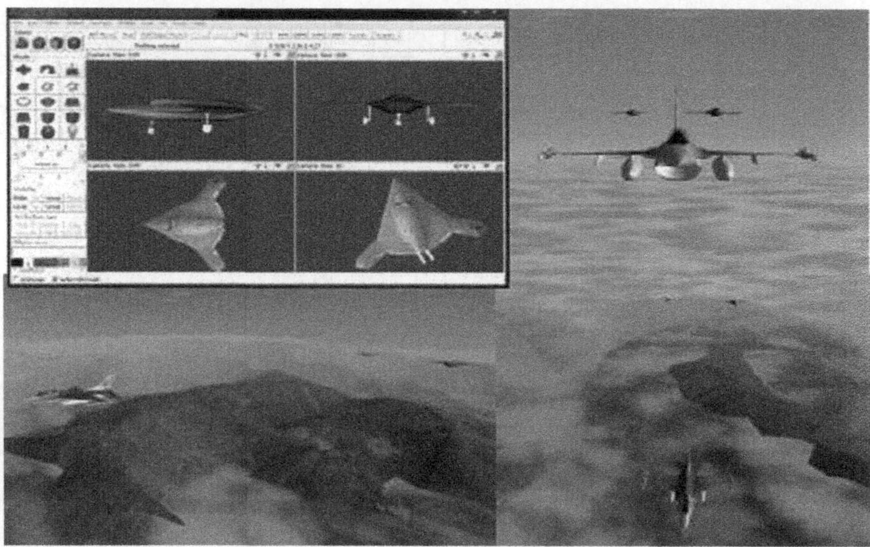

Fig. 2 The observation effect of the model production and scene

of the system independently and increase the scalability of the system simulation and the portability of the model. At the same time, the main components of FlightGear are independent of hardware and operating system, which is able to easily transfer between different operating systems. In addition, by setting the variable view and random panoramic view, the operator can observe the virtual visual scene from multi angle, large span, multi gauge distance, which brings favorable support for the operators perception. The observation effect of the model production and scene is shown in Fig. 2.

3.2 *Modularization and Integration Implement*

In order to make the simulation system have a strong expansibility, each module uses UDP communication pattern. In the deepening of the study can be off-line modify the corresponding module algorithm without need to upgrade the system through the examination of each module input and output to be location and repair, when errors or anomalies occurs. The UAV dynamics model, as shown in Fig. 3, with S function construct with external communication module, which implements data transmission between VC++ and Simulink [9]. ground stations and lead plane instructions will be transferred to wing-man mission terminals, implementing decentralized cooperative control between the frame wing-man, while flight control data that based on the PID formation flying control algorithm solver is transferred to an external data-driven and loaded model of UAV FlightGear and drives FlightGear visualization

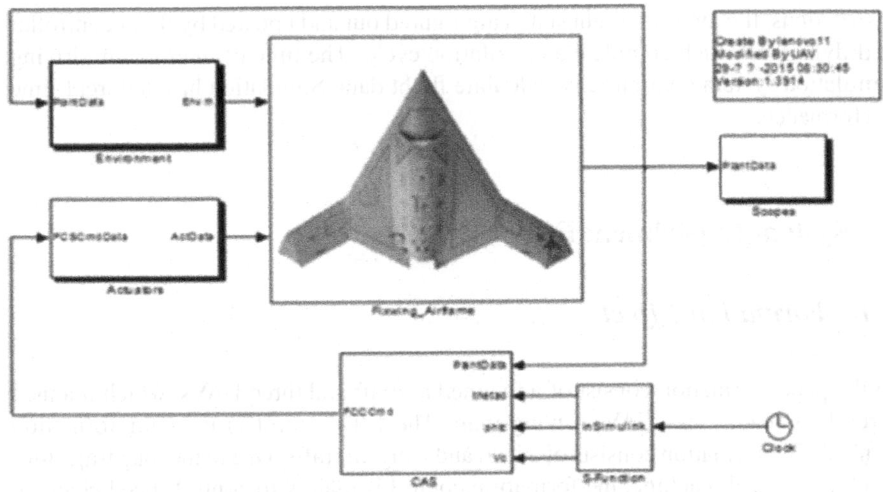

Fig. 3 The uav dynamics model

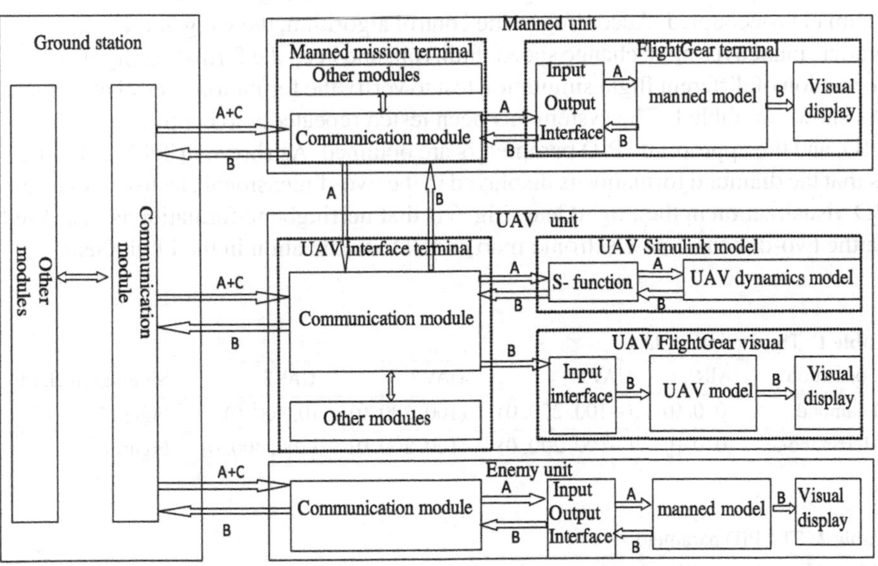

Fig. 4 System data transmission

engine, generating mixed fleet flying in formation, three-dimensional formations and combat three-dimensional visualization. There is some independent and strong data interaction coupling in the visual system for manned aircrafts/UAV. To complete the coupling data exchange in the condition of independence, the flight data is transported to leader and follower, as shown in Fig. 4. A stands for flight control and flight command data; B stands for flight data. After the ground station and leader output instructions, the aircrafts flight status are figured out and updated by flight controller and dynamic, which complete a calculation cycle. The time of intermittent driving simulation system continuously calculate flight data. Simulation has high real-time performance.

4 System Experiment Effect

4.1 Formation Effect

In the paper formation consist of a manned aircraft and three UAVs, which manned aircraft is leader and UAV is wing-man. The PID control is used for formation control. The formation consists of a lead and wing aircrafts. Due to the long trajectory tracking control machine, the formation control is mainly to control speed channel, heading channel, and altitude channel of the wing aircraft. The corresponding control amount is speed control, roll angle control and pitch angle control and the each channel is decoupled. According to the control algorithm, the wing aircraft can track leaders maneuvering to change speed, pith, roll, and keep the formation flight. We set formation of different flight simulation test to verify the formation control algorithm. As shown in Table 1. The system has been tested repeatedly to control algorithm of PID, and the appropriate PID parameters are obtained. As shown in Table 2. Figure 5 is that the diamond formation is displayed in the two-dimensional electronic map and 3D visualization in the FlightGear; Fig. 6 is that herringbone formation is displayed in the two-dimensional electronic map and 3D visualization in the FlightGear.

Table 1 The formation data

Formation	AIR	UAV1	UAV2	UAV3	Simulation chart
Diamond	(0, 0, 0)	(−100, 200, 0)	(100, 200, 0)	(0, 400, 0)	Figure 5
Herringbone	(0, 0, 0)	(−100, 200, 0)	(100, 200, 0)	(200, 400, 0)	Figure 6

Table 2 The PID parameters

Channel	PID Name	P	I	D
Speed	Incremental PID	49.8	0.024	7
Speed	Position PID	0.0296	0.00000042	0.0
Heading	Incremental PID	1.66	0.00088	3.42
Heading	Position PID	0.006	0.0	0.0
Altitude	Incremental PID	0.129	0.0000000014	0.00037

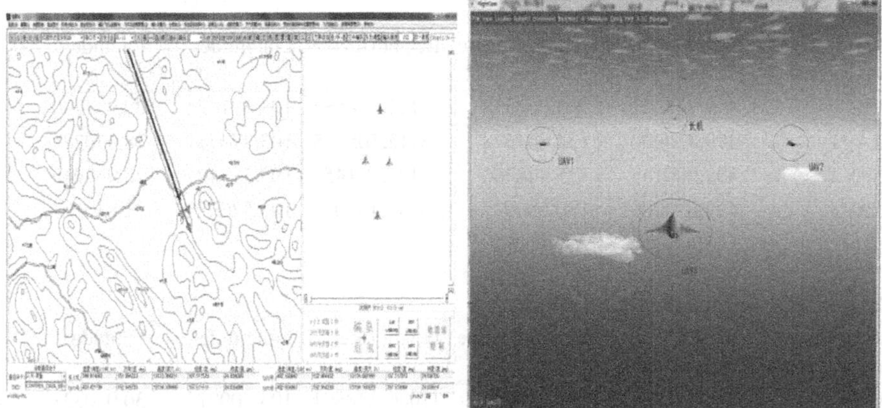

Fig. 5 The diamond formation

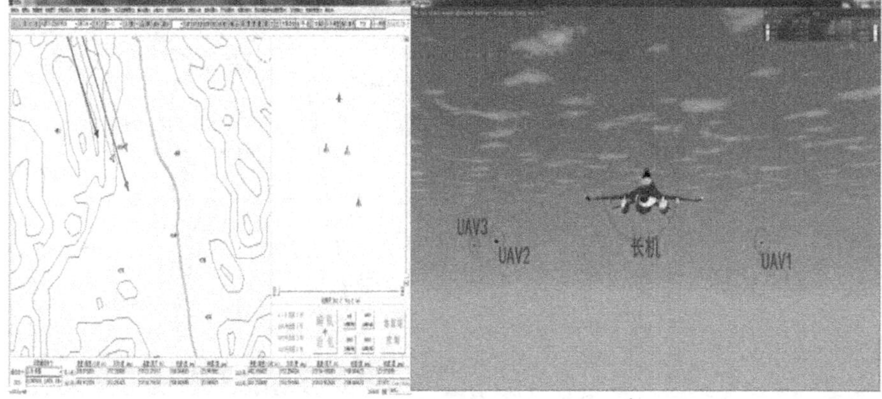

Fig. 6 The herringbone formation

4.2 Formation Effect

The initial position of our aircrafts is set in accordance with the instruction of operator. Our aircrafts have been cruising. When our aircrafts receive attack occupying command, the position of this moment is initial position. The position is shown in Table 3. The War position is initial position in the moment, as shown in Table 4. The manned aircraft orders UAV1 and UAV2 to occupy, the consequent is shown in Fig. 7.

Table 3 Our aircraft position information

Combat aircraft name	Speed (miles/hour)	Heading angle	Height (feet)	Longitude	Latitude
AIR	499.731231	103.664578	13122.634588	103.671254	29.981230
UAV1	499.830000	134.664578	13112.505715	103.374561	28.970286
UAV2	499.900145	135.448956	13122.611453	103.374402	28.968946
UAV3	499.870000	190.580095	13137.463075	103.520032	28.830107

Table 4 The war aircraft position information

Combat aircraft name	Speed (miles/hour)	Heading angle	Height (feet)	Longitude	Latitude
WAR	400.731231	160.250051	11080.320801	103.70001	30.020831

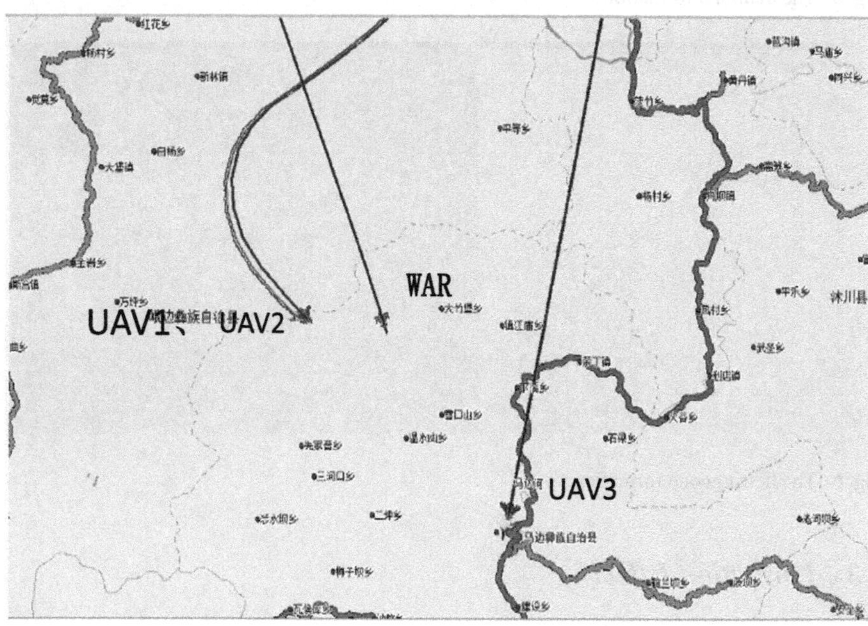

Fig. 7 The uav formation occupying

5 Conclusions

With the in-depth study on the hybrid formation flight control and air combat, the need of comprehensive verification hybrid formation control idea is becoming important and urgent in the paper replaceable aircraft model, real-time FlightGear scene update, flight data real-time display and mixed formation flight 3D visualization display can give experimenter deeply immersed and sense of participation, overcome the shortcoming of existing literature on system development, and provides the effective verification platform for the research on the related technology of the hybrid formation cooperative operation.

Acknowledgments This work is partially supported by the Program for Liaoning Excellent Talents in University (No. LJQ2014018) and the Scientific Research General Project of Education Department of Liaoning Province, China (No. L2014066). The authors also gratefully acknowledge the helpful comments and suggestions of the reviewers, which have improved the presentation.

References

1. Torens C, Adolf F (2014) Certification and Software Verification Considerations for Autonomous Unmanned Aircraft. vol 10, no 11, pp 649–664
2. Rew DY, Lee WR, Koo CH (2013) HILS approach in the virtual flight test of the Korean lunar lander demonstrator. In: Proceedings of the AIAA modeling and simulation technologies (MST) Conference. Boston, MA, pp 1013–1015
3. Tian YL, Liu F (2013) Virtual simulation-based evaluation of ground handling for future aircraft concepts. J Aerosp Inf Syst 5(10):218–228
4. Olthoff C, Schnaitmann J (2013) Development status of V-SUIT-The virtual space suit simulation software. In: Proceedings of the 43rd international conference on environmental systems, Vail, CO
5. Sullivan B, Malsom S (2002) Development of a real-time virtual airspace simulation capability for air traffic management research. In: Proceedings of the AIAA modeling and simulation technologies conference and exhibit, Monterey, California
6. Li W, Zhang H (2009) Simulation platform design and realization for unmanned aerial vehicle formation control, J Syst Simul, 21(03):691–694
7. Laughlin M, Briceno DS (2014) A virtual experimentation platform enabling the design, testing, and verification of an unmanned aerial vehicle through cyber-physical, component-based design. In: Proceedings of the 14th AIAA aviation technology, integration, and operations conference, Atlanta, GA. pp 2720–2731
8. Luo M, Kamel AE (2011) UML-based design of intelligent vehicles virtual reality platform, systems, man, and cybernetics (SMC), pp 115–120
9. Zhou T (2013) Research on network communication between Simulink and VC++ based on Sfunction. Mod Electron Tech, 13(36):108–111

Sectioned Restoration of Blurred Star Image with Space-Variant Point Spread

Tianlang Liu, Kaimin Sun and Chaoshan Liu

Abstract In order to ensure the accuracy of star sensor in high dynamic conditions, the sectional processing scheme is proposed for space-variant motion blurs. The degradation of star image may be characterized by a space-variant point spread function (PSF), and the PSF of blurred star spot was formulated by its nonuniform velocity across the image plane. Then, the image is divided into several rectangular sections and each contains an interesting star spot. The center of section is the location of star predicted by the attitude measurement equation, and its size is determined by the velocity of star spot and exposure time. Further, the PSF in each section can be approximately space-invariant. The scheme is illustrated by performing attitude determination against a restored space-variant blurred star image.

Keywords Star sensor · Space-variant psf · Sectioned image restoration · High dynamic conditions

1 Introduction

The star sensor is one of the most promising attitude measurement devices used in vehicles due to its extremely high accuracy [1]. However, high dynamic performance is one of its constraints. Due to the large angular rate, the star moves on the image plane during the exposure time, and limited starlight energy distribute in the more pixels, resulting in decreased SNR. These blurred images with high angular velocities make star extraction and identification difficult or impossible. Several solutions

T. Liu · K. Sun (✉) · C. Liu
State Key Laboratory of Information Engineering in Surveying, Mapping, and Remote Sensing, Wuhan University, Wuhan 430079, China
e-mail: sunkm@whu.edu.cn

T. Liu
e-mail: tianlang_liu@163.com

C. Liu
e-mail: chaoshan_liu@163.com

V.E. Balas et al. (eds.), *Information Technology and Intelligent Transportation Systems*, Advances in Intelligent Systems and Computing 455,
DOI 10.1007/978-3-319-38771-0_17

can be employed to improve the dynamic performance of the star sensor. One of solutions is adopting a new type of higher sensitivity or enhancing image sensor. Another solution is changing image sensor internal time sequencing and data processing algorithm such as that Pasetti used the over sampling technology and dynamical binning method to improve the sensor detection sensitivity and enhance the robustness of star trackers to high angular rates [2, 3]. However, the most convenient and efficient solution is recovering the blurred star spot with image processing algorithm, in which the estimation of Point Spread Function (PSF) characterized by the extent and direction of blurred star image algorithm is crucial. Samaan [4] discussed a predictive centroiding method with the effect of image smearing. Wu [5] discussed the Multiple blur of star spot and corresponding PSFs, and she adopted the constrained least square filter for star image restoration. Sun [6] discussed the partial image differentiation method for motion-blurred image, and obtained the motion parameters from only one image of the star sensor, which takes reference from the solution by Y. Yitzhaky [7]. G. Fasano [8] proposed the optical flow technique to compute the displacement field (or the velocity vector field) between successive star images in pixels. After the motion degradation has been described mathematically, another more challenge task is that the PSF is related to the motion direction and location of star spot across the image plane under high dynamic conditions, which is a problem of space-variant motion blur [9]. Previous work on space-variant star image restoration is relatively sparse, so it is necessary issue to study the space-variant PSFs to restore.

2 The Space-Variant PSF Estimation for Star Spot

2.1 Coordinate Frames

The derivation will be performed in an orthogonal reference frame, which is defined as follows. The focal plane coordinate frame $(o - yz)$ origin lies at the geometric center of the focal plane. In the focal plane coordinate system, the y-axis is parallel to one of the sides of the focal plane while the z-axis is parallel to the other side. The star sensor coordinate frame $(o_s - x_s y_s z_s)$ origin lies at the center of the star sensor mass. The x_s-axis is perpendicular to the focal plane and points toward the star camera optics, the y_s-axis is parallel to one of the sides of the focal plane, which is assumed to be rectangular, and the z_s-axis completes the right-handed system. The body coordinate frame, which is a convenient coordinate system, has its origin at the vehicle center of mass. Its orientation is similar to that of the star sensor coordinate system.

2.2 The Velocity Model of Star Spot

The measurement model of the star sensor can be considered as a pinhole imaging system, as shown in Fig. 1. The star spot velocity can be derived from the attitude equation of the vehicle hosting the star sensor. This section shows how this derivation can be done and it examines the problems that arise from the non-uniformity of the star spot velocity field on the focal plane. It is one of the important parameters to the restoration algorithm. As shown in Fig. 1, u_c and w_s represent the cataloged vector in the inertial frame $o_c - x_c y_c z_c$ and the measurement direction vector in the star sensor frame $o_s - x_s y_s z_s$ at the time of t, respectively. In theory, $\boldsymbol{u}_c$, $\boldsymbol{w}_s$, and the attitude matrix $\boldsymbol{M}$(t) of the star sensor satisfy

$$\boldsymbol{w}_s = \boldsymbol{M}(t) \cdot \boldsymbol{u}_c \tag{1}$$

where

$$\boldsymbol{w}_s = \frac{1}{\sqrt{f^2 + y(t)^2 + z(t)^2}} \begin{bmatrix} f \\ -y(t) \\ -z(t) \end{bmatrix} \tag{2}$$

and

$$\boldsymbol{u}_c = \begin{bmatrix} cos(\delta)cos(\alpha) \\ cos(\delta)sin(\alpha) \\ sin(\delta) \end{bmatrix} \tag{3}$$

In Eqs. (2) and (3), f represents the focal length of the star sensor and (y, z) represents the observed star location on the image plane. (α, δ) represents the right ascension and declination of the associated guide star on the celestial sphere. The derivative with respect to time t of Eq. (1) is

$$\frac{d\boldsymbol{w}_s}{dt} = \frac{\mathbf{d}\boldsymbol{M}(t)}{dt} \boldsymbol{u}_c \tag{4}$$

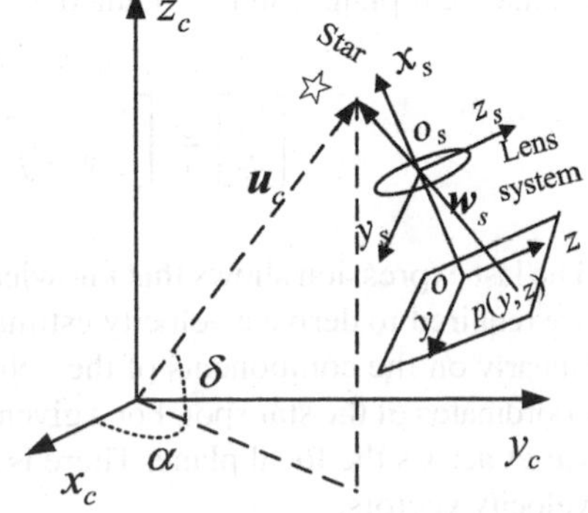

Fig. 1 The attitude determination model of star sensor

Consider the exposure time is short, take a first-order approximation of the attitude matrix $\boldsymbol{M}$:

$$\boldsymbol{M}(t + \Delta t) = [I - \tilde{\omega}\Delta t]\boldsymbol{M}$$

$$\frac{\mathbf{d}\boldsymbol{M}(t)}{dt} = -\tilde{\omega} \cdot \boldsymbol{M} \tag{5}$$

Here $\tilde{\omega}$ is the cross product matrix to $\boldsymbol{\omega}$.

$$\tilde{\omega} = \begin{bmatrix} 0 & -\omega_z & \omega_y \\ \omega_z & 0 & -\omega_x \\ -\omega_y & \omega_x & 0 \end{bmatrix} \tag{6}$$

In Eq. (6), the three components of the star sensor-referenced angular velocity $\boldsymbol{\omega} \equiv [\omega_x, \omega_y, \omega_z]$, where ω_i is the angular velocity component along the i axis of the star sensor frame and i=1, 2, 3 for x, y, z, respectively. Substituting Eq. (5) to Eq. (4) yields

$$\frac{d\boldsymbol{w}_s}{dt} = -\tilde{\omega} \cdot \boldsymbol{w}_s \tag{7}$$

substituting Eqs. (2) and (6) to Eq. (7),

$$\begin{aligned} &\frac{1}{(f^2 + y(t)^2 + z(t)^2)^{3/2}} \begin{bmatrix} -(z^2 + f^2)v_y + yzv_z \\ -(y^2 + f^2)v_z + yzv_y \end{bmatrix} \\ &= \frac{1}{\sqrt{f^2 + y(t)^2 + z(t)^2}} \begin{bmatrix} f\omega_z + z\omega_x \\ -f\omega_y - y\omega_x \end{bmatrix} \end{aligned} \tag{8}$$

Rewriting Eq. (8) as follows:

$$\begin{bmatrix} v_y \\ v_z \end{bmatrix} = (f^2 + y(t)^2 + z(t)^2) \begin{bmatrix} z^2 + f^2 & -yz \\ -yz & y^2 + f^2 \end{bmatrix}^{-1} \begin{bmatrix} f\omega_z + z\omega_x \\ -f\omega_y - y\omega_x \end{bmatrix} \tag{9}$$

By obtaining an inverse matrix equation in Eq. (9), the velocity of star spot moving on the focal plane can be obtained:

$$\begin{bmatrix} v_y \\ v_z \end{bmatrix} = \begin{bmatrix} z & -yz/f & f + y^2/f \\ -y & -f - z^2/f & yz/f \end{bmatrix} \begin{bmatrix} \omega_x \\ \omega_y \\ \omega_z \end{bmatrix} \tag{10}$$

The last expression shows that knowledge of angular velocity and the location of star are required to derive a velocity estimate of star spot. The star spot velocity depends linearly on the components of the vehicle angular velocity and quadratically on the coordinates of the star spot. For a given vehicle angular velocity, the star spot velocity varies across the focal plane. There is therefore a space-variation field of star image velocity vectors.

Under dynamical conditions, the star spot slides along the image plane at a velocity v, limited star energy will disperse into more pixels, and smearing of the star point occur during exposure time. The extent of motion blur of star spot along the y-axis and z-axis on the image plane is

$$\Delta y = v_y \cdot \Delta t, \Delta z = v_z \cdot \Delta t \tag{11}$$

$$L = \sqrt{\Delta y^2 + \Delta z^2} \tag{12}$$

Here L denotes the blur extent of the star spots. Let β is the angle between and the positive direction of y-axis, we obtain

$$\beta = \begin{cases} tan^{-1}\frac{v_z}{v_y}, \; in\ I\ or\ IV\ quadrant \\ \pi + tan^{-1}\frac{v_z}{v_y}, \; in\ II\ or\ III\ quadrant \end{cases} \tag{13}$$

Based on the degradation model of the uniform linear motion blurred image, the PSF expression of $h(y,z)$ in the time domain can be obtained:

$$h(y, z) = \begin{cases} 1/L, 0 \le y \le Lcos(\beta) \\ 0, else \end{cases} \tag{14}$$

As seen in Eq. (12), linear motion blur depends on two parameters: motion blur extent L and motion direction β. These two parameters can be computed simply by the velocity vector obtained with Eq. (10) in each location of star spot. It is also shown that the PSF $h(y,z)$ is the function of vehicle velocity (angular velocity) and the location of star in the image plane. Obviously, it is a space-variant motion blur problem.

3 Sectioned Restoration of Star Image with Space-Variant PSF

3.1 *Star Position Estimation*

Once the star sensor completes the initial star map recognition with the aid of the strap inertial navigation system (SINS) or with the lost in space algorithm, it can make correction and compensation for the drift of SINS. Due to high data rate of SINS, certain motion information can be derived from SINS, including the vehicle angular velocity, location of star spot in the image plane. Let (y_{t_i}, z_{t_i}) and $(y_{t_{i+1}}, z_{t_{i+1}})$ are the location of star spot in the adjacent frames at time t_i, t_{i+1} then rewriting Eq. (10) as

$$y_{t_{i+1}} = y_{t_i} + v_y(t_{i+1} - t_i)$$
$$z_{t_{i+1}} = z_{t_i} + v_z(t_{i+1} - t_i) \tag{15}$$

However, star image of adjacent frame changes quickly under high dynamic conditions, some distinguished stars left the image plane, and some undistinguished stars enter the image plane in next frame. The location of star spot predicted only by Eq. (15) cause the probability of mismatch increased, so it should replaced that method of Eq. (15) with the navigation star prediction for the next frame to improve the success rate of star tracking [10]. Firstly, using the angular velocity data, the next frame potential attitude $\boldsymbol{M}(t + \Delta t)$ was estimated at the current frame using the attitude matrix $\boldsymbol{M}(t)$, and following linear approximation Eq. (5). Secondly, predicting stars within the field of view and their coordinates at the next time by the predicted attitude. The potential bore-sight direction vector of star sensor is calculated with $\boldsymbol{M}(t + \Delta t)$, so the sub-partitions of the star catalogue is seek out, which center vector is closest to the bore-sight direction vector. Their inertial reference vectors $\boldsymbol{u}_{ci}$ ($i = 1, 2, \ldots$, n) to stars imaged in next frame are obtained, the next frame ideal star location in the image plane was calculated according to the potential attitude. Given the attitude matrix $\boldsymbol{M}^T(t + \Delta t) \equiv (\mathbf{M}_1, \mathbf{M}_2, \mathbf{M}_3)$ at time $(t + \Delta t)$, then

$$y_i(t + \Delta t) = y_0 - f\frac{{M_2}^T \cdot u_{ci}}{{M_1}^T \cdot u_{ci}},\ z_i(t + \Delta t) = z_0 - f\frac{{M_3}^T \cdot u_{ci}}{{M_1}^T \cdot u_{ci}} \tag{16}$$

Here y_0, z_0 are the boresight offsets. The angular velocity $\boldsymbol{\omega}$ also can be only estimated from the star sensor, which are independent of star identification and attitude information [12, 13]. The performance parameters of star sensor and the motion parameters were substituted into Eq. (10) to Eqs. (14) and (16), the corresponding image motion PSFs for each star spot can be calculated.

3.2 The Sectional Approach

For the considered frame, once the star locations are predicted, and then these locations become the center of star spots used for the real image. Based on the star motion parameters in Sect. 2.2, the location and size of blurred star spots can all be predicted. In real imaging system, the space-variant PSF characterized by the orientation and extent of motion blur changes continuously. So it is difficult to accurately restore the blurred star spot at different location by this continuous PSF. However, we can assume that the motion is locally uniform in a region of reasonable size (e.g. a small rectangular section). This assumption is reasonable since neighboring points will usually be moving at similar velocity. It is clear that the discontinuities in the velocity field will result in similar discontinuities in the motion PSF. Therefore, the PSF in every small rectangular section can be approximately space-invariant, and conventional deconvolution of image restoration methods for space-invariant linear motion blur can be used in every small rectangular section, such as iterative Lucy-Richardson or non-iterative Wiener algorithms and least squares algorithms or more complex algorithm. Furthermore, the distribution of star spots in the image plane is rather sparse, i.e. the image is composed of a stable intensity background and a small part of the star spot distribution. Therefore, the sectional processing scheme is very suitable for the blured star image of spatial variation [14, 15]. Considering the angular velocity as $(5, 1, 1)^0/s$, the blur extents of star spot at (500, 500) pixel along y, z-axis all are 34 pixels by Eq. (11). Considering the error of star spot center and orientation and extent of motion blur predicted, as far as possible to make the star spot included in the section, then, the size of the section is taken as 1.5 times the size of star spots, only the region of $34 \times 1.5 \times 34 \times 1.5$ (equal to 2600) pixels need to be scanned. On the whole star image, a region of $2600 \times n$ pixels need to be scanned, where n is the number of interested stars spot the image plane. It is important to underline that all the subsequent calculations are applied only to the section including the detected star spots and not to the whole image. If a part of star spot is beyond the boundary of the image plane, the position of the section need to be carefully adjusted. At the same time, only star spot centroid position in every section is concerned, did not deal with the sections stitching and their quality after the restoration of image section, and other various problems caused by splicing. Also, the sectional processing scheme can reduce the dimension of image section to be restored and greatly short the time of restoration, and enables a dramatic improvement in resolution.

3.3 The Constrained Least Squares Estimation

In summary, the velocity of each point in whole section is same as that of the center of section, obviously, the PSF involving the orientation and the extent of blur motion is the space-invariant in each section. The sectional image characterized by the intensity distribution *g(y,z)* was created by convolving a real sectional image *f(y,z)* with a point-spread function PSF and possibly by adding noise.

$$g(y,z) = f(y,z) * h(y,z) + \eta \tag{17}$$

The goal of image restoration is to reconstruct the original scene from a degraded observation. When the PSF in each section is known, as mentioned before, many techniques are available for estimating *f(y,z)* from *g(y,z)*. Constrained least squares (CLS) estimation is a technique for solution of integral equations of the first kind, and is applicable to image restoration. This method has been improved by Hunt [15], and the CLS filter is based on finding a direct filter solution using a criterion *C*, which ensures optimal smoothness of the image restored. Therefore the filter construction task is to find the minimum of criterion function, *C*, defined as

$$C = \sum_{y=0}^{M-1} \sum_{z=0}^{N-1} [\nabla^2 f(y, z)]^2 \tag{18}$$

subject to the constraint

$$\|g - H\hat{f}\| = \|\eta\|^2 \tag{19}$$

In the frequency domain the solution to this problem can be written as follows

$$\hat{F}(u, v) = \frac{H^*(u, v)G(u, v)}{|\ (H^*(u, v))\ |^2 + \gamma\ |\ (P(u, v))\ |^2} \tag{20}$$

where γ is the parameter that has to be adjusted to fulfill the constraint *C*, and *P(u,v)* is the Laplacian operator in the frequency domain, Typical choice is

$$P = \begin{bmatrix} 0.00 & -0.25 & 0.00 \\ -0.25 & 1.00 & -0.25 \\ 0.00 & -0.25 & 0.00 \end{bmatrix} \tag{21}$$

where,*G(u,v)*and *H(u,v)* denote the DFTs (Discrete Fourier Transform) of *g(y,z)* and *h(y,z)*. Given the estimated PSF, it is easy to obtain an estimate of the true image by Eq. (20). However, for a real motion-blurred image, the PSF is generally unknown. It is then necessary to estimate a PSF to carry out the restoration, and methods have been developed for estimating PSFs from the blurred images. In our case, the image motion PSF is well formulated by the attitude determinate equation during the exposure time in Sect. 2. The PSF for image motion can be estimated from numerical calculations, hence the degraded star image can be restored. The performance parameters of star sensor and the motion parameters were substituted into Eq. (10) to Eq. (14), the corresponding image motion PSFs in every rectangular section unit can be calculated. Then, the sectioned restoration technique can be adopted to restore the blurred star spot. The scheme can obviously both accelerate the deconvolution speed and improve the image restoration result. After CLS, a threshold technique is applied in which a $\mu + 3\sigma$ threshold is applied to identify the illuminated pixels, with and being, respectively, the intensity mean and standard deviation computed over the blocks of image. All the pixels with intensity below the noise threshold are set equal to zero. A labeling technique is applied to distinguish the different stars detected on the focal plane. Within this phase, stars whose dimension is smaller than three pixels are discarded to increase algorithm accuracy.

4 Numerical Simulation and Star Image Restoration

4.1 Star Image Simulation

The right ascension and declination corresponding to the bore-sight direction are $(292.4627, 53.1458)^0$. Select the 2514 stars brighter than $5.5M_V$ in Skymap 2000 as the guide star catalog. The exposure time is 0.3 s. The simulation takes into account the FOV, focal length, pixel size, the magnitude of the star, the noise of the image, and other parameters of the star sensor to simulate the star image as realistic as possible. The stellar magnitude model refers to Liebes conclusion. The model is reliable and frequently-used in star image simulation. The blurred star image is simulated with Eq. (10) to Eqs. (14) and (16) in which the smear parameters have little difference on the same angular velocity. The blurred star image simulation at angular velocity (1, 5, 1), and the dashed rectangle section contained the predicted star spot, are shown in Fig. 2. With a rotation around the boresight axis, velocity field simulated by our algorithm shown as Fig. 3 is more direct visual under angular velocity $(5, 1, 1)^0/s$. Velocity components and and their equivalent contours are shown as Fig. 4, the complexity of star spot movement is very obvious from the components of velocity and their equivalent contours distribution in the plane under above three angular velocities.

4.2 Simulation Data Analysis

According to the sectional approach in Sect. 3.2, the PSF in every star spot can be approximately space-invariant, however, the star spot at different location has a different PSF. In order to more clearly see the need for the introduction of the space-variant PSF to restore blurred star image, two kinds of PSF are adopted here, one

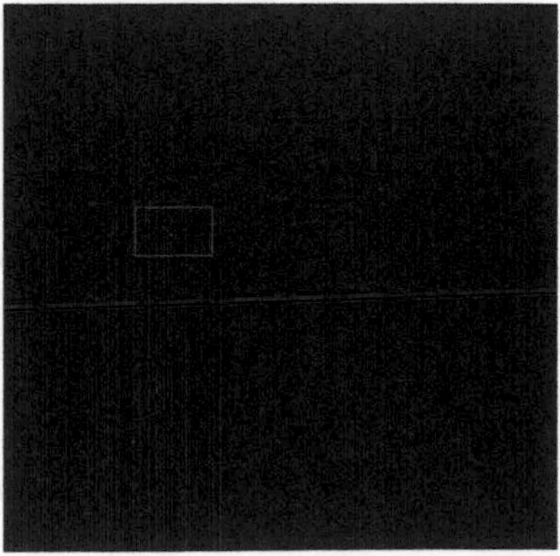

Fig. 2 Simulated blurred star image

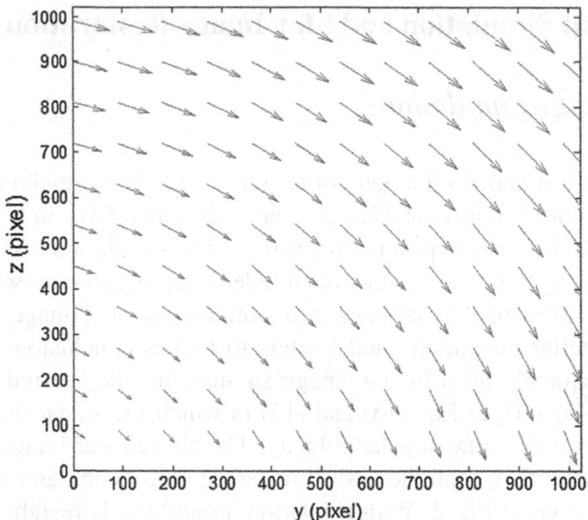

Fig. 3 Velocity field simulated under angular velocity $(5, 1, 1)^0/s$

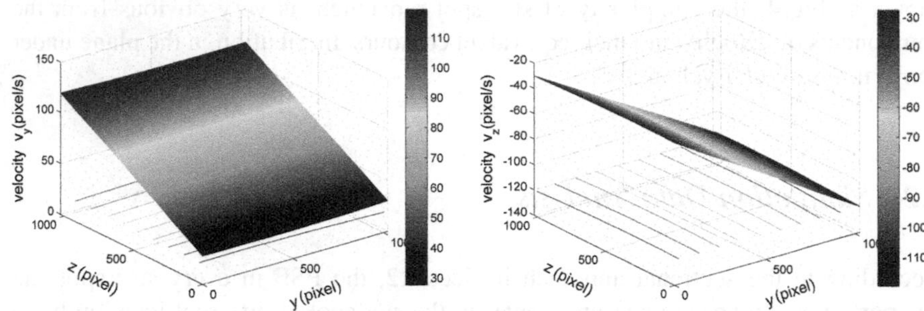

Fig. 4 Velocity distribution corresponding to angular velocity $(5, 1, 1)^0/s$

is the space-variant, and another is the space-invariant for all star spots in a single image, to study the attitude accuracy of star image under different angular velocity. For simplicity, the average values of stars PSF in the focal plane are recommended for that of the space-ivariant. Solid and dash lines are the attitude errors of SVPSF and SIPSF respectively, the consistency of attitude data SVPSF obviously gets better than those of SIPSF, almost are independent of the angular velocity and the distribution of star spot in the image plane, shown as Fig. 5. When the changes that star sensor rotates around the boresight, the PSFs at different location exist more difference, so the SIPSF method led to 3, 7 stars recognition failure under angular velocity $(3, 1, 1)^0/s$ and $(4, 1, 1)^0/s$ respectively, and led to more attitude error, and fail to solve the attitude under $(5, 1, 1)^0/s$, shown as Fig. 5. When ω_y or ω_z changes and other components unchanged, Compared with the SVPSF method, the attitude data and consistency only get slightly worse, shown as Fig. 5.

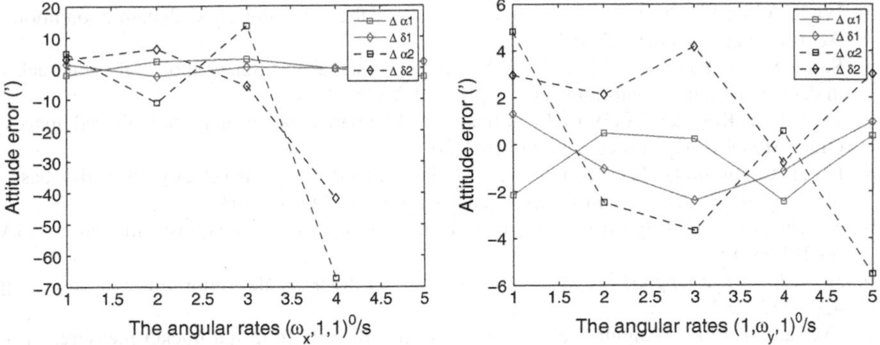

Fig. 5 The attitude errors estimated with SVPSF and those with SIPSF

5 Conclusion

The sectional processing scheme was proposed for space-variant motion blurs of star sensor under high dynamic conditions. The PSF was formulated by the velocity of star spot in the image plane. It is shown that the PSF is the function of vehicle angular velocity and the location of star in the image plane, when the manner of movement is different (three different components of angular velocity change), the extent and direction of velocity of star spot at different location is different. The sectioned block masks each star spot pixel, the center of section was predicted by vehicle attitude determinate equations, and the PSF is approximately space-invariant in each section. The normal restoration technique can be adopted to restore the blurred star spot in each region. The sectional processing scheme can reduce the dimension of processing image and greatly shorten the time of restore algorithm, and improve the accuracy of attitude.

Acknowledgments This paper is supported by National Natural Fund of China (NSFC) (No. 41471354).

References

1. Fang JC (2006) Spacecraft autonomous celestial navigation principle and method. National Defence Industry Press, People's Republic of China
2. Pasetti A, Habinc S, Creasey R (1999) Dynamical binning for high angular rate star tracking. In: Proceedings of the fourth ESA international conference on vehicle guidance, navigation and control systems. Netherlands
3. Liu CS, Hu LH, Liu GB, Yang B, Li AJ (2015) Kinematic model for the space-variant image motion of star sensors under dynamical conditions. Opt Eng 54(6):063104
4. Samaan MA, Pollock TC, Junkins JL (2002) Predictive centroiding for star trackers with the effect of image smearing. J Astronaut Sci 50:1–14

5. Wu XJ, Wang XL (2011) Multiple blur of star spot and the restoration under dynamic conditions. Acta Astronautica 68:1903–1913
6. Sun T, Xing F, You Z, Wei MS (2013) Motion-blurred star acquisition method of the star tracker under high dynamic conditions. Opt Express 21:20096–20110
7. Yitzhaky Y, Kopeika NS (1997) Identification of blur parameters from motion blurred images. Graph Models Image Process 59(5):310C320
8. Fasano G, Rufino G, Accardo D, Grassi M (2013) Satellite angular velocity estimation based on star images and optical flow techniques. Sensors 13:12771–12793
9. Sawchuk A (1974) Space-variant image restoration by coordinate transformation. J OSA 60(2):138C44
10. Jin Y, Jiang J, Zhang GJ (2013) Highly dynamic star tracking algorithm. Infrared Laser Eng 42(1):212–217
11. Crassidis JL (2002) Angular velocity determination directly from star tracker measurements. J Guid Control Dyn 25:1165–1168
12. Liu HB, Yang JC, Yi WJ, Wang JQ, Yang JK, Li XJ, Tan JC (2012) Angular velocity estimation from measurement vectors of star tracker. Appl Opt 51:3590–3598
13. Trussell HJ, Hunt BR (1978) Image restoration of space-variant blurs by sectional methods. IEEE Trans ASSP 26(10):608C9
14. Trussell HJ, Fogel S (1992) Motion blurs in sequential images identification and restoration of spatially variant. IEEE Trans Image Process 1(1):123–126
15. Hunt BR (1973) The application of constrained least squares estimation to image restoration by digital computer. IEEE Trans Comput C–22:805C812

Numerical Solutions of Nonlinear Volterra–Fredholm–Hammerstein Integral Equations Using Sinc Nyström Method

Yanying Ma, Jin Huang and Changqing Wang

Abstract In this paper, a numerical method is presented for solving nonlinear Volterra–Fredholm–Hammerstein integral equations. The proposed method takes full advantage of Nyström method and Sinc quadrature. Nonlinear integral equations is converted into nonlinear algebraic system equations. Error estimation is derived which is shown to has an exponential order of convergence. The accuracy and effectiveness of the proposed method are illustrated by some numerical experiments.

Keywords Nonlinear integral equations · Volterra–Fredholm–Hammerstein · Sinc function · Nyström method

1 Introduction

This paper is focused on proposing the numerical method for Volterra–Fredholm–Hammerstein integral equations (NVFHIE) as follows

$$u(x) + \lambda_1 \int_a^b K_1(x,t)F_1(t,u(t))dt + \lambda_2 \int_a^x K_2(x,t)F_2(t,u(t))dt = g(x), \quad (1)$$

This work was supported by the National Natural Science Foundation of China (11371079).

Y. Ma (✉) · J. Huang
School of Mathematical Sciences, University of Electronic Science and Technology of China, Chengdu 611731, People's Republic of China
e-mail: ma_yan_ying@126.com

J. Huang
e-mail: huangjin12345@163.com

C. Wang
School of Automation Engineering, University of Electronic Science and Technology of China, Chengdu 611731, People's Republic of China
e-mail: wangchangqing199008@gmail.com

V.E. Balas et al. (eds.), *Information Technology and Intelligent Transportation Systems*, Advances in Intelligent Systems and Computing 455,
DOI 10.1007/978-3-319-38771-0_18

where functions $K_1(x,t)$, $K_2(x,t)$, $F_1(t,u(t))$, $F_2(t,u(t))$, $g(x)$ and parameters λ_1, λ_2 are known and $u(x)$ is unknown with $x, t \in [a, b]$.

A variety of numerical modelling can be transformed into Eq. (1) in many fields such as the particle transport problems of astrophysics, continuum mechanics, electricity and magnetism, geophysics, antenna synthesis problems, mathematical economics, population genetics, fluid mechanics, etc.

There are some research works on developing numerical methods for Eq. (1). For example, collocation method [1–3], Talyor method [4], modified decomposition method [5] are among the most popular ones. For more details on approximating methods, see [6]. Generally, some of them employ techniques by means of an expansion in terms of some basis functions or employ some quadrature formulas, and the convergence rate of these methods are usually of polynomial order. On the other hand, Rashidinia et.al used Sinc method for solving integral equations and the convergence rate was exponential in [7]. Sinc quadrature play a prominent role in various areas of mathematics. This function have been frequently used in the solution of integral equations, differential equations and approximation theory see [8–13].

Therefore, in the present paper, we apply the Sinc quadrature to solve Eq. (1). Equation (1) is transformed to a set of nonlinear algebraic equations. The convergence rate of this method is exponential. Therefore, the current method improves conventional polynomial convergence rate.

In this paper, the basic ideas are organized as follows. In Sect. 2, we give some definitions and preliminary results about Sinc function and Sinc quadrature. In Sect. 3, Sinc Nyström method for Eq. (1) is presented. In Sect. 4, we provide the order of convergence of the scheme using a new approach. In Sect. 5, the efficiency of this method will be illustrated by comparison with other existing methods. Finally, a conclusion is given in Sect. 6.

2 Preliminaries

In this section, a summery of the basic formulations of the Sinc function are described. The Sinc function is defined on the whole real line by

$$Sinc(t) = \begin{cases} \frac{\sin(\pi t)}{\pi t}, & t \neq 0, \\ 1, & t = 0. \end{cases} \tag{2}$$

Sinc approximation for a function f on the entire real axis can be expressed in the truncated finite terms form

$$f(x) \approx \sum_{i=-N}^{i=N} f(ih)S(i,h)(x), \qquad x \in \mathbb{R}, \tag{3}$$

where the basis function $S(i,h)(t) = \frac{\sin[\pi(t/h-i)]}{\pi(t/h-i)}$ and h is a step size appropriately selected depend on $N \in \mathbb{Z}^+$ and $i \in \mathbb{Z}$. Let the function $J(j,h)$ is defined by $J(i,h)(t) = \left\{\frac{1}{2} + \frac{1}{\pi} Si[\pi(t/h - i)]\right\}$ where $Si(\mu) = \int_0^t \frac{\sin\mu}{\mu} d\mu$.

From the above, it is clear that the approximation of Eq. (3) is valid on $\mathbb{R}$, whereas the equation Eq. (1) is defined on finite interval $[a,b]$. The Eq. (3) can be suited for infinite interval using smoothing transformation. Thus, we consider conformal map of functions $\phi(z)$ is utilized.

The function $\phi(x)$ and its inverse are described as following, respectively,

$$\begin{aligned} t = \phi(z) &= \log\left(\frac{z-a}{b-z}\right), & z \in (a,b), \\ z = \{\phi\}^{-1}(t) &= \frac{a+be^t}{1+e^t}, & t \in \mathbb{R}. \end{aligned} \tag{4}$$

In order to facilitate the error analysis, the strip domain $D_d = \{z \in \mathbb{C} : |Imz| < d\}$ for some $d > 0$ is introduced. The transformation maps (a,b) onto $\mathbb{R}$, and maps D_d onto region $D = \left\{z \in C : \left|\arg(\frac{z-a}{b-z})\right| < d \le \frac{\pi}{2}\right\}$. Here, it is obvious that the range of $\{\phi\}^{-1}(t)$ on the real line as $\Gamma = \{\psi(v) = \phi^{-1}(v) \in D : -\infty < v < \infty\}$. The points $x_k = \{\phi\}^{-1}(kh) = \frac{a+be^{kh}}{1+e^{kh}} \in (a,b)$ denote Sinc grid points.

Definition 2.1 Let $L_\alpha(D)$ denotes the family of all analytic functions f and there exists a constant $C > 0$, such that $|f(z)| \le C\frac{|\rho(z)|^\alpha}{[1+|\rho(z)|]^{2\alpha}}$ for all z in D where $\rho(z) = e^{\phi(z)}$.

Theorem 2.1 *Let $f(x) \in L_\alpha(D)$ with $d > 0$ and let h be selected by the formula*

$$h = \sqrt{\frac{\pi d}{\alpha N}}, \tag{5}$$

where $N \in \mathbb{Z}^+$. Then, there exists a constant C_1, independent of N such that

$$\max\left| f(x) - \sum_{i=-N}^{i=N} f(x_i) S(i,h) \circ \phi(x) \right| \le C_1 e^{-\sqrt{\pi d \alpha N}}. \tag{6}$$

Theorem 2.2 *Let $\frac{f}{\phi'} \in L_\alpha(D)$ with $d > 0$ and let h be selected by (5). Then, there exists a constant C_2, independent of N such that*

$$\max\left| \int_a^{x_k} f(t)dt - h\sum_{i=-N}^{i=N} J(i,h)(kh)\frac{f(x_i)}{\phi'(x_i)} \right| \le C_2 e^{-\sqrt{\pi d \alpha N}}. \tag{7}$$

Theorem 2.3 *Let $\frac{f}{\phi'} \in L_\alpha(D)$ with $0 < d < \pi$ and let h be selected by (5). Then, there exists a constant C_3, independent of N such that*

$$\max\left|\int_{\Gamma} f(t)dt - h\sum_{i=-N}^{i=N}\frac{f(x_i)}{\phi'(x_i)}\right| \le C_3 e^{-\sqrt{\pi d\alpha N}}. \tag{8}$$

3 Sinc Nyström Method for Eq. (1)

According to Theorems 2.2 and 2.3, Sinc definite and indefinite integration can be directly applied to the second and third term kernel integral of Eq. (1), respectively, the integral can be accurately approximated as

$$\int_a^b K_1(x,t)F_1(t,u(t))dt \approx h\sum_{i=-N}^{i=N}\frac{K_1(x,t_i)}{\phi'(t_i)}F_1(t_i,u_i),$$

$$\int_a^x K_2(x,t)F_2(t,u(t))dt \approx h\sum_{i=-N}^{i=N}\frac{K_2(x,t_i)}{\phi'(t_i)}J(i,h)(\phi(x))F_2(t_i,u_i), \tag{9}$$

where u_i denotes an approximate value of $u(t_i)$, $t_i = \phi^{-1}(ih)$ and the mesh h is chosen by formula (5). The Nyström is exploited to Eq. (1)

$$u_N(x) + \lambda_1 h\sum_{i=-N}^{i=N}\frac{K_1(x,t_i)}{\phi'(t_i)}F_1(t_i,u_i)$$
$$+\lambda_2 h\sum_{i=-N}^{i=N}\frac{K_2(x,t_i)}{\phi'(t_i)}J(i,h)(\phi(x))F_2(t_i,u_i) = g(x). \tag{10}$$

In order to determine these $2N+1$ unknown values of Eq. (10), we choose Sinc points $x_j = \phi^{-1}(jh)$ as the collocation quadrature points. By replacing x_j into x in Eq. (10), we eventually obtain the following nonlinear system of equations

$$\mathbf{u}_N + \mathbf{K}_{1N}\mathbf{F}_1(\mathbf{u}_N) + \mathbf{K}_{2N}\mathbf{F}_1(\mathbf{u}_N) = \mathbf{g}_N. \tag{11}$$

This nonlinear system of equations can be solved by the Newton iteration method. Then, the approximate solution $u_N(x)$ ($\forall x \in [a,b]$) of Eq. (1) can be written as

$$u_N(x) = g(x) - \lambda_1 h\sum_{i=-N}^{i=N}\frac{K_1(x,t_i)}{\phi'(t_i)}F_1(t_i,u_i)$$
$$-\lambda_2 h\sum_{i=-N}^{i=N}\frac{K_2(x,t_i)}{\phi'(t_i)}J(i,h)(\phi(x))F_2(t_i,u_i), \qquad x\in[a,b]. \tag{12}$$

4 Error Analysis

Throughout this section, we shall provide the error analysis of the proposed method, which displays approximate solutions have exponential convergence order. For this result, consider the following theorem.

Theorem 4.1 *Assume that $K_1(x, .) \in L_\alpha(D)$, $K_2(x, .) \in L_\alpha(D)$ for all $x \in \Gamma$ and the nonlinear functions F_1, F_2 are bounded which satisfy the generalized Lipschitz condition with constants L_1, L_2, respectively. Then, as $|1 - (\lambda_1 L_1 M_1 + \lambda_2 L_2 M_2)| \neq 0$, there exist $N_0 \in \mathbb{Z}^+$ and a constant C such that for $N > N_0$,*

$$\max_{a \le x \le b} |u(x) - u_N(x)| \le \frac{C}{|1 - (\lambda_1 L_1 M_1 + \lambda_2 L_2 M_2)|} e^{-\sqrt{\pi d \alpha N}}. \tag{13}$$

where M_1, M_2 represent meaning as follows.

Proof According to the assumptions and above Theorems for Volterra and Fredholm type integral and for a fixed $x \in (a, b)$ we have

$$\begin{aligned}
|u(x) - u_N(x)| &= \lambda_1 \left| \int_a^b K_1(x,t) F_1(t, u(t)) dt - h \sum_{i=-N}^{i=N} \frac{K_1(x, t_i)}{\phi'(t_i)} F_1(t_i, u_i) \right| \\
&+ \lambda_2 \left| \int_a^x K_2(x,t) F_2(t, u(t)) dt - h \sum_{i=-N}^{i=N} \frac{K_2(x, t_i)}{\phi'(t_i)} J(i,h)(\phi(x)) F_2(t_i, u_i) \right| \\
&\le \lambda_1 h \sum_{i=-N}^{i=N} \left| \frac{K_1(x, t_i)}{\phi'(t_i)} \right| |F_1(t_i, u(t_i)) - F_1(t_i, u_i)| + O(\exp^{-\sqrt{\pi d \alpha N}}) \\
&+ \lambda_2 h \sum_{i=-N}^{i=N} \left| \frac{K_2(x, t_i)}{\phi'(t_i)} J(i,h)(\phi(x)) \right| |F_2(t_i, u(t_i)) - F_2(t_i, u_i)| \\
&\le \left(\lambda_1 h L_1 \sum_{i=-N}^{i=N} \left| \frac{K_1(x, t_i)}{\phi'(t_i)} \right| + \lambda_2 h L_2 \sum_{i=-N}^{i=N} \left| \frac{K_2(x, t_i)}{\phi'(t_i)} J(i,h)(\phi(x)) \right| \right) \\
&\times |u(x) - u_N(x)| + O(\exp^{-\sqrt{\pi d \alpha N}}).
\end{aligned}$$

On the other hand, due to the conditions on the kernels and transformation, there exist constants M_1, M_2, such that

$$h \sum_{i=-N}^{i=N} \left| \frac{K_1(x, t_i)}{\phi'(t_i)} \right| \le M_1, \; h \sum_{i=-N}^{i=N} \left| \frac{K_2(x, t_i)}{\phi'(t_i)} J(i,h)(\phi(x)) \right| \le M_2. \tag{14}$$

Therefore, we can conclude that

$$\max_{a \le x \le b} |u(x) - u_N(x)| \le \frac{C}{|1 - (\lambda_1 L_1 M_1 + \lambda_2 L_2 M_2)|} e^{-\sqrt{\pi d \alpha N}}. \qquad (15)$$

So, the proof of this theorem is completed.

5 Numerical Examples

In this section, we show the results obtained for two examples using Sinc Nyström method described in the previous sections. In order to observe the convergence behavior and validate the above results of theory, we solve each equation for several values of N. Specifically, we select $N = 2, 4, 8, 16, 32, 64$ and define the maximum error on Sinc point

$$E_N = \max_{-N \le i \le N} |u(x_i) - u_N(x_i)| \quad and \quad \rho_N = \log_2 \left(\frac{E_N}{E_{2N}} \right).$$

Example 5.1 Consider the following nonlinear Volterra–Fredholm–Hammerstein integral equation

$$u(x) = \int_0^1 xtu^2(t)dt + \int_0^x xte^{u(t)}dt + lnx - \frac{x}{4} - \frac{x^3}{6}, \qquad x \in [0, 1],$$

with the exact solution $u(x) = lnx$.

Example 5.2 Consider the following nonlinear Volterra–Fredholm–Hammerstein integral equation

$$u(x) = \int_0^1 (x + t)u(t)dt + \int_0^x (x - t)u^2(t)dt - \frac{x^6}{30} + \frac{x^4}{3} - x^2 + \frac{5}{3}x - \frac{5}{4}, x \in [0, 1],$$

with the exact solution $u(x) = x^2 - 2$.

Table 1 obviously show that the exponential convergence rate of the proposed numerical method. By increasing the value of N, the errors have been decreased. Table 2 represents the numerical results of the present method and the Bernstein operational matrices method [3] for Example 5.1. It is clear that our method is more efficient than the method of [3]. By comparing our method with Haar functions method [2], it is obvious that the proposed method more applicable than method of [2] for Example 5.2 in Table 3. In Fig. 1, the values of exact solutions and the approximate solutions with $N = 1$ and $N = 3$ for our method are provided. The figure clearly shows the accuracy of the proposed method.

Table 1 The numerical results for Examples 5.1 and 5.2

Example	$N = 2$	$N = 4$	$N = 8$	$N = 16$	$N = 32$	d$N = 64$
Example 5.1 E_N	3.94e−03	6.94e−04	4.09e−05	5.10e−07	1.33e−09	2.93e−012
ρ_N	*	2.56	4.08	6.33	8.59	8.82
Example 5.2 E_N	6.04e−02	6.80e−03	6.35e−04	2.79e−05	3.73e−07	9.32e−010
ρ_N	*	3.31	3.26	4.51	6.23	8.64

Table 2 The numerical solutions and exact solutions for Example 5.1

x	Present method N = 4	Method of [3] n = 16	Exact solution
0	$-\infty$	$-\infty$	$-\infty$
0.2	−1.60944	−1.61169	−1.60944
0.4	−0.91661	−0.92358	−0.916291
0.6	0.51117	−0.52270	−0.510826
0.8	−0.22340	−0.23604	−0.223144
1	−0.00056	−0.00719	0

Table 3 The numerical solutions and exact solutions for Example 5.2

x	Present method N = 8	Method of [2] n = 16	Exact solution
0	−1.9996	−1.995	−2
0.2	−1.9598	−1.965	−1.96
0.4	−1.8399	−1.841	−1.84
0.6	−1.6401	−1.643	−1.64
0.8	−1.3601	−1.359	−1.36
1	−1.0003	−0.994	−1

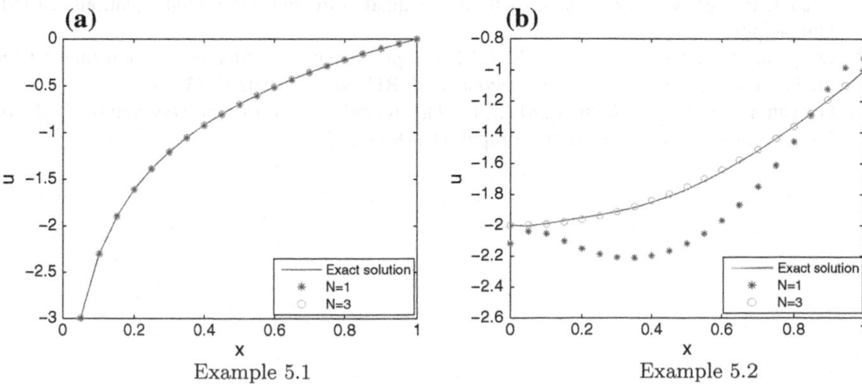

Fig. 1 The approximate solutions and exact solutions for Examples 5.1 and 5.2

6 Conclusions

In this paper, properties of the Sinc function are employed to convert Eq. (1) into some nonlinear algebraic equation and then we derive the numerical results with high accuracy. Our method is shown to be of good convergence, easy to program. In future work, we will utilized the proposed method to deal with the system of nonlinear Volterra–Fredholm–Hammerstein integral equations of the second kind.

References

1. Yousefi S, Razzaghi M (2005) Legendre wavelets method for the nonlinear Volterra–Fredholm integral equations. Math Comput Simul 70:1–8
2. Ordokhani Y, Razzaghi M (2008) Solution of nonlinear Volterra–Fredholm–Hammerstein integral equations via a collocation method and rationalized Haar functions. Appl Math Lett 21:4–9
3. Maleknejad K, Hashemizadeh E, Basirat B (2012) Computational method based on Bernstein operational matrices for nonlinear Volterra–Fredholm–Hammerstein integral equations. Commun Nonlinear Sci Numer Simul 17:52–61
4. Yalcinbas S (2002) Taylor polynomial solutions of nonlinear Volterra–Fredholm integral equations[J]. Appl Math Comput 127:195–206
5. Bildik N, Inc M (2007) Modified decomposition method for nonlinear Volterra–Fredholm integral equations. Chaos, Solitons Fractals 33:308–313
6. Mirzaee F, Hoseini AA (2013) Numerical solution of nonlinear Volterra–Fredholm integral equations using hybrid of block-pulse functions and Taylor series. Alex Eng J 52:551–555
7. Rashidinia J, Zarebnia M (2007) Convergence of approximate solution of system of Fredholm integral equations. J Math Anal Appl 333:1216–1227
8. Lund J (1983) Sinc function quadrature rules for the Fourier integral. Math Comput 41:103–113
9. Bialecki B (1989) A modified sinc quadrature rule for functions with poles near the arc of integration. BIT Numer Math 29:464–476
10. Kress R, Sloan IH, Stenger F (1997) A sinc quadrature method for the double-layer integral equation in planar domains with corners. University of New South Wales, School of Mathematics, Sydney
11. Lund J, Bowers KL (1992) Sinc methods for quadrature and differential equations. SIAM, Philadelphia
12. Okayama T, Matsuo T, Sugihara M (2011) Improvement of a Sinc-collocation method for Fredholm integral equations of the second kind. BIT Numer Math 51:339–366
13. Okayama T, Matsuo T, Sugihara M (2015) Theoretical analysis of Sinc-Nyström methods for Volterra integral equations. Math Comput 84:1189–1215

Research of Gas Purge Syringe Needle Micro Extraction System

Jun Cai, Hao Tian, Huijie Li, Donghao Li and Xiangfan Piao

Abstract Solid phase micro extraction(SPME), which is solventless, flexible, inexpensive and sensitive, plays an important role in pretreatment. Stainless steel wire can overcome the breakable defects of traditional quartz fiber, so it has been widely used as SPME fiber matrix. In order to develop a sample preparation instrument which use less organic solvent, own high enrichment quality and automated, this paper studied and developed a gas purge syringe micro extraction system, which only use stainless steel needle as extraction phase. The stainless steel needle is etched by hydrofluoric acid, without any coating adsorbent material. It is an automated and open system which contains heating sample matrix, condensing extraction phase and inert gas purging. Under optimized parameters, extraction experiments were performed, for enriching volatile and semi-volatile target compounds from the sample of 15 PAHs standard mixture. GC-MS analysis results indicated that a high enrichment factor was obtained from the system. The results demonstrate that system is potential in determination of volatile and semi-volatile analyzes from various kinds of samples.

Keywords MCU · PAH · SPME · Syringe needle micro extraction

J. Cai · H. Tian · H. Li · D. Li (✉) · X. Piao (✉)
Yanbian University, Yanbian, China
e-mail: pxf@ybu.edu.cn

D. Li
e-mail: dhli@ybu.edu.cn

J. Cai
e-mail: 2013010477@ybu.edu.cn

H. Tian
e-mail: 2012010464@ybu.edu.cn

H. Li
e-mail: 2012010389@ybu.edu.cn

V.E. Balas et al. (eds.), *Information Technology and Intelligent Transportation Systems*, Advances in Intelligent Systems and Computing 455,
DOI 10.1007/978-3-319-38771-0_19

1 Introduction

Generally, a high temperature promotes release of target compounds from the sample and accelerates the discussion of them into the gas phase, which speeds up the entry of target compounds into extraction solvent. Moreover, a low temperature of extraction phase is beneficial to enrichment, because a low temperature is advantageous for target compounds in the headspace gas phase to dissolve in extraction phase since the extraction of target compounds is an exothermic process [1–4]. So, The ideal extraction condition is the sample matrix states at high temperature, extraction phase states at low temperature. Furthermore, based on the theory of ideal gases, it is concluded that the amount of chemicals in the gas phase increases with increasing gas volume under a given temperature and pressure [5].

In sample pretreatment method development, various advanced electronic technology and computer technology are adopted to achieve better sensitivity, accuracy and automation level, which constantly promote the development of analytical chemistry and environmental science. In the conventional headspace method, hot water bath and flow of cooling water are often used for heating the sample matrix and cooling the extraction phase, respectively. These methods result in large size of device and difficulty of temperature control. This paper adopts the PTC ceramic piece and semiconductor refrigeration piece for heating sample matrix and cooling extraction phase respectively, the mass flow controller is used for regulates the gas purging.

The proposed approach is based on solid phase extraction, used stainless steel syringe needles that etched by hydrofluoric acid as extraction phase, also heater for sample matrix, condensation for extraction phase, and inert gas purging controller incorporated into organic whole, which realized the automation of sample pretreatment. Extraction experiments were performed with polycyclic aromatic hydrocarbons (PAHs) standard mixture using, then analyzed by GC-MS. The results represents advantages of simplicity of the operation, automation, and accuracy of the control in extraction conditions, and rapidness of the extraction.

2 Experimental

2.1 Materials and Chemicals

Polycyclic aromatic hydrocarbons (PAHs), and [$2H10$] phenanthrene were purchased from Supelco (Bellefonte, PA, USA). The purity of standards was higher than 99 %. Dichloromethane was purchased from Caledon (Georgetown, Ont., Canada). Stock standard solutions of PAHs (20 mg $L-1$) were prepared in methanol. The internal standard ($2H10$-phenanthrene) was spiked into hexane (200 μg $L-1$) when it was used as an extracting solvent for PAHs. The standard solutions and extracting solvent were stored in the dark at 0–4 °C till used. The gas mass flow controller and digital monitor were obtained from Beijing Metron Instruments Co., Ltd. (China).

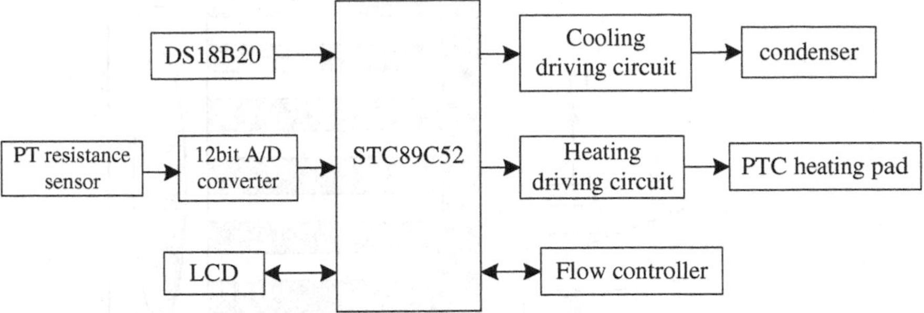

Fig. 1 Block diagram of the system

2.2 Apparatus and Mechanism

The principle of the system.

Block diagram of gas purge syringe needle micro extraction system (Fig. 1).

The principle of the system is that, first of all, setting the heating temperature, heat-ing time, condensation time and nitrogen purge flow rate with the touch screen. Then the execution module get control commands from MCU, including the mass flow controller, ceramic PTC constant temperature heating module, the semiconductor refrigeration module, In this process, condensational temperature detected by the digital temperature sensor DS18B20, heating temperature detected by the PT resistance sensor, both of two temperature compared with set temperature, it using PID control method. The touch screen is used to select the experimental parameters and display the temperature of the heater, the end time and the gas flow rate.

In the extraction process, through heating the sample matrix and inert gas purge that can accelerate the process of the volatilization of object from the sample matrix. And condensation of the injection needle can promote the target gas to liquefy around the inner wall of the needle, which strengthen the enrichment ability of needle.

The system heating the sample matrix, condensate extraction phase and dynamic purge stream to come together, the system is a combination of separated, concentration and purification, and automate the entire extraction process, greatly shorten pre-treatment time and reduce the complexity of the extraction operation. Installation drawing of gas purge syringe needle micro extraction system (Fig. 2).

The gas purge syringe needle micro extraction system is composed of Puncture needle, condenser, injection pads, glass wool, quartz sample tube, heater, nitrogen shunts. And the heater composition of the heater body, PTC ceramic heating thermostat chip, platinum resistance temperature sensor, injection pads, the sample cell, gas connection interface. Semiconductor refrigeration piece and DS18B20 is the core of Condenser, further comprising a copper heat sink, heat sink and fan.

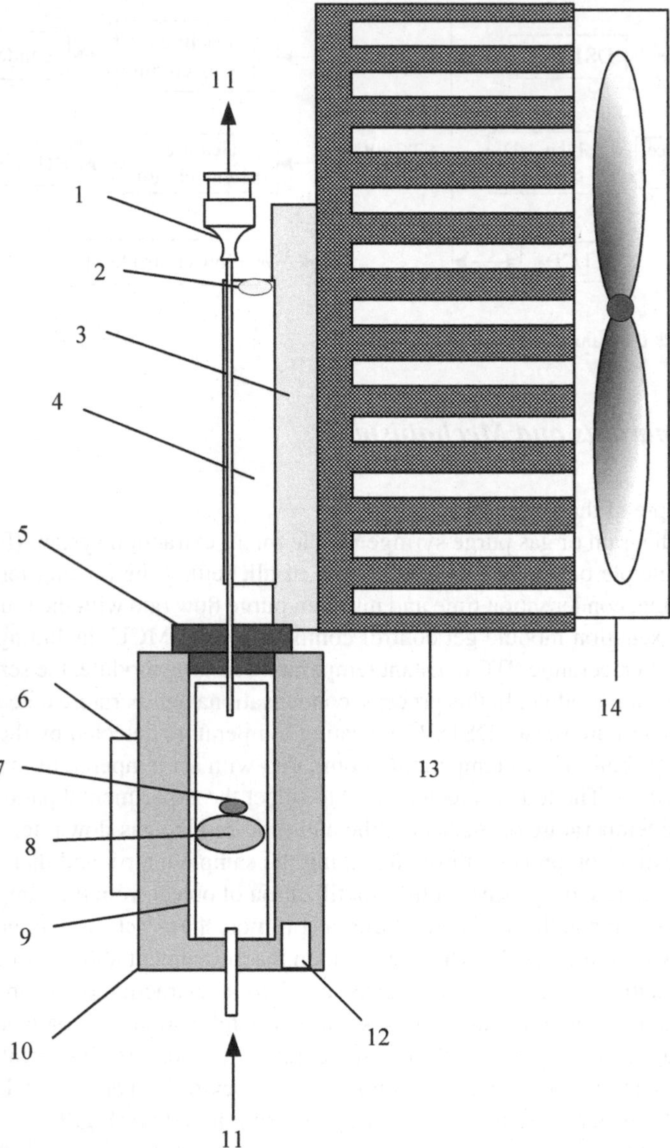

Fig. 2 Installation drawing of the system *1* needle, *2* PT resistance sensor, *3* semiconductor refrigeration piece, *4* copper plate, *5* sample introduction pad, *6* heating plate, *7* sample, *8* glass wool, *9* quartz glass tube, *10* copper heating body, *11* inert gas, *12* temperature sensor, *13* heat sink, *14* fan, *15* exhaust gas

2.3 Experimental Procedure

GP-HS-LPME system is used in extraction experiments, Proceed as follows:

(1) The actual sample or standard sample was on the glass wool layer in sample tube, Set the sample tube into the copper base with curved beak forceps, installed injection pads.
(2) The etching needle is inserted into the sample pad, glass opening out into the upper end position approximately 3 mm.
(3) Setting the purge gas flow rate, condensate and the heating temperature, extraction time on the touch screen, start the extraction process.
(4) After the end of the set extraction time, remove the needle, using a suitable solvent elution needle wall, then analysis with GC-MS.

3 Results and Discussion

Optimization of the system parameters, the heating temperature: 270 °C, extraction time: 7 min, the nitrogen flow rate: 1.5 mL/min, condensing temperature: −3 °C. In order to estimate the extraction performance of gas purge syringe needle micro extraction system to volatile and semi-volatile analyzes. First, 15 kinds of PAHs was chose for standard sample to extraction experiments, The experiment was repeated three times, Detected by gas chromatography—mass spectrometry Experimental conditions were as follows

Table 1 Recovery and RSD of PAH of the system

Target	Recovery (%)		Average recovery (%)		RSD (%)
Acenaphthylene	76.61	78.89	84.94	80.15	5.36
Acenaphthene	78.89	81.6	79.76	80.08	1.72
Fluorene	84.76	84.08	85.21	84.68	0.67
Phenanthrene	90.3	92.46	91.33	91.36	1.18
Anthracene	91.74	95.71	94.95	94.13	2.24
Fluoranthene	83.32	86.35	85.21	84.96	1.80
Pyrene	87.21	83.66	88.59	86.49	2.94
Beazo[a]fluoranthene	92.14	96.93	93.64	94.24	2.60
Chrysene	92.26	96.02	93.44	93.91	2.04
Beazo[b]fluoranthene	92.8	98.94	93.72	95.15	3.48
Benzo[k]fluoranthene	88.23	97.27	91.05	92.18	5.02
Benzo[a]pyrene	80.12	93.53	91.53	88.39	8.18
Indeno(1,2,3-cd)pyrene	84.42	96.56	88.30	89.76	6.91
Dibenz(a,h)anthracene	81.87	83.67	82.75	82.76	1.09
Benzo(ghi)perylene	87.04	86.51	82.76	85.44	2.73

(1) target sample: $1\,ppm$ PAH standard sample 20 uL;
(2) eluting solvent: 40 μL methane slowly be pushed;
(3) the internal standard: 200 ppb deuterium;

Mass spectrometer conditions are as follows: the chromatographic column was a DB-5 fused silica capillary column, carrier gas was helium, flow rate was 1.2 mL/min, and splitless sampling. The sample volume were 2 μL. The temperature of sample inlet and detector were 280 °C. The temperature of ion source was 200 °C.

It can be concluded from the Table 1 that the recycling rate of the system to PAH standard sample is satisfactory, the recycling rate of the substances are more than 80 % and most of the material recycling rate reached 90 %, which is consistent with the standard of the recycling rate for trace target object. It is proved that it is feasible to extract PAH in the system. Generally, the RSD of the extract experiment to PAH in the system is less than 10 %. It is proved that the reliability of the method.

4 Conclusion

The micro-controller (MCU) was used as the control chip in the system, which is equipped with semiconductor condensers and PTC ceramic heater. While achieving the heating of the sample matrix and condensate extraction needle, by using a gas flow controller for inert gas purge rate control and to achieve the extraction process automation, it reduced the system size. The operation is simple and extraction time is short. The method of solid phase extraction reduce the amount of the organic solvent. Through the standard sample extraction experiments, the target object of volatile and semi-volatile have a high enrichment efficiency. Therefore, the system in the field such as food chemistry, biochemistry, environmental chemistry in various sample matrices, and different target compounds enriched a broad application prospect.

References

1. Yang C, Piao XF, Qiu JX (2010) Gas purge microsyringe extraction for quantita-tive direct gas chromatographic-mass spectrometric analysis of volatile and semivolatile chemicals. J Chromatogr A 1218(12):1549–1555
2. Piao XF, Bi JH, Yang C (2011) Automatic heating and cooling system in a gas purge microsyringe extraction. Talanta 86:142–147
3. Bi JH, Yin Z, Piao XF (2011) Research and development of ME-101 multifunc-tional micro extraction system. J Yanbian Univ 37(2):115–118
4. Yushang J (2011) Simultaneously analysis of organochlorine pesticides and organophosphorus pesticides residues from ginseng using gas purge micro syringe extraction coupled with gas chromatography-mass spectrometry. Master degree thesis of Yanbian University, pp 10–15
5. Yi JJH, Junxiong H (1998) Solid-phase microextraction in the pretreatmentof environmental samples. Prog. Chem. 10(1):74–82

A Text-Independent Method for Estimating Pronunciation Quality of Chinese Students

Guimin Huang, Huijuan Li, Rong Zhou and Ya Zhou

Abstract In this paper, a novel approach is proposed for text-independent pronunciation quality assessment of Chinese students. We call the proposed method as double-models pronunciation scoring algorithm, which separates recognition from assessment stage. It can solve low recognition performance of standard method and score mismatch of nonstandard one. Applying the combination of Maximum Likelihood Linear Regression and Maximum A Posteriori adaptation achieves good recognition results for speech of Chinese students. Adjustment of scoring features signifies further improvement in correlation between machine scores and human judgment. The experimental results showed the proposed double-models technique reached good outcome for text-independent pronunciation quality assessment of Chinese students.

Keywords Pronunciation assessment · Text-independent · Speech recognition · Maximum likelihood linear regression · Maximum A Posteriori

1 Introduction

Computer Assisted Language Learning (CALL) system has been researched for several decades. As a major part of CALL, the pronunciation evaluation technologies are used for self-directed language learning to improve language ability of learners. Actually, the pronunciation evaluation technologies are based on speech recognition. There are a series of pronunciation rating features of measuring pronunciation quality

G. Huang (✉) · H. Li (✉) · R. Zhou · Y. Zhou
Guilin University of Electronic Technology, Guilin 541004, China
e-mail: 1047269234@qq.com

H. Li
e-mail: 499251602@qq.com

R. Zhou
e-mail: jessie_105@sina.com

Y. Zhou
e-mail: ccyzhou@guet.edu.cn

V.E. Balas et al. (eds.), *Information Technology and Intelligent Transportation Systems*, Advances in Intelligent Systems and Computing 455,
DOI 10.1007/978-3-319-38771-0_20

in different aspects, such as log likelihood, segment duration, segment classification, rate of speech. These features are used to test pronunciation standardness, completeness and fluency effectively [1]. Franco et al. [2] proposed posterior probability which is more robust than log likelihood in the case of standardness. Posterior probability is less affected by external factors and correlates better with human score. The Goodness of Pronunciation (GOP) is a variant of posterior probability, which was presented by Witt and Young [3]. These approaches are based on standard acoustic models which are also called golden models and trained by native corpus. So they are called as standard methods. However, these standard methods have low recognition performance for nonnative speech, especially in terms of text-independent condition.

In this paper, research on how to further improve recognition rate and assessment robustness for Chinese students under the condition of text independent is investigated. An algorithm of double models is proposed, that is one model for identification, another for evaluation. In this paper, we use Hidden Markov Model (HMM) acoustic model both in recognition and evaluation. In the acoustic model, each HMM with three states represents an English phone. Each state of a HMM is represented by eight Gaussian mixture components. The proposed method separates recognition from evaluation stage, removing the contradiction between the two processes. The combination of Maximum likelihood linear regression (MLLR) and Maximum A Posteriori (MAP) [4] is adopted to increase recognition accuracy. Finally, the correlation between machine and human scores is further enhanced through scoring features adjustment and combination. Experiment showed the proposed solution achieved good result for pronunciation measurement of Chinese students under the condition of text independent.

This paper is organized as follows. Section 2 details the designed method. Experiments and analysis are showed in Sect. 3. Conclusions are given in Sect. 4.

2 Method

Speech recognition is a speech-to-text decoding process. For the text-independent pronunciation evaluation of Chinese students that we discuss here, because the text corresponding to speech is unknown, speech recognition becomes a considerable challenge. If we only adopt conventional solutions, both recognition and assessment can not achieve good result. In this paper, MLLR and MAP adaptation methods are both introduced to enhance recognition performance. The advantage of the proposed algorithm is that it can maximize recognition rate and ensure golden character of acoustic model. The double-models method separates recognition from assessment, both there is no impact. It also makes adaptation easier without considering the difference of adaptation methods between traditional speech recognition and pronunciation assessment. So the better correlation between machine and human scores can be achieved. Furthermore, adjustment is made for scoring features to further improve evaluation performance. The processing of the method is shown in Fig. 1, which consists of recognition processing and assessment processing. First, speech

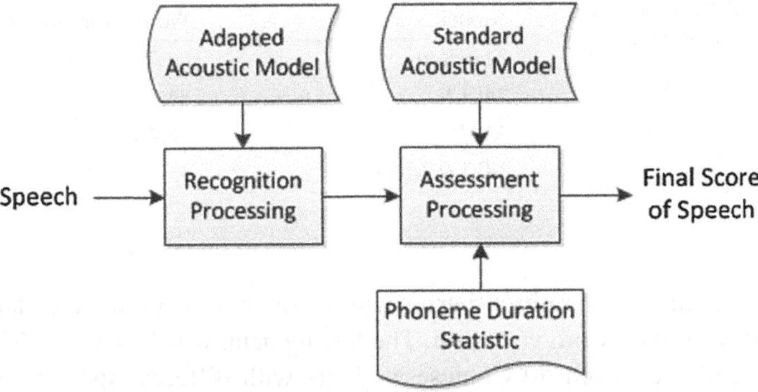

Fig. 1 The processing of double-models method

signal is recognized by a HMM recognizer. Second, scoring features are calculated from recognition result. Then the final score is output by adjustment and combination of scoring features.

2.1 *Standard Acoustic Model*

Standard acoustic model is trained by Wall Street Journal (WSJ) American English speech corpus [5]. Its a HMM-GMM model built by tri-phones.

2.2 *Adapted Acoustic Model*

Adapted acoustic model based on the standard one is trained by a lot of nonnative speech data. All range of data with different pronunciation proficiency levels is used for training.

MLLR and MAP are effective speaker adaptation methods. They have different advantages. In MLLR, phonetic parameters standing for analogous acoustic phenomenon are tied to same regression class, which will be transformed by same matrix. The number of regression class can be varied with the size of adaptation data. The more the data is, the more the number of regression class will be. But those matrices dont always have strong relevance for each parameter, especially when there are only a little data. In MAP, phonetic parameters can be updated through making full use of the details of adaptation data. But those parameters that are not covered by data cant be adjusted. So a MLLR-MAP concatenated algorithm is employed for model adaptation. At first, MLLR-adapted model is achieved by MLLR adaptation adjustment. Then, conducting MAP adaptation on MLLR-adapted model gets MLLR-MAP adapted model.

Table 1 Word error rate for different acoustic models

Model type	Word error rate (%)
Standard	86.2
MLLR	84.7
MAP	36.4
MLLR-MAP	35.7

The results about recognition performance are showed in Table 1. The performance is measured by word error rate. The testing data, which contains 6502 words in total, is collected from 50 Chinese students with different speaking ability in English.

2.3 *Phoneme Duration Statistic*

The statistic of phone duration is collected from Arctic corpus, which is used for calculating phone duration scores. The database consists of 1132 carefully selected sentences, which ensures that it is phonetically balanced and phoneme coverage is 100 %. So the database is appropriate for collecting phoneme duration statistic.

2.4 *Recognition Processing*

In recognition processing, the sphinx4 HMM recognizer is used to decode speech. Sphinx4 is developed jointly by researchers at Carnegie Mellon University. The recognition processing adopts Mel-Frequency Cepstral Coefficient (MFCC) as the feature sequence of speech signal.

2.5 *Assessment Processing*

In assessment processing, the scoring features in terms of standardness and fluency are shown in Table 2.

In assessment processing, there are three steps. They are calculating scoring features, adjusting scoring features and combining scoring features. The following introduces the calculating, adjusting and combining for these scoring features.

Calculating Log-Likelihood As described above, adapted model is just employed for recognition. Recognition results should be compared with parameters in standard model to determine how close they are to standard utterances. So when calculating Log-likelihood Probability (LP), phones mapping replacement should

Table 2 Scoring features

Scoring features	Description
Log-likelihood	Similarity comparison between test speech and standard speech
Goodness of pronunciation	Confidence measure of test speech
Rate of speech	Number of phones per frame
Phone duration	Distance of phone segment duration between test speech and standard speech
Pause duration	Total pause time in test speech

be conducted first. First, we define some symbols, $O^{p_{i-ad}}(o_1^{p_{i-ad}}, o_2^{p_{i-ad}}, \ldots, o_T^{p_{i-ad}})$ represents acoustic segment corresponding to the phone p_{i-ad}, p_{i-ad} is phone p_i in adapted model, p_{i-st} is phone p_i in standard model. The mapping replacement is $(O^{p_{i-ad}}, p_{i-ad}) \rightarrow (O^{p_{i-ad}}, p_{i-st})$. Then log-likelihood is calculated by the replaced results. The parameters both in p_{i-ad} and p_{i-st} describe the same phone p_i. The difference is that parameters in p_{i-st} are more standard than p_{i-ad}'s. For each $O^{p_{i-ad}}(o_1^{p_{i-ad}}, o_2^{p_{i-ad}}, \ldots, o_T^{p_{i-ad}})$, the double-models log-likelihood probability $DLP(O^{p_{i-ad}}, p_{i-ad})$ can be defined as the following formula (1):

$$DLP(O^{p_{i-ad}}, p_{i-ad}) = DLP(O^{p_{i-ad}}, p_{i-st}) = \frac{1}{d_i}\left\{\log\left[b\left(o_1^{p_{i-ad}}|s_1\right)\right] + \sum_{t=1}^{T-1} log\left[a(s_{t+1}|s_t)b\left(o_{t+1}^{p_{i-ad}}|s_{t+1}\right)\right]\right\} \tag{1}$$

In formula (1), phone model replacement from adapted model to standard model is conducted. $O^{p_{i-ad}}$ represents the acoustic segment $O^{p_{i-ad}}(o_1^{p_{i-ad}}, o_2^{p_{i-ad}}, \ldots, o_T^{p_{i-ad}})$, p_{i-ad} is phone p_i in adapted acoustic model, p_{i-st} is phone p_i in standard acoustic model, d_i is the duration of p_{i-ad}, s_t is the state of phone p_i at time t in HMM model, $a(s_{t+1}|s_t)$ is the HMM transition probability and $b(o_t^{p_{i-ad}}|s_t)$ is the HMM output probability of state s_t.

The sentence-level score is the average of individual scores over the N phones in the sentence and it is defined as the following formula (2):

$$DLP = \frac{1}{N}\sum\nolimits_{i=1}^{N} DLP(O^{p_{i-ad}}, p_{i-ad}) \tag{2}$$

$O^{p_{i-ad}}$ represents the acoustic segment $O^{p_{i-ad}}(o_1^{p_{i-ad}}, o_2^{p_{i-ad}}, \ldots, o_T^{p_{i-ad}})$, p_{i-ad} is phone p_i in adapted acoustic model.

Calculating Goodness of Pronunciation Goodness Of Pronunciation (GOP) is generally used for measuring standardness as well. It has higher robustness. Compared with previous method, the difference is that phones mapping replacement should be done before calculation. For each $O^{p_{i-ad}}(o_1^{p_{i-ad}}, o_2^{p_{i-ad}}, \ldots, o_T^{p_{i-ad}})$, the double-models

$DGOP(O^{p_{i-ad}}, p_{i-ad})$ can be defined as the following formula (3):

$$DGOP(O^{p_{i-ad}}, p_{i-ad}) = \frac{1}{d_i} log(P(p_{i-st}|O^{p_{i-ad}})) = \frac{1}{d_i} log\left[\frac{P(O^{p_{i-ad}}|p_{i-st})P(p_{i-st})}{\sum_{q \in Q} P(O^{p_{i-ad}}|q)P(q)}\right] \tag{3}$$

$O^{p_{i-ad}}$ represents the acoustic segment $O^{p_{i-ad}}(o_1^{p_{i-ad}}, o_2^{p_{i-ad}}, \ldots, o_T^{p_{i-ad}})$, p_{i-ad} is phone p_i in adapted acoustic model, p_{i-st} is phone p_i in standard acoustic model, Q is the full set of phones in standard acoustic model, d_i is the duration of p_{i-ad}, $P(p_{i-st})$ and P(q) are prior probability. The sentence-level GOP is defined as the following formula (4):

$$DGOP = \frac{1}{N} \sum_{i=1}^{N} DGOP(O^{p_{i-ad}}, p_{i-ad}) \tag{4}$$

where $O^{p_{i-ad}}$ represents the acoustic segment $O^{p_{i-ad}}(o_1^{p_{i-ad}}, o_2^{p_{i-ad}}, \ldots, o_T^{p_{i-ad}})$, p_{i-ad} is phone p_i in adapted acoustic model, N is the number of phones in the sentence.

Calculating Rate of Speech Rate Of Speech (ROS) [6] can also be used for assessment. Generally speaking, the more like native the speaker is, the faster the utterances are. ROS is obtained through dividing number of phones in utterance by its length.

Calculating Phone Duration Phone Duration (PD) [6] has been considered to be a good index. Phone duration statistic is collected from Arctic corpus, and ROS is used for normalization. The duration distribution of phone pi is defined as the following formula (5):

$$P(t|p_i) = \frac{1}{\sqrt{2\pi \sigma_{p_i}{}^2}} exp\left[-\frac{(t - \mu_{p_i})^2}{2\sigma_{p_i}{}^2}\right] \tag{5}$$

σ_{p_i} is the variance of phone duration of phone p_i, μ_{p_i} is the mean of phone duration of phone p_i, t is the normalization phone duration of phone p_i. The sentence-level phone duration log-score is defined as the following formula (6):

$$PD = \frac{1}{N} \sum_{i=1}^{N} log[P(ROS * d_i|p_i)] \tag{6}$$

ROS is rate of speech. d_i is the phone duration of phone p_i. N is the number of phones in the sentence.

Calculating Pause Duration If tester doesnt know exactly a words pronunciation, he may need time to think about it, which may cause pauses. So total PAUse Duration (PAUD) [5] within a sentence can be regarded as a feature.

Adjusting Scoring Features In this paper, the polynomial transformation is used to obtain a more accurate scoring feature. It is defined as the following formula (7).

$$s^* = \alpha_0 + \alpha_1 s + \alpha_2 s^2 + \cdots + \alpha_n s^n \tag{7}$$

where,s^* is a transformed scoring feature corresponding to an original s.

Combining Scoring Features Comprehensive score for utterances can be obtained by combination of several scoring features. In this paper, Multi-Regression (MR) is introduced to get overall score. The final score can be defined as follows formula (8):

$$Score = \sum_i \alpha_i s_i + \beta \tag{8}$$

where Score is the final score, α_i is the weight corresponding to score feature s_i, β is offset item. α_i and β can be estimated by using least squares method.

3 Experiments and Analysis

3.1 Experimental Environment and Database

All experiments were conducted in a PC which is installed the win7 operating system and has an Inter® Pentium® CPU, 4 GB RAM and 500 GB hard disk. Sphinx4 was used for speech recognition, which is a HMM recognizer.

In order to verify the proposed double-models method, nonnative speech database was constructed for pronunciation assessment. Speech was recorded from 50 Chinese students of Guilin University of Electronic Technology. Each student was asked to read 15 phonetically rich sentences from Arctic corpus.

3.2 Criteria of Assessment

Human score is used to assess the performance of scoring algorithm. Five English professors were invited to be as raters for scoring speech data of Chinese students on a scale of 1 to 5. To measure raters consistency, 30 percent of speech data was selected as the set of consistency assessment. This subset was repeatedly rated three times by each rater. Three kinds of correlations, which are inter-rater, intra-rater and between-rater correlation [1], were adopted.

As it is shown in Table 3, the consistency between raters is acceptable uniform. The speaker-level correlations are higher than those at sentence level. In Table 4, the correlations within each rater (intra-rater correlation) are good. It shows the raters stability. The averages of inter-rater correlations at sentence and speaker level, which

Table 3 Correlation between raters at sentence and speaker level

Rater	1	2	3	4	5
1	1.0/1.0	0.662/0.764	0.627/0.718	0.72/0.912	0.689/0.815
2	0.662/0.764	1.0/1.0	0.607/0.692	0.712/0.878	0.653/0.741
3	0.627/0.718	0.607/0.692	1.0/1.0	0.613/0.706	0.618/0.712
4	0.72/0.912	0.712/0.878	0.613/0.706	1.0/1.0	0.69/0.806
5	0.689/0.815	0.653/0.741	0.618/0.712	0.69/0.806	1.0/1.0

Table 4 Inter-rater and intra-rater correlation

Correlation type	Rater					Mean
	1	2	3	4	5	
Inter-rater (sentence)	0.765	0.729	0.626	0.791	0.742	0.731
Inter-rater (speaker)	0.882	0.806	0.716	0.937	0.852	0.839
Intra-rater (sentence)	0.925	0.852	0.762	0.937	0.863	0.868

are 0.731 and 0.839, suggest the upper bound of the level of correlation between human and machine scores.

3.3 Discussion

In this paper, experiments on the comparison of performance between previous methods and the proposed one are made in terms of standardness scores. Experiments on the scoring features adjustment, and combination of several features are made as well.

Comparison of Standardness Scores Log-likelihood and GOP scores are defined as SLP, SGOP, NLP, NGOP, DLP and DGOP separately. In the proposed one, two kinds of standard models are introduced respectively. One is trained through WSJ corpus. The other one is trained by a small amount of extremely proficient data of Chinese students and WSJ corpus.

Table 5 shows that the double-models algorithm has the highest correlation in terms of standardness scores (DLP2, DGOP2) for pronunciation measurement of Chinese students under the condition of text independent. Traditional standard method has the lowest correlation because of its poor recognition performance for nonnative speech. Standardness scores (NLP, NGOP) of nonstandard method has better results than DLP and DGOP which are the results of the first double-models method in which standard model is trained by all native data. This phenomena accounts for that

Table 5 Correlation between standardness scores and human scores at both levels

Feature type	Correlation	
	Sentence	Speaker
SLP	0.073	0.091
SGOP	0.124	0.145
NLP	0.117	0.215
NGOP	0.311	0.489
DLP	0.107	0.193
DGOP	0.231	0.44
DLP2	0.119	0.229
DGOP2	0.358	0.618

Table 6 Adjustment of features

Feature type	Before adjustment		After adjustment	
	Correlation		Correlation	
	Sentence	Speaker	Sentence	Speaker
DLP2	0.119	0.229	0.167	0.352
DGOP2	0.358	0.618	0.395	0.741
ROS	0.239	0.501	0.311	0.587
PD	0.131	0.44	0.23	0.483
PAUD	-0.35	-0.367	0.37	0.483

the difference of characteristics or acoustic channel between native and nonnative speakers does have a significant effect on pronunciation measurement. So when we use nonnative standard model which is trained by a small amount of extremely proficient data of Chinese students and WSJ corpus, the effectiveness of double-models method is apparent.

Feature Adjustment Our aim of adjusting scoring features is to obtain an increasing correlation between machine and human judgment. Features are produced before and after changes respectively. In this experiment, nonnative database is divided into two sets. One, accounting for 70 %, is used to train parameters of polynomial transformation; the other for testing, accounting for 30 %. The result is shown in Table 6 at sentence and speaker level.

Combination Combination of Several high performance features can achieve better acceptable result against single feature. We investigate different feature combinations to find the optimum manner. In this experiment, nonnative database is divided into two sets. One is used to train parameters of Multi-Regression, which accounts for 70 %; the other for testing, accounting for 30 %. All scoring features are transformed. The results are shown in Table 7 at sentence and speaker level.

Table 7 Combination Of various features

Combination type	Correlation	
	Sentence	Speaker
DGOP2 + ROS	0.437	0.768
DGOP2 + ROS + PAUD	0.482	0.783
DGOP2 + ROS + PAUD + PD	0.511	0.795
DGOP2 + ROS + PAUD + PD + DLP2	0.518	0.799

Table 7 shows that combination of scoring features can further improve correlation.

4 Conclusion

This paper has shown the feasibility of the proposed double-models scoring algorithm. The double-models technique handled the problems of low recognition performance of nonnative speech and score mismatch existed in previous attempts. Because we havent obtained the optimum correlation, great effort needs to make to get better outcome. In the future, we will focus on how to improve recognition rate for nonnative speech and explore some new scoring features.

Acknowledgments This work is supported by the Research Foundation of Humanity and Social Science of Ministry of Education of China, No.11YJAZH131, and is supported by the Foundation of Key Laboratory of Cognitive Radio and Information Processing, Ministry of Education (Guilin University of Electronic Technology, No. CRKL150105)and the Innovation Project of GUET Graduate Education, No. YJCXS201543.

References

1. Neumeyer L, Franco H, Weintraub M, Price P (1996) Automatic text-independent pronunciation scoring of foreign language student speech. In: Proceedings of the international conference on spoken language processing, vol 4, no 2, pp 1457–1460
2. Franco H, Neumeyer L, Kim Y, Ronen O (1997) Automatic pronunciation scoring for language instruction. In: IEEE international conference on acoustics, speech, and signal processing(ICASSP) 2, pp 1471–1474
3. Witt SM, Young SJ (2000) Phone-level pronunciation scoring and assessment for interactive language learning. Speech Commun 30:95–108
4. Lee CH, Gauvain J (1993) Speaker adaptation based on MAP estimation of HMM parameters. In: Proceedings of the IEEE international conference on acoustics speech and signal processing ICASSP, 558C561
5. Cincarek T, Gruhn R, Hacker C, Noth E, Nakamura S (2009) Automatic pronunciation scoring of words and sentences independent from the non-natives first language. Comput Speech Lang 23:65–88

6. Neumeyer L, Franco H, Digalakis V, Weintraub M (2000) Automatic scoring of pronunciation quality. Speech Commun 30:83–93

Sparse Projection CT Image Reconstruction Based on the Split Bregman Less Iteration

Jun-nian Gou and Hai-ying Dong

Abstract Sparse angle projection CT image reconstruction in medical diagnosis and industrial non-destructive testing has important theoretical significance and practical application value. In the paper, L1 norm was introduced as the CT images of regular constraint and optimization reconstruction model, and the method to solve it was presented based on the Split Bregman algorithm. Shepp-Logan numerical simulation experiments show that the image reconstructed by the traditional algebraic reconstruction algorithm of ART for sparse projection CT is poor. The Split Bregman may solve L1 regularization constraint model of sparse projection of CT with less number of iterations, fast reconstruction and good reconstruction quality. For the splitting factor of the algorithm, in a numerical range, the greater the reconstruction quality is better.

Keywords Split Bregman · The regular optimization · CT image reconstruction · Sparse projection · Less number of iterations

1 Introduction

CT imaging technology, on its excellent performance perspective in medical diagnosis and non-destructive testing and other fields, has a wider range of applications [1, 2]. At present, the medical CT in order to reduce the radiation dose to the patient, urgently needs to reduce radiation dose of CT scanning; Industrial CT, in many cases, also hope to be able to undertake sparse projection scanning, thus to reduce the time of CT scanning and imaging reconstruction.

J.-n. Gou (✉) · H.-y. Dong
School of Automation and Electrical Engineering, Lanzhou Jiaotong University, Lanzhou, People's Republic of China
e-mail: junnian@mail.lzjtu.cn; dannis_gou@163.com

H.-y. Dong
e-mail: hydong@mail.lzjtu.cn

V.E. Balas et al. (eds.), *Information Technology and Intelligent Transportation Systems*, Advances in Intelligent Systems and Computing 455,
DOI 10.1007/978-3-319-38771-0_21

In recent years, the use of sparse angle projection CT reconstruction has become the mainstream in the reconstruction work. Research shows, however, the data of the sparse projection data don't satisfy Tuy–Smith [3, 4] complete condition. To sparse projection data, the analytical reconstruction algorithm such as Filter Back Projection (FBP), cannot get better reconstruction quality, and the reconstructed image often exist serious artifacts. When the missing projection angle is larger, the traditional iterative reconstruction algorithm such as Algebraic Reconstruction Techniques (ART) [5] is also hard to get high quality reconstructed image.

Compressed sensing theory [6] pointed out that if the reconstructed image can carry on the sparse representation in some areas, it can be reconstructed by using incomplete (truncated) projection data in large probability. Iterative algorithm based on Bregman Distance [7, 8] is an efficient algorithm which arose in recent years. This paper will apply it to CT image reconstruction of sparse projection under the regular optimization of L1 constraint.

2 The Iteration Model of CT Image Reconstruction

Iterative reconstruction of CT images can be done by solving system of linear equations. For the discretization of two-dimensional case, a two-dimensional grid is overlaid on the image function $u(x, y)$ [9].

Assumes that the function value $u(x, y)$ within the small grid is constant, and the constant value within a small grid is presented by a single subscript u_j, and $j = 1, 2, \ldots, N_{image}$). With the total number N_{image} of grid, generally, $N_{image} = n \times n$, n for the length and width of the image. For image processing, u_j is called pixel value. Because of the detector unit itself is discrete, the projection data itself is discrete. Also single subscript b_i presents the ith projection. After the image and projection data are discretized, the relationship between the pixel u_j and the projection ray b_i can be described by using Eq. 1.

$$b_i = \sum_{j=1}^{N_{image}} a_{ij} u_j, i = 1, 2, \ldots, N_{data} \tag{1}$$

In Eq. 1, N_{data} the length of the vector b_i, each under the projection angle of the total number of projection data; a_{ij} is the element of system matrix A, the i th projection ray interacting with the pixel u_j. In this paper, the value of a_{ij} is defined as the length of the i th ray through the pixel u_j; the system matrix A (i.e., projection matrix) is a $N_{data} \times N_{image}$ matrix. As shown in Eq. 2. CT image reconstruction of discretization model is aimed at fan beam X-ray source in the paper.

$$Au = b \tag{2}$$

CT image reconstruction is detected from the input A and output data b to get the internal structure characteristics of the object u. In Eq. 2, u is the unknown parameters of the system model, also is the goal need to be reconstructed; system matrix A is related to the imaging geometry, usually a large sparse matrix. At this point, the solution of Eq. 2 belongs to the typical type of inverse problem of discomfort. Regularization method is used to define an objective function, by minimizing the objective function to solve the problem can be converted to a class of constrained extreme value problem. For the industrial CT detection object, the type of the composition material generally is less, and this characteristic makes the reconstruction image grey value has obvious sparse feature of fragmentation. Based on this characteristic, L1 norm is introduced to measure the sparseness of the image, and the optimal solution of Eq. 2 is got by minimizing the L1 norm approximation. In order to speed up to solve the convergence speed, the Split Bregman algorithm is introduced to get an approximate reconstruction.

3 Solution to the Reconstruction Model

3.1 Bregman Distance

Definition 1. Given a differentiable convex function $J(u) : R^n \to R$, define the distance between two points u and v, which belong to the space R^n:

$$D_J^p(u, v) = J(u) - J(v) - < p, u - v > \tag{3}$$

In Eq. 3, p is the sub-gradient of function J at point v. When the function $J(u)$ is a Bregman function, its corresponding distance $D_J^p(u, v)$ is called Bregman distance. Due to the convex function properties, for arbitrary point w between the middle of u, v, there are $D_J^p(u, v) \geq 0$ and $D_J^p(u, v) \geq D_J^p(w, v)$. Considering the convex energy functional, J and L, both of them belong to the space R^n, and there is $\min_{u \in R^n} L(u) = 0$. The corresponding solving unconstrained minimization problem is

$$\min_u |\Phi(u)| + \mu L(u) \tag{4}$$

In Eq. 4, $|\cdot|$ represents L1-norm.

3.2 Bregman Iteration Method

For the study, the images of CT image reconstruction can be expressed in Eq. 5 by using penalty function form, and Eq. 5 can be further converted into Eq. 6.

$$u^{k+1} = \min D_J^p(u, u^k) + \frac{\mu}{2} \parallel Au - b^k \parallel_2^2 \tag{5}$$

In Eq. 5, μ is a balance between fidelity and sparse conditions parameter. If Eq. 3 is drag into Eq. 5, and apply the optimality conditions, after deformation the basic Bregman iteration model can be gotten, as shown in Eq. 6

$$\begin{aligned} u^{k+1} &= \min_{f} J(u) + \tfrac{\mu}{2}||Au - b^k||_2^2, \\ b^{k+1} &= b^k + b - Au^k \end{aligned} \tag{6}$$

In Eq. 6, $J(u) = \| u \|_1$.

3.3 The Split Bregman Iteration Method

When the two parts in Eq. 4 are considered separately, then Eq. 7 can be got.

$$\min_{u,d} |d| + L(u)\ such\ that\ d = \Phi(u) \tag{7}$$

Reference [10] gives the Split Bregman iteration method. In order to solve Eq. 7 it is necessary to convert it to unconstrained problem Eq. 8.

$$\min_{u,d} |d| + L(u) + \frac{\mu}{2}||d - \Phi(u)||_2^2 \tag{8}$$

According to the Ref. [8], assume $J(u, d) = |d| + L(u)$, then Eq. 8 is the specific application of Eq. 5. In order to strengthen the constraint condition, use Bregman iterative formula as two parameters, and Eq. 9 can be got.

$$(u^{k+1}, d^{k+1}) = \min_{u,d} D_J^p(u, u^k, d, d^k) + \frac{\mu}{2}||d - \Phi(u)||_2^2 \tag{9}$$

By applying of simplified form of Eq. 6, Split Bregman iterative algorithms Eq. 10 with two parts can be got.

$$(u^{k+1}, d^{k+1}) = \min_{u,d} |d| + L(u) + \frac{\mu}{2}||d - \Phi(u) - b^k||_2^2 \tag{10}$$

In order to effectively minimize the effect, Eq. 10 functional can be splitted into L1 part and L2 part, which are respectively on the minimization of parameters u and d. In order to solve the above problem, Eqs. 11 and 12 are needed.

$$u^{k+1} = \min_{u} L(u) + \frac{\mu}{2}||d^k - \Phi(u) - b^k||_2^2 \tag{11}$$

$$d^{k+1} = \min_{d} |d| + \frac{\mu}{2}||d - \Phi(u^{k+1}) - b^k||_2^2 \tag{12}$$

Based on the above description of the equations, the general Split Bregman algorithm can be described by Algorithm 1.

- Algorithm 1 Split Bregman Algorithm

$$\begin{aligned}
&Initialization : b^0, u^0, \mu, N, tol, lambda\\
&while||u^k - u^{k-1}||_2 > tol\\
&\quad for\ n = 1\ \ to\ N\ \ do\\
&\qquad u^{k+1} = \min_u L(u) + \tfrac{\mu}{2} \parallel d^k - \Phi(u) - b^k \parallel_2^2\\
&\qquad d^{k+1} = \min_d |d| + \tfrac{\mu}{2}||d - \Phi(u^{k+1}) - b^k \parallel_2^2\\
&\quad end\ for\\
&\quad b^{k+1} = b^k + (\Phi(u^{k+1}) - d^{k+1})\\
&end\ while
\end{aligned}$$

In Algorithm 1, two regularization factors are contained, namely the regularization parameter μ and the regularization splitting factor *lambda.* In Sect. 4, not only the effect of the regularization method for CT image reconstruction is verified, but also the influence of *lambda* for reconstruction effects are compared.

4 Numerical Experiments

The classic Shepp-Logan image is used to do experiment, and to compare the traditional ART algorithm with the Split Bregman regularization algorithm from view of reconstruction quality of the image. Fan beam of X-ray source was used in experiments, and the projection of sparse coverage of target was done. Assume that the image center and the center of rotation are the same point, the angle range of $0 \sim 360°$ are used to get 64 groups of projection data sampling for image reconstruction. The image size is 64×64. In this paper, the computer for experiment is Intel CPU 4 core processor, with frequency of 3.0 GHz and 4 GB of memory. In the process of actual projection and image reconstruction, noise is inevitable. So 5 % Gaussian random noise was added in the projection data. Set the iteration stop condition $\parallel u^{k+1} - u^k \parallel \leq 0.001$, and the reconstruction results are shown in Fig. 1.

To intuitive observation of Fig. 1, graph (a) is Shepp-Logan original image; Graph (b) is the result of reconstruction using classical ART algorithm (regular factor = 1); Graph (c) and (d) are the results of alternative Split Bregman iterative reconstruction algorithm. Graph (c), the splitting factor *lambda* is 0.35, (d) of the splitting factor *lambda* is 0.6. Obviously, the effect of graph (c) is better than graph (b), and graph (d) is the most close to graph (a). In the aspects of number of iterations, graph (b), the ART of the number of iterations is 100 times, and then both the number of iterations for the later two are only about six times. In order to more clearly compare the quality of the convergence reconstruction of all algorithms, Fig. 2 shows the norm residual of the above three kinds of reconstruction image after six times iteration. In Fig. 2,

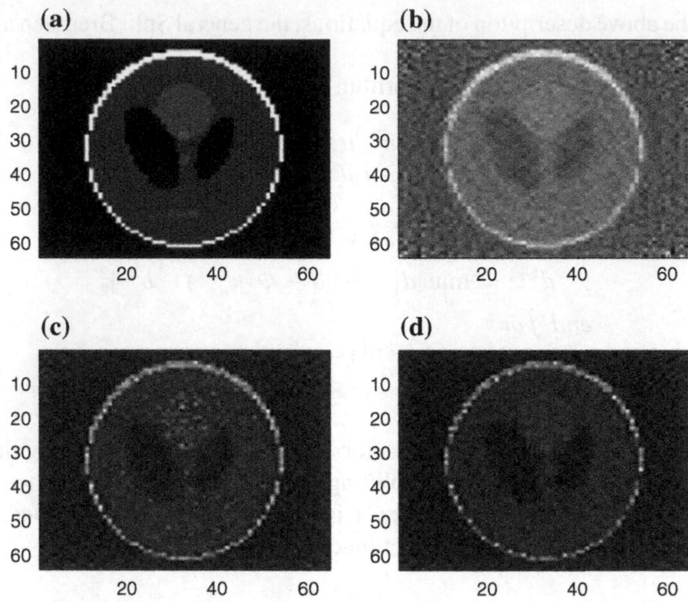

Fig. 1 Reconstruction images with noise. **a** Original image. **b** ART. **c** Split (lambda = 0.35). **d** Split (lambda = 0.6)

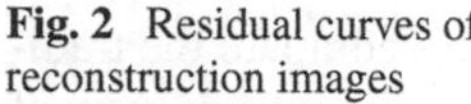

Fig. 2 Residual curves of reconstruction images

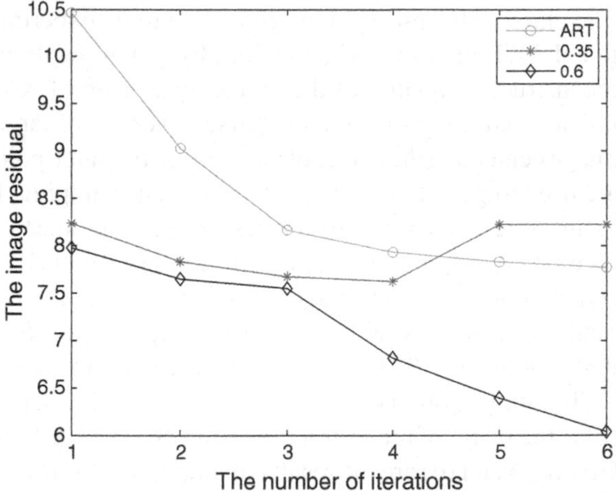

the horizontal axis is the number of iterations of the reconstruction algorithm, and the vertical coordinate shows the image Residuals relative to the real value.

As can be seen from Fig. 2, the residual of ART image reconstruction algorithm are still bigger after 6 times iteration; The convergence speed of residual for splitting factor 0.6 is far greater than that of the splitting factor 0.35. The regularization factor for the Split Bregman algorithm is 0.5.

5 Conclusion

Based on CT iterative reconstruction model and the actual needs of sparse projection reconstruction, this paper introduces "Split Bregman" framework to solve the L1 regular optimization problem of image reconstruction. Numerical experiments show that Split Bregman iteration method can be implemented within a few iterations of CT image reconstruction, and compared with the traditional algorithm of ART, the Split Bregman iterative reconstruction method may achieve good effect within a few iterations of CT image reconstruction with less artifacts. The Algorithm can effectively solve the similar L1 regular optimization problem.

Acknowledgments This research are financially supported by the Science and Technology Plan Projects of Lanzhou city, P. R. China (Grant No.214160) and the Youth Science Fund of Lanzhou Jiaotong University, P. R. China (Grant No.2012035).

References

1. Sidky EY, Kao CM, Pan X (2006) Accurate image reconstruction from few-views and limited-angle data in divergent-beam CT. J X-ray Sci Technol 14:119–139
2. Sidky EY, Pan X (2008) Image reconstruction in circular cone-beam computed tomography by constrained, total-variation minimization. Phys Med Biol 53:4777–5807
3. Tuy HK (1983) An inversion formula for cone-beam reconstruction. SIAM J Appl Math 43(3):542–546
4. Smith BD (1985) Image reconstruction from cone-beam projections:necessary and sufficient conditions and reconstruction methods. IEEE Trans Med Imag 4(1)
5. Gordon R (1974) A tutorial on ART (algebraic reconstruction techniques). IEEE Trans Nucl Sci 21(3):78–93
6. Donoho DL (2006) Compressed sensing. IEEE Trans Inf Theory 52(4):1289–1306
7. Bregman LM (1967) The relaxation method of finding the common point of convex sets and its application to the solution of problems in convex programming. USSR Comput Math Math Phys 7:200–217
8. Goldstein T, Osher S (2009) The split Bregman method for L1-regularized problems. SIAM J Imag Sci 2(2):323–343
9. Avinash CK, Slaney M (2001) Principles of Computerized Tomographic Imaging. Society for Industrial and Applied Mathematics, Philadelphia
10. Wang Y, Yin W, Zhang Y (2007) A fast algorithm for image deblurring with total variation regularization. CAAM Technical Reports (2007)

A Program Threat Judgement Model Based on Fuzzy Set Theory

Xiaochuan Zhang, Jianmin Pang, Yujia Zhang and Guanghui Liang

Abstract In the age of information security, whether a program is threatening or not is a crucial problem to solve. In this paper, a novel threat program judgment model based on fuzzy set theory is proposed. In the model, we derive a new evaluation function from multi-factor determined fuzzy synthetic function. Using the function, the program threat is evaluated by the membership of programs with great threat. Furthermore, we realize the judgment model and the experiment data shows its feasibility and effectiveness.

Keywords Program threat · Fuzzy synthetic functions · Fuzzy set theory

1 Introduction

Currently, two major problems at the technical level in the evaluation of threat program are as follows: (1) the data source is single, making it hard to describe the threat program comprehensively. The traditional method to evaluate a threat program is mainly based on the disassembly instructions features of the code or the function call sequences features. Due to the various forms that a malicious program manifests and the lack of comprehensive data sources indispensable for the threat evaluation, it is difficult for the data from very few sources to characterize the features

X. Zhang (✉) · J. Pang · Y. Zhang · G. Liang
State Key Laboratory of Mathematical Engineering and Advanced Computing, Advanced Computing and Parallel Processing Lab, Zhengzhou Information Science and Technology Institute, Zhengzhou, China
e-mail: zhangxiaochuan@outlook.com

J. Pang
e-mail: jianmin_pang@hotmail.com

Y. Zhang
e-mail: zhang52580684@vip.qq.com

G. Liang
e-mail: lghray1987@163.com

V.E. Balas et al. (eds.), *Information Technology and Intelligent Transportation Systems*, Advances in Intelligent Systems and Computing 455,
DOI 10.1007/978-3-319-38771-0_22

of the program; (2) the evaluation algorithm is so simple that the features of threat program cannot be depicted clearly. A threat program has its own characteristics. For instance, a malicious behavior can be implemented in different ways. In general, the threats brought by these different ways are not the same in the sense of the detect ability and cannot be simply accumulated, and thus, the traditional evaluation algorithm, unable to distinguish the different threats brought by the different implementations, can hardly demonstrate the comprehensive features of threat program.

To detect the malicious program, many efforts have been done [1–5]. In this paper, we focus on fuzzy pattern recognition. A lot progress has been made in this field. In [6], Zhang B et al. used fuzzy pattern recognition to detect unknown malicious code. They calculated Euclid degree of similarity between test file behaviors and malicious code behaviors. In [7], fuzzy pattern recognition was first used to abstract malicious characteristics of a program. Then the maliciousness of this program was quantitatively analyzed by probability method and whether the program is malicious was determined approximately using a fuzzy reasoning algorithm. However, the rate of false positives and false negatives can be further reduced.

This paper greatly improves the fuzzy synthetic function by introducing multi-factors determined fuzzy synthetic function. To solve the mentioned problems in the first paragraph, this paper presents a novel threat program judgment model based on fuzzy set theory which can completely distinguish malicious program. In specific, we derive an evaluation function is derived from two multi-factor determined fuzzy synthetic function, and a multi-source information collection framework for program behavior is built. First, this model collects multi-source data rather than single-source data. The multi-source information collection framework for program behavior including static analysis and dynamic analysis will collect full-scale information of a threat program. Second, the model, by exploiting the fuzzy set theory, establishes an evaluation function, which uses weighted average decomposition method and multi-factors determined decomposition method to gradually decompose the fuzzy concept of threat to the concrete behavior. So, we can evaluate a threat program more accurately. Finally, to verify the validity of the judgment model, we use it to distinguish some malicious programs and benign ones. The experimental data shows that our judgment model can completely make a right distinction between these two categories of programs with an acceptable running time. As far as the accuracy to detect the malicious program is concerned, our model is the best so far.

The paper is organized as follows. A brief review of some related work is provided in Sect. 2. The concept of classical fuzzy set theory is introduced and a multi-factor determined fuzzy synthetic function derived from it is presented in Sect. 3. A new threat program judgment model is proposed and the details of its design, solution and realization are given in Sect. 4. The experiments designed to test the validity of the model is presented in Sect. 5. Finally, we conclude the paper in Sect. 6.

2 Related Work

Static analysis technology of malicious code is based on the features of the code of disassembly instructions. Hiran V. Nath and Babu M. Mehtre [1] compared the static detection technology of malicious code which uses machine learning. Meanwhile, they pointed out that the bottlenecks of static detection techniques which use machine learning, data mining and other methods lie in the lack of samples, and difficulties in unpacking. Table 1 in [1] compares different methods of static analysis of malicious code.

With the wide use of packing technology and code obfuscation techniques, static analysis technology faces a lot of challenges. Dynamic analysis technique has thus emerged.

Generally, dynamic analysis puts the malicious code into a virtual environment to execute and acquires its behavior characteristics for further analysis. A dynamic analysis tool named TTAnalyze, which was developed by Bayer, Ulrich et al. [5], is able to acquire the behavior characteristics of executable program. The results of manual analysis show that TTAnalyze provides more accurate information than documentation [8] provided by Kaspersky Lab. But TTAnalyze is unable to identify the specific value of new-built registers.

M. Egele et al. compared a number of mainstream dynamic malicious code analysis tools, which is shown in Table 1 in [9]. It can be seen that these tools focus on distinct program behavior respectively. For instance, Anubis focuses on API calls, system calls, creation of processes or threads and the registry operations, etc., while Ether focuses on system calls and instruction sequences.

However, more and more attackers are using this increasingly popular technology to determine whether malicious code is executed in virtual environment, and thus the malicious code can evade the dynamic analysis. This whittles down the powers of dynamic analysis techniques.

The aforementioned methods collect information of programs by various means, and based on the information, determine whether they are malicious codes. But there still exist problems regarding the single source of data and natural disadvantages of the analysis method that are difficult to overcome. Meanwhile the accuracy of the analysis is expected to be further improved. Therefore, to solve the problems above, after integrating the advantages of static analysis with dynamic analysis, we design a program of multi-source information collection framework combining static and dynamic analysis, present and implement a program threat judgment model based on fuzzy set theory.

3 Fuzzy Set Theory

In this section, we introduce some related concepts and definitions to fuzzy set theory based on which we build a multi-factor determined fuzzy synthetic function.

3.1 Classical Fuzzy Set Theory

Fuzzy sets and membership Ordinary set is the collection of objects sharing certain attributes. The concept expressed by such objects is clear and well-defined. Therefore, the subordination of each object to the collection is also clear. However, there are many semantics of natural language terms which are not clear, and the attributes of many objects cannot be simply described by the answer of "yes" or "no". Thus, there goes the following definition: Suppose U is a collection of objects denoted generically by x and

$A = < x, \mu_A(x) > |x \in U \Lambda \mu_A(x) \in [0,1] \wedge \forall y(y \in U \wedge x = y \rightarrow \mu_A(x) = \mu_A(y))$.

Then we call A a fuzzy set in U, and $\mu_A(x)$ a membership function which associates each object in U with a real number in the interval $[0, 1]$. The value of $\mu_A(x)$ represents the degree of membership of x in A.

Fuzzy synthetic evaluation model To evaluate a specific object properly, we should evaluate all the factors which affect the object firstly, and then integrate them. Now we focus on threat program judgment model based on this idea. From the above observations, the threat of the programs analyzed should firstly be decomposed into "Starting performance", "Hiding performance" and "Self-protection performance" and other aspects. Each aspect should be evaluated and then integrated.

We first define a useful function.

Definition: assume a function $\mu : [0,1]^n \rightarrow [0,1]$, satisfy the following conditions:

(1) Regularity: if $\mu_1 = \mu_2 = \cdots = \mu_n = k$, then $\mu(\mu_1, \mu_2, \ldots, \mu_n) = k$;

(2) Monotonically increasing: $\mu(\mu_1, \mu_2, \ldots, \mu_n)$ is monotonically increasing on all variables. That is to say, for arbitrary i, if $\mu_i^1 \leq \mu_i^2$, then

$\mu(\mu_1, \ldots, \mu_(i-1), \mu_i^1, \mu_{i+1}, \ldots, \mu_n) \leq \mu(\mu_1, \ldots, \mu_{i-1}, \mu_i^2, \mu_{i+1}, \ldots, \mu_n)$

(3) Continuity: $\mu(\mu_1, \mu_2, \ldots, \mu_n)$ on all variables are continuous.

Then we call μ is an n-ary fuzzy synthetic function.

The weighted average fuzzy synthetic function is a frequently-used fuzzy synthetic function and has been widely used in the threat program judgment model studied in this paper.

Assume $K = (k_1, k_2, ..., k_n) \in [0,1]^n$ is a normalized weight vector ($\Sigma_(i = 1)^n k_i = 1$), and $\forall(\mu_{A_1}(x), \mu_{A_2}(x), ..., \mu_{A_n}(x)) \in [0,1]^n$. Let

$\mu_A(x) = \mu_\sigma(\mu_{A_1}(x), \mu_{A_2}(x), \ldots, \mu_{A_n}(x)) = \sigma_(i = 1)^n k_i \mu_{A_i}(x)$

where $A_1, A_2, \ldots, A_n$ are called weighted average decomposition of fuzzy set A on the base set U and μ_σ is called the weighted average of fuzzy synthetic function, the value of which represents the degree of membership of x in A. The value of $\mu_{A_i}(x)$ is the degree of membership of the ith fuzzy set which is decomposed from A, and k_i can be explained as the weight of ith fuzzy set in synthetic evaluation.

3.2 Multi-factors Determined Fuzzy Synthetic Function

The weighted average fuzzy synthetic function encounters some problems when directly applied to the threat program judgment model studied in this paper. When a fuzzy set is decomposed into several independent fuzzy sets and forms sufficient condition for the original fuzzy set, the weighted average fuzzy synthetic function including geometric average, single factor determined fuzzy synthetic function [9] will be no longer effective. For example, a program whose starting performance through modifying registry is good and starting performance through loading drivers is good can inevitably be described as a program whose starting performance is good. We assume that membership degree of program A in fuzzy set, starting performance through modifying registry is good, is 1. At this time, membership degree of program A in fuzzy set, starting performance is good, should also be 1. But the membership degree of program A in fuzzy set, starting performance through loading drivers is good, is 0, that means A doesnt load any drivers at all. If we still use weighted average fuzzy synthetic function, we will get a wrong result that the membership degree of program A in fuzzy set, starting performance is good, is between (0, 1). Hence, the weighted average fuzzy synthetic function is inappropriate in dealing with such problems. To solve this problem, this paper presents the concept of multi-factors determined fuzzy synthetic function.

The definition of multi-factors determined fuzzy synthetic function Assuming that $A, A_1, A_2, \ldots A_n$ are fuzzy sets on the base set U, where A_1, A_2, A_n are independent and are all sufficient conditions for fuzzy set A. Then we call A_1, A_2, A_n are the multi-factors determined decomposition of fuzzy set Aon the base set U. The membership function of it, is called multi-factors determined fuzzy synthetic function, is:

$\mu_A(x) = \mu_\cup(\mu_{A_1}(x), \mu_{A_2}(x), \ldots, \mu_{A_n}(x)) = 1 - \prod(i = 0)^n(1 - \mu(A_i)(x)).$

Next we explain how to derive this formula.

The foundation of multi-factors determined fuzzy synthetic function Frequency stability presented in fuzzy statistics tests can undertake the objective meaning of membership degree [9, 10]. Therefore, the degree of membership is able to be explained as the probability of the specific object belonging to a specific fuzzy set. Based on this property, we give the foundation of multi-factors determined fuzzy synthetic function as follows.

Assume: A_1, A_2, A_n are multi-factors determined decomposition of fuzzy set A on the base set U and $\mu_A(x)$ is able to be explained as the probability of $x \in A$. Obviously, according to the definition of the multi-factors determined decomposition, the probability of $x \in A$ equals to the probability of $x \in A_1 \cup A_2 \cup \ldots \cup A_n$, that is,

$\mu_A(x) = \mu_\cup(\mu_{A_1}(x), \mu_{A_2}(x), \ldots, \mu_{A_n}(x))$

Fuzzy set A^C is the complement or negation of A. For $\forall x \in U$, then [11]

$$\mu(A^C)(x) = 1 - \mu_A(x)$$
$$= 1 - \mu_{\cup}(\mu_{A_1}(x), \mu_{A_2}(x), \ldots, \mu_{A_n}(x))$$
$$= \mu_{\cap}(\mu_{A_1}(x), \mu_{A_2}(x), \ldots, \mu_{A_n}(x)) \quad (1)$$

where $A_1, A_2, \ldots, A_n$ are independent. According to the Bayes formula 12, we have

$$\mu(A^C)(x) = \mu_{\cap}(\mu_{A_1}(x), \mu_{A_2}(x), \ldots, \mu_{A_n}(x))$$
$$= \mu(A_1)^c \cap A_2^c, \ldots \cap A_n^c)(x)$$
$$= \mu(A_1)^c)(x) \cdot \mu_{A_2}^c)(x) \ldots \cdot \mu_(A_n)^c)(x)$$
$$= \Pi(i = 0)^n \mu_{A_i}^c)(x) \quad (2)$$

Combine (1) and (2) and we obtain the result,
$\mu_A(x) = 1 - \Pi(i = 0)^n(1 - \mu_{A_i}(x))$
Notably, the foundation of multi-factors determined fuzzy synthetic function contains intersection and union operations in fuzzy set theory. In the classical fuzzy set theory presented by L.A. Zadeh, intersection operation is defined as $\mu_(A \cap B)(x) = min?(\mu_A(x), \mu_B(x))$, and union operation is defined as $\mu_(A \cup B)(x) = max?(\mu_A(x), \mu_B(x))$. Because fuzzy and random events are all uncertain events, there are many similarities between them. The difference is that for a random event, the event itself is an uncertain event, while for a fuzzy event, the concept of it is not clearly defined, whether the object belongs to this concept is difficult to determine. But this does not prevent using the knowledge of probability theory for solving the fuzzy sets problems. Combining the knowledge of probability theory, we give a new intersection and union algorithm in fuzzy set theory (in 5 [13] was defined as the algebraic product and sum). On this basis, this paper deduces new intersection and union algorithms, and provides an alternative way which can be used for reference in fuzzy set theory. The formula of multi-factors determined fuzzy synthetic function also fits the principle of inclusion for fuzzy sets presented by Kukov, Mria and Mirko Navara [14].

4 Program Threat Judgment Model

On the basis of fuzzy set theory, to solve the problem in program threat judgment, this paper presents a threat program judgment model which is called FSTJ (Fuzzy Set theory based Threat program Judgment model).

4.1 Design of FSTJ

Selection of base set and fuzzy set In FSTJ, the base set denoted by U is taken as the set of all programs being tested. The concept of "program with great threat" takes as a fuzzy set, which is denoted by A, on the base set U.

Inputs and outputs of FSTJ The inputs of FSTJ are the data sets of collected program behaviors. These data sets of program behaviors are collected from the program behaviors collection framework and can reflect the threat of program.

The output of FSTJ is the degree of membership μ_A of the program being tested in the fuzzy set Awhich is defined as the concept of "program with great threat". Here, U is the base set of all the programs being tested.

Decomposition of fuzzy set A For the malicious code, its main behavior characteristics include starting, hiding, self-protection and destroying. Each aspect contains many implementations. In starting method, for example, it can be achieved by modifying the configuration files, registry, ActiveX, or setting malware itself as a system service or a device driver, etc. Also, in every way they can achieve it with a variety of techniques, such as API calls through ring 3 level or kernel function calls though ring 0 level, etc. Therefore, we should decompose the fuzzy set Athrough weighted average decomposition and multi-factors determined decomposition to a specific behavior which can be obtained by the program behaviors collection framework. If a fuzzy set can be decomposed to several specific behaviors, and these behaviors are independent and form sufficient conditions for the original fuzzy set, this decomposition should be multi-factors determined decomposition. In FSTJ, schematic diagram of fuzzy sets decomposition is shown in Fig. 1. Arrow pointing to the right represents the weighted average decomposition, while arrow pointing to the left represents the multi-factors determined decomposition.

4.2 Solving FSTJ

According to the description in Sects. 3.1 and 3.2, the algorithm calculating the membership μ_A of a certain program tested in fuzzy set A, "the program with large threat", is shown in Fig. 2.

4.3 Achieving FSTJ

JSTP (Judgment System of Threat Program), is a judging system of program threat based on FSTJ. Its framework is shown in Fig. 3. This section will detailedly present some key technical issues of the system.

The basis of threat program judgment model is the behavior characteristics of the program, so we need a complete program behaviors collection framework which is

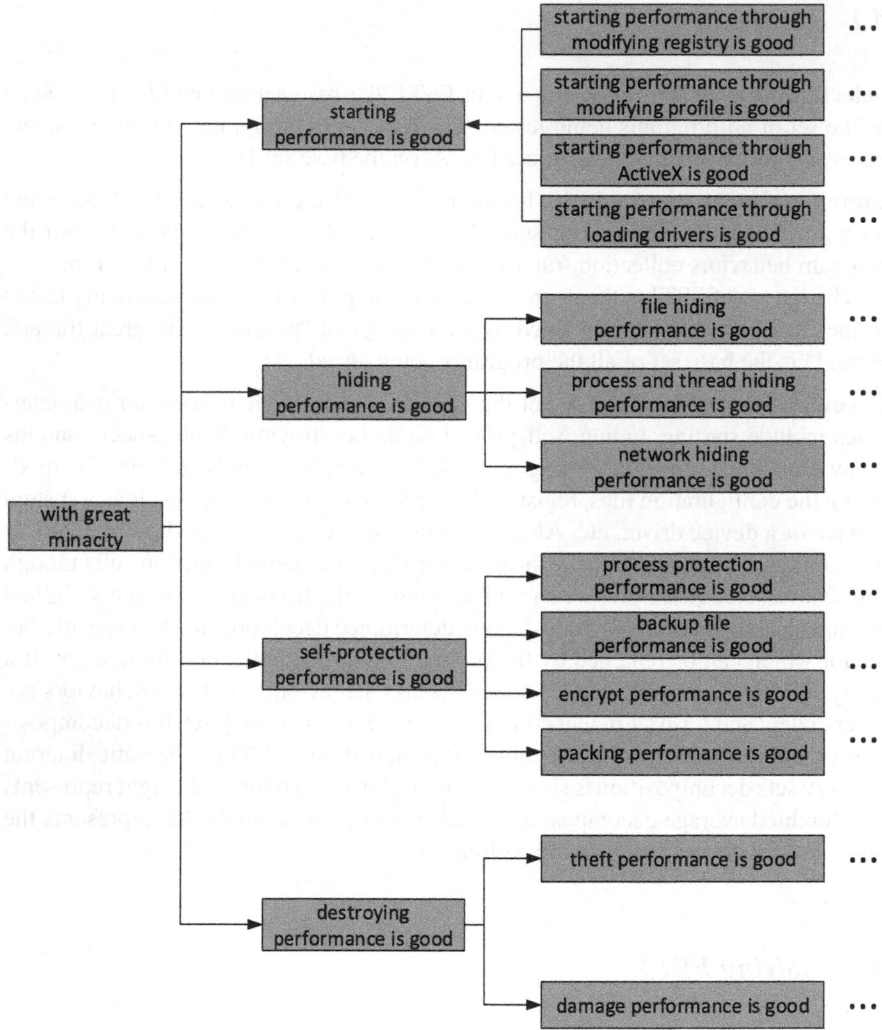

Fig. 1 Schematic diagram of fuzzy sets decomposition in FSTJ

adapted to the threat judgment. The establishment of JSTP is based on the multi-source information collection framework of program behaviors. This framework is divided into static analysis module and dynamic analysis module.

Static analysis module gives a comprehensive evaluation of the programs, mainly focusing on their self-protection ability. Static analysis module first disassembles the program being tested, then, analyzes control flow and data flow, and gets comprehensive and accurate description of it. At the same time, the capabilities of self-protection can be detected and capabilities against static analysis could be verified.

```
Func CountWeightedAverage
    Input: fuzzy set F being counted and program behaviors set B
    Output: the grade of membership μ_F of the program being tested in the fuzzy set F
{
    decompose F into F_1,F_2,…,F_n through weighted average decomposition;
    for each F_i in F_1, F_2, … , F_n
    {
        if(F_i can be decomposed through weighted average decomposition)
            result_i= CountWeightedAverage(F_i);
        if( F_i can be decomposed through multi-factors determined decomposition)
            result_i= CountMulti-factorsDetermined(F_i);
    }
    result = Σ_{i=1}^{n} k_i · result_i;
    return result;
}
```

```
Func CountMulti-factorsDetermined
    Input: fuzzy set F being counted and program behaviors set B
    Output: the grade of membership μ_F of the program being tested in the fuzzy set F
{
    if (fuzzy set F can be decomposed to a specific behavior which can be got by the program behaviors collection framework through multi-factors determined decomposition)
        result = 1 − Π_{i=1}^{n} (1 − μ_{F_i}(x));
    else
    {
        decompose F into F_1 , F_2 , … , F_n through multi-factors determined decomposition;
        for each F_i in F_1, F_2, … , F_n
            if(F_i can be decomposed through weighted average decomposition)
                result_i = CountWeightedAverage(F_i);
            if( F_i can be decomposed through multi-factors determined decomposition)
                result_i = CountMulti-factorsDetermined(F_i);
        result = 1 - Π_{i=1}^{n}(1 − result_i);
    }
    return result;
}
```

Fig. 2 Calculate the membership μ_A

Dynamic analysis module is mainly used to detect a variety of characteristics shown by the tested program in the run time. For the implementation of this module, we need to collect execution characteristics from several hierarchies, including network, the host application layer, the kernel layer and IRP sequences. Dynamic analysis framework is shown in Fig. 4.

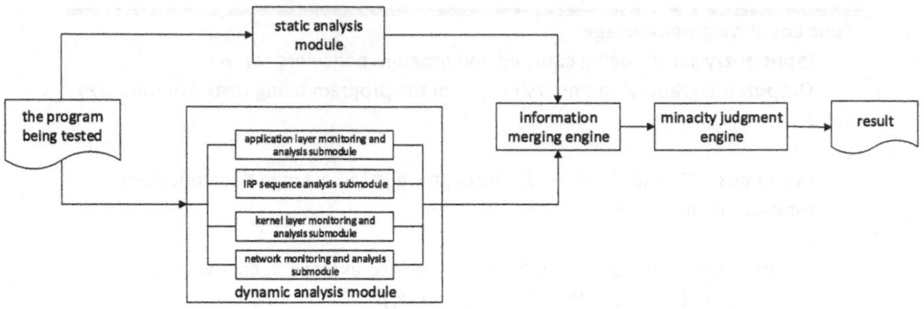

Fig. 3 Framework of JSTP

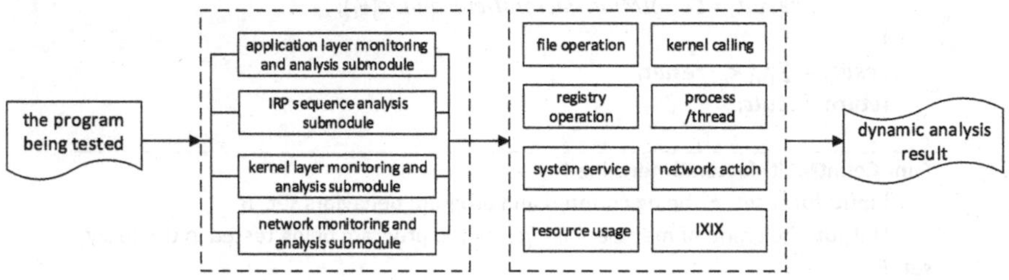

Fig. 4 Dynamic analysis framework

After static analysis and dynamic analysis, we send the data collected by the two modules to the information merging engine for synthetic analysis. We merge the data from different sources and with different format in the information merging engine, and then send the merged data to threat judgment engine. Finally, we get result from threat judgment engine according to fuzzy set theory.

4.4 *Case Study*

We take a Trojan horse, named Final Fantasy, as a case study. We detect several potential malicious behaviors of Final Fantasy by JSTP, which is shown in Table 1.

We decompose Final Fantasys threat as shown in Fig. 5. The numbers on connectors are their weights of target fuzzy sets to the origin fuzzy sets. The green boxes represent specific behaviors. And numbers in red boxes behind green boxes are the degrees of memberships of these behaviors in corresponding fuzzy sets, representing degree of specific behaviors belong to corresponding fuzzy sets.

Now, we can calculate the degree of Final Fantasys threat. We should calculate each degree of membership of Final Fantasy in all fuzzy sets in Fig. 5. The results is shown in parentheses at blue boxes in Fig. 5.

Thus, Final Fantasys minacity equals $0.3 \times 0.8 + 0.2 \times 0.81892 + 0.20.4 + 0.3 \times 0.44 = 0.615784$.

Table 1 Potential malicious behaviors of Final Fantasy

load driverFinal Fantasy_Service
inject remote thread to Internet Explore
hide fileC:\ WINDOWS \ system32\ 6to4ex.sys
delete fileC:\ WINDOWS\ system32\ Drivers\ beep.sys
delete fileC:\ DOCUME 1\ ADMINI 1\ LOCALS 1\ Temp\\ release.tmp
delete itself
hide network protocol via 80 port
move file: FinalFantasy.exe from "C:\ DOCUME 1\ ADMINI 1\ Desktop" to "C:\ WINDOWS"
use Base64 as well as CRC32 algorithm to encrypt
steal file: C:\ DOCUME 1\ ADMINI 1\ DOCUME 1\ 1.docx
turn on the camera

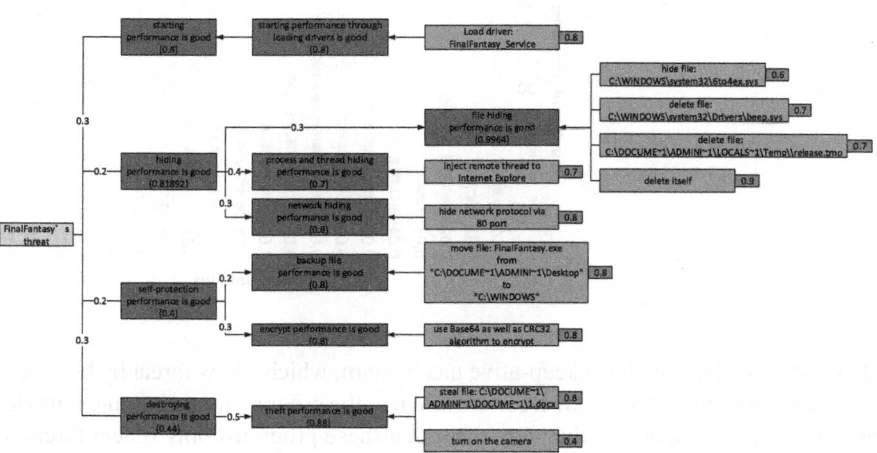

Fig. 5 Final Fantasys threat decomposition

5 Experiments

There are 468 samples, 146 of which were malicious programs gleaned from the Internet and the remaining 322 were benign programs. Experiment platform is Windows 7 ultimate SP1 32 Bit. Test results for threat judgment of test samples are shown below.

It can be seen from Fig. 6 that membership of malicious programs in fuzzy sets, "program with great threat", is larger than 0.5, which reflects the threat of malicious programs is not only shown in a single aspect, but usually in multiple aspects.

And it can be seen from Fig. 7 that test results of ordinary benign programs are less than 0.2, and most appear between 0.04 and 0.11. This is because some of these benign programs will modify registry or load service, making it run automatically at boot time; or the program itself is an installation file which will release new files; or

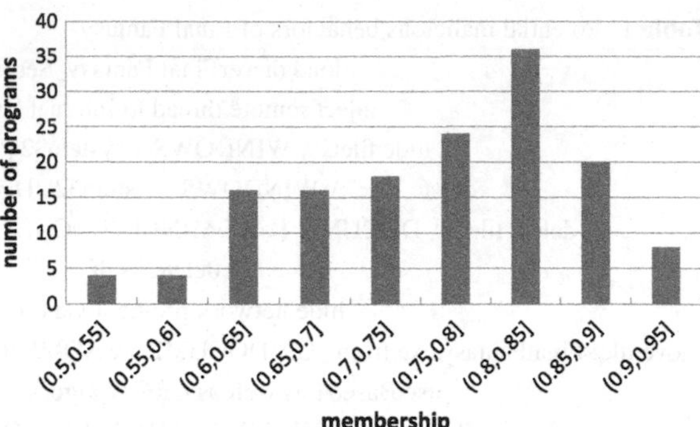

Fig. 6 Test results for malicious programs

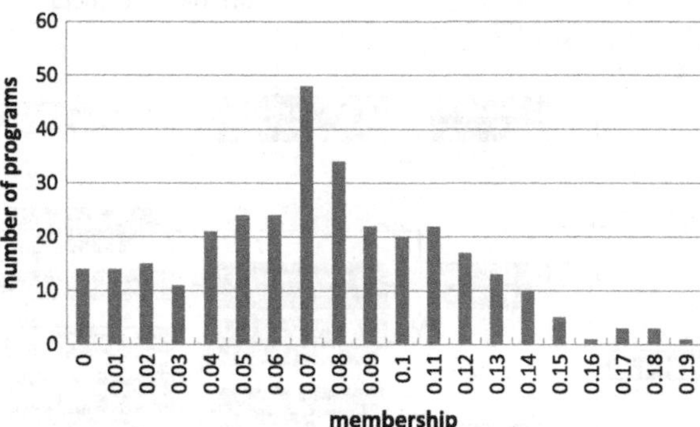

Fig. 7 Test results for benign programs

chat softwares like QQ have keep-alive mechanism, which throw threat by heartbeat behavior. All of these have shown certain threat in the program threat judgment model presented in this paper, but due to the fact that these programs only reflect threat in a small number of aspects, the memberships obtained of their synthetic judgment were not high. This caused an obvious distinction between malicious programs and benign programs.

To sum up, if we choose any of these values of results between [0.3, 0.4] as a standard to judge whether a program is malicious or not, test results show that both false positives and false negatives of JSTP are zeros in our judgment model. Therefore, JSTP is an objective and accurate model in malicious program judgment, and the experiment verified the feasibility and effectiveness of FSTJ.

6 Summary

Because of the advantages of fuzzy set theory in dealing with ambiguous and imprecise concepts, this paper presents and realizes a threat program judgment model FSTJ based on fuzzy set theory, which solves the problems that exists in the application of traditional fuzzy synthetic functions by defining multi-factors determined fuzzy synthetic function. The experimental results show that the JSTP is objective and accurate in the program threat judgment and also prove the feasibility and effectiveness of FSTJ. The future work may introduce machine learning to FSTJ for better result. The decomposition methods of the fuzzy set presented in this paper can also be applied to other areas, such as pattern recognition, medical diagnostics, as well as intelligence-gathering, etc.

Acknowledgments This work is supported by National Natural Science Foundation of China under Grant No. 61472447, and also supported by Shanghai Commission of Science and Technology Research Project under Grant No. 13DZ1108800.

References

1. Nath H, Mehtre BM (2014) Static malware analysis using machine learning methods recent trends in computer networks and distributed systems security. Springer, Heidelberg
2. Feng Y, Anand S, Dillig I (2014) Apposcopy: semantics-based detection of android malware through static analysis. In: Proceedings of the 22nd ACM SIGSOFT international symposium on foundations of software engineering. ACM, pp 576–587
3. Kruegel C (2014) Full System Emulation: achieving successful automated dynamic analysis of evasive malware. In: Proceeding of BlackHat USA security conference
4. Vinod P, Jain H, Golecha YK, Gaur SM, Laxmi V (2010) MEDUSA: MEtamorphic malware dynamic analysis usingsignature from API. In: Proceedings of the 3rd international conference on security of information and networks. ACM, pp 263–269
5. Bayer U, Moser A, Kruegel C, Kirda E (2006) Dynamic analysis of malicious code. J Comput Virol 2(1):67–77
6. Zhang B, Yin J, Hao J (2005) Using fuzzy pattern recognition to detect unknown malicious executables code., Fuzzy systems and knowledge discovery. Springer, Heidelberg
7. Fu W, Wei B, ZHAO R, Pang J (2010) Fuzzy reasoning model for analysis of program maliciousness. J Commun, vol 31(1)
8. Kaspersky Lab: antivirus software (2006). http://www.kaspersky.com/
9. Egele M, Scholte T, Kirda E, Kruegel C (2012) A survey on automated dynamic malware-analysis techniques and tools. ACM Comput Surv 44:1C49
10. Baoqing Hu (2010) Fundamentals of fuzzy theory. Fuzzy systems and mathematics, 2nd edn. pp 45–45
11. Zadeh Lotfi A (1965) Fuzzy sets. Inf control 8(3):338–353
12. Thomas B (1763) An essay towards solving a problem in the doctrine of chances. By the late Rev. Mr. Bayes, FRS communicated by Mr. Price, in a letter to John Canton, A.M.F.R.S. Philosophical Transactions, Giving Some Account of the Present Undertakings, Studies and Labours of the Ingenious in Many Considerable Parts of the World 53:370C418
13. Zadeh LA (1968) Probability measures of fuzzy events. J Math Anal Appl 23(2):421–427
14. Kukov M, Mirko N (2013) Principles of inclusion and exclusion for fuzzy sets. Fuzzy Sets Syst 232:98–109

Implementation of Adaptive Detection Threshold in Digital Channelized Receiver Based on FPGA

Leilei Jin and Haiqing Jiang

Abstract Aimed at the problem of adaptive detection threshold in the digital channelized receiver, this paper introduces a threshold generating method based on sorting statistics, which utilizes the magnitude vector in small serial number after ranking output amplitude of sub-channels from small to large order as a noise sample to form the threshold, but sorting operation of this method takes up a large number of FPGA resources. When intercepting and receiving the sky signal, this paper puts forward a simplified and improved method based on characteristics of clean sky signal environment and pulse signal environment. The sliding adaptive threshold method takes the forepart signal of the same channel as a noise sample to form the threshold.

Keywords Digital channelized receiver · Sliding adaptive threshold · FPGA

1 Introduction

The Future war is information warfare, and the digital channelized receiver is a class of the most widely used electronic warfare receivers currently [1]. In the digital channelized receiver, signal detection is accomplished in sub-channels after channelization. The traditional signal detection method always adopts fixed detection threshold. Although fixed detection threshold is easy to be carried out, it has disadvantages of poorer adaptability to signal environment in practical application [2]. To compensate for this shortcoming, people propose the changeable threshold to detect the signal. The adaptive threshold method can automatically generate a detection threshold for subsequent signal detection based on the current signal characteristics, which can be better adapted to the practical changing signal environment [3]. The literature [4] describes a way to use the "silent" channel to estimate the noise

L. Jin · H. Jiang (✉)
School of Information and Electronics, Beijing Institute of Technology, Beijing, China
e-mail: haiqingd99@bit.edu.cn

L. Jin
e-mail: jinlei330@126.com

V.E. Balas et al. (eds.), *Information Technology and Intelligent Transportation Systems*, Advances in Intelligent Systems and Computing 455,
DOI 10.1007/978-3-319-38771-0_23

variance, but this method is biased estimation and has a greater error. Aimed at this problem, a noise variance estimation method based on sorting statistics achieves the unbiased estimation and has been applied in practice; Further on, when intercepting and receiving sky signal, according to the characteristics of the clean sky signal environment and pulse signal environment, this paper proposes a sliding adaptive threshold method which proves an effective way in practice.

2 The Adaptive Threshold Based on Sorting Statistics

2.1 *Basic Idea of the Method*

It is considered that as long as we divide the sub-channels into odd and even groups, the sub-channel outputs in each group are uncorrelated [5]. For the same sub-channel output, extract output data in time domain, and when the decimation factor is large enough to make no overlapping between the data blocks, the output data from the same sub-channel is completely independent to each other. In the following discussion, it is assumed that the odd and even grouping and time-domain extraction have been used in sub-channels, namely that independent samples are obtained. And take $K/2$ sub-channels grouped odd (or even) for example. The parameter $k^{'}$ and the parameter $m^{'}$ represent the signal channel number after grouping and sampling time after extraction, respectively.

The amplitude matrix of the channel outputs is defined as

$$\mathbf{Z} = \begin{bmatrix} Z(1,1), Z(1,2)\cdots Z(1,m')\cdots \\ Z(2,1), Z(2,2)\cdots Z(2,m')\cdots \\ \vdots \\ Z(K/2,1), Z(K/2,2)\cdots Z(K/2,m')\cdots \end{bmatrix} \tag{1}$$

For each m', after sorting the column vector of Z by the magnitude from small to large order, we can get a new matrix composed of sorting statistics which is defined as

$$\mathbf{Z}' = \begin{bmatrix} Z^1(1), Z^1(2)\cdots Z^1(m')\cdots \\ Z^2(1), Z^2(2)\cdots Z^2(m')\cdots \\ \vdots \\ Z^{K/2}(1), Z^{K/2}(2)\cdots Z^{K/2}(m')\cdots \end{bmatrix} \tag{2}$$

In Eq. (2), $Z^i(m')$ is the output amplitude of the sub-channel with serial number i after sorting all odd (or even) sub-channel outputs by the magnitude from small to large order at the moment m'. The row vector of Z' is composed of the sub-channel outputs with serial number i at every moment, and it has the form as follows:

$$\mathbf{Z^i} = [Z^i(1), Z^i(2) \cdots Z^i(+\infty)] \tag{3}$$

After channelization, SNR significantly increases, making the signal samples with higher probability greater than the noise samples. So Z^i with a smaller numberi is made up of noise samples, and we can use the noise samples in these vectors to estimate noise variance when signal is existing.

2.2 *Implementation of the Method*

Since the front-end ADC sample rate is above 2 GHz, FPGA needs to adopt multiplexing structure in the process of achieving channelization. For instance, the digital channelized receiver with 256 channels can use the structure of the 32 channels multiplexing eight times. In the part of Fast Fourier Transformation, output channel numbers are all odd or even at each beat, which ensures the 32channels of each beat are uncorrelated. Take 32 channels of each beat as a group, so all channels are divided into eight groups. First of all, find the minimum value of each group respectively, and then get a sum of N points and figure out their mean respectively. Next select the maximum from the results of eight groups. At last multiply the maximum by the adjustment factor and get the signal detection threshold. We divide 256 channels into eight groups instead of two groups for the reason that the analog front end is not always white noise. To a certain extent, it has a better performance in amplitude fluctuation. Another reason is that it is easy to accomplish multiplexing structure. Its implementation process is shown in Fig. 1.

Determination of multiplying factor: Firstly, collect the inputs of noise amplitude in this module. Secondly save the data into a data file, and figure out the standard deviation of the group data sequence by mathematical software. Finally divide the standard deviation by the mean calculated by FPGA, and the quotient is multiplying factor. The multiplying factor can be slightly larger to guarantee that the noise will not be detected.

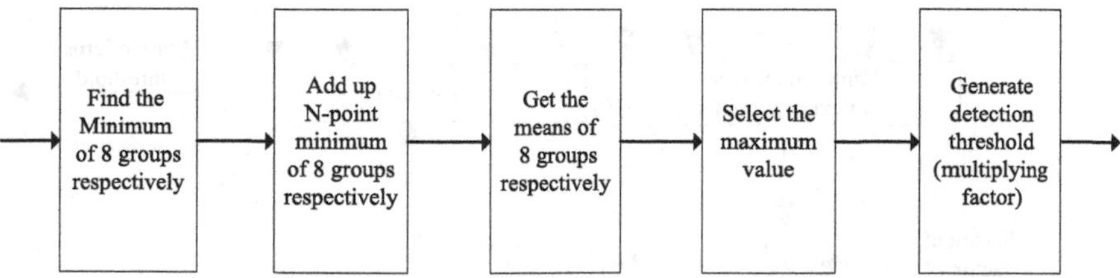

Fig. 1 Processing flow of adaptive threshold generation

3 The Sliding Adaptive Threshold

3.1 Basic Idea of the Method

The sorting operation of the adaptive threshold based on sorting statistics takes up a large number of FPGA resources. When intercepting and receiving the sky signal, the sky signal environment is relatively clean and signal form is pulse, so we can always find a piece of signal with noise only. Therefore we can eliminate the sorting process and directly use the noise sequence of this channel to generate the threshold [6]. The basic principle of sliding adaptive detection threshold is to take the forepart signal of the same channel as a sample to calculate a detection threshold which is used to detect the subsequent signal. In order to prevent from using the threshold calculated by signal amplitude as detection threshold, we need to filter results according to some rule. Figure out the mean of every L points within N duration continuously, make a comparison among the N mean, and select the minimum value. Multiply the minimum value by the adjustment factor and get the product. Then add the product to the offset compensation factor and obtain the detection threshold. This detection threshold is used for the signal detection of next N duration, at the time of signal detection, take the average of every L points within this N duration and work out the detection threshold for next N duration. Slide the sample constantly like this, and get new detection threshold constantly. Where the size of L is determined by the PRI lower limit of the system and the size of N is determined by L and maximum pulse width of the system. Specifically in order to ensure that noise samples do not contain the pulse signal, L is less than twice of radar pulse width. In limiting case the duty ratio of radar pulse signal is not more than 20 %, namely that L is less than 40 % of PRI. The size of N is not less than $(PW_{max}/L) + 2$ (Fig. 2).

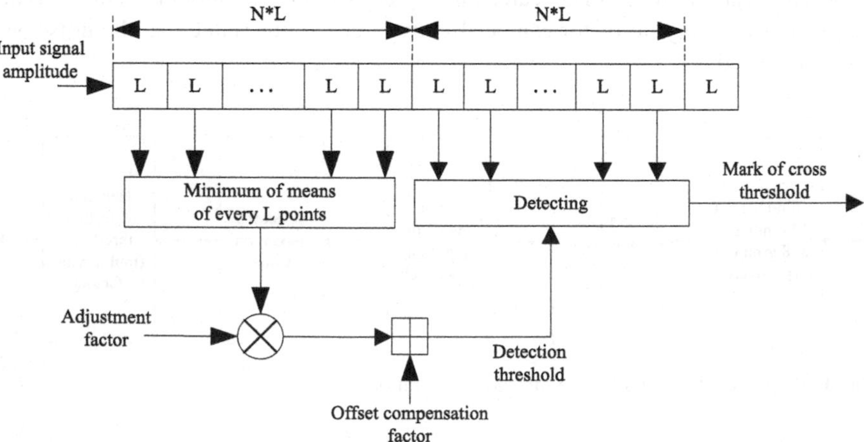

Fig. 2 Sliding adaptive threshold generating and signal detecting

We could calculate sliding adaptive detection threshold DT_a by using the following equation:

$$DT_a = k \times \bar{x} + \delta \quad (4)$$

Where, $\bar{x}$ is the mean value of the sample sequence, k is the adjustment factor, and δ is the offset compensation factor.

3.2 Implementation of the Method

Similarly, we assume that there are 256 channels in the digital channelized receiver, and when implemented in FPGA, it can use the structure of the 32 channels multiplexing 8 times. Sliding adaptive threshold module adopts the same multiplexing structure. First of all, extract the instantaneous amplitude of each channel receptively. Then get a sum of L points of each sub-channel and figure out the mean receptively. Next wait for the mean of the following L points, and get the smaller one of the two mean values. Make a comparison between the smaller one and the mean of the third L points and get the smaller one until N duration. Finally obtain the smallest mean value among N duration and get the detection threshold based on Eq. (4).

In FPGA implementation, solve the count of L from the equation $L = F_{clk} \times 40\,\% \times PRI_{min}$, where F_{clk} is the system clock and PRI_{min} is the system PRI lower limit. Considering the division in FPGA, it needs to amend the counter value as 2^M. And it needs to cut off M bits of the sum to get the mean. The size of N is determined by the maximum of pulse width and L, and N is calculated by the equation $N = (PW_{max} \times F_{clk}/L) + 2$, where PW_{max} is the maximum of pulse width.

Determination of adjustment factor k: Firstly, collect the inputs of noise amplitude in this module and save the data into a data file. Secondly figure out the standard deviation of the group data sequence by mathematical software. Finally divide the standard deviation by the mean calculated by FPGA, and the quotient is the adjustment factor. For the determination of offset compensation factor, the front-end tunable microwave module of the system will increase the attenuation to guarantee that the ADC does not overflow in high SNR environment. Therefore the noise floor is relatively lower. In order to prevent false-alarm probability over high, it needs to set the offset compensation factor.

4 Comparison of Two Methods

Two methods are validated by the engineering practice, and method 2 is simplified and improved on the basis of method 1 according to the characteristics of the sky signal environment. The specific characteristics of the two methods are as follows:

(1) Method 1 uses the sequence of sub-channel minimum values after channelization as a noise sample to estimate the threshold, which can adapt to the continuous wave signal and pulse signal environment, so method 1 is not restricted by signal environment. Method 2 is searching noise samples in their respective sub-channels, and only under the pulse signal environment it is possible to obtain noise samples to generate the detection threshold. So the method 2 is only applicable to the pulse signal environment;
(2) The sorting operation of method 1 takes up a large number of FPGA resources. It uses space for time of parallel sorting method in sorting algorithm with FPGA. If the total data number required sorting is N, the comparator required in parallel sorting algorithm is N*(N-1). Each comparator occupies 5 logic resources in the FPGA design, and then the required logical units for sorting are 5*N*(N-1) [7]. Method 2 has no sorting operation, so its implementation is simple;
(3) All sub-channels of method 1 are unified with one threshold, but in practical engineering, this threshold may not apply to every subchannel for the reason that the analog front end is not always white noise. While in method 2, each sub-channel uses its own threshold, which has good adaptability to the amplitude fluctuation between sub-channels.

5 Conclusion

This paper studies the problem of adaptive detection threshold in digital channelized receiver, and introduces an adaptive threshold method based on sorting statistics. According to the characteristics of the sky signal environment, this paper proposes a simplified and improved method, the sliding adaptive threshold has been applied in practice, which proves to be an effective way.

References

1. Xiang S, Qixiang F, Li Y (2014) The hotpots and development trend of wideband digital channelized receiver technology. Aerosp Electron Warf 1:32–35
2. Huang W (2010) Study on signal detection and adapative threshold in digital receiver. Chongqing University, Chongqing, pp 24–25
3. Yucheng W, Chen N, Gao S (2007) Adapative threshold signal detection method for burst communication. J Electron Inf Technol 29(12):2896–2898
4. Taherpour A, Gabor S, Nasiri-Kenari M (2008) Wideband spectrum sensing in unknown white gaussian noise. IET Commun 6(2):736–771
5. Burke BT (2000) Subband channelized radar detection and bandwidth estimation for FPGA implementation. University of Illiions at Urbana-Champaign, Illinois
6. Zou G (2015) Design and implementation of signal reconnaissance and sampling processor of ultra-wideband radar. Beijing Institute of Technology, Beijing
7. Shi Y, Jin C (2013) The parallel comparison sorting algorithm based on FPGA. Digit Technol Appl 10:126–127

Design and Implementation of Wireless Sensor Network Gateway with Multimode Access

Huiling Zhou, Yuhao Liu and Yang Xu

Abstract Wireless sensor network has the characteristics of large-scale, ad-hoc network and wireless communication, it is one of research hotspots. The gateway plays a very important role as a conversion device between different protocols in the network. A kind of gateway with Ethernet, WI-FI, GPRS three access modes has been designed and realized in this paper, it realized the intelligent management of wireless sensor network and the conversion between sensor network protocol to TCP/IP. In the practical application test of nearly one year, gateway worked normally and performed stably. It can meet the storage, medical, household and other different application requirements.

Keywords Wireless sensor network · Gateway · Ethernet · WI-FI · GPRS

1 Introduction

Wireless sensor network (WSN) is composed of a large number of sensor nodes that are densely deployed either inside the phenomenon or very close to it [1]. These sensor nodes can sense, measure, and gather information from the environment and, based on some local decision process, they can transmit the sensed data to the user [2]. The composition of its network is shown in Fig. 1. Application scenarios of WSNs

---

This work is partially supported by the 2013 National Science and Technology support program (2013BAD17B06).

H. Zhou · Y. Liu (✉) · Y. Xu
Automation School, Beijing University of Posts
and Telecommunications, Beijing, China
e-mail: liuyuhaobupt@163.com

H. Zhou
e-mail: huiling@bupt.edu.cn

Y. Xu
e-mail: 549309321@qq.com

V.E. Balas et al. (eds.), *Information Technology and Intelligent Transportation Systems*, Advances in Intelligent Systems and Computing 455,
DOI 10.1007/978-3-319-38771-0_24

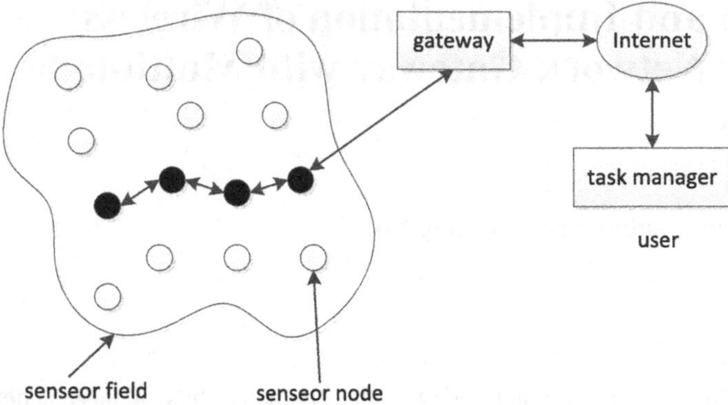

Fig. 1 Components of WSN

include military, industrial, household, medical, marine and other fields, especially in natural disasters monitoring, early warning, rescuing and other emergency situations [3].

As the coordinator in the WSN, gateway is responsible for the establishment and management of network and data collection. Communication protocol of WSN is generally Zigbee and so on, while the Internet uses TCP/IP. So gateway needs to repack data collected from sensor nodes through the protocol conversion, and transfer them to the user's server over the Internet for display and processing. Gateway also needs to receive and process the management control commands sent by the terminal users.

At present, many universities, companies and research institutions have attached great importance to the research of gateway related technology, especially the design of the new gateways, and have made some achievement [4–10]. In the above researches, gateways usually adopt GPRS or Ethernet one way only when choosing Internet access mode. As is known to all, in different application fields, the specific application environment of WSN tends to vary widely. In the practical deployment, gateway can only be used in the scenarios that satisfy the current access mode by using a single access pattern, so that the diversity of network applications greatly reduces.

Therefore, in this paper, a new type of WSN gateway with multi-mode access is designed and implemented, and it is aimed at solving the problem of the single gateway access pattern. This gateway which meets functions of data gathering, network management, protocol conversion, management control and so on has realized Ethernet, WI-FI, GPRS three kinds of Internet access. By considering the specific application requirements and the network environment factors, such as the difficulty of the Ethernet routing, WI-FI coverage and GPRS traffic fee, the most suitable access mode can be selected in the current scenario.

2 Hardware Design of Gateway

The hardware of the gateway consists of MCU module, memory module, wireless RF module, passive wake-up module, Ethernet module, Wi-Fi module and GPRS module (Fig. 2).

MCU module as the core part of the gateway, is mainly responsible for networking, sensor data collection, protocol conversion and other functions, so it needs strong processing capacity. Based on this consideration, STMicroelectronicssStm32f103ZET6 has been selected as the MCU processor in this paper. Stm32f103 ZET6 is based on ARM Cortex-M3, its clock frequency is up to 72 MHZ, with 64 KB RAM and 512 KB Flash. Chip contains five serial ports, two I2C interfaces and three SPI interfaces. The peripheral driver library of chip is complete and easy to develop.

The memory module is composed of EEPROM and NandFlash. EEPROM is used to store network access mode, server IP address and other configuration information, its capacity requirements is small, so AT240C whose capacity is 256 bytes has been chosen. NandFlash is used for non-volatile data storage that are collected in the WSN when the gateway and data server disconnect the connection. In order to prevent data loss, ST NAND128W3A with large capacity has been collected, its capacity is 128 MB and the largest erase cycles is up to 100000 times.

Wireless RF module is responsible for data communications between sensor nodes and gateway. In this paper, AT86RF212 designed and produced by ATMEL company has been selected as wireless RF chip, its receiving sensitivity is −110 dBm, the maximum output power can be up to +10 dBm. AT86RF212s sleep state power consumption is only 0.2 A, it has excellent performance and low power consumption. Usually sensor nodes will enter low-power mode to save power when sensor data have been collected and uploaded. During this period, the gateway is unable to communicate with sensor node until the them wake up in the next wake cycle. This message model restricts the application of WSN, such as in emergent cases the need for temporary data collection. So the passive wakeup module needs to be introduced into the gateway, which will wake up the sensor nodes in the deep sleep

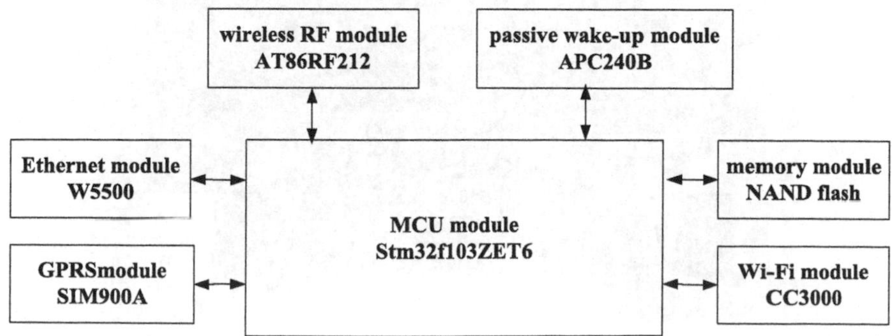

Fig. 2 The gateway hardware structure

by electromagnetic wave wake technology. APC240B wireless transmission module that is designed and manufactured by Shenzhen, PR company has been chosen in this paper. It is integrated with highspeed communication module and SCM. In addition, it adopts the advanced error-correction algorithm, and has higher anti-interference ability. At the same time, the power consumption of this module in the receiving state is only 3.2 mA.

W5500 embedded Ethernet controller of WIZnet company has been selected in the Ethernet module. W5500 hardware integrates TCP/IP protocol stack, and this can save the resources of the microprocessor processing network protocol. Users just need to call the socket interface function to quickly complete network programming. The chip supports up to 8 socket communications.

WI-FI module uses CC3000 wireless network processor launched by TI, it works in 2.4 GHz band and supports 802.11 b/g agreement. Chip contains the complete embedded IPV4 TCP/IP protocol stack and WI-FI driver, its firmware is only 6 KB, memory is only 3 KB. Its firmware driver provides the standard socket interface for users, which greatly reduces software requirements of the host microcontroller, and the maximum emission current is only 275 mA.

GPRS features with high-speed transmission, always online, billing by flow, automatic voice data switch, fast login, etc. its more suitable for the application of wireless sensor network with low data traffic. GPRS module adopts SIM900A dual-band GSM/GPRS that is specifically launched for Chinese market by SIMCom company, this module which owns the embedded TCP/UDP protocol supports standard AT instruction set, so it is very convenient to debug and develop and can connect remotely data server. The circuit board of gateway after hardware design is shown in Fig. 3.

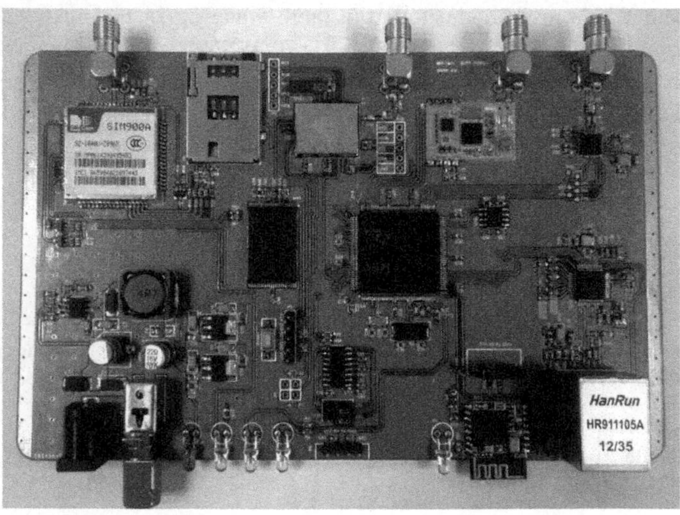

Fig. 3 The circuit board of gateway

3 Gateway Software Design

3.1 Software Task Description

There are seven tasks running on the gateway, which are startup (starting task), system (system maintenance task), debug (debug configuration task), WSN (wireless sensor network maintenance task), TCP (network maintenance task), collect (WSN data acquisition task) and flush (data push task). Startup task is responsible for the initialization of mailbox and semaphore and creating five other tasks, it is performed only once, and then deletes itself after execution. System task monitors system time and sends e-mail massages to collect task at the time of acquisition. System task also monitors the network connections status with data servers, and establishes flush task on each connection, sends data that is saved in NANDFlash for temporary storage to the data servers. Debug task configures the data queue informations of serial port (data received by the bluetooth module will be sent through the serial port to MCU) one by one, and handles them timely. TCP task polls the data queue informations of network (Ethernet, WIFI and GPRS network interface), and processes them timely. WSN task provides the scheduling interfaces of protocol stack state machine, and completes all the maintenance work related to the sensor network. Collect task executes the acquisition of WSN data, the information of waiting for mailbox will be hung up after this task has been set up, and it can be set to the ready state and executed by system tasks (timing acquisition), debug tasks and TCP tasks (real-time acquisition). Flush task is created by the system task only after the network connection has been established, and then deletes itself after execution (Fig. 4).

Data transition is the core function of the gateway, showing the process of sensor data from the collection to the protocol conversion, and is uploaded to the server through the Internet. Then collect task, TCP task and the design of communication protocol between gateway and servers will be emphatically introduced.

3.2 Collect Task

There are two ways of data collection. One is the timing acquisition, according to the preset time to perform; the other is real-time collection, users can use the client to issue a collection instruction according to their own needs. they are all completed by collect task. After collect task receives the gathering information from the mailbox, the gateway first send a pilot code of 2 s to wake up all sensor nodes by the passive sensei module. Then time slot is allocated according to the network topology, the slot table is broadcasted to all nodes, and the message is also used as the network synchronization signal, and the starting slot time of the sensor nodes is aligned with the time of receiving the data. Gateway wait for the data aggregation in the network according to the total time of allocating slots, and analysis the sensor node data

Fig. 4 Tasks of gateway

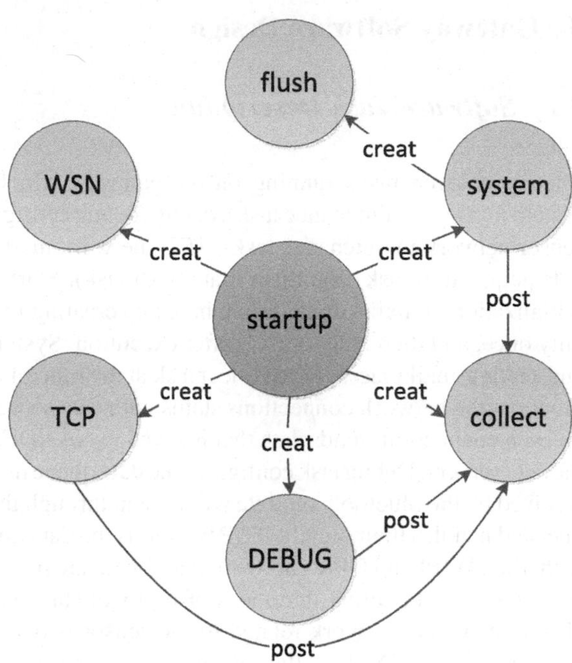

after waiting, and then package the parsed data into a data packet. Then the gateway check the server connection status, if not connected, it will be directly written in NANDFlash for backup, if the connection is normal, it will send data packets to the server, and wait for the server to confirm the frame, if the waiting time is overtime, it will back up the data to NANDFlash, if receiving acknowledgment frame will suspend, waiting for the next collection command. The program execution flow is shown in Fig. 5.

3.3 Software Design of TCP Task

TCP task is responsible for connecting and communicating with internet servers, and it is the software implementation of multi mode access. TCP/IP protocol is all used in the communication between Ethernet module, Wi-Fi module, GPRS module and the servers. In the process of network communication, TCP task is abstracted as the socket client. So there is no difference in the software level, they share the same process logic.

The data server assigns a fixed port for each gateway. Considering the reliability of data transmission, the server and gateway use TCP protocol to communicate. After the server is started and the socket is established, the listen () function will be called

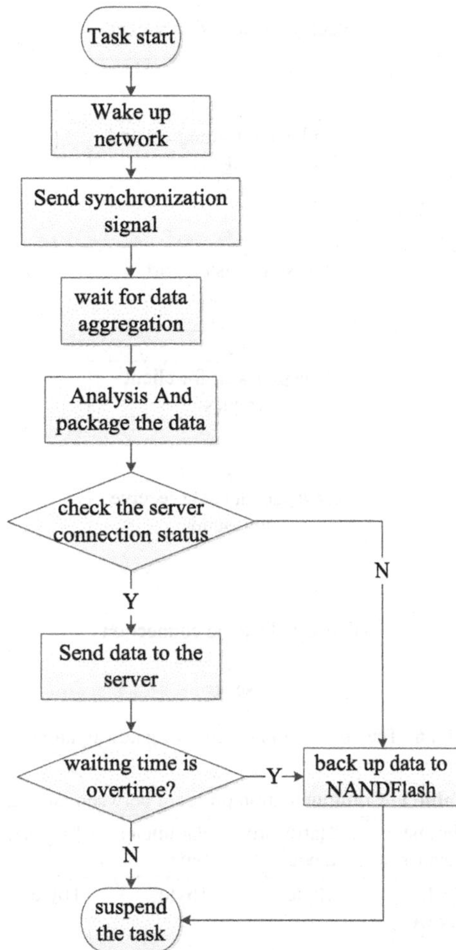

Fig. 5 Collect task flow

to monitor ports, waiting for gateway to connect. After gateway initialize the socket, the connect () function will be called to connect with the specified server port, and the two successfully establish the communication after three handshakes through TCP, then the send () and recv () function are used to transmit data. If either party wants to disconnect, the close () function will be called to interrupt the connection with the socket after the fourth handshake with TCP. The process of using the socket function to communicate between gateway and data servers is shown in Fig. 6.

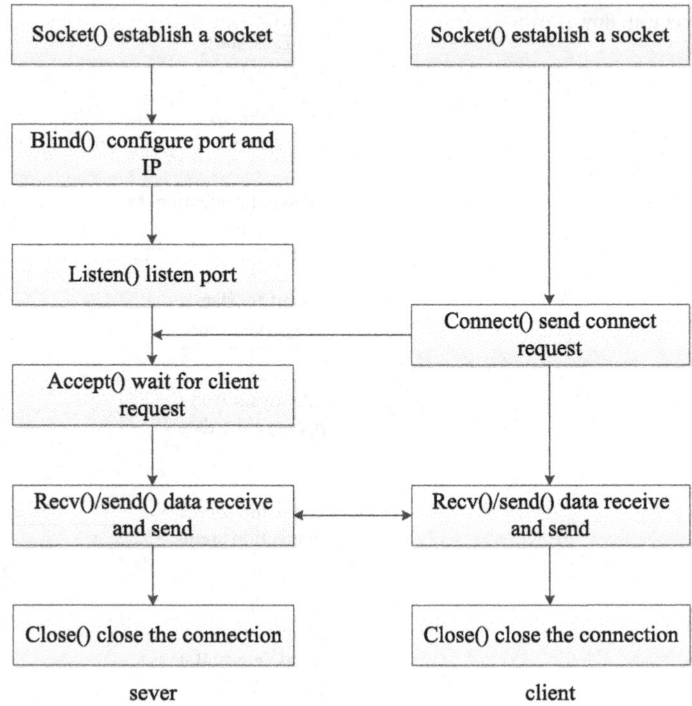

Fig. 6 The connection process of gateway and socket server

Table 1 Communication protocol between gateway and server

Frame header	Hardware device	Parameter type	Instruction	Data length	Data	Frame footer
0x42 0x3A	1byte	1byte	1byte	2byte	2048bytes	0x45 0x2E

3.4 *Design of Communication Protocol Between Gateway and Data Server*

In order to guarantee the data transmission reliable and unify the configuration of the parameters, the communication protocol between gateway and server has been worked out, its format is shown in Table 1.

Hardware device refers to the functional modules of the gateway, such as system equipment, network equipment, wireless sensor equipment, serial configuration device and wireless sensor data. Parameter type refers to the parameters that can be configured in hardware device, such as there are gateway name, gateway ID, running time, and timing acquisition cycle and so on under system equipment. The instructions include reading, writing, requesting three types. The data length is represented by two bytes, the first byte is high eight bits, and the second byte is low eight bits. Data refers to the data in the frame, and the length is specified by the length of data. Head frame and tail frame defines the start and end position of the frame.

4 Livedeployment Test

The gateway designed and implemented in this paper were installed in the three tobacco alcoholization warehouses in a city of Sichuan in January 2014, they monitored the environment, temperature and humidity of the goods in real time. Through more than a year of actual operation, the gateway nodes worked stable and reliable. Figure 7 is one of the warehouse equipment deployment diagram. The hardware devices deployed in this warehouse are a gateway and twenty-two sensor nodes. Each sensor node is connected with a temperature and humidity sensor, and the interval is 4 h.

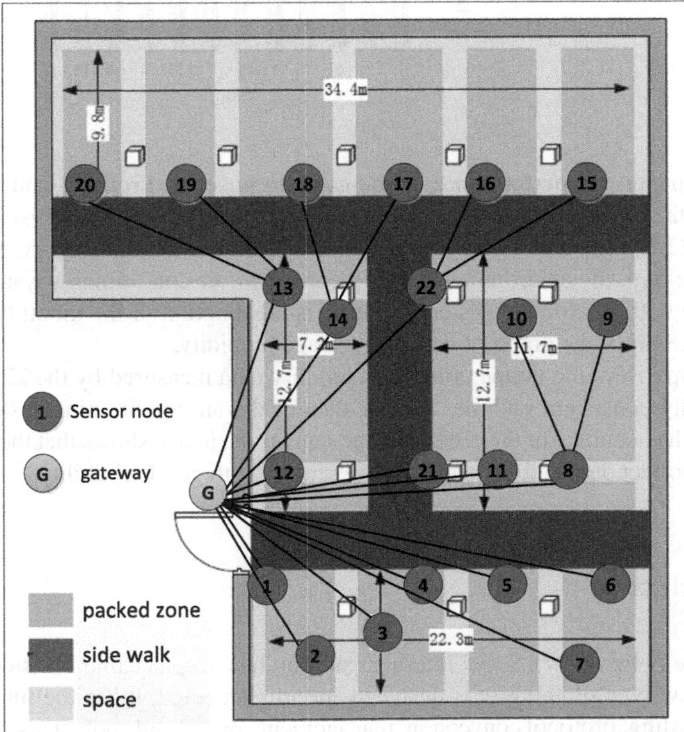

Fig. 7 Node deployment topology

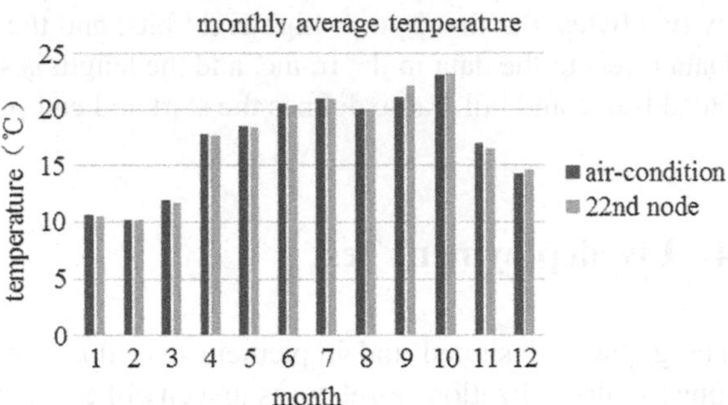

Fig. 8 Monthly average temperature

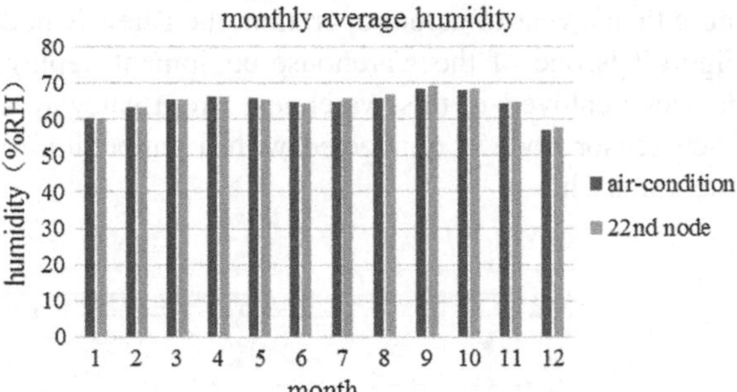

Fig. 9 Monthly average humidity

In this paper, the performance of the gateway is verified reliable and stable by analyzing the continuity and accuracy of the collected data during the system operation. The 22nd node data of environmental temperature and humidity were selected to compare with the environmental data recorded by air-conditioning system in the warehouse with the storage system. Figure 8 is the diagram of the monthly average temperature Fig. 9 the graph of monthly average humidity.

By comparison, the temperature and humidity data measured by the 22nd nodes are basically consistent with the average data of the air conditioning system. That can prove the accuracy of these data. At the same time, it also shows that the gateway node can collect data and transmit data accurately with good reliability.

5 Conclusion

As a bridge between WSN and Internet, gateway has irreplaceable function. In this paper, a new type of gateway is designed and implemented, it has the functions of data forwarding, protocol conversion, management control and soon, it also has three kinds of Internet access modes, such as Ethernet, WI-FI, and GPRS, that can greatly improve the adaptability to different application scenarios. Live deployment tests show that it has good reliability and stability. It can be widely used in warehousing, logistics, smart home, environmental monitoring and other fields.

References

1. Akyildiz IF, Su W, Sankarasubramaniam Y et al (2002) A survey on sensor networks. IEEE Commun Mag 40(8):102–114
2. Yick J, Mukherjee B, Ghosal D (2008) Wireless sensor network survey. Comput Netw 52(12):2292–2330
3. Han G, Xu H, Duong TQ et al (2013) Localization algorithms of wireless sensor networks: a survey. Telecommun Syst 52(4):2419–2436
4. Liu L, Meng Z, Shi B et al (2012) Research on wireless sensor network sink node based on embedded Linux. J Hefei Univ Technol Nat Sci 35(4):499–502
5. Cheng K, Zhang H, Zhang Y et al (2014) Design of CC2530 based gateway node for wireless sensor network. J Northwest A F Univ (Nat Sci Ed) 5:029
6. He S, Chen X (2012) Gateway design for urban lighting monitoring system based on GPRS and ZigBee. Microcontroller Embed Syst 12(1):27–29. doi:10.3969/j.issn.1009-623X.2012.01.013
7. Zhao L, Shao S, Zhu J et al (2012) Design and implementation of a multi core embedded gateway for linking WSN and internet. J Northeast Univ Nat Sci 33(1)
8. Wang Y, Si L, Liang S (2014) The design of wireless sensor network gateway based on ZigBee and GPRS. Int J Future Gener Commun Netw 7(2):47–56
9. Lv Y, Tian Y (2010) Design and application of sink node for wireless sensor network/industrial and information systems (IIS). 2nd international conference on IEEE, vol 1, pp 487–490
10. Deng HN, Zhou C, Zhao C et al (2012) Design and realization of domestic smart gateway based on wireless sensor network. 2nd international conference on electronic and mechanical engineering and information technology. Atlantis Press

Research on Monitoring Technology of Traditional Chinese Medicine Production Process Based on Data Opening

Lu Pei, Zhou Guangyuan and Wang Nana

Abstract There is useful data directly related to the end products quality, which is implied in the process data of traditional Chinese medicine (hereinafter referred to as TCM) production. Rapid process data collection, open data management and scalable monitoring technology, have great significance to achieve the direct quality control. Aiming at the procedure of TCM, it directly collects process date based on OPC technology and meanwhile stores real time data in the SQL database, and then develops monitoring software by using programming language C (C#). The result shows that the minimum monitoring system based on C# can improve the data acquisition and storage rate, and realize data open and sharing, with advantages of small memory footprint, simplified structure, and highly expansibility.

Keywords Data open · Traditional Chinese medicine production · OPC technology · The monitoring technology

1 Introduction

In production of TCM, it may cause significant differences to the intermediates, and affect the final quality [1] directly when the change of the various processes involved in each operation parameter. Xiuli Shao [2] and others find the association relationship between the temperature, the amount of the liquid, the concentration of the liquid etc., and the product quality, by association analysis the extraction and concentration process data of TCM. Aimed at the quality consistency problem for different batch of TCM, Haoshu Xiong [3] and others research the method

L. Pei (✉) · Z. Guangyuan · W. Nana
Tianjin University of Science and Technology, Tianjin, China
e-mail: 1728134316@qq.com

Z. Guangyuan
e-mail: 978407115@qq.com

W. Nana
e-mail: 1223879607@qq.com

V.E. Balas et al. (eds.), *Information Technology and Intelligent Transportation Systems*, Advances in Intelligent Systems and Computing 455,
DOI 10.1007/978-3-319-38771-0_25

for monitoring the intermediates quality from all production processes of TCM. Pei Lu [4] and others identity the process parameters of the quality factors and study the relationship between the extraction process and the extraction rate by analyzing the production data of TCM. On the one hand, these researches demonstrate the relationship of process parameters and production quality, which provides the basis for production optimization. On the other hand, those also can change the process parameters in real time through online quality analysis, in order to achieve the purpose of quality control and demonstrates the broad prospect. But its firstly need to improve the real-time data acquisition and opening rate, in order to ensure the real-time analysis and optimization.

With the implementation of our new Good Manufacturing Practice (GMP), automatic monitoring of the production has been achieved gradually among TCM manufacturing enterprises. Currently, seeing from the structure of the monitoring system, it mainly includes: Distributed Control System (referred to DCS) and IPC + PLC control system [5–7]. The monitoring software mainly develops by adopting configuration, which means that TCM production data is stored in an internal database of the configuration software. If needs data open, or namely accesses to production data from external, it needs to establish an external database and writing VB script, by collecting data via event-trigger mechanism. But if use script frequently, it may result in a waste of industry system resources and reduction operating speed. At the same time, the real-time data cant be transmitted and the analysis of real-time performance will be reduced when data open. Therefore, to research the technology of fast data acquisition and data open, and the monitoring technology during production based on data open, is a big need to the TCM manufacturing enterprises development, and also has a great importance to achieve online real time quality control.

In recent years, the using of OPC [8] makes it possible to solve industrial control in the way of simpler system structure, longer life and lower price. Actually, OPC provides a mechanism, through which the system can obtain data from the server and pass it to any client application. OPC realizes the fast data transfer between the application and the industrial control equipment with high access efficiency. That is a foundation to realize the production data open. OPC technology has been successfully applied in many fields such as electric power, oil [9, 10] and so on. This technology realizes real time data sharing, which makes the whole monitoring system have better openness, lower development cost, easy to upgrade and maintain ea. However in the relevant literature, it has not been reported that there is monitoring technology research for direct data open by combining with OPC technology in the field of TCM production.

Combined with OPC technology, the monitoring system can collect and store the data of the production process of traditional Chinese medicine in real time and it can be stored in the relational database SQL directly, which realizes the data of unification, open and efficient transmission. And Visual Studio Microsoft 2010 development platform using C# programming language to develop the monitoring software, the development of the monitoring interface is more friendly, take up system resources is smaller, while the user interaction. To realize the opening and real time of production data is the basis for the real time quality control, which is very important to improve the quality of Chinese traditional medicine.

2 Data Acquisition and Storage Scheme of TCM Production

2.1 Based on OPC Technology, It Can Access to Data Directly and Improve Acquisition Rate

TCM enterprises need to read the real-time data in production, as real-time data is needed whether in online optimization of control strategy and the process parameters of the real-time curve in, or in the achievement of quality control directly in production. According to this real-time production data, the extraction rate of TCM production can be calculated based on the active ingredient proportion in the extract. Therefore, it needs a unified data access method to achieve the efficient real-time data transmission.

The traditional data transfer way in intelligent equipment, PLC and the host computer monitoring software is thorough device driver. The monitoring software of host computer of some automation manufacturers, such as Siemens, MITSUBISHI, develops some typical communication drivers; but those cannot solve the communication problem in all the hardware and software. Thus, there are following features in monitoring software development: repeat development; different driver in different developers; unfit for changes of hardware features.

After using OPC technology, the data collection is more effective and convenient. It can make flexible and efficient data read between the TCM production monitoring system and the intelligent equipment PLC. With OPC technology, the delay of data transmission is greatly reduced, and the performance and synchronization of real-time data transmission is greatly improved.

2.2 Stored the Date Directly to the Open SQL Database, Lays the Foundation to Realize Quality Control of Real Time

For TCM enterprises, not only need to collect real-time data, but also need to open that. Only to open the production data can analyze the online optimization, and control online quality in production.

In the general TCM monitoring system, the production data is stored in internal database of configuration software in WINCC and others. It needs to create an external database and write the script to access production data stored in the internal database, but this cannot achieve real-time data. By using OPC technology, it can access the data stored in the CP5611 card through the OPC client and also store the data in the SQL database. This solves the problem of real-time data dump from the configuration software to the standard database.

Server SQL is a widely used database management system with the advantages of ease of use, and suitable for distributed organization, used for data warehouse decision support, and so on. So, it is more convenient to analyze the production data, after storing the date in the relational database (Server SQL). And it also provides the conditions for the network transmission, for the global data sharing and for the development of the monitoring software.

2.3 *Using Programming Language C (C#) to Develop Monitoring Software, to Construct the Minimum Monitoring Software System to Improve the Extensibility*

With the development of process optimization and quality control, monitoring software must upgrade the software functions; but some of the on-line optimization analysis functions cannot develop through configuration software. Thus, integration and extension of the monitoring system cannot be realized by using configuration software.

If adopts OPC technology, TCM production monitoring system, which develops by object-oriented C#, can easily add embedded data correlation analysis module, on-line optimization analysis function module, and on-line quality control function module, etc. Therefore, by combining with OPC, TCM production monitoring system is easy to upgrade technology, is more extensible, and is more conducive to the sustainable development.

In conclusion, this paper builds a data flow monitoring system, as shown in Fig. 1, based the example that Siemens PLC is widely used in TCM manufacture enterprises. Parameters must be monitored in TCM production include extraction temperature, solvent quantity, concentrate density parameters. Extraction temperature, for example, affects the data change in production. The signal generating by scene testing equipment is adopted by PLC315-2 DP controller. Field bus uses the protocol of Profibus-DP, and production data is cached on the CP5611. Using OPC technology, the data communicate between OPC server and the CP5611. Data can be collected and stored in a SQL database by writing OPC client and accessing OPC server. Thus,

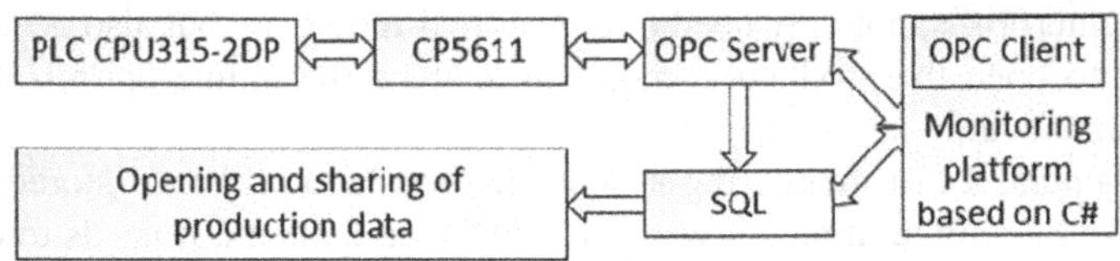

Fig. 1 Production data flow chart

it is easy to access data in the database, and realize human-computer interaction by using C# to develop the monitoring software.

3 The Realization of the Monitoring Technology in TCM Production

3.1 The Realization of OPC Asynchronous Communication

Firstly, it needs to add the reference of OpcRcw.Comn and OpcRcw. Da in development project, and defines the OPC related variables: OpcRcw.Da.IOPCServer ServerObj; Opc-Rcw.Da.IOPCGruopStateMgt IOPCGroupStateMgtObj=null;int[] ItemServerHandle; In the realization of OPC clients, it use Asynchronous communication is used in the process of realization of OPC client. It allocates memory queue for the defined read and write; then checks if it is effective. If yes, these items will be sent into the queue and operated. It will check the activity variables group. If finds a variable, it will perform a callback function, to realize action that is needed to implement when the variable changes. Delegate can be used to implement this feature (Fig. 2).

Following are a few key functions.

(1) The connecting function of the OPC server. To connect different OPC server only needs to change the server name and computer name OPC server. The connecting function of the OPC server is btnStartConnection ().
(2) The real-time monitoring about data change functions. Firstly it needs to declare a variable, which can trigger events to get the real-time data. When the data changes in the Group of OPC server, the data can automatically refresh in the corresponding client, according to the update cycle. Use the C# event handling mechanism and register DataChange to the event. Once the server-side

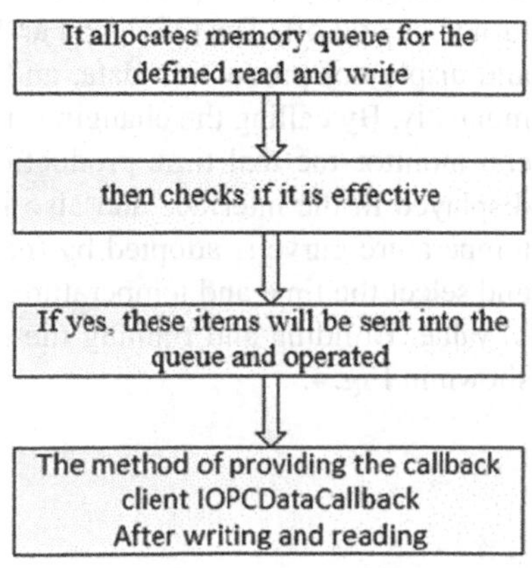
It allocates memory queue for the defined read and write
then checks if it is effective
If yes, these items will be sent into the queue and operated
The method of providing the callback client IOPCDataCallback After writing and reading

Fig. 2 The mechanism of OPC asynchronous communication

data have changed, it automatic triggers the process. Trigger function is chkGroupActive_CheckedChanged ().

(3) The data read and write function. The reading data function is btnRead (). Data can read and write in OPC server through DataChange events trigger in the group. The events have several parameters and parameter ItemValues stores the data of data items. Parameter TimeStamps stores the time of data change. Parameter ClientHandles is a tag index for the data. The value of the OPC tag is ItemValues [1]. And only when the data is changed, the event is triggered.

3.2 The Realization of Storage the Production Data in the SQL Database

Connecting with SQL database, when the triggering event OnDataChange () occurs, SQL insert statement will store the change date in database. The key code to store the production date is sqlStr="server=ADMIN-PC \\ SQLEXPRESS; database = OPC;uid=sa;pwd=********"; SqlConnection mySqlConnection = new SqlConnection (sqlStr); // mySqlConnection.Open(); // str = string.Format(@"insert into wenducaijibiao(Item_value,Item_quality, Datetime) values('" + strAReadVal.Text. Substring (0,5) + "','" + strAReadQuality.Text + "','" + System. DateTime. Now. ToLongTimeString ().ToString ()+ "')"); SqlCommand command = new SqlCommand(str, mySqlConnection); command. ExecuteNonQuery (). Figure 3 shows the data storage.

3.3 Monitoring Software Development

Embedded the OPC client in the monitoring system and connected the OPC server, realizes production data collection and display, and achieves production data monitor and storage. By calling the link and disconnect function, the OPC server can be turned on and off. The OPC read and write data function is to achieve the collection and display of production data, and the display data can be stored in the database manually. By calling the changing function of real time monitoring data, the system can monitor the real time production data. Once the data is changed, the data is displayed in the interface and also inserted into the database. The development of temperature curve is adopted by the chart control. It can bind data in the database and select the time and temperature data in the table and regard the two fields as X, Y value. Building and running the production data shows monitoring interface, as shown in Fig. 4.

ADMIN-PC\SQLEXP... - dbo.wendubiao

ID	value	quality	Datetime	time
10	31.72	Good	2015-10-16 1...	16:59
12	35.59	Good	2015-10-16 1...	17:01
14	39.53	Good	2015-10-16 1...	17:03
15	41.84	Good	2015-10-16 1...	17:04
17	45.52	Good	2015-10-16 1...	17:06
19	48.82	Good	2015-10-16 1...	17:08
21	51.86	Good	2015-10-16 1...	17:10
22	53.21	Good	2015-10-16 1...	17:11
24	55.77	Good	2015-10-16 1...	17:13
26	58.15	Good	2015-10-16 1...	17:15
28	60.11	Good	2015-10-16 1...	17:17
31	62.54	Good	2015-10-16 1...	17:20
33	63.88	Good	2015-10-16 1...	17:22
35	65.14	Good	2015-10-16 1...	17:24
36	65.75	Good	2015-10-16 1...	17:25
38	66.71	Good	2015-10-16 1...	17:27
41	67.88	Good	2015-10-16 1...	17:30
43	68.57	Good	2015-10-16 1...	17:32
45	69.09	Good	2015-10-16 1...	17:34
47	69.66	Good	2015-10-16 1...	17:36
48	69.92	Good	2015-10-16 1...	17:37
50	70.31	Good	2015-10-16 1...	17:39
52	70.7	Good	2015-10-16 1...	17:41
53	70.78	Good	2015-10-16 1...	17:42
55	71.13	Good	2015-10-16 1...	17:44

Fig. 3 Data stored in the SQL database

4 Conclusion

Using the OPC technology, storing the production data in the relational database SQL, as well as using the C# to develop the monitoring software in the Visual Studio Microsoft 2010 development platform, realizes the data- oriented open of TCM production monitoring.

(1) Based on OPC technology, direct access to the TCM production improves the data collection and transmission rate; achieves efficient and convenient data read and write. In the research process, the development C# program of the OPC client reduces the complexity of the data command transmission process, improves the efficiency of data transmission in the industrial process, and improves the

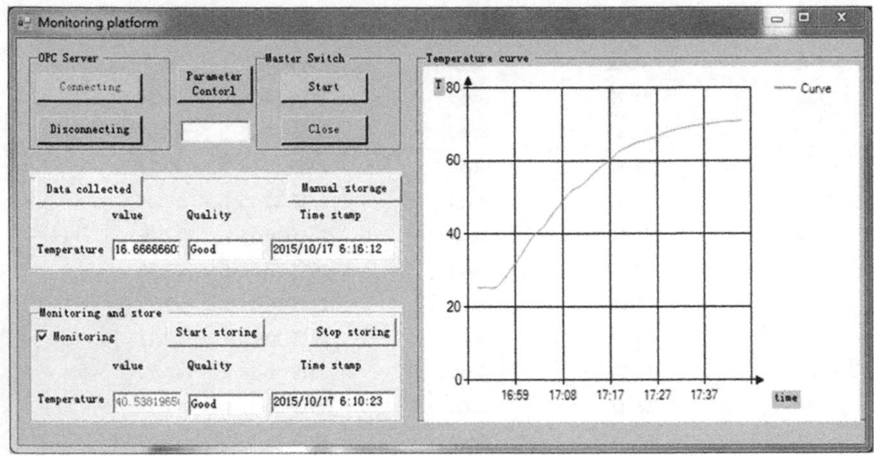

Fig. 4 The monitoring interface of the production data

real-time performance. So, improving the real-time data acquisition and open rate are very important to real-time data analysis and optimization.

(2) Stored the data directly in the open SQL database, lays the foundation for the realization of the real time quality control. The production data can be collected and stored in the relational database SQL. And the open data is good for online optimization analysis and quality control.
(3) Using C# to develop the monitoring software, it can build the smallest monitoring system with advantages of small memory, simplified structure and great expansibility. TCM production monitoring system, which is developed by C#, can easily add embedded data correlation analysis module, on-line optimization analysis function module, and on-line quality control function module, etc. Therefore, this monitoring system is easy to upgrade technology, is more extensible, and is more conducive to the sustainable development.

References

1. Haibin Q, Yiyu C, Yuesheng W (2003) Some engineering problems on developing production industry of modern traditional chinese medicine. China J Chin Materia Med 28(10):904
2. Xiuli S (2010) Research and implementation of data mining analysis of traditional Chinese medicine production. J Nankai Univ Nat Sci Ed 5:46–51
3. Haoshu X (2012) Method for statistical quality control of multi process and multi index of traditional Chinese medicine production. China J Chinese Materia Med 37(13):1935–1941
4. Pei L, Meng J (2014) The research on monitoring system of Chinese medicine extraction process based on the information integration. Tianjin University of Science and Technology, Tianjin
5. Bilgic A et al (2011) Low-power smart industrial control design automation & test in Europe conference & exhibition (DATE) 2011. IEEE

6. Jiexian S, Xiaoping S (2006) Automatic control technology in the extraction of Chinese herbal medicine. Chinese Herbal Med 29(9):984–986
7. Liu Q, Bintang X, Gaobin, Jin L, Cuiming W (2010) The digital control system of traditional Chinese medicine extraction. China instrument (Suppl):183–185
8. OPC Foundation. Data Access Custom Interface Standard Version 2.04. 5 September 2000
9. Xue L (2012) The research on the application of OPC technology in substation monitoring system, Shandong university
10. Luan X (2007) The research on the application of OPC technology in oil field monitoring and controlling system. Micro Comput Inf 13:109–110

Regenerative Braking Control Strategy of Electric Truck Based on Braking Security

Shiwei Xu, Ziqiang Tang, Yilin He and Xuan Zhao

Abstract In order to improve the energy efficiency of electric vehicles, a regenerative braking control strategy of electric truck was developed to improve energy recovery based on braking security. After the prerequisites, the restrictions of ECE regulations, battery and motor were completed to ensure the braking security, the regenerative braking force allocation strategy was designed. Then a co-simulation of Cruise and Matlab of this control strategy was executed in Japan1015 operating cycle to evaluate the strategy effects. Simulation results show that the strategy proposed in this paper can recover as much as 11.48 % braking energy in Japan1015 cycle under braking security requirements. So this regenerative braking control strategy can significantly improve the economic performance for electric vehicles.

Keywords Automotive engineering · Electric truck · Regenerative braking · Braking stability · Co-simulation

1 Introduction

According to recent studies of electric vehicles, regenerative braking seems to be a most promising technology to improve the driving range because it can recover the energy wasted on braking energy. So far main focus has been placed on factors that affect the electric power on recovering the braking energy, such as motor and battery. However, there is not enough attention paid to the braking stability. Since

S. Xu (✉) · Z. Tang · Y. He · X. Zhao
School of Automobile, Chang'an University, Xi'an 710064, China
e-mail: xushiweide2008@163.com

Z. Tang
e-mail: tangzqa@126.com

Y. He
e-mail: 553571576@qq.com

X. Zhao
e-mail: bluesky_xuan@163.com

V.E. Balas et al. (eds.), *Information Technology and Intelligent Transportation Systems*, Advances in Intelligent Systems and Computing 455,
DOI 10.1007/978-3-319-38771-0_26

the capacity of regenerative braking is restricted by velocity, maximum torque of motor and maximum charge current of battery, it is crucial to make a suitable division percentage of motor braking and mechanical braking for regenerative braking. More recently, the division of braking force has been investigated in some studies. Hellgren [1] conducted a research on the effect of motor, battery and structure of braking system; however, the division of braking force was not mentioned. References [2, 3] conducted studies on distribution of braking force under considering the restriction of electric power; however, the application of these theories are needed to improve because all of these theories were proposed on ideal situation. Therefore, in order to recover the braking energy as much as possible, the issues, the principle of braking force distribution strategy and maintaining the braking stability warrant further attention.

In this paper, considering the security requirements, a regenerative braking control strategy of electric truck was proposed. The structure of the paper is organized as follows: Sect. 2 introduces the requirement of braking security restrictions, which are consist of ECE R13, battery and motor, Sect. 3 describes the regenerative braking control strategy, Sect. 4 presents the co-simulation, simulation results and the analysis, Sect. 5 is the conclusion.

2 Constraints of the Regenerative Braking Under Braking Security Requirements

2.1 ECE R13 Brake Regulations

I curve is the ideal braking forces distribution curve applied on front and rear wheels. However, in practice, the braking force distribution generally cant be employed as '*I* curve'. ECE R13 brake regulations enacted by Economic Commission have proposed clear requirements for the front and rear braking forces of biaxial vehicle [5]: For truck, the adhesion utilization coefficient of front wheels should exceed the adhesion utilization coefficient of rear wheels with various loading conditions; when the friction coefficient φ is between 0.2 to 0.8, the requirements of the braking intensity should meet $z \geq 0.1 + 0.85(\varphi - 0.2)$. According to this requirement, rear wheels must have a certain braking force if the front wheels locked to maintain vehicle stability and high braking efficiency. The related curve between the smallest ground braking force of rear wheels and that of front wheel is called *M* curve, and the braking forces distribution curve should be higher than the *M* curve, *M* curve determined by formula (1) and (2).

$$F_{br} = \frac{z + 0.07}{0.85} \frac{G}{L} (L_a - zh_g) \tag{1}$$

$$F_{bf} = G_Z - F_{br} \tag{2}$$

From the ECE brake regulations, it can be concluded that the ideal security range of braking force distribution should distribute between the I curve and M curve, however, even if the braking force allocation curve is above I curve but below the ECE regulations rear axles upper limit [4], the truck braking security will also be guaranteed. So considering these restrictions, the actual regenerative braking force allocation curve should be located in the region between M curve and ECE regulations rear axles upper limit curve.

2.2 Battery

For regenerative braking system, maximum charging current, maximum charging power, the battery state of charge(SOC) are the major restriction factors. Both of the charging current and charging power of regenerative braking cannot exceed the maximum charging current and the maximum allowable charging power of batteries. Battery SOC cannot operate beyond the active battery charging region. Here, the active battery charging region is from 30 to 90 % [5]. For most batteries, the charging power P_{bat} can be expressed as

$$P_{bat} = (U_{OC} + IR)I \tag{3}$$

Where U_{OC} is the open circuit voltage in V, and R is the internal resistance of the battery in Ω. I is the current of the charging current in A, which is calculated as:

$$I = I_0 e^{-\sigma t} \tag{4}$$

Where I_0 is the maximum initial charging current in A, while σ is the attenuation coefficient, which is also called as the charge acceptance ratio. The battery charge power calculated by the formula (3) and (4) limits the maximum regenerative braking force, which is:

$$F_{reg1} \leq \frac{P_{bat}}{v\eta_t \eta_m \eta_b} \tag{5}$$

Where F_{reg1} is the maximum motor regenerative braking under the limitation of the battery charging power in N; v is the velocity in km/h; η_t is mechanical transmission efficiency; η_m is the power generation efficiency of motor; η_b is battery charging power.

2.3 Motor

As motor is a main factor affecting the energy recovery, the regenerative braking torque provided by it is affected by the motor torque characteristics, motor speed characteristics, vehicle velocities and other factors [6]. When the motor is working in the generator state, the torque output characteristics are basically similar to the output characteristics of the motor state, which can be expressed as:

$$T_{reg1} = \begin{cases} 9550P_N/n_b & n \leq n_b \\ 9550P_N/n & n > n_b \end{cases} \tag{6}$$

Where T_{reg1} is the motor regenerative braking torque in N.m; P_N is the motor rated power in kW; n_b is the motor base speed in r/min.

When the vehicle braking at a low velocity, the regenerative braking capacity will decrease with the vehicle velocity decreasing due to the lack of kinetic energy. In order to ensure safety, the regenerative braking force provided by motor is set as zero when the motor speed drops to 500 r/min, so the amendment formula is turned into:

$$T_{reg} = \lambda(n)T_{reg1} \tag{7}$$

Where λ (n) is the correction factor associated with the motor speed:

$$\lambda(n) = \begin{cases} 0 & n \leq 500r/min \\ 1 & n > 500r/min \end{cases} \tag{8}$$

From the above analysis, the maximum motor regenerative braking force on driving wheel under the generator power limits is obtained as:

$$F_{reg2} = \frac{T_{reg}i}{r}\eta_t \tag{9}$$

Where i is the total ratio of power train; r is the wheel radius in m. During the braking, in order to ensure the brake safety, the braking force distribution ratio needs to be adjusted when the maximum motor regenerative braking force can't satisfy the required braking force on the drive wheels.

In summary, the motor maximum regenerative braking force is determined by battery charge power and motor generator power, which is:

$$F_{regmax} = min(F_{reg1}, F_{reg2}) \tag{10}$$

3 Regenerative Braking Control Strategy

3.1 Braking Force Allocation Strategy

In order to maximize the recovery of braking energy, the braking force allocation strategy is established within the range of the safety brake force distribution [7].

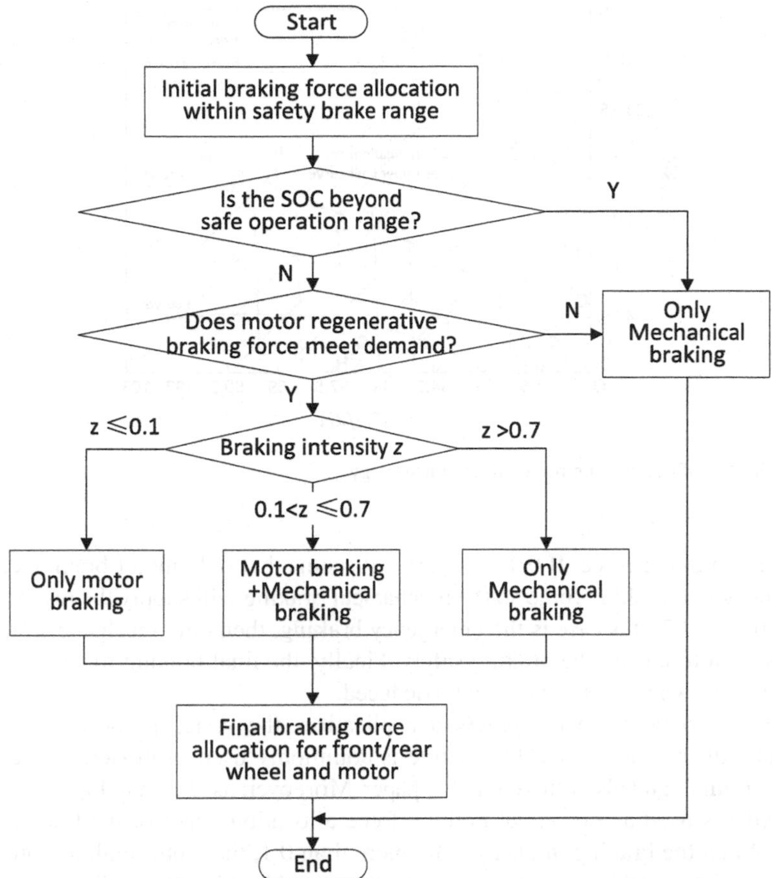

Fig. 1 Control flow of the braking force distribution

One way to improve braking energy recovery rate is to increase motor braking force in conjunction with a decrease of mechanical braking force [8]. The braking force allocation strategy are shown in Fig. 1.

Specific steps are presented as follows: Firstly, the initial braking force distribution ratios of front wheel and rear wheel are gotten within the braking safety range. Secondly, a judgment, whether the battery SOC is beyond safe operation range or not, should be made. If yes, it means the regenerative braking will damage the battery, so the regenerative braking should be stopped and only mechanical braking supplies the whole braking force. Otherwise, the flow enters the next step. Thirdly, we should judge if the motor regenerative braking meets demand or not. If not, the motor braking force will be too small to supply enough braking force, so the whole braking force also should be supplied by mechanical braking solely. Otherwise, the flow enters the next step. Fourthly, braking intensity z should be classified. If $z \leq 0.1$, it means the motor braking can meet the whole braking force so that there is only motor braking

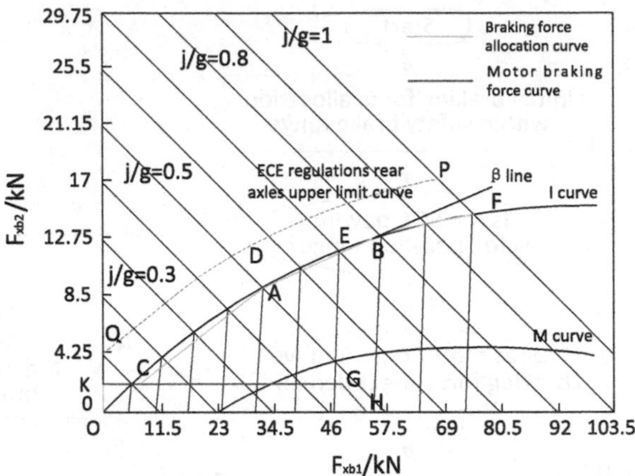

Fig. 2 Sketch of the braking force allocation strategy

supplying braking force. If $0.1 < z \leq 0.7$, it seems that only motor braking cannot meet the whole braking force so the mechanical braking will supply the rest braking force. If $z > 0.7$, this case is the emergency braking, then only mechanical braking is employed to ensure the braking safety. Finally, the final braking force allocation for front/rear wheel and motor can be deduced.

In order to facilitate the practical application, the varied proportional valves hydraulic distribution curve like β line is commonly used in modern vehicles to replace *I* curve and also utilized in this paper. Moreover, as shown in Fig. 2, the Line OKCABF is used as the actual braking force allocation curve of the truck in this paper. When the braking intensity is no more than 0.1, the motor braking force can meet the whole braking force and thereby the actual braking force allocation curve is coincident with motor braking force curve. Then, when the braking intensity is between 0.1 and 0.7, the whole braking force consists with motor braking force and mechanical braking force, and the curve is varying along with β line to get a good braking security. Then again, when the braking intensity exceeds 0.7, ABS or other braking security electric equipment are activated and the curve will follow the idea *I* curve.

3.2 Regenerative Braking Control Process

Regenerative braking control system structure were shown in Fig. 3. From the structure, it can be revealed that during braking process, the total target braking force was obtained by the pedal travel, and the motor braking force was determined by velocity, battery SOC and so on. Then the actual regenerative braking force and mechanical braking force were determined in the braking force allocation strategy module, and these signals were sent to the corresponding control module. All signals were transported through the CAN bus between modules.

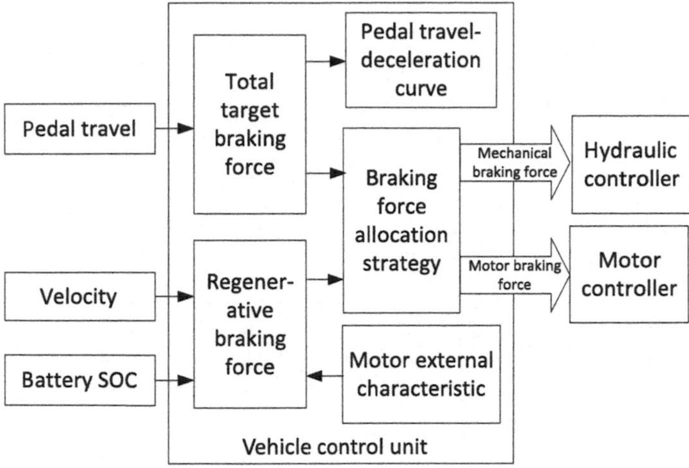

Fig. 3 Regenerative braking control system structure

4 Simulation Results and Analysis

To evaluate the results of regenerative braking control strategy in this paper, the co-simulation of Cruise and Matlab is executed in Japan1015 operating cycle [9], and the results were compared with non-braking energy recovery strategy. In the operation cycle, all of the initial battery SOC were set to 80 %. The parameters of the electric truck are shown in Table 1.

Figure 4 shows the speed track curve at Japan1015 cycle, which suggests that the braking strategy developed in this paper can meet the operation cycle conditions. In Figs. 5 and 6, the current and SOC change under regenerative braking strategy are compared with non-braking energy recovery strategy. In Fig. 5, the current value was positive in some braking conditions with the regenerative braking strategy, which implies that parts of electric energy were recovered. Figure 7 shows that the battery SOC can decrease more slowly under regenerative braking strategy owing to the

Table 1 Parameters of the electric truck

Parameters	Value
Length × Length × height (mm)	7830 × 2470 × 2760
Front body over-hang (mm)	1480
Rear body over-hang (mm)	1830
Wheel base (mm)	4600
Curb weight (kg)	10800
Gross mass (kg)	16000
Maximum velocity (full load) (km/h)	80

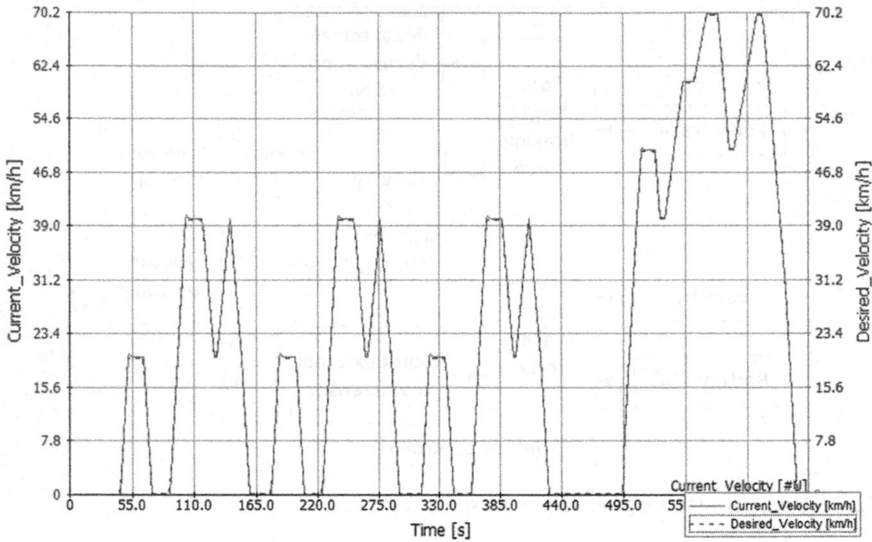

Fig. 4 Speed track curve at Japan 1015 cycle

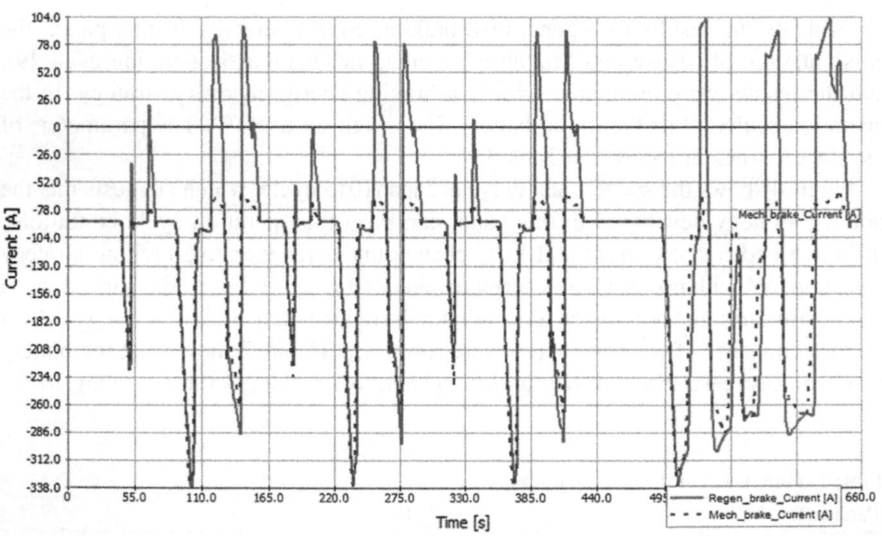

Fig. 5 Current change curves at Japan 1015 cycle

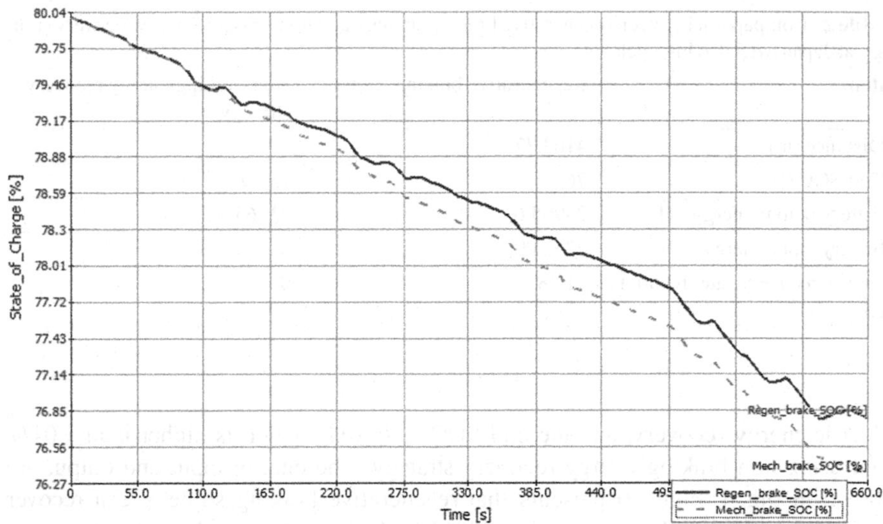

Fig. 6 Battery SOC change curves at Japan 1015 cycle

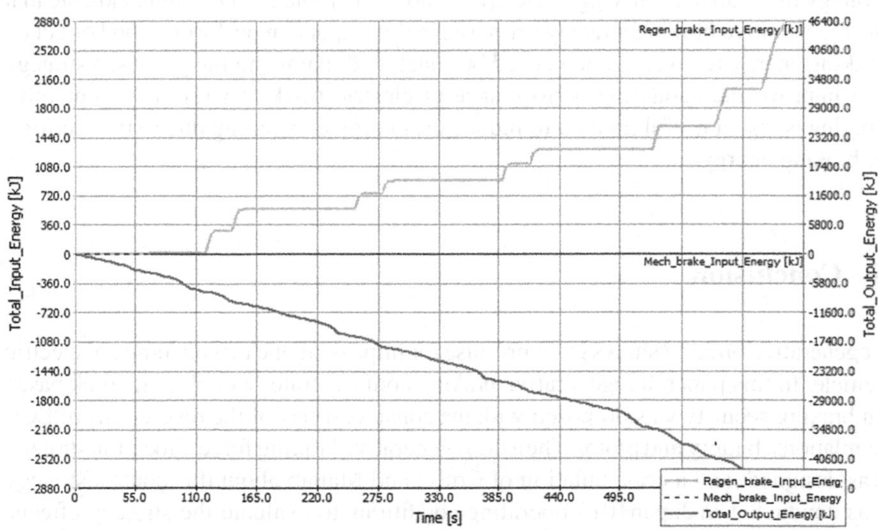

Fig. 7 Energy input and output curves at Japan1015 cycle

Table 2 Comparison between regenerative braking strategy and non-braking energy recovery strategy at Japan1015 driving cycle

Item	Regenerative braking strategy	Non-braking energy recovery strategy
Distance (m)	4163.77	
End SOC (%)	76.6	75.59
battery output energy (kJ)	23965.6	23965.6
battery input energy (kJ)	2752.09	0
energy recovery rate in battery (%)	11.48	0

electric energy recovery, and the end SOC is 76.6 %, which is higher than 1.01 % that using non-braking energy recovery strategy. The energy input and output are shown in Fig. 6. Figure 6 presents that regenerative braking strategy can recover 2752.09 kJ energy while consuming 23965.6 kJ energy in the whole driving cycle, and the energy recovery rate in battery can reach to 11.48 %. The comparison of performance between regenerative braking strategy and non-braking energy recovery strategy in Japan1015 driving cycle are as shown in Table 2. The results indicate that the regenerative braking strategy proposed in this paper can perform a good effect on braking energy recovery, and it could be concluded that using the proposed strategy can improve the economic performance of electric truck. Therefore, regenerative braking is also a useful method to reduce the energy consuming due to the recovery of braking energy.

5 Conclusion

Regenerative braking shows great promise for improving the driving range of electric vehicle. In this paper, a regenerative braking control strategy of electric truck based on braking security was proposed with the considerations of the restrictions of ECE regulations, battery and motor. Then the regenerative braking force allocation strategy was designed, and a co-simulation of Cruise and Matlab about this control strategy was conducted in Japan1015 operating conditions to evaluate the strategy effects. Simulation results show that the regenerative braking control strategy proposed in this paper can recover as much as 11.48 % braking energy in the operating cycle. Consequently, it is affirmed that this regenerative braking control strategy has a great ability to improve the economic performance of electric trucks. In the future, how to recycle more energy under the large braking intensity needs increasing attention.

Acknowledgments This work has been supported by NSFC (51507013) and Fundamental Research Funds for the Central Universities (310822151025).

References

1. Hellgern J, Jonasson E (2007) Maximisation of brake energy regeneration in a hybrid electric parallel car. J Int J Electr Hybrid Veh 1:95–121
2. Ko J, Ko S, Son H (2015) Development of brake system and regenerative braking cooperative control algorithm for automatic-transmission-based hybrid electric vehicles. J. IEEE Trans Veh Technol 64:431–440
3. Junzhi Z, Yutong L, Chen L (2014) New regenerative braking control strategy for rear-driven electrified minivans. J. Energ Convers Manag. 82: 135–145 (in Chinese)
4. Yimin G, Ehsani M (2001) Electronic braking system of ev and hev-integration of regenerative braking. In: Future transportation technology conference, SAE International, Costa Mesa, pp. 2001–2478
5. Han J, Park Y (2014) Cooperative regenerative braking control for front-wheel-drive hybrid electric vehicle based on adaptive regenerative brake torque optimization using under-steer index. J Int J Automot Technol 15:989–1000
6. Hongwei G, Yimin G, Ehsani M (2001) A neural network based srm drive control strategy for regenerative braking in EV and HEV. In: Electric machines and drives conference 2001 (IEMDC 2001), IEEE International, IEEE Press, Cambridge, pp 571–575
7. Yin G, Jin J (2013) Cooperative control of regenerative braking and antilock braking for a hybrid electric vehicle. J. Math Probl Eng 7:1–9
8. Maia R, Silva M, Arajo R (2015) Electrical vehicle modeling: a fuzzy logic model for regenerative braking. J. Expert Syst Appl 42:8504–8519
9. Farhad S, Vahid E, Hassan N (2008) Effect of different regenerative braking strategies on braking performance and fuel economy in a hybrid electric bus employing CRUISE vehicle simulation. In: 2008 SAE International Powertrains, Fuels and Lubricants Congress, Shanghai, pp 2008-01-1561

RS-Coded Signal-Carrier Faster-than-Nyquist Signaling in Fading Channels

Xing Liu, Rong Liu, Meng Wen and Xiaohu Liang

Abstract Nowadays, the technology of Faster than Nyquist (FTN) signaling has become one of the most developed technologies. Traditionally, FTN transmission is investigated for a point-to-point AWGN link. The demand to transmit FTN signal in frequency-selective fading channels is still growing dramatically. Rayleigh fading channels are the foundation of all frequency-selective fading channel models. The FTN signal transmits based on Reed Solomon (RS) codes and LMMSE algorithm in Rayleigh fading channel is surveyed in this paper.

Keywords Faster-than-Nyquist (FTN) · Fading channel · Rayleigh channel · Reed Solomon (RS) code

1 Introduction

The way of using mobile and wireless communication would be changed because of societal development. With the number of communicating machines continuing to rise, some essential communications will become more mobile. Although 4G (sometimes called 3.9G), also named as Long-Term Evolution (LTE) could be applied in some countries, the current LTE-Advanced can not accommodate such high traffic without scarifying user experience which can emerge in about 2020. Although 5G would be studied to support the predicted increase in the mobile data volume and

X. Liu (✉) · M. Wen · X. Liang
PLA University of Science and Technology, Nanjing 210007, China
e-mail: m18511170901@163.com

M. Wen
e-mail: wenmeng0507@163.com

X. Liang
e-mail: liangxiaohu688@163.com

R. Liu
Institute of China Electronic System Engineering Corporation, Beijing 100141, China
e-mail: rliu56@aliyun.com

V.E. Balas et al. (eds.), *Information Technology and Intelligent Transportation Systems*, Advances in Intelligent Systems and Computing 455,
DOI 10.1007/978-3-319-38771-0_27

respond to the new and diverse requirements such as mission-critical control, logistics application through a effective combination of evolved existing methods and new radio concepts, the efficient support of a broad range of data rates going from low-rate sensor application up to ultra-high rate multi-media services is a great challenge at the physical layer. The future mobile radio system could consider filterbank based multi-carrier schemes which could allow for the efficient use of fragmented spectrum sharing with other applications as the potential new waveform forms. A potential solution for these challenges is the non-orthogonal transmission schemes to improve the spectral-power efficiency available. Now, The study of Faster-than-Nyquist (FTN) transmission is popular as a technique to improve the spectrum by increasing the data rate.

FTN comes at the cost of a higher complexity of the receiver design. Modern wireless receivers do in reality encounter the different communication channels. Most often, the channel fade. The frequency selective channel varies over time because of the mobility of the transmitter, the receiver, or both. And the frequency selective fading presents varying amplitudes and phases for the different subcarriers used in wireless system [1]. The properties of the fading channel must be learned or evaluated at the receiver for recovering the transmitted signal. To know about the characteristics of the FTN transmission in the fading channel is helpful to the study of wireless communication.

However, FTN signaling in fading channel has not be researched widely. There are some work which is mainly on channel equalization and detection. In [2] there assumed two static multipath channel, these channels in a line of sight component for conditions to implement . In [3] there evaluates the performance of FTN signaling for fading environments to suppose the channels satisfy some assumptions. For instance, the channels are realized by random generation of multipath components, each multipath component is assumed to be IID Gaussian with zero mean. The number of multipath components can vary depending on the environment where the receiver is operating. And the time variance of channel is supposed to be static during receiving the block of information. The extension to fading channel presented in some fading environments, the more in-depth researches about large Doppler and delay spreads are required in the future.

In this paper, we will focus on the performance of FTN transmission in the fading channel. We test the transmit performance of FTN signaling to frequency-selective fading channels which has been presented by Fig. 1. Tests are given for some scenarios. We are concerned with the performance of FTN transmission through the fading channel, so we are interested in the effects of the FTN factor, the roll-off factor and the message length on error performances. We also compare the performance of FTN transmission between the Rayleigh fading channel and the AWGN channel to deepen our understanding of the big difference between the ideal communication and the actual applications.

This paper is organized as follows: Section describes the Faster-than-Nyquist's concept and the model of FTN signal, Section describes the comprehensive system model, Section gives numerical results showing the expression of FTN signal transmission in the fading channel, Section analyzes the results.

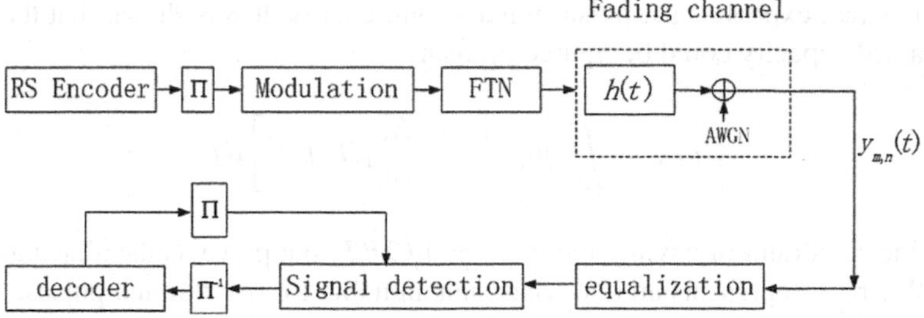

Fig. 1 System model

2 FTN Signaling

Generally digital data is most often sent by trains of pulses [4, 5]. More specifically, a signal would be expressed as

$$s(t) = \sqrt{E_S} \sum_n b_n h(t - nT) \tag{1}$$

Each pulse has energy E_S, T seconds is symbol time and the shape of one pulse is $h(t)$. The Nyquist criterion would be interpreted as

$$\sum_{k=-\infty}^{\infty} \left| H\left(f - \frac{k}{T}\right) \right|^2 = T \tag{2}$$

with transmit pulse shape $h(t)$ and receive matched filter $h^*(-t)$. Where $H(f)$ is just the Fourier transform of $h(t)$. When referred to an FTN signal, the $h(t)$ is no longer orthogonal. Symbol time becomes τT, $\tau < 1$. So the signal becomes

$$s(t) = \sqrt{E_S} \sum_n b_n h(t - n\tau T) \tag{3}$$

The τ can be named as FTN factor, which also could be thought of as a time acceleration factor. If a filter matched to $h(t)$, whose sample at $n\tau T$, is used at the receiver. The intersymbol interference (ISI) could be contained. Shannon defined the capacity of an AWGN channel [6, 7] as

$$C = \int_0^B \log_2 \left[1 + \frac{2E_s}{N_0} |H(f)|^2 \right] df \tag{4}$$

where B is the one sided bandwidth of the signal, E_s means the average power of the signal, N_0 indicates the white noise spectral density, $H(f)$ is the Fourier transform of $h(t)$, which expresses the pulse signal of unit energy. It was shown that the FTN constrained capacity could be figured up using

$$C_{FTN} = \int_0^B \log_2 \left[1 + \frac{E_s}{N_0}|H(f)|^2\right] df \quad (5)$$

When the constraint of having optimal $\tau = 1/2BT$ that prevents the aliasing influence. Where B represents an i.i.d. Gaussian alphabet for bandlimited pulses. When the Nyquist capacity is expressed as

$$C_N = \int_0^{1/2T} \log_2 \left[1 + \frac{2E_S T}{N_0}\right] df \quad (6)$$

Equation (5) is equal to the AWGN capacity equation (4). The excess pulse bandwidth makes the FTN capacity is larger than the Nyquist capacity.

Furthermore, We are focus on the similar degree of the two signals, that is, the two signals that are most easily confused with each other. These are the two that lie closest in terms of Euclidean minimum distance [3]. For linear modulation signals, determine their error rate is a basic parameter, the minimum distance, $d_{\min}$. The $d^2_{\min}$ is the least of the square Euclidean distance between any such pair, which can be express as

$$1/2E_s \int_{-\infty}^{\infty} \left|s_i(t) - s_j(t)\right|^2 dti \neq j \quad (7)$$

To any pulse shape, the $d^2_{\min}$ in (7) with binary orthogonal pulses is always 2. $d^2_{\min}$ would be smaller when the pulses are non-orthogonal in a general way. However, Mazo noticed that $d^2_{\min}$ would not be change for τ in the range [0.802, 1], despite the ISI [3]. So in transmission of an uncoded binary phase-shifting keying (BPSK) sequence, It doesn't decrease the minimum Euclidean distance when sending the sinc pulse up to $1/0.802 \approx 25\,\%$ faster showed by Mazo.

Mazo's result was taken into account after 1990, which was shown that the same phenomenon occurred with root raised cosine (RRC) pulses [8], which was the most commonly used in applications.

3 System Model

Figure 1 shows a general transmit-receive block diagram of the system model. This section extends the earlier development of FTN signaling to frequency-selective fad-ing channels. Reed Solomon (RS) coded system with BPSK modulation would be considered for reasonable energy efficiency with suitably good coding gains. Matched filtering and LLR computation are also used to the fading case. The following two parts respectively introduce the system.

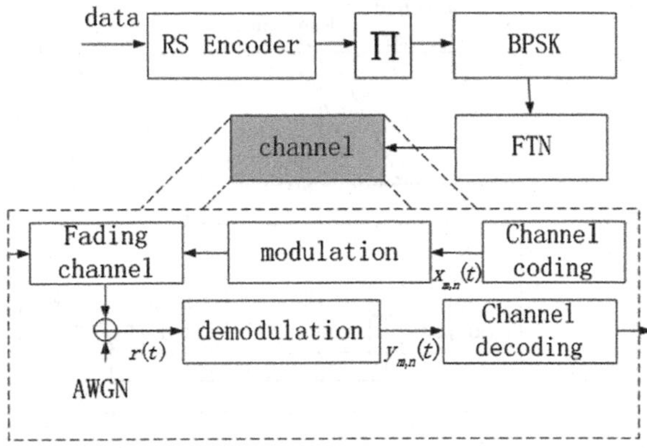

Fig. 2 Transmission system model

3.1 Transmission System Model

Reed Solomon (RS) codes are a particularly important kind of block error correction coding. Figure 2 shows a plain block diagram of RS coded modulation scheme. They are employed in this systems to enhance system robustness against burst errors.

Consider the FTN transmission over the Rayleigh fading channel as shown in Fig. 2. If $r(t)$ is the received signal, the inputs to the channel equalizer $y_{m,n}(t)$ will be

$$y_{m,n}(t) \equiv \int r(t)\psi^{*}_{m,n}(t)dt \tag{8}$$

We can also use a frequency-domain representation of the Eq. (8) such as

$$Y_{m,n} = X_{m,n} \cdot H_m + \eta_{m,n} \tag{9}$$

This frequency domain model is a simplified model of the channel [9, 10] which expresses the channel coefficient at subcarrier. In order to make the model presented in Eq. (9) hold true, we assume the channel parameters from the realizations in this paper remain within the limits in an orthogonal fashion.

3.2 FTN Signal Receiver Processing Through Fading Channel

Rayleigh fading channel is the base of fading channels. When the apparent path between transmitting antenna and receiving antenna exists, the envelope fading abides by

$$\gamma_i = \sqrt{x_i^2 + y_i^2} \tag{10}$$

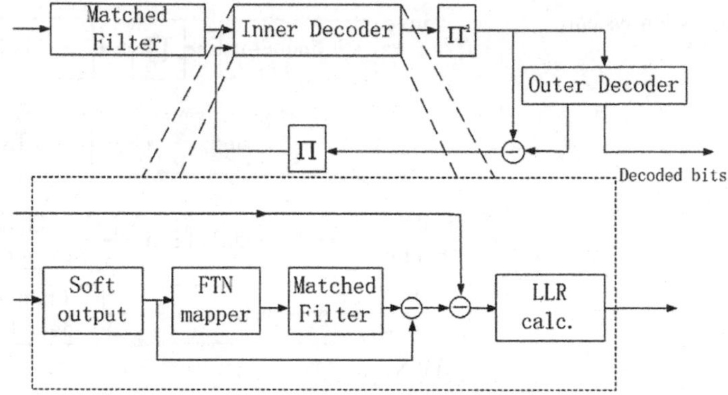

Fig. 3 Receiver system model

x_i and y_i are samples of zero-mean Gaussian random variables each with variance σ^2. The Rayleigh PDF is expressed as

$$f_{Rayleigh}(\gamma) = \frac{\gamma}{\sigma^2} \exp[-\gamma^2/2\sigma^2]\gamma \geq 0 \tag{11}$$

The phase of Rayleigh fading in accordance with uniform distribution of $[-\pi, \pi]$. Modeling the channel behavior and its properties is helpful to wireless communication.

In the presence of a frequency-selective fading channel, the receiver processing also need to know the effects of the channel. The simplified blocks of the receiver processing are showed by Fig. 3.

The processing blocks within the receiver remain similar with the AWGN channel. It's useful to reduce the traditional transmitter-receiver system with an AWGN channel to a system with discrete-time samples, at the symbol rate τT [11]. We consider the method of the whitened matched filter (WMF) receiver.

In the presence of the signals' non-orthogonality, the received filter matched to $h(t)$ is sampled each τT, followed by a noise-whitening filter. The whitening filter is discrete-time, as z-transform is $1/D(1/z^*)$. The data symbols a_n enter the filters can satisfy $r = a * g + \eta$, here $g_k = \int h(t)h(t + k\tau T)dt$, The z-transform $G(z)$ can be expressed as $D(z)D(1/z^*)$ for details see [12, 13]. The output r would become $r' = a * d + \omega$, where ω is the white Gaussian noise. And the $D(z)$ is the channel model of WMF. The WMF receiver scenario can be expressed like this: Transmit Filter $h(t)$-AWGN-Receive Filter $h(t)$-Sample at $n\tau T$-Whitening Filter-LMMSE [14].

4 Simulation and Results

In order to simulate the coded FTN, the information bits are first encoded using the RS code. The coded bits are then modulated using BPSK and transmitted at the different FTN rates. At the receiver side, after passing through WMF and sampling at the FTN rate, the output is passed in the form of LLRs to the RS decoder.

In this section, we provide our simulation results for the FTN transmission in the following scenarios.

Figure 4 shows the BER versus SNR for FTN transmission using the same FTN factor and RRC pulse shape, but with different channels. It can be verified by observing that for the same SNR, the performance of FTN transmission in fading chan-nel is strictly worse than those of FTN transmission in AWGN channel.

Figure 5 shows the data BER performance versus E_b/N_0 for FTN factors $\tau = 0.4$, $\tau = 0.5$, $\tau = 0.6$, and compares them with that of Rayleigh channel transmission.

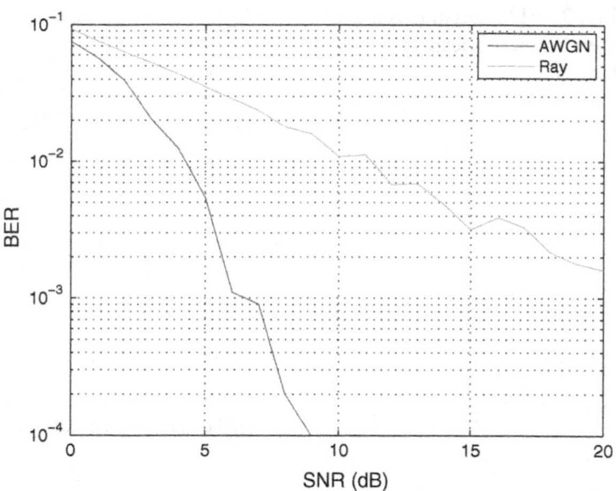

Fig. 4 FTN signal transmit through different channels

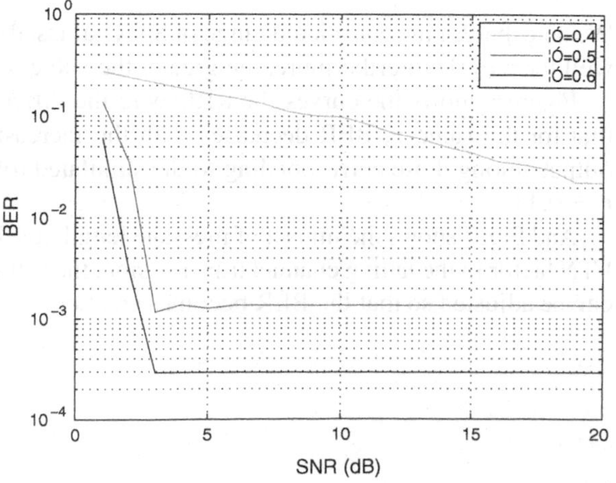

Fig. 5 FTN transmission with different FTN factors

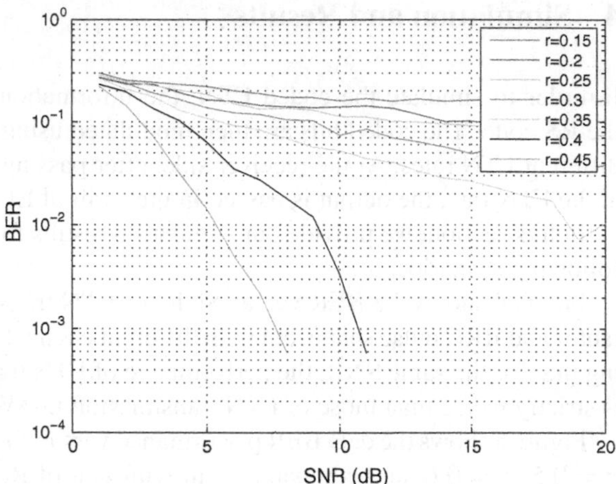

Fig. 6 FTN transmission with different roll-off factors

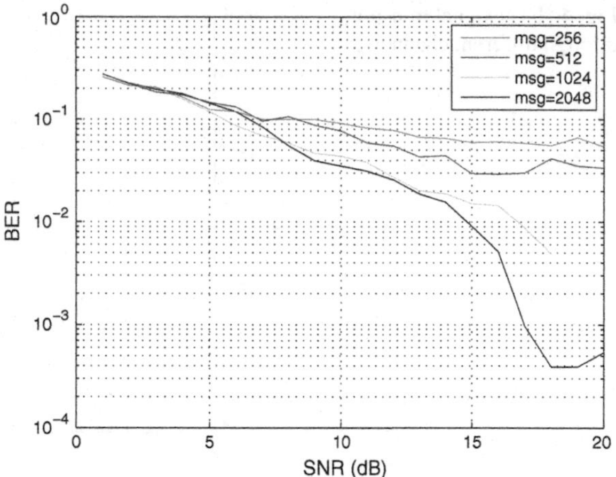

Fig. 7 FTN transmission with different message length

We note that for the considered range of SNR values, the transmission performs quite well for $\tau = 0.6$. Furthermore, the greater the value of τ, the lower the error rate.

Figure 6 shows the curves for RRC with the different roll-off factors. It is seen that for the same τ, BER decreases with the increasing roll-off factors. The best output among those corresponding to the simulated roll-off factors is achieved with $r = 0.45$.

And Fig. 7 shows the impact of message length to BER. Therefore, for each given FTN factor in the fading channel transmission, the roll-off factor and message length can be adjusted so that the BER becomes smaller.

5 Conclusion

We investigated FTN for signal carrier transmission in the fading channel. We interpreted the difference between the AWGN and Rayleigh transmission, and using the simulations, we showed that although FTN signaling transmission becomes poor in the fading channel, the BER would be decreasing with the FTN factor, roll-off factor and message length increasing. Sometimes, we can adjust those parameters to reduce the error rate. To understand the performance of FTN transmission in fading channel could have a great help for our future research.

References

1. Molisch A (2005) Wireless Communication. Wiley, New York
2. Karedal J (2009) Measurement-based modeling of wireless propagation channels-MIMO and UMB. Ph.D. Dissertation, Dept. of Electrical and Information Technology Lund University
3. Dasalukunte D, Owall V, Rusek F, Anderson JB (2014) Faster-than-Nyquist Singnaling, White Paper. Springer, Heidelberg
4. Mazo JE (1975) Faster-than-Nyquist Signaling. Bell Syst Tech J 54(8):1451–1462
5. Proakis JG, Salehi M (2008) Digital Communication, 5th edn. McGraw Hill, New York
6. Shannon E (1948) A mathematical theory of communication, reprinted with corrections. Bell Syst Tech J 27:379–423, 623–656
7. Rusek F, Anderson J (2009) Constrained capacities for faster-than-Nyquist signaling. IEEE Trans Inf Theory 55(2):764–775
8. Liveris AD, Georghiads CN (2003) Exploiting faster-than-Nyquist signaling. IEEE Trans Commun 51(9):1502–1511
9. Cavers J (1972) Variable-rate transmission for Rayleigh fading channels. IEEE Trans Comm 20(1):15–22
10. Goldsmith AJ, Chua SG (1997) Variable-rate variable-power MQAM for Rayleigh fading channels. IEEE Trans Comm 45(10):1218–1230
11. Mazo JE, Landau H (1988) On the minium distance problem for faster-than-Nyquist signaling. IEEE Trans Comm 34(6):1420–1427
12. Liveris AD, Georghiades CN (2003) Exploiting faster-than-Nyquist signal. IEEE Trans Comm 51(9):1502–1511
13. Rusek F, Anderson JB (2008) Non binary and precoded faster than nyquist signaling. IEEE Trans Comm 56(5):808–817
14. ten Brink S (1999) Convergence of iterative decoding. IEEE Electron Lett 35:806–808

Research on the Evaluation Model on the Adaptability of the Support Equipment Basing on the Entropy Weight and TOPSIS Method

Wei Zhao-lei, Chen Chun-liang, Shen Ying and Li Yong

Abstract Aiming at the status quo that the support equipment cannot well meet the support needs in the information war, the concept and connotation of the adaptability were defined basing on the thought of system engineering. And then the intrinsic relationship among support ability, the battlefield environment, support equipment, and its performance was systematically analyzed. And then the evaluation model on the adaptability of the support equipment basing on the entropy weight and TOPSIS method was constructed. And the evaluation model example of the support equipment in 9 kinds of environments was developed as the validation example. And it proved feasible and rational. At last, an application example in the performance measurement of supply and demand coordination is demonstrated.

Keywords Support equipment · Adaptability · Evaluation

Adaptability refers to the ability to realize its expected function and performance under different environments by changing system structure. And it is a time-effect dynamic characteristics [1]. And its target is to meet all environmental needs. And its main means are changing its functional structure. Now it is a hotspot of current research how to evaluate the system objectively and reasonably to optimize the structure and performance of its system. Currently, the research on the adaptability evaluation index system for the support equipment is not enough scientific and rational. And they mainly considered the adaptability between the operational personnel,

W. Zhao-lei (✉) · C. Chun-liang · S. Ying · L. Yong
Department of Technical Support Engineering,
Academy of Armored Force Engineering, Beijing, China
e-mail: zl.vi@163.com.cn

C. Chun-liang
e-mail: 303076486@qq.com.cn

S. Ying
e-mail: 957002130@qq.com.cn

L. Yong
e-mail: 975107319@qq.com.cn

V.E. Balas et al. (eds.), *Information Technology and Intelligent Transportation Systems*, Advances in Intelligent Systems and Computing 455,
DOI 10.1007/978-3-319-38771-0_28

equipment and the natural environment in different training and battlefield conditions. Or the research was developed just on the evaluation method. They all failed to consider the intrinsic relationship among support ability, the battlefield environment, support equipment, and its performance [2]. The evaluation model on the adaptability of the support equipment basing on the entropy weight and TOPSIS method was constructed to meet all the support needs in the informatization war.

1 The Connotation Analysis on the Adaptability

According to the definition of adaptability by Holland [3], the adaptability of support equipment refers to the satisfaction degree of all the actual support needs which is meted by the supportability of the support ability in the specified conditions. The higher the satisfaction degree is, the better the adaptability of support equipment is.

If the support equipment is treated as a "system", its current "environment" is the information warfare. And the environment raises the requirements to the function structure and performance by bringing up the support needs. And the action of the support is to meet the support needs by changing its functional structure and performance indicators. And eventually, the system can meet all the support needs raised by the environment in real-time. Support equipment adaptability contains the main contents of three aspects. They are the adaptability between the support equipment to the battlefield environment, the matching between the support equipment and the equipment-to-support, the implementation of its functional structure and performance indicators.

(1) The adaptability between the support equipment to the battlefield environment. As information technology and high-tech weapons are widely used, the battlefield space and damage rate are unprecedented increasing. And it raises a severe test to the information application level, battlefield survivability, adaptability in various environments of the support equipment. Therefore, it is the primary factor, whether it can meet the needs of various environment or not, to judge the overall adaptability of the support equipment.

(2) The matching between the support equipment and the support needs of equipment-to-support. The support object of the support equipment is the main battle equipment. To meet all the demand of the systematic battle based on the information system, the functional structure and performance indicators of the main battle equipment have improved in a larger range than ever before. And the decrease in the autologous supportability of the main battle equipment greatly increased its dependence on its support equipment at the same time. The matching level from its functional structure and performance indicators to the support needs of the equipment-to-support will deeply impact the maintaining and improvement of the combat effectiveness. Therefore, it is the important factor to evaluate the overall adaptability of the support equipment judging by the matching between the support equipment and the equipment-to-support.

(3) The implementation of its functional structure and performance indicators. The implementation level of good adaptability in battlefield environmental and the matching with the support needs of equipment-to-support depends on the following three factors. Firstly, the reliability, maintainability and supportability are the important foundation to ensure the implementation its function of the support equipment. Secondly, the man is the operator of the support equipment. So the rationality level of man—machine C environment design will directly impact the generation effect of the support capability. Thirdly, the economy level of the equipment support in the whole life cycle is usually restricts the functional structure and performance indicators. Therefore, the implementation level of its functional structure and performance indicators is the crucial factor to judge the overall adaptability of the support equipment.

2 The Establishment of the Adaptability Evaluation Model for the Support Equipment

2.1 The Construction of the Adaptability Evaluation Indexes System for the Support Equipment

The adaptability evaluation for the support equipment refers to the evaluation research on the matching level between the supportability in completing the support mission under specified conditions and the comprehensive needs in real support process using qualitative and quantitative methods. And the analysis process of the adaptability evaluation system for the support equipment is shown as in Fig. 1.

The scientific and comprehensive adaptability evaluation index system for the support equipment is the basis and prerequisite for the reasonable adaptability evaluation. And the construction of the adaptability evaluation indexes system involves a series of factors. Simultaneously, all the indexes will combine into an organic whole according to the relations of the hierarchical structure [4].

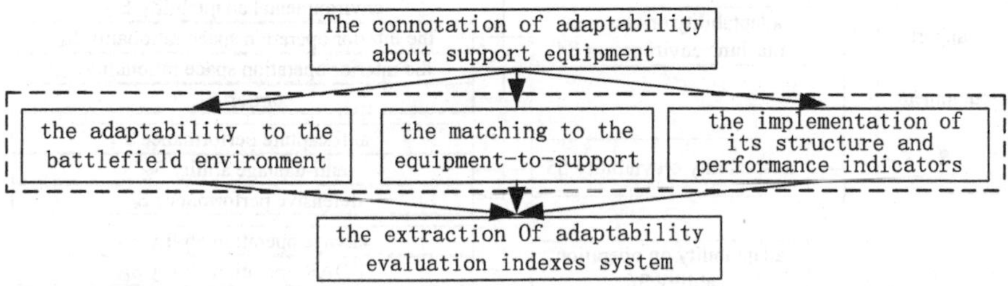

Fig. 1 The analysis process of the adaptability evaluation system for the support equipment

Basing on the principle of completeness, objectivity, independence, feasibility, comparability, the consistency, as well as the combination of quantitative and qualitative, the evaluation indexes system for the support equipment is constructed considering meeting the comprehensive support needs in wartime. And then the elementary indexes system is reduced from the angles of the importance, validity, stability by using the CVS (Concernment Estimate, validity Estimate, Stability Estimate, CVS) [5]. Ultimately, Considering about the adaptability on information application ability, adaptability on mobility, adaptability on RMS of the support equipment, adaptability on the man-machine- environment, adaptability on viability and adaptability on operation ability, the adaptability evaluation indexes system for the support equipment is established [6]. And the evaluation index system is shown as in Fig. 2.

Based on the above method, the adaptability evaluation index system of the three layer structure is constructed. Among them, the top level is the target layer of the adaptability evaluation indexes system for the support equipment. The secondary layer is the performance layer, which is used to character each sub-goal. And the performance mainly involves five indicators that is adaptability on information application ability, adaptability on mobility, adaptability on RMS of the support equipment, adaptability on the man-machine environment, adaptability on viability and adaptability on operation ability. The third layer is the characteristics layer. And it is mainly composed of the parameter to directly observe and measure every performance. In the hierarchical structure, one secondary adaptability indicator is usually

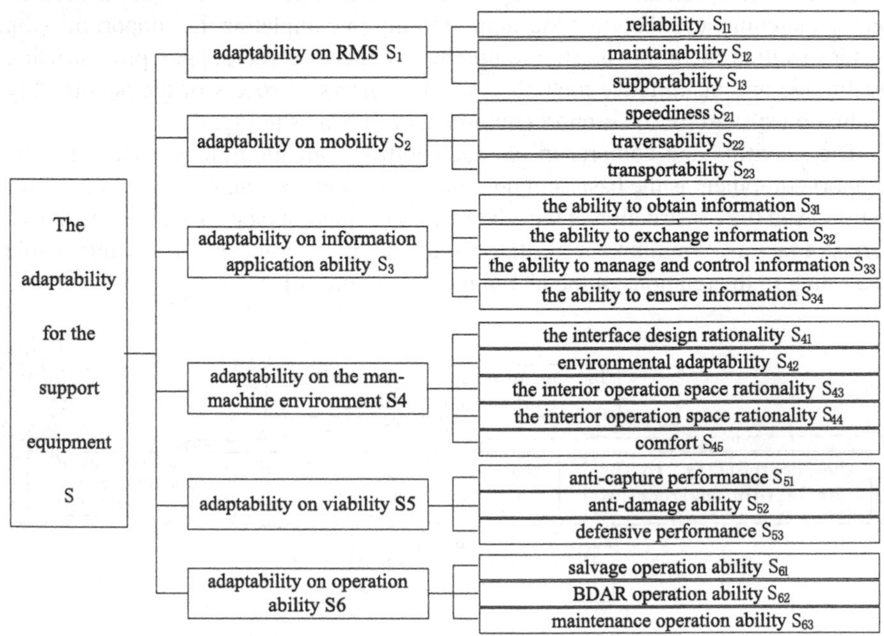

Fig. 2 The adaptability evaluation indexes system for the support equipment

held out by one or some sub-indicators. And sub-indicators maybe reflect one or some secondary adaptability indicators.

Adaptability on RMS of the support equipment refers to the capability level to reach and keep its functional design index after putting into use. It is usually characterized by reliability, maintainability and supportability parameters.

Adaptability on mobility refers to the adaptability level of mobility in the support process under the informatization battlefield. And it is usually characterized by speediness, traversability, and transportability.

Adaptability on information application ability refers to the adaptability level of information application ability in the equipment support process under the informatization battlefield. And it is usually characterized by the ability to obtain, exchange, manage and control, as well as ensure information.

Adaptability on the man-machine-environment refers to optimization combination level of the man-machine-environment aiming at the safety and comfort, high-efficiency and economy. And it is usually characterized by the man-machine interface design rationality, the interior operation space rationality, comfort, environmental adaptability.

Adaptability on viability refers to the adaptability level to achieve all its expected function, performance under various battlefield environments. And it is usually characterized by the anti-capture performance, anti-damage ability, defensive performance etc.

Adaptability on operation ability refers to the adaptability level to complete the equipment support mission by support equipment in the specified condition. And it is usually characterized by salvage operation ability, BDAR operation ability, and maintenance operation ability.

2.2 The Comprehensive Adaptability Evaluation on the Support Equipment Basing on the Entropy Weight and TOPSIS Method

Aiming at the plain area, sunlight area, alpine region, plateau region, high humidity area, high temperature area, high salt-spray area, complex terrain, and high sandy area, the adaptability evaluation is conducted taking a certain type of support equipment as example. The primary data is got from the questionnaires from four sets of professional experts in different fields related to the equipment support. And then the data is corrected and calibrated basing on the comprehensive analysis of the historical data. Finally, the reasonable raw data for the corresponding parameters is obtained. Supposing that the adaptability evaluation will be conducted in m kinds of environments, and the number of the evaluation index is n. The value of evaluation index $D_j(j = 1, 2, \ldots, n)$ under the environment $M_i(i = 1, 2, \ldots, m)$ is M_{ij}.

The comprehensive adaptability evaluation on the support equipment basing on the entropy weight and TOPSIS method is conducted taking evaluation on the RMS of the support equipment as example. And other indexes follow the same steps. So the raw evaluation data matrix M of the RMS of the support equipment is shown as follows.

$$\mathrm{M} = \mathrm{M_{ij}n} \times \mathrm{m} = \begin{pmatrix} 52 & 31.5 & 40.25 & 62 & 55.25 & 24 & 10 & 9.75 & 16.5 \\ 0.125 & 0.1125 & 0.1375 & 0.2875 & 0.075 & 0.1125 & 0.05 & 0.0375 & 0.0625 \\ 386400 & 220000 & 126100 & 484900 & 264917 & 94200 & 58910 & 35800 & 13816 \end{pmatrix}$$

Using the formula $\mathrm{P_{ij}} = \mathrm{M_{ij}} / \sum_{i=1}^{j} \mathrm{M_{ij}}$, the raw data matrix M is standardized. And then the matrix P is obtained.

$$\mathrm{P} = \begin{pmatrix} 0.1726 & 0.1046 & 0.1336 & 0.2058 & 0.1834 & 0.0797 & 0.0332 & 0.0324 & 0.0548 \\ 0.125 & 0.1125 & 0.1375 & 0.2875 & 0.075 & 0.1125 & 0.05 & 0.0375 & 0.0625 \\ 0.2293 & 0.1306 & 0.0748 & 0.2877 & 0.1572 & 0.0559 & 0.035 & 0.0212 & 0.0082 \end{pmatrix}$$

Employing the formula $ej = -\frac{1}{\ln m} \sum_{i=1}^{m} pij \times \ln pij,\ aj = (1 - ej) / \sum_{j=1}^{n} (1 - ej)$ the entropy weight and weight of the three secondary indexes for RMS is determined [7, 8]. And the weight matrix X is shown as follows.

$$X = \begin{pmatrix} a_1 & & & \\ & a_2 & & \\ & & \cdots & \\ & & & a_j \end{pmatrix} = \begin{pmatrix} 0.2013 & & \\ & 0.2413 & \\ & & 0.5574 \end{pmatrix}$$

Employing the formula $S = X \times (pij)n \times m = (aj \times pij)n \times m,\ S = X \times (pij)n \times m = (aj \times pij)n \times m$ the matrix S can be calculated.

$$\mathrm{S} = \begin{pmatrix} 0.0418 & 0.0253 & 0.0324 & 0.0499 & 0.0444 & 0.0193 & 0.0080 & 0.0078 & 0.0133 \\ 0.0314 & 0.0283 & 0.0346 & 0.0722 & 0.0188 & 0.0283 & 0.0126 & 0.0094 & 0.0157 \\ 0.1161 & 0.0661 & 0.0379 & 0.1457 & 0.0796 & 0.0283 & 0.0177 & 0.0107 & 0.0041 \end{pmatrix}$$

Employing the formula $\begin{cases} \mathrm{Sj}^{+} = \max(\mathrm{S1j, S2j, \ldots, Smj}) \\ \mathrm{Sj}^{-} = \min(\mathrm{S1j, S2j, \ldots, Smj}) \end{cases}$, $\begin{cases} di^{+} = \sqrt{\sum_{j=1}^{n} (\mathrm{Sij} - \mathrm{Sj}+)^2} \\ di^{-} = \sqrt{\sum_{j=1}^{n} (\mathrm{Sij} - \mathrm{Sj}-)^2} \end{cases}$, $\mathrm{Ci} = 100 \cdot \frac{d_i^+}{d_i^+ + d_i^-}$ the relative approximation of the plain area Ci of the adaptability on RMS under different environments can be calculated [9, 10]. And the results are shown as in Table 1.

Following to the same steps, the relative approximation and rank aiming at the overall goal of the adaptability for the support equipment can be calculated. And so

Table 1 The relative approximation of RMS in different environments

	Alpine region	High temperature area	High humidity area	High sunlight area	High salt-spray area	High sandy area	Complex terrain	Plateau region
Relative approximation of the plain area Ci.	36.281	24.193	19.071	87.733	33.580	11.135	33.719	15.171

can the relative approximation and rank aiming at the adaptability on RMS, adaptability on mobility, the adaptability on information application ability, adaptability on man-machine environment, adaptability on viability, adaptability on operation ability. And the results are shown as in Tables 2 and 3.

2.3 Analysis on Evaluation Result

Generally speaking, the plain area is where the adaptability for some support equipment under informatization battlefield in the plain area is strongest. And it is followed by high sunlight area, alpine region, plateau region, high humidity area, high temperature area, high salt-spray area, complex terrain, high sandy area. For the alpine region, the adaptability on mobility, the adaptability on information application ability, as well as adaptability on viability maybe seriously restricts the completion of the support mission.

For the high temperature area, the adaptability on RMS, adaptability on man-machine- environment, as well as adaptability on operation ability maybe very seriously restricts the completion of the support mission.

For the high humidity area, the adaptability on RMS and adaptability on mobility be strengthened. And they maybe restrict the completion of the support mission to some extent. For the high smog area, the adaptability on RMS and adaptability on viability maybe very seriously restrict the completion of the support mission.

For the high sandy areas, the adaptability on RMS and adaptability on mobility maybe seriously restrict the completion of the support mission.

For the plateau region, the adaptability of viability to some extent restricts the completion of the support mission.

For the complex terrain, the adaptability on mobility, adaptability on information application ability, and adaptability on man-machine- environment adaptability on viability maybe to some extent restricts the completion of the support mission.

Table 2 The adaptability evaluation results of the each parameter in the performance layer under different environments

	Adapt ability on RMS		Adapt ability on mobility		Adapt ability on information application ability		Adapt ability on man machine environment		Adapt ability on viability		Adapt ability on operation ability	
	Ci	Rank	Ci	Rank	Ci	Rank	Ci	Rank	Ci	Rank	Ci	Rank
Alpine region	69.979	2	6.259	8	2.432	9	21.856	5	3.850	9	17.230	6
High temperature area	20.223	6	40.067	4	9.549	7	7.550	8	15.690	5	77.443	8
High humidity area	8.629	7	21.982	6	41.228	3	19.348	6	20.052	3	38.656	3
High sunlight area	29.650	5	64.964	2	58.359	2	35.989	2	38.615	2	66.043	2
High salt spray area	5.030	8	32.190	5	26.629	5	28.351	4	7.818	8	43.664	4
High sandy area	4.095	9	3.721	9	4.181	8	3.587	9	11.692	6	6.801	9
Plain area	100	1	100	1	98.756	1	99.820	1	18.631	4	80.985	1
Plateau region	41.595	4	46.699	3	30.686	4	30.624	3	10.633	7	32.091	5
Complex terrain	49.775	3	8.063	7	25.123	6	11.213	7	92.556	1	13.540	7

Table 3 Comprehensive adaptability evaluation for the support equipment under different environment

	Alpine region	High temperature area	High humidity area	High sun light area	High salt spray area	High sandy area	Plain area	Plateau region	Complex terrain
Ci	24.983	28.420	23.947	48.937	20.268	5.679	83.032	32.055	33.378
Rank	6	5	7	2	8	9	1	4	3

3 Conclusions

Basing on the systematic ad adaptability theory, the evaluation index system is conducted. And the conduction well considers about the adaptability between the support equipment to the battlefield environment, the matching between the support equipment and the equipment-to-support, the implementation of its functional structure and performance indicators. And then the weight of each index is determined by employing the entropy weight theory. And the comprehensive evaluation on the adaptability evaluation of the support equipment is conducted by employing the TOPSIS method. Therefore the adaptability evaluation model of the support equipment is established. And the model can reduce the subjectivity caused by expert scoring to a certain extent. Whats more, it can improve the scientific and rationality of the adaptability evaluation, which can lay the technical foundation to improving the adaptability of the support equipment.

References

1. Tai-liang SONG (2008) Supportive system engineering. National Defense Industry Press, Beijing
2. Dan-yang Z (2007) Research on the adaptability evaluation of armored support material. Academy of armored force engineering
3. Fan-ke KONG, Chun-liang CHEN (2009) Thinking on supportability objective for equipment development: discussion on equipment mission-performing capability. J Acad Armor Forc Eng 23(2):1–5
4. Hong-wei W, Yun-feng Z, Shuai T (2013) On environmental worthiness evalua-tion during whole life cycle of equipment. Equip Environ Eng 10(1):70–72
5. Xu-cheng Z, Jun-weiZHU MA, Yi-tian (2014) Research on plateau environmental wor-thiness of support equipment. Equip Environ Eng 11(5):27–31
6. Li-e MA (2006) Research and analysis of ship weapon equipment suitability of environment. Ship Sci Technol
7. GJB 4239: General requirements for material environmental engineering
8. Jiulin L, Zhaolei W, Hongping P (2013) Application of entropy weight-set pair analysis method in efficiency evaluation of emergency repair. Ordnance Ind Autom 32(5):10–13
9. Zhao-lei WEI, Chun-liang CHEN, Hui LIU (2015) Research on the evaluation of equipment supportability based on ANP and MCGC. Comput Measure Control 23(4):1238–1242
10. Xiu-peng W, Chun-run Z, Ya-dong L (2012) Study of equipment support forces complex adaptability simulation experimentation. Command Control Simul 32(4):65–70

Secure Geographic Based Routing Protocol of Aeronautical Ad Hoc Networks

Song-Chao Pang, Chang-Yuan Luo and Hui Guan

Abstract Aiming at the problem that routing security for Geographic information cannot be guaranteed, a Secure Geographic based Routing Protocol of Aeronautical Ad hoc Networks (S-GRP) is proposed. Through identity-based public key cryptosystem, it makes authentication for neighbor node and establish a shared secret in the neighbor discovery phase, and then makes signcryption for data in the data forwarding phase. When in the next hop selection, it makes an integrated assessment for node location-speed factor and trust factor, selecting the optimal next hop. Security analysis and performance analysis for S-GPR show that the algorithm is effective and feasible.

Keywords Routing security · Aeronautical ad hoc networks · Geographic information · Routing protocol

1 Introduction

At present, with the rapid development of the aviation industry, especially the gradual opening of low-altitude areas proposed by the States, the various types of aircraft fly density increase rapidly, and communications support capabilities at this stage cannot meet the needs of Air Traffic Management (ATM) in the future [1]. Aeronautical Ad hoc Networks (VANETs) is applications of MANET in aerospace, and the main direction of future development in aviation network. The topology is shown in Fig. 1.

S.-C. Pang (✉) · C.-Y. Luo · H. Guan
Zhengzhou Institute of Information Science and Technology,
Zhengzhou 450001, China
e-mail: pangsongc@163.com

C.-Y. Luo
e-mail: luocyzz@126.com

H. Guan
e-mail: 83570798@qq.com

V.E. Balas et al. (eds.), *Information Technology and Intelligent Transportation Systems*, Advances in Intelligent Systems and Computing 455,
DOI 10.1007/978-3-319-38771-0_29

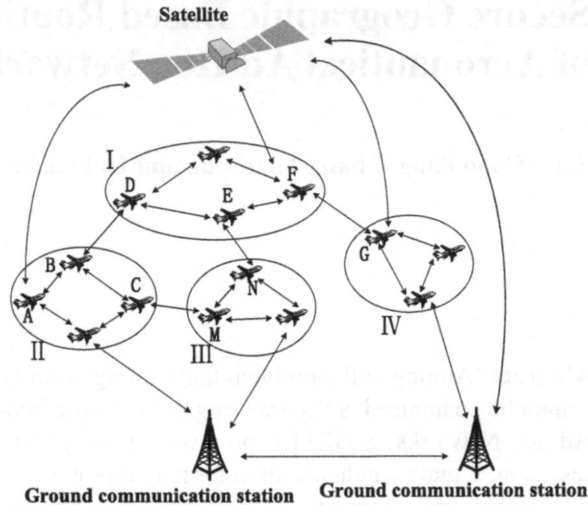

Fig. 1 Network structure of Aeronautical Ad hoc Networks

Many of the current domestic and foreign research institutions and scholars conduct study on aviation Ad Hoc Networks and achieved certain results. Ehssan et al. propose MUDOR [2], using aircraft Doppler shift to obtain relatively stable routes. On this basis, through increasing support for QoS, Ehs-san et al. propose a Qos-MUDOR [3] routing, introducing the optimal forwarding the request strategy. Greedy Perimeter Stateless Routing (GPSR) [4] uses geographical information for routing. In data forwarding phase, node uses greedy algorithm to choose the nearest neighbor to the destination node as the next hop. Its biggest drawback is without considering movement factor of nodes, which does not apply to high dynamic network. Based on GPSR, NEWSKY project propose geographic load share routing (GLSR)[5] by increasing the shortest queue principle, which is an effective solution to the congestion problem GPSR, but the emulation only considers static topology, lack of persuasiveness. AeroRP [6] protocol proposed by University of Kansas selects routes based on intercepted time between nodes, but without considering the local stability characteristics of the node.

Kumar put forward SCMRP [7], intermediate malicious nodes cannot have an impact on the path selection, effectively preventing wormhole attack, but not apply to Aeronautical ad hoc networks without infrastructure. Pathan et al. put forward SERP [8], which makes node authentication via shared key stored in advance, and authenticates the data packets by one-way hash chain during the message transmission, but it cannot meet the high dynamics of aeronautical ad hoc network and prevent internal malicious nodes attack. Reference [9] proposed an aeronautical network safety geographic information routing based on ADS-B system, which ensures the confidentiality of geographic information by cryptographic algorithms, and ensures data security by dividing the route, but it only works on Civil Aviationcan with advance route planning.

As the positioning technologies become more sophisticated, routing protocol based on geographical information can quickly and accurately obtain the network topology information, better adapted to highly dynamic, large-scale network, becoming a hot spot in aeronautical hoc routing algorithm [10]. Identity-based cryptography scheme does not require a public key certificate, and key management mechanism is relatively simple, which has a great advantage in wireless ad hoc network [11], and it is well suited for aeronautical ad hoc networks with no central node. This paper designs a Secure-Geographic based Routing Protocol (S-GRP), establishes a new next-hop node selection mechanism through the combination of node locations velocity factor and trust factor, and achieves the node authentication, message authentication and encrypted transmission through identity-based signatures and encryption.

System initialization: First, the credibility PKG makes initialization of security mechanisms for each airplane nodes within the network system. PKG selects two cyclic groups G_1 and G_2 of order q (q is a prime number greater than 160), the generator of G_1 is P, bilinear mappings $e : G_1 \times G_1 \rightarrow G_2$, given the security parameter n. KG randomly selects $k \in Z_q^*$ as the master key of system, computes the public key $P_{pub} = kP$ of system; selects $R \in G_1^*$, and computes $\theta = e(R, P_{pub})$. Define the following secure hash function: $H_0 : \{0, 1\}^* \rightarrow G_1$, $H_1 : G_1 \times \{0, 1\}^* \rightarrow Z_q^*$, $H_2 : G_2 \rightarrow \{0, 1\}^*$. PKG keeps secret master key k of system, and discloses system parameters $\{G_1, G_2, q, P, P_{pub}, e, R, \theta, H_0, H_1, H_2\}$.

Each aircraft node in the network can compute the corresponding public key of node i based on parameters and node status information ID_i disclosed by PKG, that is $k_{i+} = H_0(ID_i)$. The private key of aircraft node i is precalculated by the PKG, and transmitted to the node i via a secure channel.

2 S-GRP Protocol Description

2.1 Neighbor Discovery

Each node performs neighbor discovery by sending HELLO message periodically, and HELLO message format sent by node A is as following:

$$A \rightarrow broadcast : HELLO = hello, \{GI_A\} k_{A-}, ID_A \quad (1)$$

where, hello represent that this packet type is a neighbor discovery message, $\{GI_A\} k_{A-}$ represents using the private key k_{A-} of A to sign geographical information GI_A of A. After receiving this message, the neighbor node B first computes node A's public key k_{A+}=$H_0(ID_A)$ by ID_A to verify the signature part of HELLO message. If verification is not passed this HELLO message will be discarded. If the signature can be verified, the node B checks whether there is a shared key K_{AB} with the node A, if not K_{AB} is generated and has public-key encryption by node A, Node B uses its own private key and public key of the node A to signcrypt on C and combine into reply message, the reply message is as follows:

$$B \rightarrow A : REPLY = reply, ID_B \parallel SignCryptt_{A,B}\{K_{AB}, GI_B\} \quad (2)$$

If there is K_{AB} no longer generated, after node A receives a reply message, first make inverse signcryption for the latter part of the package by its own private key k_{A-} and public key k_{B+} of node B to obtain content and validation results for routing messages. If the message content failed legality verification, then discard the packet. If the message content through legality verification, carry out neighbor table update according to the decrypted GI_B, and save the shared key K_{AB} of the two nodes. By periodically neighbor discovery, there will be a shared key K_{ij} between the neighbor nodes, which can be used as the session key in data forwarding phase.

2.2 Next Hop Selection

Location-Speed Factor In Aeronautical Ad hoc Networks, it is assumed that aircraft node M is candidate next hop node of aircraft node i, the node D is the destination node, the communication range of the aircraft nodes is R. from the geographical information can obtain that the position coordinate of M is (x_1, y_1, z_1), the speed vector is $\overrightarrow{v}_M = (x_m, y_m, z_m)$; the position coordinate of D is (x_2, y_2, z_2), the speed vector is $\overrightarrow{v}_D = (x_d, y_d, z_d)$, as shown in Fig. 2.

The Euclidean distance between Node M and Node D is:

$$|\overrightarrow{MD}| = \sqrt{(x_2 - x_1)^2 + (y_2 - y_1)^2 + (z_2 - z_1)^2} \quad (3)$$

Angle between $\overrightarrow{MD}$ and $\overrightarrow{v_{MD}}$:

$$\angle\varphi = \arccos \frac{\overrightarrow{MD} \bullet \overrightarrow{v_{MD}}}{|\overrightarrow{MD}| * |\overrightarrow{v_{MD}}|} \quad (4)$$

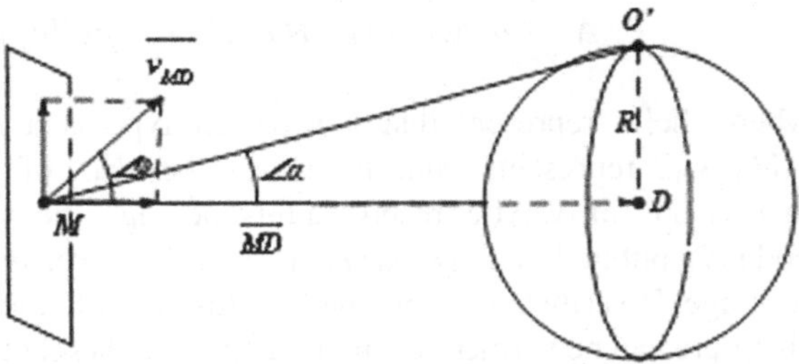

Fig. 2 A schematic view of the collision time calculation

Herein:

$$\angle\alpha = \arctan \frac{R}{|\overrightarrow{MD}|} \tag{5}$$

We use the arrival time T of the next hop moving to the communication boundary of the destination node to evaluate the merits of a next hop node. To sum up, each candidate next hop node T can be calculated:

$$T = \begin{cases} +\infty, & \angle\varphi > \angle\alpha \wedge \left|\overrightarrow{MD}\right| > R \\ \frac{|\overrightarrow{MD}|-R}{|\overrightarrow{v_{MD}} \bullet \cos\varphi|}, & \angle\varphi > \angle\alpha \wedge \left|\overrightarrow{MD}\right| > R \\ 0, & \left|\overrightarrow{MD}\right| < R \end{cases} \tag{6}$$

Define the location-speed factor as:

$$\Delta_{iM} = \frac{1}{T+1} \tag{7}$$

Trust Factor Suppose trust assessments between nodes are conducted periodically, in each survey cycle node i and node M will perform a certain number of neighbor discovery and data forwarding process. In each neighbor discovery and data forwarding process, it will be accompanied by the mutual authentication process between nodes. $\hat{T}_M(i,n)$ is defined to represent the trust value of the node i to node M after n statistics periods, and the trust value of the initial node is $\hat{T}_M(i,0) = T_{MIN}$. T_{MIN} is minimum of nodes can participate in the routing process; $T_M(i,n)$ represents trust of node i to node j in the nth period, which is calculated as follows:

$$T_M(i,n) = N_{ver}(i,n)/N_{rec}(i,n) \tag{8}$$

A and B respectively represent the number of sending packets received by node i from node M in the nth period and the number of packets through authentication or successful forwarding packets. The number of packets is counted by two-jump pass mechanism. After each statistical period, the updating rule of trust value is as follows:

$$\hat{T}_M(i,n+1) = \lambda\hat{T}_M(i,n) + (1-\lambda)T_M(i,n), 0 < \lambda < 1 \tag{9}$$

λ represents the weight of historical trust value of evaluation and current observations. The trust factor is defined as follows:

$$\nabla_{iM} = \hat{T}_M(i,n) \tag{10}$$

Based on the above analysis, comprehensive factor need to combine the location-speed information of node and trust level, and accordingly select the optimal next hop. In order to weigh the relationship between the two factors, combine with formula (7, 10) to give a comprehensive selection factor ψ as:

$$\psi(i, M) = (\varDelta_{iM} + 1)^{\nabla_{iM}} - 1 \tag{11}$$

The system set location-speed factor and trust factor threshold $\varDelta_{\min}$ and $\nabla_{\min}$ according to the level of dynamics of node and size of safety requirements, and select the next hop node with largest comprehensive factor on the basis of compliance with the minimum threshold value.

2.3 Data Forwarding

Suppose the set composed of all neighbor nodes of the source node S is S[i], S would like to send a message DATA to the destination node D. before sending packets, S first computes the comprehensive selection factor $\psi(S, i), i \in S[i]$ of all the next hop node through the next hop node selecting algorithm. Assume that $\psi(S, n) = Max[\psi(S, i)]$.

Node S first generates a random number R_{SD}, and tests the identity of destination node by way of response to the random number. The packet format is defined as follows:

$$I_D = [GI_D, ID_D, Seq]\,K_{Sn}, TTL \parallel SignCrypt_{S,D}\{DATA, R_{SD}, GI_S, ID_S\} \tag{12}$$

For the former half of the packet, only node n can decrypt the geographical information of destination node according to the shared key K_{Sn}, for routing in the next step. TTL represents the maximum number of hops, and the forwarding value reduces one per pass. When TTL $= 0$ stop forwarding, the packet is discarded. For the latter part of the package, only the destination node can perform opposite signcryption by using its own private key and public key of the source node, get the message content and make legality verification on node S.

After the node n receives a data packet, it works as the following steps.

Step 1: Check whether the TTL is greater than 0, if equal to 0, discard; if greater than 0, then the next step.

Step 2: decrypt the first half of the package through shared key K_{Sn} with node S, and check whether Seq is repeated, discarded directly if repeated; or the next step.

Step 3: according to decrypted identity information to determine whether itself is the destination node or not; if yes, use their own private key and public key of source node to opposite signcrypt the latter part of the package to give the message content and validation results, and go to Step 4; or go to Step 5.

Step 4: If the message pass the legitimacy test, the next step is performed according to the message contents of the package, and put random number R_{SD} on the second half of the message in the replying message to the source node S for the identity inspection of the node D; if the message does not pass the legitimacy test, directly discard the packet.

Step 5: node n start the next hop node selection process and choose the next hop according to the geographic location information of the destination node, assumption that the selected hop node is p. Use shared key K_{np} between node n and node p encrypt the first half of package, TTL value minus one:

$$I_D = [GI_D, ID_D, Seq]\, K_{np}, TTL \parallel SignCrypts_{S,D}\{DATA, R_{SD}, GI_S, ID_S\} \tag{13}$$

Forward this packet to the next hop.

Step 6: After each intermediate node receives a data packet, in accordance with the processing work of node n until the packet reaches the destination node.

3 Security and Performance Analysis

3.1 Security Analysis

Replay attack: packet sequence number mechanism is added into the packet, and encrypting with a shared key between nodes, which can quickly verify duplicate packets, find and discard the replayed data packets. External nodes cannot obtain the correct serial number through decryption, and therefore cannot implement effective replay attacks.

Sybil attack: in the neighbor discovery phase, each node and its neighbor nodes perform mutual authentication by identity-based signature, and establish a shared secret. Nodes with forged identity can be effectively seen through by other nodes, it is possible to effectively prevent Sybil attacks.

Confidentiality, integrity and non-repudiation of routing messages: Neighbor Discovery messages perform encryption and signature through public key system, external malicious node which cannot be verified cannot get geographical information of node, to ensure the confidentiality of geographic information. Routing packets are protected by signcryption, so external nodes cannot get the true content of the message packet, to ensure the confidentiality of data. When the content of the package has been tampered with external nodes, other nodes will identify by opposite signcryption, thus discarding the tampered package, it will not allow illegal packets to expand its influence to ensure the integrity of the data. The data packets have all passed signcryption, to ensure non-repudiation of data.

Selfish behavior of internal nodes: both neighbor discovery and routing process are accompanied with trust assessment process of nodes, which assess the trust of node by monitoring communication behavior of node. Selfish behavior of node will reduce its trust value, when less than a minimum threshold, it will be disqualified to participate in routing, which can effectively suppress selfish behavior of nodes, to ensure the efficiency of network routing.

3.2 Analysis of Computational Overhead

This section analyzes the increasing additional overhead due to the added security in the routing process. As mentioned before, the main security operation has three parts: signatures and verification in the neighbor discovery process; trust evaluation in next hop selection process; shared key encryption and decryption and signcryption operations in data forwarding process. When analyzing the cost of algorithm, the main consideration is time-consuming public key algorithm with high computational complexity, ignoring the symmetric ciphers with smaller cost. For S-GRP algorithm, main analysis are "for operation", the multiplication operation of P, G1 and index of G2 these three cryptographic operations in identity-based public key cryptography algorithm (Table 1).

Assuming in a routing process, each node needs to update the neighbor table, the routing information sent by the source node S reach the destination node D through n skip. The main operations are shown in Table 2. Neighbor discovery process between nodes is proceeding in parallel, and the consumption of time is equivalent to that of neighbor discovery. According to the current computing power of onboard computer, the computing time of the operations above is in milliseconds, and the operation do not increase additional route control packets. It is considered that the introduction of security mechanisms will not bring significant delay and increasing overhead to the routing process.

Table 1 Main cryptographic operations in S-GRP

Node	Source node	Intermediate node			Destination node
Process	S	i_1	...	i_{n-1}	D
Neighbor discovery	$2P + 2G_1 + 1G_2$	$3P + 5G_1 + 2G_2$	...	$3P + 5G_1 + 2G_2$	$1P + 3G_1 + 1G_2$
Data forwarding	$2G_1 + 1G_2$	1 symmetric encryption and decryption	...	1 symmetric encryption and decryption	$2P + 1G_2$

Table 2 Simulation parameter table

Simulation parameters	Value	Simulation parameters	Value
Simulation area size	1000 km × 1000 km	Stand-alone communication range	100 km
Node mobility model	Gauss–Markov	MAC protocol	IEEE 802.11
Simulation time	900 s	Transmission rate	100 Kb/s
Antenna type	Omnidirectional antenna		

4 The Simulation Analysis

NS2 simulation software is used to establish a dynamic simulation environment in the Linux environment, and the specific parameter settings are shown in Table 2.

Scheme I: Select GPSR as a comparative target, assuming the total number of nodes in the network is 50, including five external malicious nodes and five internal selfish nodes. Malicious nodes inject illegal packets to the network in the rate of 1 Kb/s, only the intermediate nodes can be used as selfish nodes, and randomly drop packets in probability of 0.3 0.1. Select the number of illegal packets successfully delivered and the number of legitimate packets successfully delivered two indicators as simulation. The simulation results are shown in Figs. 3 and 4.

Obtained from Fig. 3, GPSR routing algorithm does not have any resistance for external malicious node, and illegal packets injected by malicious nodes can be spread in the network basically, except for some dropped by inside selfish nodes. For S-GRP routing algorithm, since adding the authentication process of node in the neighbor discovery process, so the malicious node cannot participate in the routing process, and illegal packet cannot be received and forwarded by legitimate nodes, in the simulation results, an illegal packet is successfully delivered is the result of an illegal route between nodes.

Obtained from Fig. 4, the number of legitimate packets successfully delivered in GPSR routing algorithm is approximately 15 percent less than that in S-GRP. On the one hand, this is because the S-GRP adds node trust evaluation mechanism, selfish behavior of inside selfish nodes will cause a decline in the value of self-trust, when trust value below a minimum threshold to participate in the route will be excluded from the network; on the other hand, GPSR has no resistance for external malicious

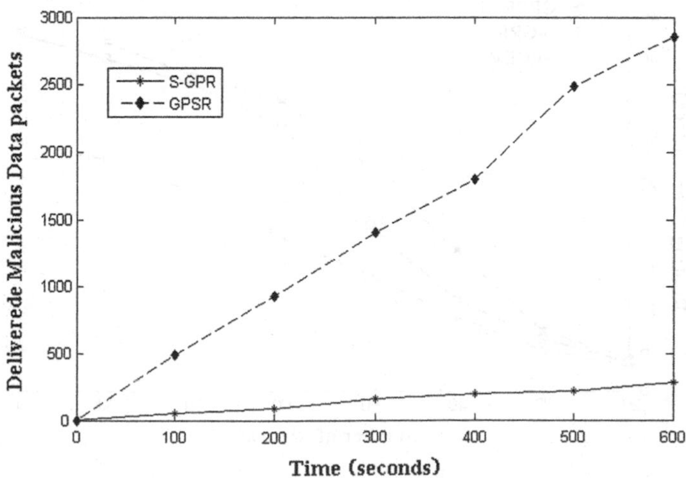

Fig. 3 Number of illegal packets successfully delivered

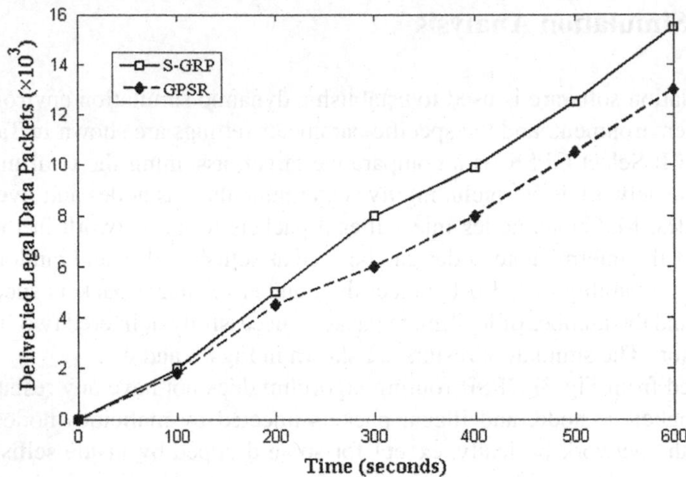

Fig. 4 Legitimate packets successfully delivered quantity

nodes, some legitimate packets is stolen, altered or discarded, resulting in normal transfer of legitimate packet is destroyed.

Scheme II: To evaluate the routing performance of S-GRP, choose GPSR and S-AODV [12] routing algorithm as contrast target, select two elements of the routing overhead, an average end to end delay. The simulation result is average of three simulation, shown in Figs. 5 and 6.

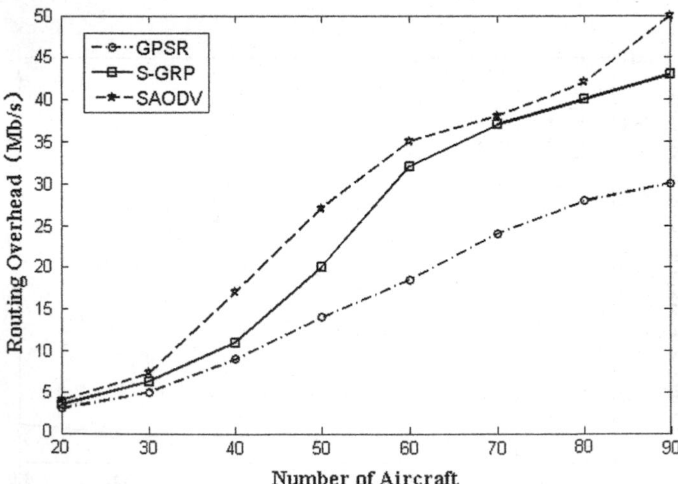

Fig. 5 Number of illegal packets successfully delivered

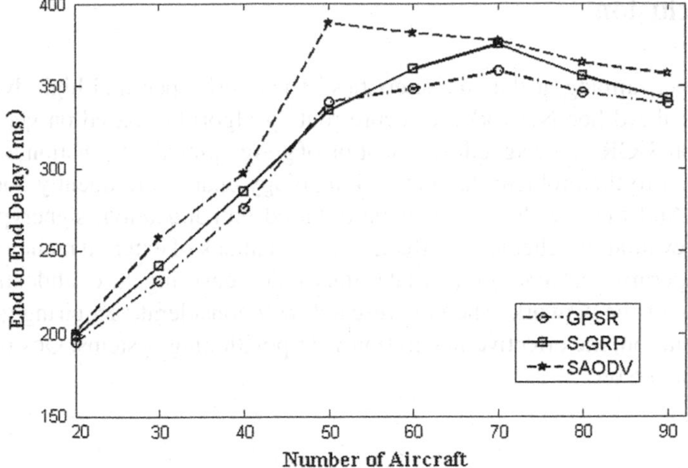

Fig. 6 Legitimate packets successfully delivered quantity

Figure 5 shows the effect of network node density on routing overhead. As can be seen from the figure, routing overhead of S-GRP is about 1.3 times as GPSR. This is because the S-GRP added node trust evaluation mechanism, using the strategy of two-jump transfer ACK packet, but on the other hand ACK packet is only a simple message acknowledgment packet, which is relatively small and will not cause routing overhead increasing dramatically. S-AODV routing algorithm generates a large amount of route discovery response, maintenance data packet in the routing establishing process, and in the topology rapidly changing network, frequently performs routing discovery, and the increase in node density will cause a rapid increase in rout-ing overhead. When the node density reaches a certain level it will flatten. S-GRP does not require a route discovery, so the routing overhead is smaller than the S-AODV.

Figure 6 shows the effect of network node density on an average end to end delay. As can be seen from the figure, the delays of S-GRP and GPSR routing algorithm are significantly less than the S-AODV, and GPSR is slightly less than the S-GRP. This is because the S-AODV routing algorithm is demand routing, in network environment where topology frequent changes, it will result in the failure of the link, routing repeat-ed discovery that increasing the end to end delay. In addition, the algorithm adds the hop authentication mechanism, which adds a certain delay. Since the S-GRP algorithm signcryption packets, it adds a certain delay, but it is more reasonable than GPSR in the next hop selection, and reduces the delay to some extent, so the total delay is slightly longer than GPSR.

5 Conclusion

In this paper, aiming at the characteristics of network open and high dynamic in Aeronautical Ad hoc Networks, a secure routing algorithm based on geo-location information S-GRP is designed. Application of geographical information is an effective solution to the problem that network topology changes frequently due to high dynamic. And the introduction of identity-based authentication, signcryption and node trust evaluation scheme can effectively resist attack of external malicious nodes and uncooperative behavior of internal selfish nodes, ensuring the confidentiality and availability of the network. The next research will considerate acquiring geographical information, the effective integration with positioning systems, Qos routing of network and so on.

References

1. Tricia GILBERT (2013) Future aeronautical communication infrastructure technology investigation. BiblioGov
2. Sakhaee E, Jamalipour A (2006) Hoc aeronautical ad networks. IEEE WCNC 1:246–251
3. Ehssan S, Abbas J, Nei K (2006) Multipath doppler routing with qos support in pseudolinear highly mobile ad hoc networks. IEEE ICC 8:3566–3571
4. Medina D, Hoffmann F, Rossetto F (2010) Routing in the airborne internet. 2010 Integrated communications navigation and surveillance (ICNS) conference. A7:1-10
5. Medina D, Hoffmann F, Rossetto F et al (2010) A cross-layer geographic routing algorithm for the airborne internet. Proc IEEE ICC 1–6
6. Peters K, Jabbar A, Cetinkaya EK, et al (2011) A geographical routing protocol for highly-dynamic aeronautical networks. Proc IEEE WCNC 492–497
7. Kumar S, Jena S (2010) SCMRP: secure cluster based multipath routing protocol for wireless sensor networks. Proceedings of the 6th international conference on wireless communication and sensor networks. 1C6
8. Pathan AS, Hong C (2008) SERP: secure energy-efficient routing protocol for densely deployed wireless sensor networks. Annals of telecommunications-annales des tcommunications. 63(9–10):529–541
9. Mohamed S, Nicolas L (2013) An ADS-B Based secure geographical routing protocol for aeronautical ad hoc networks. Computer Software and Applications Conference Workshops(COMPSACW), IEEE 37th Annual. 556–562
10. Melodia T, Pompili D, Akyildiz LF (2005) On the interdependence of distributed topology control and geographical routing in ad hoc and sensor networks. IEEE J Sel Area Commun (JSAC). 23(3):520–532
11. Eduardo DS, Aldri LDS et al (2008) Identity-based key management in mobile ad hoc networks: techniques and applications. IEEE Wirel Commun 15(5):46–52
12. Patroklos G (2005) Argyroudis, donal omahony.: secure routing for mobile ad hoc networks. IEEE Commun Surv Tutor 7(3):2–21

Research of Recycling Resource Website Based on Spring and MyBatis Framework

Yujie Guo, Ming Chen and Kanglin Wei

Abstract In order to realize the timely management of available garbage recycling and utilization in daily life, so writing code based on Spring and MyBatis framework. Through the use of role-based access control principle, the system achieve different login roles display the different function of related modules, and gives part of the implementation code. The management of available garbage prompt the user, waste recycling stations, processing treatment station and ele-commerce company which make a well-connected cycle network. The system not only be beneficial to the service provider, but also reduce the number of garbage bin, strengthen the awareness of environmental protection, finally make high quality living environment. Web application framework based on Spring and MyBatis can solve problems such as low reusability of code, hard Extensible maintainability, poor performance and etc. through the analysis of experimental results.

Keywords Recycling resource website · Spring · MyBatis

1 Introduction

In recent years, with the rising number of e-commerce sites, the online shopping rate is also rising. In order to make full use of online shopping and recyclable resources in everyday life, the system is a combination of electrical business operators, the recycle bin, the relations between and among processing treatment station, develop a recyclable resources sub-modular of ele-commerce, The recyclable resources sub-

Y. Guo (✉)
College of Science, Three Gorges University, Yichang, China
e-mail: raclen0326@163.com

M. Chen · K. Wei
College of Computer Information Technology, Three Gorges University, Yichang, China
e-mail: chenming131@163.com

K. Wei
e-mail: zeyuanwei@163.com

V.E. Balas et al. (eds.), *Information Technology and Intelligent Transportation Systems*, Advances in Intelligent Systems and Computing 455,
DOI 10.1007/978-3-319-38771-0_30

modular function can not only reasonable process the relationship among three party, but also make and receive benefits for them. At the same time, it can let people to reasonable using and recycling resources in everyday life. This article focuses on the function of the user, the recycle bin design.

2 Spring and MyBatis Framework Research

2.1 Spring Framework Review

Spring is the most popular application development framework for enterprise Java, Spring Framework is an open source Java platform and it was initially written by Rod Johnson. There are three Core components of the Spring Framework: the Core, the Context, and Beans. Without them, there is no the upper features, such as AOP, Web, etc. Spring is the Bean oriented programming (BOP, Bean Oriented Programming), based on Bean programming model promote Spring Framework to create high performing, easily testable, reusable code. Bean is encapsulated in the Object data, Context is used to find and management the relationship between the data collection. This relationship is set as the IOC container again. The Core is a tool which establish and maintain relationships between each Bean. The next focus on the Spring Framework two core functions: inversion of control (IoC) and the aspect oriented (AOP).

2.2 Introduce Spring Core Function

IoC (Inversion of Control, Control of inverted): In the process of Java development, the achievement of one logic business usually requires two or more objects of collaboration. When using partner, each Object need new Object () to complete the cooperation objects. High coupling results when an object relies on too many other objects. The Spring IoC [1] container use Java reflection mechanism to realize the interdependence between the object to create, coordinate work, it make each object more focus on business logic itself, completely decoupling between each object. Thus, Object obtain dependent objects reversed, also known as DI (DI).

AOP (Aspect oriented programming): A system usually composed of many components, each component is responsible for part of the corresponding function. These components are often additional collateral to function such as logging and transaction management and security services. These services business distribute in multiple components, causing double complexity of code. AOP lateral growth of separation in the system. It is usually aspect oriented modularized cross-cutting concerns. These cross-cutting concerns separate the application logic of business. The key unit of modularity in OOP is the class, whereas in AOP the unit of modularity is the aspect.

Whereas DI helps you decouple your application objects from each other, AOP helps you decouple cross-cutting concerns from the objects that they affect. Put it simple, its just an interceptor to intercept some processes, for example, when a method is execute, Spring AOP can hijack the executing method, and add extra functionality before or after the method execution.

2.3 *MyBatis Framework*

MyBatis is established on the basis of the iBatis. MyBatis is a Java persistence framework with support for custom SQL, stored procedures, and advanced mapping. Programmers need to manually write specific SQL. through the way of using the configuration file, Mabatis make the data layer interface associated with Pojos class, they are mapped the result to the records in the database.

MyBatis is a kind of semi-automatic ORM implementation. The ORM tool get Sessionfactory through a configuration file (usually XML configuration file), again by the Sessionfactory get Session, completed data CRUD (i.e., add and delete) and transaction commit, and other functions by the Session. The meet disconnect after the closing Session.

It separate business logic layer and data access layer separation, make the overall system level clear, easy to maintain and test. The separation of code and the SQL improve the maintainability of the system.

3 Based on Spring and MyBatis Framework of the System Level Design

The current popular J2EE framework into the MVC three layer structure. This paper designed the three layer extends into four layers: the presentation layer (View) (Controller), the business layer, control layer (Service), data processing layer (DAO).

The presentation layer uses the Jquery Easy UI framework. Based on the Jquery UI library, uses HTML + CSS 3, contains common UI components, such as layout, form, Tabs, forms, tree, toolbars, pop-up menus, and so on. It can support the commonly used browser, such as Chrome, FireFox7 +, Safari 5 + and IE 9 +, etc. The framework separate static HTML tags and dynamic Script with late system upgrade and maintenance [2].

Transfer the data processing parameters of presentation layer to the control layer. Control layer based on the URL to the corresponding method. The method pass the data to the specified business layer method.

The business layer [3] is mainly responsible for processing business logic. Business layer method only need to be concerned with the method of business process, requires no additional care about data processing problem. When the business layer method need to get the object data and information, it usually call data processing methods.

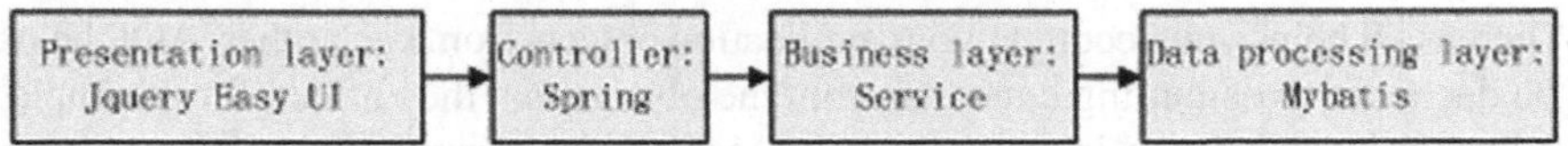

Fig. 1 Hierarchy

Data processing layer is mainly responsible for data processing. Data processing provides data services for the service layer. In the data layer, data transmission use the Entity. Because of using MyBatis technology, only by the mapping configuration file, the SQL required parameters and processing the results of the field mapping to the Entity, user can get the required results [4].

The system use the configuration file such as web, XML, the application Context.xml of Spring Framework, and Mybatis.xml, as well as the configuration files related to SQL processing. This hierarchy as shown in the system (Fig. 1):

4 Spring and MyBatis in the Application of Recycle Resources Business System

4.1 Overall System Process

The recycle bin and processing treatment station with electric business operators have certain relations of cooperation. Users will be sold to the recycle bin recycling, recycling treatment station again to sell the waste resource processing, processing treatment station to sell processed waste resources in electric business operators, the electricity business operators sell the related resources to its required businesses [5]. The overall operation implement a "sell". The overall process as shown in Fig. 2.

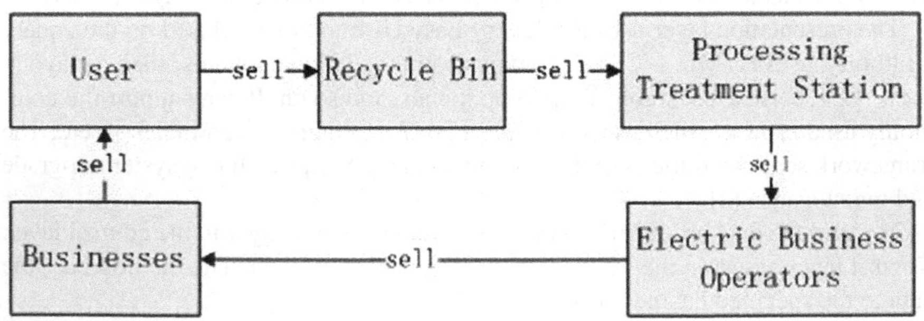

Fig. 2 Overall process

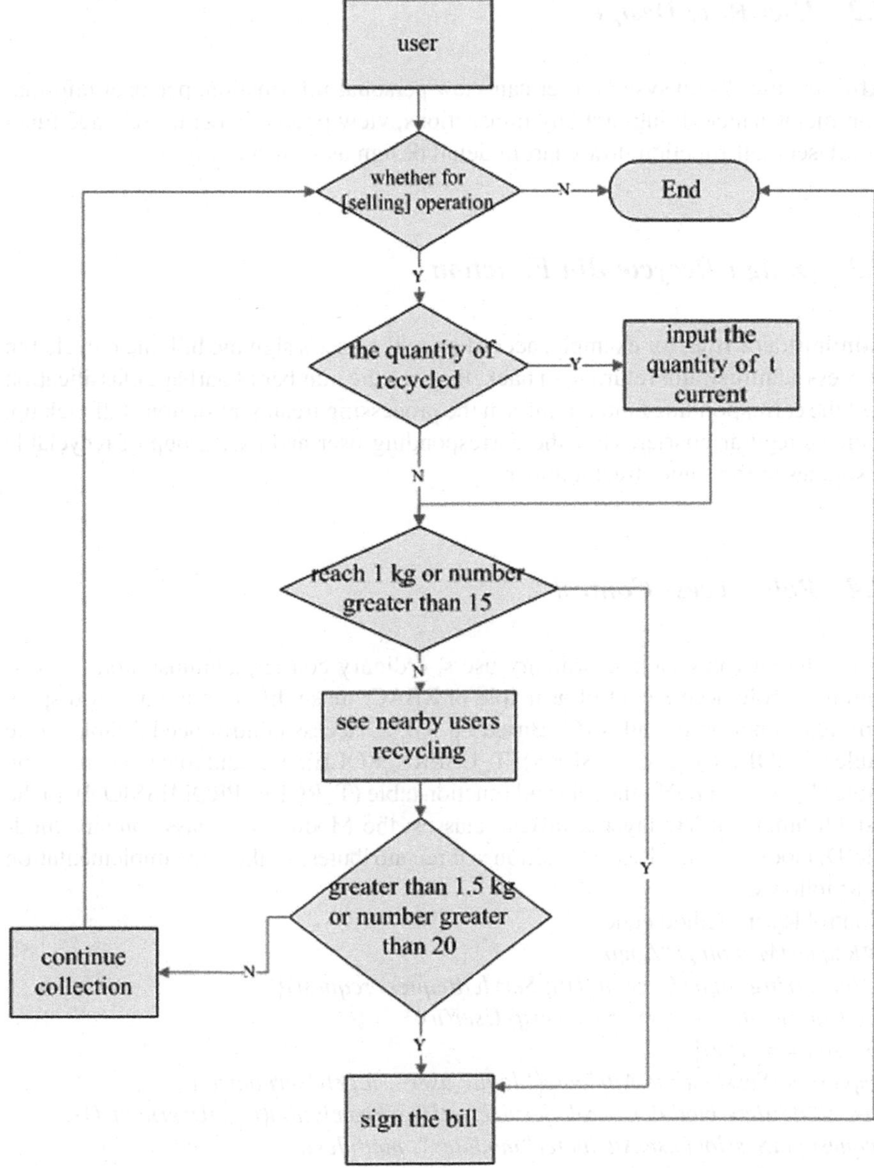

Fig. 3 User sell function

4.2 User Role Design

After the user login system, user can view personal information, personal information maintenance, bring-and-buy transactions, view points, integral exchange function.Users sell function flow chart in detail design as shown in Fig. 3.

4.3 Design Recycle Bin Function

Administrator role, for example, according to the user's sign the bill, the recycle bin express administrator return cash back, they see the number of garbage classification and the corresponding amount, inform the processing treatment station staff pick up, manage regular courier, view the corresponding user and user group of recyclable resources in the same street number.

4.4 Role Access Control

For different roles such as ordinary users, ordinary courier, administrator, the system use a role access control principle of RBAC, make different login user display different function modules [6]. Based on RBAC access control need 5 tables: role table (T_ROLE), user (T_USERS) (T_USERS_ROLE), user and role table, function table (T_FUNCTION), the role and function table (T_ROLE_PROMISSION). In the establishment of VO layer JavaBean classes: the Module, the class contains moduleID, moduleName, List <Function> three attributes. In the code implementation is as follows:

Control layer method code:

```
@RequestMapping("/login")
public String login(Users u ,HttpServletRequest request){
boolean result = usersService.verifyUser(u);
if(result == true){
request.getSession().setAttribute("login_user", u.getUsername());
List<Module> modules = roleService.getUserPromission(u.getUsername());
request.getSession().setAttribute("modules", modules);
return "redirect:/index.jsp";
}else{
return "redirect:/login.html";
}
}
```

Business layer method code:

```
public List<Module> getUserPromission(String username){
List<Module> moduleList = roleDAO.findModuleByUsername(username);
```

```
for(Module module : moduleList){
Map params = new HashMap();
params.put("username", username);
params.put("moduleId", module.getModuleId());
List<Function> functions = role-DAO.findFunctionsByModule(params);
module.setFunctions(functions);
}
return moduleList;
}
```

Data layer method code:

```
public interface RoleDAO {
public List<Module> findModuleByUsername(String user-name);
public List<Function> findFunctionsByModule(Map params);
}
```

Mybatis configuration file code:

```
<select id="findModuleByUsername" parameterType="java.lang.String"
resultMap="moduleMap">
select distinct f.function_id , f.function_name
from t_role t, t_role_promission rp, t_function f , t_users_role ur , t_users u
where t.role_id = rp.role_id and f.function_id = rp.function_id
and t.role_id = ur.role_id and ur.user_id = u.user_id
and u.username = #{username} and function_level = 1
</select>
<select id="findFunctionsByModule" parameterType="java.util.Map"
resultMap="functionMap">
select *
from t_role t, t_role_promission rp, t_function f , t_users_role ur , t_users u
where t.role_id = rp.role_id and f.function_id = rp.function_id and t.role_id =
ur.role_id
and ur.user_id = u.user_id
and u.username = #{username} and function_parent_id= #{moduleId} and func-
tion_level = 2
</select>
```

5 Conclusions

Spring and MyBatis framework is applied in this article, which is combined with a specific role access control code examples. This paper discusses the whole system-level of design and the integration between them. Spring technology [7] is applied to the system architecture of recycle resources, which can decouple the presentation layer, control layer, business layer and data interface layer, improve the performance of the system, and also improve the maintenance and scalability of the system;

MyBatis is applied to the recycle system, as a result, the correlation between the model object and database design is reduced to a minimum. Spring Framework play a overall control role in the whole system, reduce the complexity of the system. It has proved that the Spring and MyBatis framework can be full use of the advantages of both for Web development [8], and effectively improve the project's development efficiency, simplify the project's later expansion and maintenance.

References

1. Ren Yongchang, Jiang Deyi, Xing Tao, Zhu Ping (2011) Research on software development platform based on SSH framework structure. Proc Eng 15:3078–3082
2. Zhang Jingjun, Wang Lei, Li Hui, Liu Guangyuan (2011) Research Java Web framework based on OSGi. Proc Eng 15:2374–2378
3. Fogh RH, Boucher W, Vranken WF, Pajon A, Stevens TJ, Bhat TN, Westbrook J, Ionides JMC, Laue ED (2004) A framework for scientific data modeling and automated software development. Sci Math 21:1678–1684
4. Krill Paul (2004) Genuitec introduces MyEclipse IDE for the spring framework. ProQuest J 14:1178–1187
5. Montgomery K, Montgomery K, Bruyns CD (2004) Generalized interactions using virtual tools within the spring framework: cutting. Stud Health Technol Inf 85:79–85
6. Reynaud-Plantey Daniel (2005) New threats of Java viruses. J Comput Virol 1(1):32–43
7. Gupta P, Govil MC (2010) Spring Web MVC Framework for rapid open source J2EE application development: a case study. Int J Eng Sci Technol, 2(6):1684
8. Manor A, Shaojuan L, Saurav S, Marto JA (2010) Pathway palette: a rich internet application for peptide-, protein- and network-oriented analysis of MS data. Proteomics 10(9):1880–5

A Fast Video Stabilization Method Based on Feature Matching and Histogram Clustering

Baotong Li, Yangzhou Chen, Jianqiang Ren and Lan Cheng

Abstract Stable video has a significant effect on video based traffic parameters extraction system. This paper proposed a fast video stabilization method based on feature matching and histogram clustering. Firstly, the method extract ORB feature points on sub-images which are set in adjacent frames and then matches them based on Hamming Distance. Secondly, the motion consistency principle combined with histogram clustering is used to get correct matches which can be used to calculate the parameters of the affine transformation matrix. Finally, after filtering those parameters, stable frame can be obtained by warping the original frame using the affine transformation matrix. Experimental results show that the proposed method can effectively stabilize videos with low complexity in real time.

Keywords Video stabilization · Motion estimation · Feature matching · Histogram clustering

1 Introduction

Vehicle detection and tracking based on video method has advantages of easy maintenance and high flexibility, etc. Thus, it becomes one of the most useful techniques to obtain traffic parameters in Intelligent Transportation System (ITS) [1, 2]. However,

B. Li · Y. Chen (✉) · J. Ren · L. Cheng
Beijing Key Laboratory of Transportation Engineering, College of Metropolitan Transportation, Beijing University of Technology, Beijing 100 Pingleyuan, Chaoyang District, Beijing 100124, China
e-mail: yzchen@bjut.edu.cn

B. Li
e-mail: limingdao311@emails.bjut.edu.cn

J. Ren
e-mail: renjianqiang@emails.bjut.edu.cn

L. Cheng
e-mail: chenglan@emails.bjut.edu.cn.com

V.E. Balas et al. (eds.), *Information Technology and Intelligent Transportation Systems*, Advances in Intelligent Systems and Computing 455,
DOI 10.1007/978-3-319-38771-0_31

the shaking of hand or carrier not only decreases the visual quality of the recorded video significantly but also causes errors for traffic parameters, especially for aerial videos [3]. So the video stabilization technique is introduced to remove jitters from videos. Video stabilization is composed of two parts: motion estimation and motion filtering. The motion estimation module is designed to estimate the global motion vector and it includes intention motion and random jitter motion which can be eliminated by the motion filter module to obtain the stable image [4].

The contribution of this paper is that we propose a fast feature-based video stabilization method which uses the motion consistency principle combined with histogram clustering to eliminate false feature matches and by doing this we can get correct matches which are important for motion estimation. The remainder of this paper is organized as follows. The second section is a review of the video stabilization. The elaboration of proposed method is arranged in the third section. The fourth section shows some experimental results and the analysis of them. Finally, we conclude for this paper.

2 Literature Review

Motion estimation is the most important part of video stabilization technique because incorrect camera motion parameters will result in retaining unwanted motions in the stabilized videos thereby degrading the stabilization performance. According to its implementation principle, motion estimation methods can be divided into two categories: intensity based and feature based method [5]. The intensity based motion estimation methods obtain the camera motion vector by detecting intensity changes between images or sub images. They include the represent point matching [6], block matching [7], optical flow [8], gray encoded bitplane matching [9] and phase correlation method [10]. Generally speaking, intensity based methods are slower than feature-based methods because they often use all the gray information of the image. In addition, benefit from a variety of feature detection and description operators have been put forward, feature-based methods are being used widely.

The feature based motion estimation methods often are composed of three parts: feature extraction, feature description and feature matching, just as shown in Fig. 1.

Firstly, the feature extraction operator is used to detect features from F_k and F_{k+1} frame of a video sequence, and then those features are described to be feature description vectors. Finally, by judging the distance between those vectors we

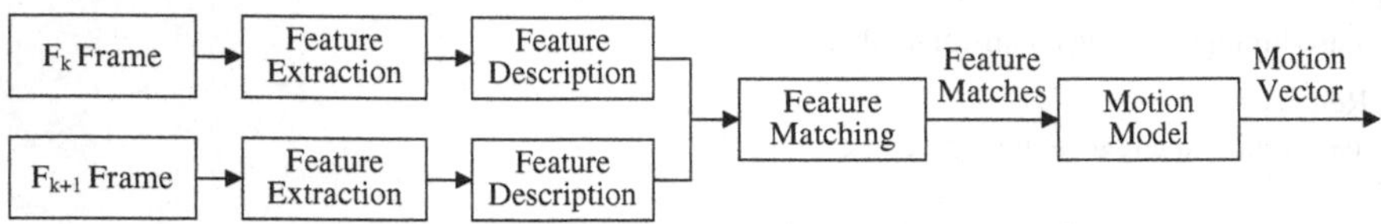

Fig. 1 The flow chart of feature based motion estimation methods

can get feature matches which can be used to calculate parameters of the camera motion model. According to the difference of selected features, feature-based methods include edge pattern matching [11], line feature matching [12], regional feature based matching [4] and feature point matching method [5, 13–16]. Among them, because of the advantages of fast speed and high robustness, methods based on feature point matching are widely concerned by researchers in recent years, especially after the SIFT (Scale Invariant Feature Transform) operator has been proposed in 2004 [13].

In [13, 14], the SIFT operator is used to extract features from video frames and estimate the global motion vector according to their trajectories. Binoy and Anurenjan P.R. [16] used SURF (Speeded Up Robust Features) algorithm which is improved from SIFT to achieve stabilized video. However, the SIFT and SURF features are robust, but because of the high dimension of feature vector, its real-time performance is poor. Then in [5], the authors proposed to use ORB (Oriented FAST and Rotated BRIEF) operator to estimate the global motion vector and it is proved to be faster than SIFT and SURF by experiments. But they extract ORB feature points from the whole frame image, which has not good effectiveness in such scenes which contain multiple moving targets. Furthermore, the removal of error matches by using RANSAC (RANdom SAmple Consensus) is time-consuming. So we put forward a method of removing the error matches based on histogram clustering and it will be described minutely in the third part of this paper.

Several techniques like motion vector integration, frame position smoothing, Gaussian filtering, Kalman filtering and extended Kalman filtering have been used to eliminate the jitter [4]. Among them, the Kalman filter has been widely used [4, 10, 14], but the performance is poor when using it to solve nonlinear or non-gauss problems. In addition, the Kalman filter is more timeconsuming than the Gaussian filter. So in order to meet the requirement of real-time processing, we select the Gaussian filter with k_t frames delay to filter motion parameters.

3 Methodology

The flow chart of the proposed method is shown in the Fig. 2, which mainly consists of 5 parts named camera motion model, sub-images setting, ORB feature extraction and matching, removal of error matches and motion filtering respectively. The details of them are introduced below.

3.1 Camera Motion Model

A necessary condition for accurate motion estimation is to select an appropriate camera motion model. Translation transformation, similarity transformation, affine transformation, homographic transformation and perspective transformation models

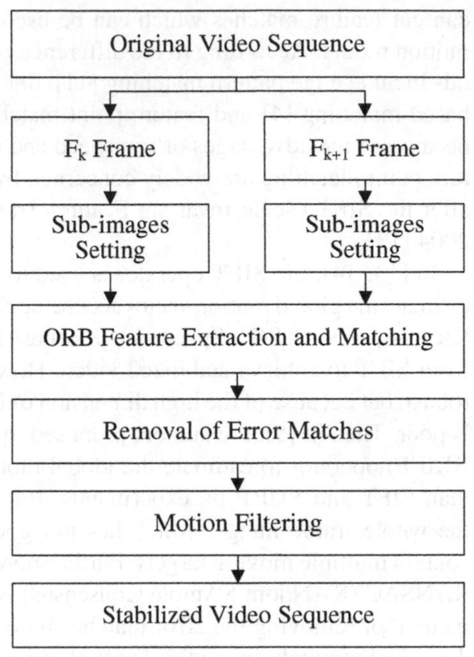

Fig. 2 The flow chart of the proposed method

are the most commonly used in recent years. The affine transformation model with six parameters can guarantee the accuracy of estimating camera motion and save time [17]. So we adopt it as the camera motion model of this paper and its formula is:

$$P_{k+1} = MP_k \Rightarrow \begin{pmatrix} x_{k+1} \\ y_{k+1} \end{pmatrix} = \begin{bmatrix} a_{11} & a_{12} & a_{13} \\ a_{21} & a_{22} & a_{23} \end{bmatrix} \begin{pmatrix} x_{k+1} \\ y_{k+1} \\ 1 \end{pmatrix} \tag{1}$$

where P_{k+1} and P_k are match points of the F_{k+1} and F_k frame and M is affine transformation matrix respectively. Parameters a_{11}, a_{12}, a_{21} and a_{22} of M are used to represent the zoom, shear and reverse transform, while a_{13} and a_{23} represent the translation transform between images. The purpose of motion estimation is to calculate the six parameters of M which represents the global camera motion. It can be seen that we need at least three correct feature point matches to calculate the six parameters from formula 1. And then our next goal is to get at least three correct feature point matches.

3.2 Sub-images Setting

Inspired by the idea of block matching, some researchers used sub-image instead of whole image to achieve motion estimation. In [10], phase correlation detection was done on four sub images located in the four corners of the whole image to obtain the global motion vector. However, it is effective only when the camera moves translationally. In [18], a method for global motion estimation is introduced by extracting SIFT-features from four sub-images. But SIFT feature extraction and description are slow which makes the method's real time performance obviously poor. Otherwise, it is important to remove error matches. This paper also uses four sub images instead of whole image and the setting of the four sub images is the same as it is done in [18]. The reasons why we use sub images are as follows:

1. Feature extraction and description on sub-images are significantly faster than on the whole image.
2. Generally speaking, moving targets in video sequences are often in the center of the frame, while the background locates in the corner.
3. The feature extraction of the whole frame is easy to make the feature set in a certain area which reduces the accuracy of the motion estimation. However, by setting up sub images in four corners can eliminate this influence.

3.3 ORB Feature Extraction and Matching

As described in Sect. 3.1, we need at least 3 feature point matches, and now there are four sub images, so we can find a correct match in each sub image to solve the problem. We first extract and describe ORB feature points from each sub image in adjacent frames. And then Hamming Distance is measured to get feature point matches. ORB algorithm is proposed by Ethan et al. in [2], compared with SIFT and SURF, it is fast and effective. ORB algorithm is a method combined with FAST (Features from Accelerated Segment Test) feature point detector and BRIEF (Binary Robust Independent Elementary Features) feature descriptor, at the same time, improving and optimizing them. Because the focus of this paper is not on the improvement of ORB operator, we just use the ORB to detect feature points and matches them, this is not going to repeat it here.

Based on the principle of ORB algorithm, the oFAST(oriented FAST) operator is used to detect feature points of each sub image and N feature points are gotten according to the Harris corner response function firstly. And then those N feature points are described by the steer BREIF operator to get feature description vectors. Lastly, we get feature point matches by measuring distance between description vectors using Hamming Distance criterion which is more suitable for the processing of binary number. While the feature description vector consists of binary numbers and this can be clearly seen in [2].

3.4 Removal of Error Matches

Feature point matches gotten in Sect. 3.3 maybe contain some error matches caused by moving objects and the similarity between feature points. The elimination of those error matches is an essential part of the process of video stabilization because they can bring great error to the motion estimation. The main contribution of this paper is also in this section. We first use the motion consistency principle to make a rough elimination, and then use the histogram clustering to get correct matches. Its specific practices are as follows.

In adjacent frames, all pixels in a sub image should have the same movement trend [10, 18]. Therefore, two corresponding sub-images with W $\times$ H pixels from adjacent frames are mosaicked to be a big image with 2W $\times$ H pixels firstly. And then we connected feature point matches which had been obtained in Sect. 3.3 to calculate slope for each match. Finally, those larger and smaller slope matchings are rejected because their movement is different from other matches that increased the probability of false matching. In addition, we have to say that our algorithm can adaptively decide how many matches to eliminate based on the numbers of matches gotten in Sect. 3.3. By doing this, we can not only reduce the calculation time of the histogram clustering but also guarantee the accuracy of motion estimation.

Although matches with larger and smaller slope have been removed, but it cannot guarantee that the rest of the matches are correct. Therefore, in order to obtain the correct matches, we proposed to use histogram clustering to classify the remaining matches. At begin, we obtain the maximum and minimum value of the slope and divide the slope into 10 equal intervals according to them. Secondly, interval statistics of all the slopes is done to obtain the highest probability of the slope range. Lastly, those feature matches responded to the highest slope range are used to calculate average position $(\overline{x}_k, \overline{y}_k)$ and $(\overline{x}_{k+1}, \overline{y}_{k+1})$. Four feature point matches can be obtained by doing above treatment for each pair of sub-images and then they can be used to calculate parameters of the affine transformation matrix.

3.5 Motion Filtering

The motion parameters we got in the previous step include intentional camera motions and undesired camera motions. A stabilized video is obtained by removing undesired motions. As assumed in [4], undesired camera motions like sudden jitter, handshake involve rapid motion variations over time, are random in nature and hence regarded as high frequency components. While intentional camera motions are assumed to be smoothed with slow inter-frame variations. So we can use a low pass filter to remove undesired motions. We select the motion filtering technique adopted by Matsushita et al. [19] which is free from accumulative error because their method locally smooths displacement from the current frame to the neighboring frames by using a Gaussian filter as described below.

The transformation from a frame to the corresponding motion compensated frame is directly computed using only the neighboring transformation matrices. The global transformation matrix T_t is smoothed using a Gaussian filter by

$$S_t = \sum_{j \in N_t} T_t^j * G(k_t) \tag{2}$$

where, $N_t = \{j | t - k_t \leq j \leq t + k_t\}$, denotes the temporal neighborhood of frame t, S_t is the smoothed corrective transformation, and

$$G(k_t) = \frac{1}{\sigma\sqrt{2\pi}} \exp\left(\frac{-k_t^2}{2\sigma^2}\right) \tag{3}$$

is a Gaussian kernel with $\sigma = \sqrt{k_t}$. The window size k_t varies over time and enables the smoothing operation to be adaptive to the different magnitudes of motion that might exist in video sequences. The smoothness of the camera motion parameters can be controlled by varying the window size kt. In this paper, we set $k_t = 6$ just as advised in [19]. Finally, a stabilized video can be obtained by warping the original video frames with S_t.

4 Experimental Result

The proposed method is verified with a large number of video sequences which are all captured from different real-life scenes. As shown in Fig. 3, video 1 and video 2 are recorded by a mobile phone while video 3 and video 4 are aerial videos. Video 1 records a general traffic scene and the violent shakes are added to it when shooting. A traffic accident occurs in video 2 which can be used to study the impact of traffic accidents on traffic flow. Video 3 records traffic situation of a highway exit ramp and traffic flow changes in the cross section of a highway can be seen in video 4. These video sequences are preprocessed to contain 300 frames and each frame composed of 640×480 pixels and then we use them to verify the proposed method from two aspects of computational cost and effectiveness.

4.1 Computational Test

The computational test process was performed offline with an Intel Core 2 Duo 2.4 GHz laptop with 2 GB of RAM and without any hardware acceleration. All algorithms of the proposed approach are written in C++ programming language based on the OpenCV library. The result is shown in Table 1.

Fig. 3 Experimental results of four video sequences and from the first row to the last are the 5th, 85th, 178th and 256th frame

Table 1 Result of computational test

Sequence	Video 1	Video 2	Video 3	Video 4
Time cost	66.3459 ms	73.3201 ms	72.7198 ms	87.9508 ms

In Table 1, the unit of time cost is millisecond per frame It is an average time of each video sequence. It can be obviously seen that the real time performance of our algorithm is good. The different time cost between each video sequence may be caused by their different scenarios. The reason why video 4 costs more time than the others maybe that its scene contains a lot of leaves which are rich in features.

4.2 Effectiveness

To evaluate the effectiveness of the proposed approach, we adopted the *ITF* (Interframe Transformation Fidelity) which has been widely used to measure the temporal smoothness for the performance of video stabilizatio [5, 12, 16]. A high *ITF* value often indicates a video with good quality. The *ITF* can be got by:

Table 2 Result of *ITF* test

Video/*ITF*	Video 1	Video 2	Video 3	Video 4
original	23.3901	21.2243	21.1176	20.6428
stabilized	25.3628	24.0865	23.6878	21.8463

$$ITF = \frac{1}{T_f - 1} \sum_{k=1}^{T_f - 1} PSNR(k) \tag{4}$$

where T_f is the total frames of video sequence and *PSNR*(k) is the peak signal-to-noise ratio between two consecutive frames (F_k, F_{k+1}) which is defined as:

$$PSNR(k) = 10\log_{10} \frac{I_{\max}}{MSE(k)} \tag{5}$$

where I_{max} is the maximum pixel intensity of a frame and it equals to 255 in 8 bit grayscale images. The *MSE (k)* (Mean Square Error) is the deviation value of corresponding pixel gray level between adjacent frames and its calculation formula is:

$$MSE = \frac{1}{mn} \sum_{i=0}^{m-1} \sum_{j=0}^{n-1} \|I_{k+1}(i, j) - I_k(i, j)\|^2 \tag{6}$$

where m and n are the height and width of the image, $I_{k+1}(i, j)$ and $I_k(i, j)$ are the gray values of frame F_{k+1} and F_k.

According to the above three formulas, we calculate the *ITF* value of the original video and the stabilized video and shows them in Table 2.

As shown in Table 2, the *ITF* score of each stabilized video is 1.2–2.8 units higher than its original video sequence. It means that a more desirable and smoothed sequence can be obtained by using the proposed method.

5 Conclusion

In this paper, we propose a fast and robust video stabilization method based on the combination of feature points matching and histogram clustering. The ORB algorithm is used for feature point detecting and matching in the four sub-images between two consecutive frames. And then error matches are eliminated by applying the motion consistency principle. After that histogram clustering are used to get correct matches which can be used to calculate parameters of the affine transformation matrix M. Finally, stabilized images are gotten by warping the original frame images with M. Experiment results have confirmed the real time performance and effectiveness of the proposed method. However, the effectiveness of our method decreases heavily when two of four sub-images have not enough features. Besides, error matches caused by moving targets cannot be completely eliminated. Then our future work is to study the collaboration between four sub-images to obtain a higher efficiency.

Acknowledgments This work was supported by the National Natural Science Foundation of China [61273006, 61573030], and Open Project of Beijing Key Laboratory of Urban Road Intelligent Traffic Control [XN070], Scientific Research Foundation of Hebei Higher Education [QN2015209], Science and Technology Support Foundation of Hebei Province [13210807].

References

1. Hsieh JW, Yu SH, Chen YS et al (2006) Automatic traffic surveillance system for vehicle tracking and classification. IEEE Trans Intell Transp Syst 7(2):175–187
2. Jazayeri A, Cai H, Zheng JY et al (2011) Vehicle detection and tracking in car video based on motion model. IEEE Trans Intell Transp Syst 12(2):583–595
3. Walha A, Wali A, Alimi AM (2014) Video stabilization with moving object detecting and tracking for aerial video surveillance. Multimed Tools Appl 1–23
4. Okade M, Biswas PK (2014) Video stabilization using maximally stable extremal region features. Multimed Tools Appl 68(3):947–968
5. Xu J, Chang H, Yang S et al (2012) Fast feature-based video stabilization without accumulative global motion estimation. IEEE Trans Consum Electron 58(3):993–999
6. Uomori K, Morimura A, Ishii H et al (1990) Automatic image stabilizing system by full-digital signal. IEEE Trans Consum Electron 36(3):510–519
7. Vella F, Castorina A, Mancuso M et al (2002) Digital image stabilization by adaptive block motion vectors filtering. IEEE Trans Consum Electron 48(3):796–801
8. Chang J, Hu W, Cheng M, Chang B (2002) Digital image translation and rotation motion stabilization using optical flow technique. IEEE Trans Consum Electron 48(1):108–115
9. Ko SJ, Lee SH, Lee KH (1998) Digital image stabilizing algorithms based on bit-plane matching. IEEE Trans Consum Electron 44(3):617–622
10. Ertürk S (2003) Digital image stabilization with sub-image phase correlation based global motion estimation. IEEE Trans Consum Electron 49(4):1320–1325
11. Paik JK, Park YC, Park SW (1991) An edge detection approach to digital image stabilization based on tri-state adaptive linear neurons[J]. IEEE Trans Consum Electron 37(3):521–530
12. Liang YM, Tyan HR, Chang SL et al (2004) Video stabilization for a camcorder mounted on a moving vehicle. IEEE Trans Veh Technol 53(6):1636–1648
13. Battiato S, Gallo G, Puglisi G et al (2007) SIFT features tracking for video stabilization, In: 14th International conference on image analysis and processing, 2007. ICIAP 2007. IEEE, pp 825–830
14. Yang J, Schonfeld D, Mohamed M (2009) Robust video stabilization based on particle filter tracking of projected camera motion. IEEE Trans Circuits Syst Video Technol 19(7):945–954
15. Peng X, Chen J, Zhang J (2011) Robust digital image stabilization based on spatial-location-invariant criterion. In: IECON 2011-37th annual conference on IEEE industrial electronics society. IEEE, pp 2250–2254
16. Pinto B, Anurenjan PR (2011) Video stabilization using speeded up robust features. In: 2011 International conference on communications and signal processing (ICCSP). IEEE, pp 527–531
17. Litvin A, Konrad J, Karl WC (2003) Probabilistic video stabilization using Kalman filtering and mosaicing. In: Electronic imaging. International society for optics and photonics, pp 663–674

18. Li C, Liu Y (2011) Global motion estimation based on SIFT feature match for digital image stabilization. In: 2011 International conference on computer science and network technology (ICCSNT), vol 4. IEEE, pp 2264–2267
19. Matsushita Y, Ofek E, Ge W et al (2006) Full-frame video stabilization with motion inpainting. IEEE Trans Pattern Anal Mach Intell 28(7):1150–1163

Time Domain Reflectometry Calculation Model of Landslides Slippage

Li Dengfeng, Zhang Baowei, Wang Gang, Yang Kebiao, Wang Leiming and Xu Xuejie

Abstract For the application of Time Domain Reflectometry (TDR) technology in landslide monitoring, a new mountain landslide slip amount calculation model based on TDR technology is put forward. This article describes the basic principles of TDR landslide monitoring and the theory of electromagnetic pulse propagation in a coaxial cable. Deduced calculation model of landslides slippage, conducted a laboratory simulation tests landslides, and discussed the landslides slippage effect on TDR waveform, get and verify specific calculation model by the test. The results showed that, TDR reflected voltage with the landslide slip amount is increasing, the different deformation point of the reflective voltage with the shift to a linear relationship. The two errors between calculation model slip and the actual slip are not more than 2 mm.

Keywords TDR · Landslide monitoring · The equivalent sampling method

L. Dengfeng (✉) · Z. Baowei · W. Gang · Y. Kebiao · W. Leiming · X. Xuejie
College of Electronics and Control Engineering, Changan University,
Xian 710064, China
e-mail: dfli@chd.edu.cn

Z. Baowei
e-mail: z_bo_123@163.com

W. Gang
e-mail: 1315703001@qq.com

Y. Kebiao
e-mail: 18202953244@163.com

W. Leiming
e-mail: thunder8415@163.com

X. Xuejie
e-mail: xu1129005165@163.com

V.E. Balas et al. (eds.), *Information Technology and Intelligent Transportation Systems*, Advances in Intelligent Systems and Computing 455,
DOI 10.1007/978-3-319-38771-0_32

1 Introduction

TDR technology is an electrical measurement technique, which has been used for many years to detect a variety of objects morphological characteristics and spatial orientation. From the 1970s, foreign researchers began to apply TDR technology to measure the water content of soil [1], structural damage detection [2], slope stability monitoring [3, 4] and so on. Due to the features of security, high efficiency, remote measurements etc., TDR technology has been rapidly developed in foreign countries [5]. And domestic theory research and practice application of TDR technology is still in infancy. At present, many scholars primarily focus on variation of the reflection coefficient caused by landslides, but the study of landslide slippage in the time domain reflection is very limited. In this paper, on the basis of previous studies, the rules of the pulse signal transmission in a coaxial cable is studied, the calculation model of landslides slippage is proposed and the calculation test model is verified.

2 TDR Landslide Monitoring Principle

A TDR landslide monitoring system is shown in Fig. 1. A coaxial cable is vertically installed in a borehole and back filled with cement grout to make it close to the mountain. The top of the cable is connected to a monitor platform. The coaxial cable which is buried in the earth will be deformed due to the action of shearing the soil and rock when landslides occurs. As a result, the characteristic impedance is changed.

TDR monitor platform can be connected by a USB to the computer, and the platform is depends on measuring voltage reflections from an electrical pulse as it pass through a coaxial cable. When TDR monitoring platforms send a serious of pulse signal, the reflection and transmission of the pulse signal at the deformation point occurs. With the change of reflection voltage on TDR waveform, the depth of the landslide and landslide displacement can be calculated.

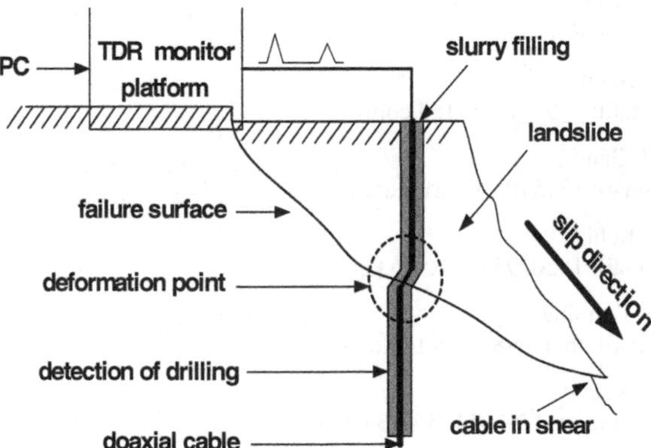

Fig. 1 TDR landslide monitoring system

The calculation formula of the landslide surface depth L is given as follows.

$$\Delta x = -0.6679 \times V_{rt}(f) + 3.69 \quad (1)$$

In Eq. (1), v is the transmission speed of the detection pulse in the test cable, δt is the time difference between the detected pulse and the reflected pulse.

3 Landslide Slip Calculation Model

According to the electromagnetic theory, when electromagnetic wave is transmitted in the coaxial cable, the attenuation and phase shift will be happened, but there is no reflection and transmission of electromagnetic wave. The propagation function of the coaxial cable is defined as follows.

$$\Delta x = -0.7451 \times V_{rt}(f) + 3.69 \quad (2)$$

In Eq. (2), providing a voltage signal $v_0(f)$ at the top of the coaxial cable and the transmission distance is L, then the voltage value is given as follows [6].

$$v_L(f) = v_0(f)e^{-\alpha(f)L - j\beta(f)L} \quad (3)$$

In Eq. (3), α is the attenuation constant, β is the phase constant, f is the frequency of the transmission pulse, L is the transmission distance.

When a coaxial cable is affected by shearing, the stress area is deformed, which causes the characteristic impedance of this area to change. The electromagnetic wave will reflect and transmit in this area, as shown in Fig. 2. Section A–A to Section B–B is the output of the signal generator to the start of the coaxial cable, its electrical impedance is Z_1. Section B–B to Section 1–C is normal cable, the impedance is Z_0. Section 1–C to Section D–D is deformation of the cable, the impedance is Z_2. V_0 is incident signal voltage at cross section B–B, VL_i is the voltage at cross section C–C, VL_r is reflected signal voltage for the first time at cross section C–C, VL_t is

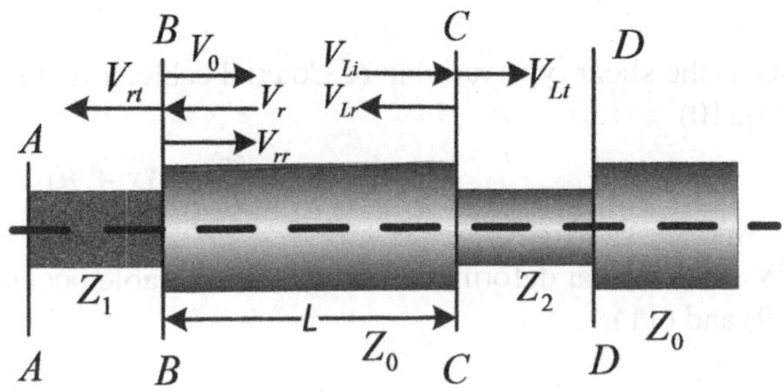

Fig. 2 The sketch of pulse signal reflection and transmission

the transmission signal voltages for section C–C, V_{rr} is the reflection voltage at the cross section of the reflective signal V_r, V_{rt} is the transmission voltage at the section A–A of the reflected signal V_r, and they can be directly measured by the system.

By Eq. (3):

$$V_{Li}(f) = V_0(f)e^{-\alpha(f)L - j\beta(f)L} \tag{4}$$

$$V_r(f) = V_{Lr}(f)e^{-\alpha(f)L - j\beta(f)L} \tag{5}$$

The reflection coefficient ρ_1 at the set of section B–B is provided by Eqs. (4) and (5) as follows.

$$\rho_1 = \frac{V_{Lr}(f)}{V_{Li}(f)} = \frac{V_r(f)}{V_0(f)}(e^{\alpha(f)L + j\beta(f)L})^2 \tag{6}$$

The reflection coefficient ρ_0 at the set of section A–A is provided, the transmission voltage V_{rt} in the cross section of A–A is shown as follows.

$$V_{rt}(f) = V_r(f)(1 + \rho_0) \tag{7}$$

The voltage reflection coefficient in the deformation is got by Eq. (6) with (7) as follows.

$$\rho_1 = \frac{V_{rt}(f)}{V_0(f)(1 + \rho_0)}(e^{\alpha(f)L + j\beta(f)L})^2 \tag{8}$$

The calculation formula of reflection coefficient is shown in Eq. (8), the reflection coefficient can reflect the status of the coaxial cable of deformation condition.

Foreign researchers Dowding, C.H. et al. (1988) equality the shear deformation of the coaxial cable into an equivalent capacitance in the equivalent circuit [7], the other calculation formula of reflection coefficient is

$$\rho = \Delta C Z_0/(2t_r) \tag{9}$$

In Eq. (9), δC is equivalent capacitance, t_r is pulse rise time. Dowding, c. et al. verified the proportional relationship between the shear coaxial cable shape at the same time, variables and the equivalent capacitance finite element method is

$$\Delta \mathrm{C} = k\Delta x + b \tag{10}$$

δx is the shear type variable of Coaxial cable, k and b are constant. By Eq. (9) with Eq.(10)

$$\rho = \frac{Z_0}{2t_r}(k\Delta x + b) \tag{11}$$

When the shear deformation of the coaxial cable occurs, it could be obtained by Eqs. (9) and (11).

$$\Delta x = \frac{\left(e^{\alpha(f)+j\beta(f)}\right)^{2L} V_{rt}(f)}{(Z_0/2t_r)k\,V_0(f)(1+\rho_0)} - \frac{b}{k} \tag{12}$$

In Eq. (12), the parameters $e\alpha(f) + j\beta(f)$, $Z_0/2t_r$, k, $V_0(f)$, $1 + \rho_0$, b/k are constant. Equation (12) is the theory calculation model of slippage of the TDR landslide monitoring system. Thus it can be seen, when it is determined that the position of coaxial cable deformation, the coaxial cable deformation Δx has a linear relationship with the reflected voltage amplitude V_{rt}(f). The value L can be determined by Eq. (1). Therefore, in the process of landslide detection, TDR technology obtained the landslide surface depth L and slippage through analyzing the change of the reflected signal, determined the landslide hazard degree by the landslide surface depth, and estimates the stability of landslide by landslide slip trends.

4 The System Test and Analysis

The experiment chooses the TDR landslide monitoring system and the shearing device of the coaxial cable for the Simulation test. TDR landslide monitoring system includes TDR hardware system and PC applications. TDR landslide monitoring system parameters are shown in Table 1.

In the landslide simulation tests, coaxial cable shear device make test cable produce shear deformation, the shearing process of coaxial cable is shown in Fig. 3. When both block in a stable state, coaxial cable is not affected by shearing action;

Table 1 TDR landslide monitoring system parameters

Signal type	Rise time	Voltage amplitude	Sampling frequency
Single pulse	1.85 ns	912 mv	20 GHz

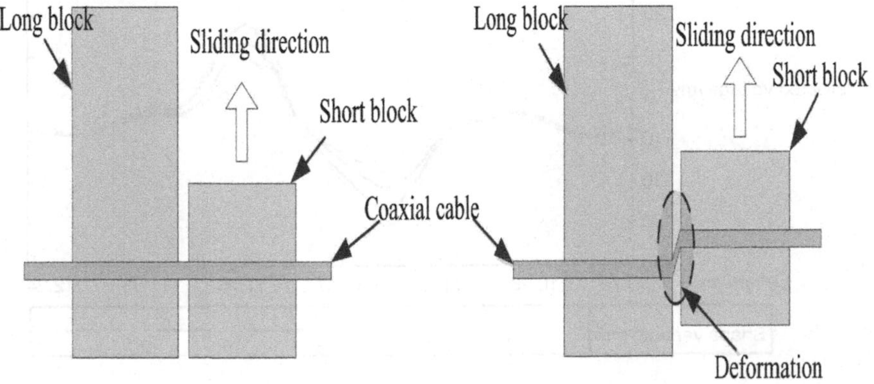

Fig. 3 Coaxial shear diagram

When short block slides, coaxial cable is affected by sliding surface shear force and lead to shear deformation of the sliding surface of coaxial cable.

Through several groups of test obtaining the constant term in Eq. (12), the transmission speed of the pulse signal in the coaxial cable is 2.574 × 108 m/s, Attenuation factor of the coaxial cable is $e^{\alpha(f)+j\beta(f)} = e^{0.00693}$, the transmission coefficient in the incoming end of coaxial cable is $1 + \rho_0 = 0.73$, constant term is $Z_0/2t_r \sim 2.03 \times 10^{10}$, $k = -1.22 \times 10^{-13}$, $b/k = -3.69$, obtained by Eq. (12).

$$\Delta x = \frac{e^{0.01386L}}{-1.65} V_{rt}(f) + 3.69 \tag{13}$$

In Eq. (12), δx is a slip rate, L is the depth of the slide surface, and $V_{rt}(f)$ is transmission voltage which reflection signal at the starting end of a coaxial cable.

In the test, Intervals of 1 mm to record a TDR waveform and save the waveform data, the maximum test form variables is 25 mm. Test coaxial full length 15.9 m, Fig. 4 is coaxial cable at 11.2 m shearing action occurs, the reflected waveform with the deformation of the coaxial cable and a partially enlarged waveform changes.

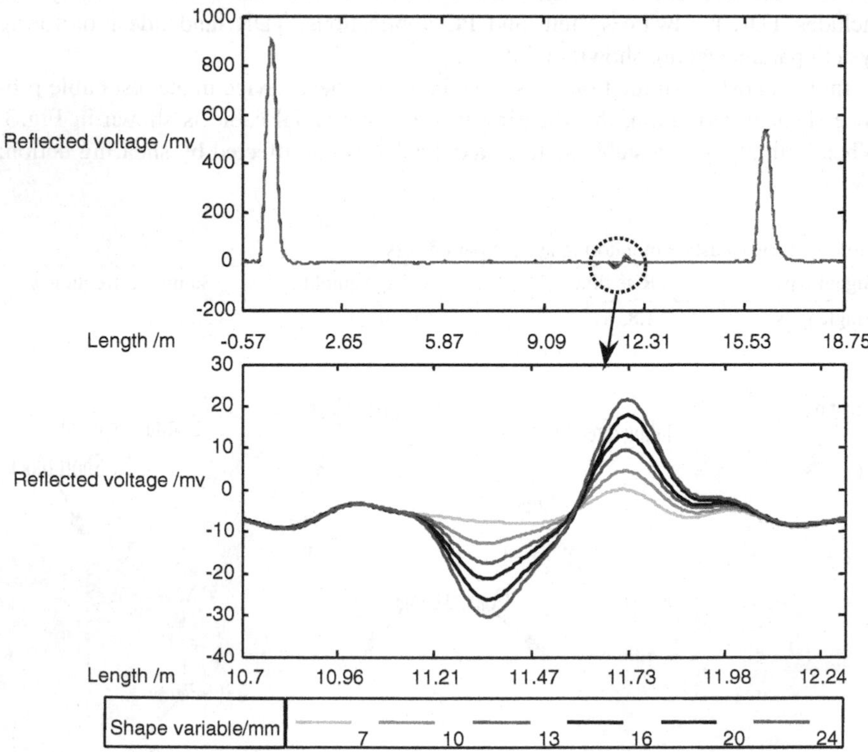

Fig. 4 11.2 m shear deformation process and partial enlarged waveform

Among them, the first pulse signal pulse for launch, the second pulse signal to the terminal reflection pulse, dotted circles for the coaxial cable of the deformation in the reflected wave. Figure 4 below for deformation of waveform of partial enlargement, down from the top of the 7 mm respectively, 10, 13, 16, 20 and 24 mm form variables of the waveform. It is clear that with the increase of coaxial cable shape variable, TDR waveform of the gradual emergence of a negative a positive two spikes, wherein the negative spike is the first change in the impedance plane CC of the reflected signal, the second signal is positive spike impedance change of the reflected signal DD surfaces [8]. In the theoretical analysis to choose the first surface impedance change of B–B reflected signal voltage is calculated. When the coaxial cable of the shape variable is changed from 0 to 25 mm, the peak amplitude of reflection signals is changed from 0.14 to −28.44 mv.

Figure 5 for coaxial cable in the 3.1, 7.0, 11.2 and 14.9 m in deformation occurs, the first peak pulse voltage amplitude changes with coaxial cable shape variables of the measured curve [9, 10]. Visible, coaxial cable shape variable changes from 0 to 4 mm, basic reflected voltage amplitude remains the same, when the coaxial cable of the type variable changes from 4 to 25 mm, reflected voltage amplitude increase with the increase of coaxial cable shape variable, two deformation point reflected voltage curve of a certain distance between each other, and the coaxial cable of the reflected voltage amplitude and form a linear relationship with variable approximation.

Combining the measured data and the deformation location L with Eq. (13), obtained the calculation formula of deformation: (1) In 3.1 m calculation formula of deformation point form variables:

$$\Delta x = -0.6327 \times V_{rt}(f) + 3.69 \quad (14)$$

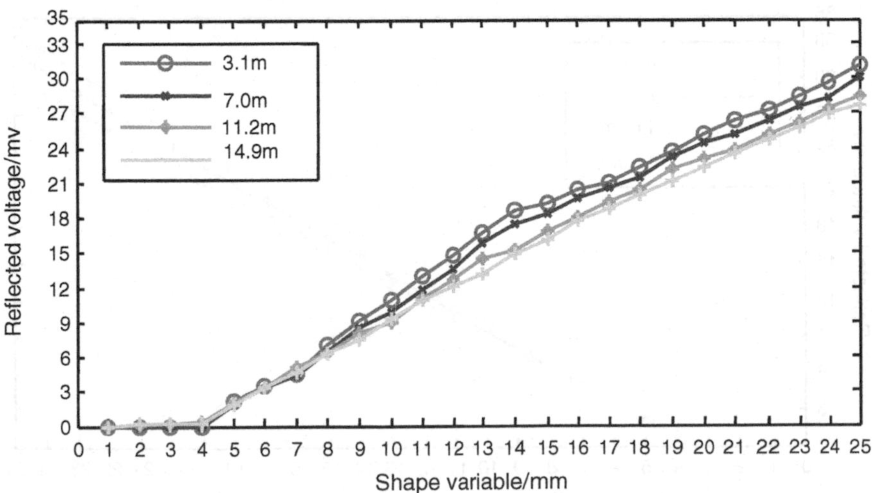

Fig. 5 Reflected voltage with variable changes in the measured *curve*

(2) In 7.0 m calculation formula of deformation point form variables:

$$\Delta x = -0.6679 \times V_{rt}(f) + 3.69 \tag{15}$$

(3) In 11.2 m deformation shape variable calculation formula:

$$\Delta x = -0.7078 \times V_{rt}(f) + 3.69 \tag{16}$$

(4) In 14.9 m calculation formula of deformation point form variables:

$$\Delta x = -0.7451 \times V_{rt}(f) + 3.69 \tag{17}$$

By the Eqs. (14)–(17) can obtain the coaxial cable of the shear type variables and a linear relationship with reflected voltage [11], all kinds of drawing into curve is shown in Fig. 6.

As shown in Fig. 6, from above in turn, is 3.1, 7.0, 11.2, 14.9 m of the curve. By Eq. (13), the further away from the coaxial cable between deformation point location, the greater the attenuation of the pulse signal, the smaller of the reflected voltage amplitude. The Fig. 6 shows, in the same deformation points, a reflection of 3.1 m, the highest voltage amplitude, the second is 7.0, 14.9 m minimum reflected voltage amplitude. With Eqs. (14)–(17) image of calculated value and the actual variable comparison, get the error of calculated value and the actual shape variable, is shown in Table 2.

As can be seen from Table 2, the absolute value of the calculation error is no more than 2 mm. From Figs. 5 and 6, in the form of variable 10–16 mm, and deformation

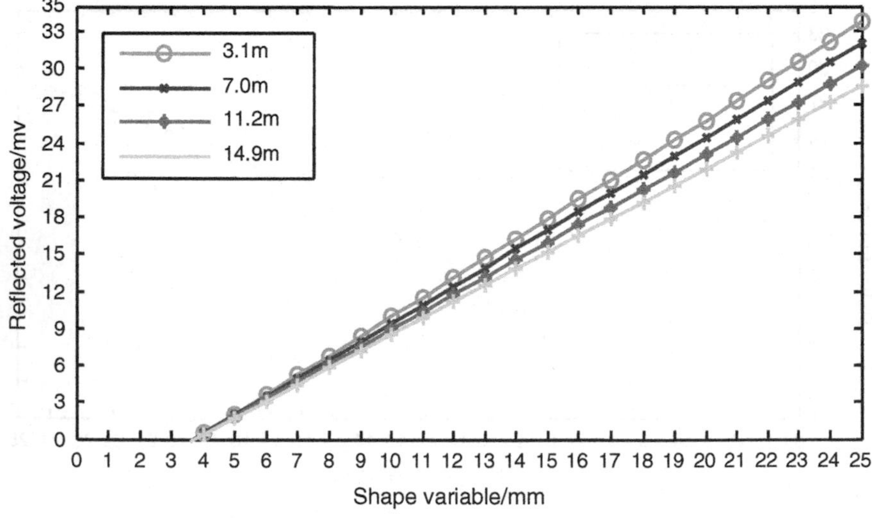

Fig. 6 Reflected voltage with the calculated *curve* of the variable changes

Table 2 Calculated value and the actual error of the shape variables

Actual deformation shape variable calculation error/mm				
	3.1 m	7.0 m	11.2 m	14.9 m
4 mm	0.31	0.32	0.08	0.09
7 mm	0.43	0.43	−0.26	−0.17
10 mm	−0.59	−0.30	0.00	−0.65
13 mm	−1.28	−1.22	−0.91	−0.46
16 mm	−0.57	−0.80	−0.47	−0.89
20 mm	0.40	0.04	0.00	−0.33
24 mm	1.55	1.43	0.91	0.23

point curve upward projections between 3.1 and 7.0 m. Therefore, in the interval range, error values are larger.

5 Conclusion

(1) The TDR monitoring system for landslide monitoring using coaxial cable, the monitoring of the dead zone, dead zone for [0 mm, 4 mm].
(2) Coaxial cable of the by shearing action, within the range of deformation/ 4, 25 mm, TDR coaxial cable of the reflected voltage amplitude and form of a variable into a linear relationship.
(3) Application of slippage calculation model of reflected voltage and slippage relation curve and measured curve of error values are not more than 2 mm.

References

1. Zhao J, Wang D, Chen J (2015) Experimental study on slope sliding and debris flow evolution with and without barrier. Water Sci Eng 8(1):68–77
2. Lin MW, Thaduri J (2005) Structural damage detection using an embedded ETDR distributed strain sensor. Subsurf Sens Technol Appl 6(4):315–336
3. Dennis ND, Ooi CW, Wong VH (2006) Estimating movement of shallow slope failures using time domain reflectometry. In: Proceedings of TDR
4. Dowding CH, Dussud ML, Kane WF et al (2003) Monitoring deformation in rock and soil with TDR sensor cables. Geotech News 21(2):51–59
5. Cortez ER, Hanek GL, Truebe MA et al (2009) Simplified users guide to time-domain-reflectometry monitoring of slope stability
6. Chen Y, Chen R-P, Chen Y-M (2003) TDR slope monitoring system calculation model and experimental study. Ind Constr 33(8):37–41
7. Dowding CH, Su MB, O'Connors K (1988) Principles of time domain reflectometometry applied to measurement of rock mass deformation. Int J Rock Mech Min Sci Geomech Abstr Pergam 25(5):287–297

8. Martinez P, Marciano JJS Jr (2014) A simulation study of a time domain reflectometry (TDR) cable with non-standard cross-section. In: TENCON 2014-2014 IEEE Region 10 Conference. IEEE, pp 1–6
9. Lin CP, Tang SH, Lin WC et al (2009) Quantification of cable deformation with time domain reflectometryimplications to land-slide monitoring. J Geotech Geoenviron Eng 135(1):143–152
10. Singer J, Thuro K, Sambeth U (2006) Development of a continuous 3d-monitoring system for unstable slopes using time domain reflectometry. Felsbau 24(3):16–23
11. Singer J, Festl J, Thuro K (2010) Application of time domain reflectometry (TDR) as a monitoring system for subsurface deformations. Geol Act 2459–2465
12. Chen Y, Chen R-P, Chen Y-M (2009) TDR The application of the test technology in geotechnical engineering. Engineering survey, vol 2
13. Lin C, Xiao-Ping L (2013) Based on the technology of TDR slope monitoring and early warning. Autom Subgrade Eng 1:120–125
14. Tritium LY, Xie L (2011) The coaxial cable of the TDR reflection characteristic under the pure shear state study. J Geotech Found 25(1):49–52
15. China TH, Fu HL (2010) TDR technology application in the highway slope monitoring test. Rock Soil Mech 31(4):1331–1336

Image Classification Based on Modified BOW Model

Gaoming Zhang, Jinfu Yang, Shanshan Zhang and Fei Yang

Abstract Image classifications the basis to solve visual tracking, image segmentation, scenes understanding and other complex visual tasks. Bag of words (Bow) model is initially applied in text classification area, introduced in image proceeding and recognition on account of its simple and effective. This paper follows the standard bag-of-words pipeline, but substitutes original SIFT descriptor with DSP-SIFT(domain-size pooling sift). Then, taking account of the truth that an DSP-SIFT was developed for gray images which limits its performance with regard to some colored objects, we present to add CN(color-name) descriptor to collect the color information to form visual words of image. The experimental results over fifteen scene categories and Caltech 101 datasets indicate that our method can achieve very promising performance.

Keywords Bow model · Visual word · DSP-SIFT descriptor · CN descriptor

1 Introduction

Image classification is the basis to solve visual tracking, image segmentation, scene understanding and other complex visual tasks. Bag of words(Bow) [1] model is initially applied in text classification area, the document is represented as the unordered

G. Zhang (✉) · J. Yang · S. Zhang · F. Yang
Department of Control and Engineering, Beijing University of Technology,
No, 100 Chaoyang District, Beijing 100124, China
e-mail: zhanggaoming003@126.com

J. Yang
e-mail: jfyang@bjut.edu.cn

S. Zhang
e-mail: dkzhang@emails.bjut.edu.cn

F. Yang
e-mail: yangfei199217@163.com

V.E. Balas et al. (eds.), *Information Technology and Intelligent Transportation Systems*, Advances in Intelligent Systems and Computing 455,
DOI 10.1007/978-3-319-38771-0_33

combination of key words, achieving the goal of classification by counting the frequency of occurrence of key words. Due to the property of simple and effective, Bow model was introduced in computer vision by Li Fei-Fei obtaining nice classification performance. BOW method mainly contains four key steps: feature extraction, code-book training, feature coding, classifier training. In feature extracting section of image, classical approach is to choose Scale Invariant Feature Transform(SIFT). The local feature descriptor, proposed by David G. Lowe in [2] for the first time, has been modified and perfected in 2004. SIFT descriptor can handle nuisance variability of illumination, view, scale and affine to a certain extent by detecting the stable key points in scale space. Convolutional neural networks CNN has drawn greater attention in recent years, using large amount labeled images, it can also be trained to learn away nuisance variability. [3, 4] show that the CNN outperforms SIFT, albeit with a much larger dimension. [5] presented a modification of SIFT, called domain-size pooled SIFT or DSP-SIFT, obtained by pooling gradient orientations across different domain sizes(scales) and spatial locations, outperforms the best CNN.

This paper follows the standard Bow pipeline, but substitutes original sift descriptor with DSP-SIFT(domain-size pooling sift). Then, taking account of the truth that an DSP-SIFT develops for gray images which limits its performance with regard to some colored objects. However, color information is a significant attribute for human vision. Color names are linguistic labels humans use to communicate the colors in the world. Van de Weijer [6] proposed a method to automatically learn the eleven basic color names of the English language from Google images. They are black, blue, brown, grey, green, orange, pink, purple, red, white, and yellow. Then, an eleven dimensions local color descriptor can be deduced simply by counting the occurrence of each color name over a local neighborhood. Considering that color names were found to compare favorably against photometric invariant descriptions on several computer vision application, such as image classification and object detection. On the basis of above, Rahat Khan [7] put forward an approach of discriminative color feature computation, which clusters color values together based on discriminative power on a training data set.

To be specific, we propose to use DSP-SIFT local descriptor to extract shape information. Meanwhile, In view of the above-mentioned facts, we choose to use CN descriptor to collect the color information to form visual words of image. Subsequently, visual words of Bow model can be gained after the clustering of DSP-SIFT and CN descriptors respectively. As a result, a image can be represented by the shape representation words and color representation words. The goal of classification by counting the frequency of occurrence of key words. We conduct several image classification experiment on universal datasets, the results show that our method is superior to original Bow method.

In the following sections we describe related work (in Sect. 2), clarify our Bow model of modified visual words (in Sect. 3) and explain DSP-SIFT and CN descriptor in detail, and evaluate our approach by some experiments (in Sect. 4).

2 Relate Work

For many visual recognition tasks, a critical problem is to discover robust image expressions (features). In recent years, the research over designing local descriptors has attracted much attention since the success of using SIFT as local features in retrieval, detection and recognition. The idea of Bow model, taking dense-sampling sift as local feature, achieved best performance for object classification in pascal VOC 2007 challenge. The features are invariant to image scale and rotation, and are shown to be robust across a substantial range of affine distortion, change in 3D viewpoint, addition of noise, and change in illumination. Inevitably, it also exists drawbacks, such as the number of matching points are too low, key points are not evently distributed, feature computing is very slow. A bunch of descriptors are proposed to improve original SIFT. PCA-SIFT [8] performed principal component analysis (PCA) on the gradient patches, which helps to reduce the noise in the descriptors. To improve sift feature computing efficiency, Bay [9] proposed SURF(Speed Up Robust Features), adopting rectangle filter other than Gaussian second-order filter, altering the size of rectangle instead of the size of image to get scale space and employing integral image to speed up convolution operation. Local extremum acquired by SIFT/SURF using hessian matrix are utterly stable. However, it may rely on gradient direction of pixels in local region too much when determine major orientation. Unfortunately, feature extracting as well as matching rely heavily on the major orientation in follow step make matching failed easily in spite of a slight major orientation deviation. Jingming Dong [5] came up with a modification of SIFT, obtained by pooling gradient orientations across different domain sizes (scales), in addition to spatial locations, addressed the problem of major orientation inaccuracy. On the whole, DSP-SIFT improves SIFT by a considerable margin. Since SIFT descriptor and its variants mentioned above utilize the shape information of an image only, they perform poorly when classify objects that have similar shapes but different colors in recognition tasks. Notwithstanding that there are some attempts in the literature which have been introduced to make use of the color information inside the SIFT descriptors. For example, in [10], the normalized RGB model has been used in combination with SIFT to achieve partial illumination invariance besides its geometrical invariance. The color invariance of this approach is still limited because of the primitive color model used. In [11], a multi-stages recognition approach has been developed in order to achieve both color and geometrical invariance. In the first stage, a color classifier is used to label the different image regions. Then, the SIFT descriptors are augmented by adding the color labels. In spite of the good performance of this approach, its need for colored learning instances limits its performance in several applications. The built Colored SIFT (CSIFT) [12] is more robust than the conventional SIFT with respect to color and photometrical variations. However, CSIFT descriptor reserves color information around key points merely, leading to consider all the points in the image for object description is not feasible.

Inspired by the outstanding attribute of DSP-SIFT and CN descriptor, following classic Bow model, we present a approach of adopting DSP-SIFT and COLORNAME as local features to construct visual words.

3 Bow Model of Modified Visual Words

On account of merely shape and color information can not represent image factually and adequately, we propose a novel image classification method based on Bow model to be specific, we use DSP-SIFT feature to extract the shape information of the image and then get visual words after clustering. Meanwhile, color information also is utilized by CN feature, and then we can get visual words after clustering in the same way. As a result, the image can be represented by the shape representation words and color representation words.

For our method, we follows the standard bag-of-words pipeline. For feature detection, we use a combination of domain with interest point detectors. For shape we use the DSP-SIFT descriptor. A visual vocabulary is constructed for shape representation. For color, we use the CN descriptor and construct a visual vocabulary. The vocabularies are constructed using standard K-means and the histograms are constructed using hard assignment. To represent an image we use the spatial pyramid representation as in [13]. For classification, we use the non-linear SVM using histogram intersection kernel [14].

3.1 DSP-SIFT Descriptor

As described in [15, 16], formula bellow denotes a single un-normalized cell of SIFT.

$$h_{SIFT}(\theta \left| I, \hat{\sigma} \right.)[x] = \int \mathrm{N}_{\varepsilon}(\theta - \angle \nabla I(y)) \mathrm{N}_{\hat{\sigma}}(y - x) d\mu(y) \tag{1}$$

where I is a square image, deduced by applying a difference-of-Gaussian(DOG) operator across all scales and locations(SIFT detections), $d\mu(y) = \|\nabla I(y)\| \, dy$, and θ denotes angle variation range $(0, 2\pi)$, ε is an orientation histogram bin of size, and $\hat{\sigma}$ is the spatial pooling scale. The Convolution kernel N_{ε} is separable bilinear of size ε and $\mathrm{N}_{\hat{\sigma}}$ is bilinear of size $\hat{\sigma}$, whereas they could be substituted for a Gaussian filter with standard deviation and an angular Gaussian with dispersion parameter ε. Computing 16 cells illustrated in (1) at locations $x \in \{x_1, x_2, \ldots, x_{16}\}$ on a 4×4 lattice Λ, normalizing, and then we get SIFT descriptor after concatenating them.

Because of the spatial pooling scale $\hat{\sigma}$ is deduced from the response of a DoG operator on the (single) image, and that both of $\hat{\sigma}$ and image domain size are both tied to the photometric characteristics of the image. Using fundamental principles of classical sampling theory, [5] propose to untie the size of the domain where the descriptor is computed, other than the idea of scale selection. As a consequence, DSP-SIFT achieve robustness to domain size changes due to occlusions.

If SIFT is expressed as (1), thus DSP-SIFT can be written as

$$h_{DSP}(\theta \mid I \ , \sigma)[x]\varepsilon_s(\sigma)d\sigma x \in \Lambda \tag{2}$$

where $S > 0$ is the size-pooling scale and ε is an exponential or other unilateral density function. Unlike SIFT, that is computed on a scale-selected lattice, DSP-SIFT is computed on a regularly sampled lattice Λ.

3.2 CN Descriptor

It is difficult to describe color since a series of accident events may influence its measurement duo to the presence of variations of scene geometry, viewpoint, illuminant and shadows. That is to say, it is important to weight all of the factors affecting color when designing color representations. Color names are linguistic labels which humans use to communicate the colors in the real world. Van de Weijer [6] have proposed an approach that the eleven basic color names of the English language are automatically learnt from Google images. As a consequence of this learning, the color space is partitioned into eleven regions. Then, simply by counting the occurrence of each color name over a local neighborhood region, an eleven dimensions local color descriptor can be deduced.

To learn a discriminative color representation, Rahat Khan et al. [7] apply divisive information-theoretic feature clustering (DITC), proposed by Dhillon et al. [17] to find connected clusters in L*a*b* space, to achieve the aim of finding discriminative Color Representation.

DITC algorithm The idea of DITC algorithm is to divide feature clusters into a smaller set of clusters, clustering process is conducted so as to minimize the decrease of mutual information of the new more compact feature representation. If dataset has L classes, they are $c1,\ldots,cL$. Each class can be represented by color words histogram. Mutual information formula below shows the discriminative power of the color words W among the classes C:

$$I(C, W) = \sum_i \sum_t p(c_i, \omega_t) \log \frac{p(c_i, \omega_t)}{p(c_i)p(\omega_t)} \tag{3}$$

where $p(\omega_t)$ the priors knowledge $p(c_i)$ and the joint distribution $p(c_i)p(\omega_t)$ can be determined empirically from the dataset.

The mutual information shows the information that the words W contain about the classes C. Then, dividing the words W into k clusters $W^C = \{W_1, W_2, \ldots, W_K\}$

which are invariant with respect to some physical variation. Each cluster contains W_i a set of words. Then Dhillon et al. [17] proved that the decline of mutual information caused by clustering a word W_t to cluster W_j meis equal to:

$$\Delta i = \pi_t KL(p(C|\omega_t), p(C|W_j)) \tag{4}$$

where the Kullback–Leibler (KL) divergence is given by

$$KL(p_1, p_2) = \sum_{x \in X} p_1(x) \log \frac{p_1(x)}{p_2(x)} \tag{5}$$

And $\pi_t = p(\omega_t)$ is the word prior knowledge.

The total decline of mutual information caused by clustering the words as follows:

$$\Delta I = \sum_j \sum_{\omega_t \in W_j} \pi_t KL(P(C|\omega_t), p(C|W_j)) \tag{6}$$

Therefore the clusters W that we seek are those which minimize the KL divergence between all words and their assigned cluster. In our case the words represent $L*a*b*$ bins of the color histogram.

Minimizing Eq. 4 is equal to joining bins from the $L*a*b*$ histogram in such a way as to minimize the ΔI. $L*a*b*$ bins which have similar, $P(C|\omega_t)$ are joined together. We adopt an algorithm similar to EM algorithm to optimize the objective function 4, the algorithm contains two steps (1). Compute the cluster means with

$$p(C|W_j) = \sum_{\omega_t \in W_j} \frac{\pi_t}{\sum_{\omega_t \in W} \pi_t} p(C|\omega_t) \tag{7}$$

(2). Assign each word to nearest cluster based on

$$w_t^* = \arg\min KL(p(C|w_t), p(C|w_j)) \tag{8}$$

w_t^* provides a new cluster index for word W_t. The two steps is repeated until converges. The DITC algorithm has been studied in the context of joining color and shape features into so-called Portmanteau Vocabularies by Khan et al. [7] use the DITC algorithm for a distinct purpose, i.e to learn discriminative color features automatically.

Discriminative Color Representations However, primitive DITC clustering algorithm does not consider the location of the words in the L*a*b* color space. As a result, the algorithm can join non-connected bins. Hence Eduard make the words to be connected when learning photo-metric invariants in [18]. Beyond that, connectivity has several conceptual advantages: it allows for comparison to photometric invariance, comparison with color names (CN), semantic interpretation and comparison with human perception. Therefore [7] propose to adapt the DITC algorithm to ensure that the cluster are connected in space. As a second adaptation we enforce smoothness of the clusters which prevents them from over fitting to the data.

4 Experiment and Result

Our experiments were performed on a Dell PC with an Intel Core i3 CPU M380 @2.53GHz, and a4GB random access memory. We report our results on two different datasets: fifteen scene categories dataset [13], and Caltech 101 [19].

The fifteen class scene dataset is a dataset of fifteen natural scene categories that expands on the thirteen category dataset released in [1]. The two new categories are industrial and store added by Oliva [20]. The sizes of images in the dataset are about 250×300 pixels, with 210 to 410 images per class. This dataset contains a wide range of outdoor and indoor scene environments. In Caltech-101 dataset, pictures belong to 101 categories with 40 to 800 images per category. Most categories have about 50 images, collected in September 2003 by Li Fei-Fei [19]. The size of each image is roughly 200×300 pixels. Some examples for every dataset are shown in Fig. 1.

We tested the proposed modified Bow method using fifteen scene categories dataset, and Caltech 101 dataset respectively. In the first experiment, we randomly choose 50 images per class from fifteen scene category dataset to assess classification accuracy. In our experiments, we evaluate the performance of the modified Bow model in scene recognition. DSP-SIFT method outperforms original SIFT and CSIFT, with regard to recognize the object images in the Caltech 101 dataset, we randomly chose 15 training images per class and 5 testing images per class from Caltech-101dataset. The results are shown in Tables 1 and 2

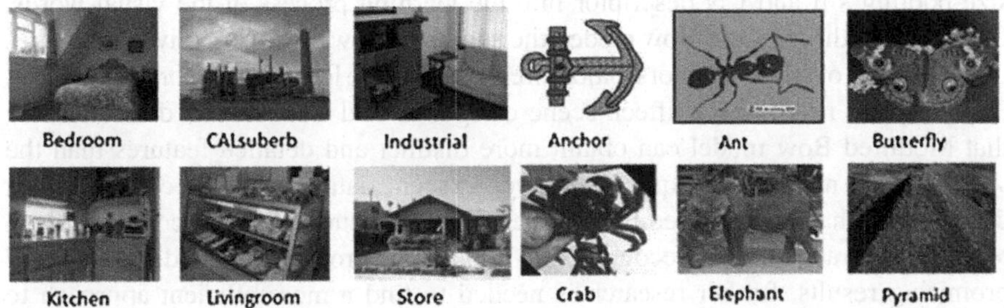

Fig. 1 Some examples of the fifteen scene categories, and Caltech 101. The *left group* is six types of images sampled from the fifteen categories including indoor and outdoor. The *right one* is six types of images sampled from Caltech 101

Table 1 Classification accuracy over fifteen scene categories dataset

Approaches	Recognition rate (%)
SIFT(Bow)	77.5
CSIFT(Bow)	78.2
DSP-SIFT(Bow)	85.5
DSP-SIFT + CN(Bow)	86.8
Zhang et al. [21]	90.2

Table 2 Classification accuracy over Caltech-101dataset

Approaches	Recognition rate (%)
SIFT(Bow)	56.3
CSIFT(Bow)	62.7
DSP-SIFT(Bow)	65.8
DSP-SIFT + CN(Bow)	76.3
Zhang et al. [21]	73.6
Gao et al. [22]	75.6

The proposed modified Bow performs better than original SIFT but worse than Zhang et al. [21], the reason may be the major images in Fifteen scene categories dataset are gray image.

In Table 2, the proposed modified Bow performs better than original SIFT and other methods [22].

5 Conclusion

In this paper, we have presented a Modified Bow model by introducing the domain-size-pooling sift and CN descriptor into the learning process of the visual words. Compared to the standard Bow model, the Modified Bow model effectively remedies the instability of sift descriptors major directions and the loss of the color information. Experimental results over fifteen scene categories and Caltech 101 datasets show that Modified Bow model can obtain more distinct and detailed features than the standard Bow model. The experiments over 15 scene dataset and Caltech 101 dataset demonstrate that the Modified Bow model performs better than their counterparts in terms of accuracy of scene recognition. Although the proposed method has achieved promising results, further research is needed to find a more efficient approach to feature extraction and scene recognition.

Acknowledgments This work is partly supported by the National Natural Science Foundation of China under Grant no. 61201362, 61273282 and 81471770, Graduate students of science and technology fund under no. ykj-2004–11205.

References

1. Fei Fei L, Perona P (2004) A bayesian hierarchical model forlearning natural scene cate-gories. In Proceedings of IEEE CVPR
2. Lowe DG (2004) Distinctive image features from scale-invariant keypoints. Int J Comput Vis
3. Dosovitskiy A, Tobias J (2014) Unsupervised feature learning by augmenting single images. arXiv:1405.5769
4. Fischer P, Dosovitskiy A (2014) Thomas, Brox.: Descriptor matching with convolutional neural networks: a comparison to sift. arXiv:1405.5769
5. Dong J, Soatto S (2015) Domain-size pooling in local descriptors DSP-SIFT. In: Proceedings of IEEE CVPR
6. Van J, De Weijer Cordelia (2009) Schmid: learning color names for real-world applications. IEEE Trans Imag Process 18(7):1512–1523
7. Khan R, Van de Weijer J (2013) Damien muselet discriminative color descriptors. In: Proceedings of IEEE CVPR
8. Ke Y, Sukthankar R (2004) PCA-SIFT: A more distinctive representation for local image descriptors. In: Proceedings of the IEEE conference on computer vision and pattern recognition (CVPR)
9. Bay H, Tuytelaars T (2008) Speeded Up Robust Features. In: Computer vision and image understanding (CVIU), vol 110, pp 346–359
10. Brown M, Lowe D (2002) Invariant features from interest point groups. BMVC
11. Mikolajczyk K, Schmid C (2005) A performance evaluation of local descriptors. IEEE Transactions on PAMI, vol 27 p 10
12. Alaa E, Hakim A, Farag AA (2006) CSIFT: A SIFT descriptor with color invariant characteristics. In: Proceedings of IEEE CVPR
13. Lazebnik S, Schmid C, Ponce J (2006) Beyond bags of fe atures: spatial pyramid matching for recognizing natural scene categories. In: Proceedings of IEEE CVPR. 2169C2178
14. Maji S, Dosovitskiy A, Malik J (2008) Classification using intersection kernel support vector machines is efficient. In: Proceedings of IEEE CVPR
15. Dong J, Karianakis N, Davis D (2014) Multi-view feature engineering and learning. In: Proceedings of IEEE CVPR
16. Vedaldi A, Fulkerson B (2010) Vlfeat: an open and portable library of computer vision algorithms. In: ACM multimedia
17. Dhillon IS, Mallela S, Kumar RV (2003) A divisive information-theoretic feature clustering algorithm for text classification. J Mach Learn Res
18. Vazquez E, Joost V, Weijer D (2011) Describing reflectances for color segmentation robust to shadows, highlights, and textures. In: IEEE Transactions on Pattern Analysis and Machine Intelligence
19. Feifei Li, Fergus R, Perona P (2004) Learning generative visual models from few training examples: an incremental Bayesian approach tested on 101 object categories. In: Proceedings of IEEE CVPR
20. Oliva A, Torralba A (2001) Modeling the shape of the scene: a holistic representation of the spatial envelope. Int J Comput Vis
21. Zhang C, Liu J, Tian Q (2011) Image classification by non-negative sparse coding, low-rank and sparse decomposition. In: Proceedings of IEEE CVPR
22. Gao J, Yang J, Li M (2015) A novel feature extraction method for scene recognition based on centered convolutional restricted boltzmann machines. arXiv:150607257

A Novel Method for Scene Classification Feeding Mid-Level Image Patch to Convolutional Neural Networks

Fei Yang, Jinfu Yang, Ying Wang and Gaoming Zhang

Abstract Scene classification is an important task for computer vision, and Convolutional Neural Networks, a model of deep learning, is widely used for object classification. However, they rely on pooling and large fully connected layers to combine information from spatially disparate regions; these operations can throw away useful fine-grained information, and in natural scenes, there are many useless information which will increase computation cost. In this paper, mid-level discriminative patches are utilized to pre-process the full images. The proposed method which combines mid-level discriminative patches for preprocessing with CNN for feature extraction improved the efficiency of computation and are more suitable for classifying scenes. Firstly, full images are divided into discriminative parts. Then utilize these patches to go through CNN for feature extraction. Finally, a support vector machine will be used to classify the scenes. Experimental evaluations using MIT 67 indoor dataset performs well and proved that proposed method can be applied to scene classification.

Keywords Convolutional natural networks · Mid-level discriminative patch · Scene classification · Deep learning

1 Introduction

Scene classification is an important task for computer vision, since most scenes are collections of entities organized in a highly variable layout. And it has been a driving intention for research in computer vision for many years which is basic for

F. Yang (✉) · J. Yang · Y. Wang · G. Zhang
Department of Control and Engineering, Beijing University of Technology,
No. 100 Chaoyang District, Beijing 100124, People's Republic of China
e-mail: yangfei199217@163.com

J. Yang
e-mail: 354003021@qq.com

Y. Wang
e-mail: 15650750952@163.com

V.E. Balas et al. (eds.), *Information Technology and Intelligent Transportation Systems*, Advances in Intelligent Systems and Computing 455,
DOI 10.1007/978-3-319-38771-0_34

machine learning. Scene classification is widely applied in many fields, including robot navigation, location, and map construction. In recent studies, there are many researches about scene classification. Much of these methods have appeared by the development of robust image features, such as SIFT and HOG [1, 2], even a spatial pyramid on Bag of visual Words (BoW) [3–5] which is low-level image descriptors as well as deformable part models [6] which is high-level semantic features. However, the low-lever features are based on statistics, they can represent the descriptor of images, but failed to represent the understanding of images for human. And both features are just local information of images, they will neglect some hidden elements of images. Not all the images can include the extracted features. So a novel method-convolutional neural net works(CNN) is performed to solve the problem.

CNN was firstly proposed by LeCun et al. [7] and was firstly applied in hand-writing recognition. Recently, with the rise of GPU computing, convolutional neural networks (CNN) has achieved impressive state of the art results over a number of areas, including scene recognition [8–13], object and semantic segmentation [14–17], object detection [18–20], Contour detection [21, 22] and so on. Most of them get excellent results. However, although deep CNN extract features from low-level edges information to high-level semantic information layer by layer, it will still loss some important information when it is trained. Since they rely on pooling and large fully connected layers to combine information from spatially disparate regions; these operations can throw away useful fine-grained information. In natural images, there are a lot of useless information for classification, such as a cow in a large grass, a bird in the sky. Training CNN with the whole image will sometimes increase computation cost. So we need pre-process the whole images before going through CNN.

For the human visual system, which features are the best primitive is a very important problems for artificial intelligence. For low-level pixels, there are not enough information to contain more useful features. What if we use the whole image to ex-tract features as a extreme? The Method is not feasible since it requires a great many of training data and a lot of semantic entity which is ambiguous and less discriminable. So we adapt the mid-level image patches as our primitive. There are many prior works on mid-lever features [23–26]. But for one image, there are a lot of image patches and the majority of them are useless for our classification. Therefore we aim to find image patches more representative and discriminative.

In this paper, we will firstly find the representative and discriminative image patches, and then take them as convolutional neural networks inputs to extract features. Finally, a support vector machine will be used to classify the scenes. The main contributions of this paper are as follows:

(1) The introduced method based on CNN can be used to extract features for images.

(2) The representative and discriminative mid-level image patches are used for pre-process the images. This paper will be organized as follows. In Sect. 2, we introduce the structure and algorithm of convolutional neural networks. A novel method for finding mid-level patches will be introduced in Sect. 3. Section 4 will show the

experiments about our method on MIT-67 Indoor datasets. Finally, we will conclude our paper in Sect. 5.

2 Related Work

There are many prior works on scene classification. Quattoni et al. [27] firstly explained that we can regard scene as a combination of parts, which is represented as regions of interest (ROI). More and more researchers agreed with him, and caused a boom that complete scene classification using the component model [28–32]. Li et al. [28] proposed the Object Bank, which is a high-level image representation, they take an image as a scale-invariant response map to combine image representations, then use simple off-the-shelf classifiers to classify scene. Singh et al. [29] proposed an unsupervised discovery of mid-level discriminative patches. They adapt discriminative clustering to train SVM, and use obtained patches to complete scene classification. Its worth noting that both of them still ignore the adverse impact of false response. To solve this problem, Lin et al. [30] presented a joint model for scene recognition, which used important spatial pooling regions (ISPRs) combining part appearance information to decrease the effect of false responses. However, both representations are just local information of images, they will neglect some hidden elements of images. So we utilize CNN to achieve image global representation.

There are also some methods for scene classification using CNN. For example, Cheng et al. [31] proposed to use a coarse-to-fine framework which is an unsupervised and supervised single-hidden-layer neural networks respectively to train coarse sparselets, and complete object detection and scene classification. He et al. [13] proposed a spatial pyramid pooling combined with CNN, and there is no requirements for the size of input image, they used the proposed method for image classification and object detection. However, both of them changed the structure of network which make the training process more complicated. Instead, we propose a joint method for scene classification, which combined mid-level patches with CNN.

3 Convolutional Neural Networks

Convolutional neural networks (CNN) is the further step of multilayer neutral network. It introduces the idea of deep learning to the neutral networks by computing the convolution to extract features from the shallower to the deeper. At the same time, acquire parameters of convolutional kernel automatically at the training process. Finally the whole network will produce more exact representations for classification. CNN are mainly composed of convolution layer, sub-sampling layer, and full connection layer (see Fig. 1).

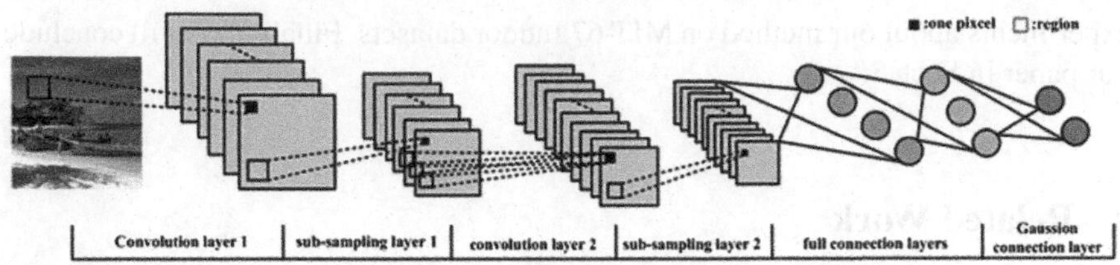

Fig. 1 The structure of CNN. It consist of convolutional layer, sub-sampling layer and full connection layer. The last layer is Gaussian connection layer which is used as classifier

3.1 Feed-Forward Pass of Traditional CNN

As shown in Fig. 1, a natural image passes the network, firstly, it will go through the first convolutional layer. And it is convolved with the convolutional kernel which is learned itself. The next output will use the activation function to obtain the first layer feature maps. These feature maps go through several similar structure layers and finally obtain more comprehensive feature representations.

The whole procedure obtains features from the contour information to high semantic information. The following is a feed-forward propagation process. Firstly, given a set of images, they go through continuous convolutional and sub-sampling layers to get final feature maps. We take as current layer, then the output of this layer is defined as:

$$X_j^l = f(u^l)\, with\, u_j^l = W_j^l x^{l-1} + b^{l-1} \tag{1}$$

where W_j^l is the weight of the layer, j represents the jth convolution kernel, b^{l-1} is bias of the layer, $f(.)$ is usually the function of sigmoid or tanh. Then the output go through the sub-sampling

$$X_j^l = f(\beta_j^l down(x_j^{l-1}) + b_j^l) \tag{2}$$

where $down()$ represent a function of sub-sampling. We adopt the max pooling, which have the effect that reduce the dimension.

Full connection layers can be seen as special convolution layers. The difference between them is the dimension. Full connection layer is a one-dimensional vector. At the end of the network, there is a classifier which is called softmax classifier.

3.2 Back Propagation Pass

We have known the feed-forward pass, but how does CNN train itself to update the weight and bias? In this section, we will introduce the training process. The training process is the same as traditional Error Back Propagation (BP). We define the squared-error loss function with N training examples.

$$E^N = \frac{1}{2}\sum_{k=1}^{c} ||t_k^n - y_k^n||^2 \tag{3}$$

where t^n represents labels with the nth training example, y^n represents the output. Define δ as sensitivities of every neuron, for the error of background propagation can be seen as follows:

$$\frac{\partial E}{\partial b} = \frac{\partial E}{\partial b}\frac{\partial u}{\partial b} = \delta \tag{4}$$

Here ${}^{\partial u}/_{\partial b} = 1$, so the sensitivities of bias is equal to the derivative of error E to the whole input U. The derivative is to make the error backing propagation from top to the basis of the network. Back propagation use the following formula:

$$\delta^l = (W^{l+1})^{T\circ} f^{'}(u^l) \tag{5}$$

o denotes element-wise multiplication. Then update weights using δ. Specifically, for the layer l, every derivative of error for the weights of the layer is multiplication cross of δ and input of this layer. Then we can get the updates of weight:

$$\Delta W^l = -\eta\frac{\partial E}{\partial W^l} with \frac{\partial E}{\partial W^l} = x^{l-1}(\delta^l)^T \tag{6}$$

The whole process is applied to bias b. Next we will introduce the training of different layers. η is the learning rate.

Convolution layer Noting that the convolution layer is followed by a sub-sampling (see Fig. 1). To get sensitivities for layer l, we must compute the sensitivities and weight of $l+1$ layer (δ^{l+1}, W^{l+1}) from (5). And from (6), we can get ΔW. However, one pixel of sampling layer corresponds one unit of former convolution layer output. So, one node of one map for l layer is just connected with one node relevant to that map for $l+1$ layer.

To compute sensitivities of l layer, we need up-sample the sensitivities of map corresponding down-sampling layer. Each pixel of feature map corresponds to a sensitivities which consists of one map too. Then take advantage of formula (5), we can get δ_j^{l+1}

$$\delta_j^{l+1} = \beta_j^{l+1}(f^{'}(u_j^l \circ up(\delta_j^{l+1}))) \tag{7}$$

where $up(.)$ represents a up-sampling. Here we define it as $up(x) \equiv x \otimes 1_{n\times n}$, n is a factor of down-sampling.

Given a feature map j, we can compute the gradient of bias for layer l as follows.

$$\frac{\partial E}{\partial b_j} = \sum_{u,v}(\delta_j^l)_{uv} \tag{8}$$

Also we can compute the gradient of convolution kernel weight from Eq. (6).

Sub-sampling layers Here, we aim to compute sensitivity maps. From these maps, we can get parameter of n and b. As we can see in Fig. 1, sub-sampling is followed by convolution layer. If convolution layer is full connected with sub-sampling, then we can compute sensitivity maps of sub-sampling.

Suppose there are M feature maps in sub-sampling layer, we find which patch correspond one pixel of the next layer sensitivity map. Take use of Eq. (5), finally, we will get the sum of sensitivity maps δ^l.

Now we are ready to compute parameter of β and b from (8) and (9)

$$\frac{\partial E}{\partial \beta_j} = \sum_{u,v} (\delta_j^l \circ d_j^l)_{uv} with\ d_j^l = down(x_j^{l-1}) \tag{9}$$

where d_j^l is computed by down-sampling of feed-forward process.

4 Discovery of Mid-Level Discriminative Patches

In this section, we will introduce a novel method for finding a certain number of discriminative patches. There are many other prior works on finding good mid-lever image patches, and most of them rely on the original visual words. For example, J et al. [32] proposed clustering a sparse key-point detections in SIFT space to get representation units of visual meaning. Although this method can get high-level object parts, but it always capture some simple corners which is simple coding. To avoid this, G et al. [33] proposed a approach that discovered highly probable regions of object instance. The whole process consists of two procedures: one is choosing the exemplar sets, the other is using the first step to refine the ROIs of each image with a circulation. Singh et al. [29] adapted the same thought to propose a method to find image patches. Suppose that there are many mid-level patches, K-means can not get a better cluster for K-means use the low-level measure(such as Euler distance, $L1$, cross-correlation). Usually not similar patches in vision will be clustered together. So they use linear SVM to produce more appropriate similarity measure to avoid that, which is called discriminative clustering. We utilize the method to produce a certain amount of patches as inputs to CNN, which refrain from large computation consuming and information losing.

4.1 Total Framework of Our Method

Given a set of images, we randomly sample for each image. Therefore, for one single image, there are hundreds or thousands of patches. We must process these patches and make them more different from each other. The typical solution is to cluster them in an unsupervised way. Reference [29] adopt common K-means. However, K-means as well as some other clustering method just use a low-level distance measure, which is not proper clustering for mid-level patches. But if we find that patches are visually similar, we can use a classifier to cluster the similar patches to solve the problem. So we take use of liner SVM to act as classifier to produce similarity metric and cluster large amount of patches. Note that only good clusters can produce good similarity, and good similarity can produce the good clusters. So the method involves the problem of iteration just like the method proposed by G et al. [33].

The method above is still infeasible, since if we want represent the entire visual world, there must be many clusters. To solve the problem, we turn the classification step of discriminative clustering into a detection step, taking each cluster as a detector, trained to find other patches like those it already has. Therefore, we propose a natural world dataset N. This method makes each cluster discriminative enough against the other clusters in the discovery dataset D and the rest of dataset N.

4.2 Details of Our Method

We aim to discover top n discriminative patches. Firstly, we need two datasets, one for discovery dataset D, the other is natural world set N. To avoid overlapping, we adapt cross-validation. So divide both of them into two average parts, denoted as $D1$, $D2$, $N1$ and $N2$. Then compute HOG descriptors at 7 different scales. At the same time, select patches from randomly, discarding highly overlapping patches and those with no gradient information(e.g. sea patches) and run K-means in HOG space all the same. Since there will some problems about K-means to mid-level patches mentioned before, we set k quite high($k = S/4$) to produce tens of thousands of clusters, most with very few members. Finally, prune out clusters which are less than 3 patches and proceed to the next iterative step.

Now, we are ready to train liner SVM. Taking k clusters as positive examples, all the patches in $D1$ as negative examples, start to train SVM. Then find top m new members in $D2$ using trained classifier, here we set $m = 5$ to make sure the purity of clusters. Swap the two sets and continue this training process utilizing the updated classifier and k clusters until the clusters are converged. Finally, compute purity scores of obtained patches and select top n patches.

5 Experimental Evaluations

Our algorithm is evaluated on the MIT 67 indoor dataset for scene classification. The MIT dataset consist of 67 scene categories, which contains 15,620 images and are divided into subway, bathroom, closet and so on. Our experiment consists of two main parts and is performed on one single GPU 980Ti.

5.1 Discovery of Mid-Level Discriminative Patches

We must pre-process the images before going through CNN, which will add the miss-ing information CNN may lose and reduce compute consuming.

The Choice of Experiment Data We first conduct our experiment to find discriminative patches. For every category, we select 80 images as D for one category, and divide it into two equal, non-overlapping parts. One for $D1$, the other is $D2$. And we treat other images in the dataset as natural word set N, separate it two average parts in the same way of D. Both of this way are prearranged for cross-validation. Finally, conduct experiment using our method on these data. We adapt HOG feature as our descriptor, The parameter C of linear SVM we use is equal to 0.1, and we choose 12 as our iterations of hard negative mining. Figure 2 shows some of our discriminative patches by our methods. Note that the extracted patches show the typical information of one specific scene.

5.2 Extracting Features and Classification

After retaining a small amount of patches from Sect. 5.1, we then extract feature using convolutional neutral networks. And then a Support Vector Machine is trained to classify the scene. Our model consists of 7 layers, five convolution layers followed by three pooling layers, and two connection layers. We extract features from 7th layer

Fig. 2 A sample for mid-level discriminative patches in MIT 67 indoor scene dataset. Its worth noting that our patches can capture the main information of a typical scene

Table 1 Classification for using features from different layers

Method	Accuracy (%)
Conv2	50.32
Conv5	53.8
Pool3	57.28
Fc6	58.76

Table 2 Accuracy over MIT-67 indoor dataset

Method	Accuracy (%)	Method	Accuracy (%)
ROI [27]	26.05	Sparselets [31]	59.87–64.36
DPM [6]	30.4	Mode Seeking [10]	65.1
Object Bank [28]	37.6	MS+IFV [10]	66.87
Patches [29]	38.1	ISPR+IFV [30]	68.5
ISPR [30]	50.1	Semantic FV [11]	72.86
SPP [13]	56.3	Ours	71.34

which is representative for image patches and conduct an experiment for comparing classification of extracted features from different layers. Finally we use SVM to classify the scene. The result is in Table 1. Compared with features from other layers, our fc7 feature gain a better classification effect. For reducing the over-fitting of connection, we use a latest regularization method dropout.

Table 2 lists some results of previous methods for scene classification on MIT 67 indoor, also with our method. Its worth noting that our method acquires a relatively better performance except for semantic fisher vector. However, the performance of semantic FV is resulting from two mapping, one is feature maps from image to feature, another is mapping from feature maps to semantic maps, our method just has one mapping from images to feature, which will of course be better than us. But compared with other methods, our approach is much better.

6 Conclusions

In order to supplement the possible information that CNN may lose, and reduce the computation cost, we propose a new method to pre-process the full images, which obtain a better performance than many of other methods for scene classification. Firstly, we use a trained discriminative classifier to get mid-level patches, which is proved to be more representative and discriminative for a specific scene. Then we use a 7 layers CNN to extract features from patches which is already got former. Finally use a SVM to categorize scenes. The results on MIT-67 indoor dataset demonstrate that our proposed method get a better performance.

Acknowledgments This work is partly supported by the National Natural Science Foundation of China under Grant no. 61201362, 61273282 and 81471770, Graduate students of science and technology fund under no. ykj-2004-11205.

References

1. Dalal N, Triggs B (2015) Histograms of oriented gradients for human detection. In: CVPR
2. Lowe DG (2003) Distinctive image features from scale-invariant keypoints
3. Csurka G, Dance C, Fan L, Willamowski J, Bray C (2004) Visual categorization with bags of keypoints. In: ECCV workshop
4. Lazebnik S, Schmid C, Ponce J (2006) Beyond bags of features: spatial pyramid matching for recognizing natural scene categories. In: CVPR
5. Perronnin F, Sanchez J, Mensink T (2010) Improving the fisher kernel for large-scale image classification. In: ECCV
6. Pandey M, Lazebnik S (2011) Scene recognition and weakly supervised object localization with deformable part-based models. In: ICCV
7. LeCun Y, Bottou L, Bengio Y, Haffner P (1998) Gradient-based learning applied to document recognition. Proc IEEE 86(11):2278–2324
8. Gong Y, Wang L, Guo R, Lazebnik S (2014) Multi-scale orderless pooling of deep convolutional activation features. In: ECCV
9. Zhou B, Lapedriza A, Xiao J, Torralba A, Oliva A (2014) Learning deep features for scene recognition using places database. In: NIPS
10. Doersch C, Gupta A, Efros A (2013) Mid-level visual element discovery as discriminative mode seeking. In: NIPS
11. Dixit M, Chen S (2015) Scene classification with semantic fisher vectors. In: CVPR
12. Liu L, Shen C, van den Hengel A (2015) The treasure beneath convolutional layers: cross-convolutional-layer pooling for image classification. In: CVPR
13. He K, Zhang X, Ren S, Sun J (2014) Spatial pyramid pooling in deep convolutional net-works for visual recognition. In: ECCV
14. Dai J, He K, Sun J (2015) Convolutional feature masking for joint object and stuff segmentation. In: CVPR
15. Ciresan DC, Giusti A, Gambardella LM, Schmidhuber J (2012) Deep neural networks segment neuronal membranes in electron microscopy images. In: NIPS, pp 2852–2860
16. Farabet C, Couprie C, Najman L, LeCun Y (2013) Learning hierarchical features for scene labeling. IEEE transactions on pattern analysis and machine intelligence
17. Hariharan B, Arbelaez P, Girshick R, Malik J (2014) Simultaneous detection and segmentation. In: european conference on computer vision (ECCV)
18. Pinheiro PH (2014) Recurrent convolutional neural networks for scene labelling. In: ICML
19. Sermanet P, Eigen D, Zhang X, Mathieu M, Fergus R, LeCun Y (2014) Overfeat: integrated recognition, localization and detection using convolutional networks. In: ICLR
20. Girshick R, Donahue J, Darrell T, Malik J (2014) Rich feature hierarchies for accurate object detection and semantic segmentation. In: CVPR
21. Shen W, Wang X, Wang Y (2015) DeepContour: a deep convolutional feature learned by positive-sharing loss for contour detection. In: CVPR
22. Uijlings JRR, Ferrari V (2015) Situational object boundary detection. In: CVPR
23. Albaradei S, Wang Y (2014) Learning mid-level features from object hierarchy for image classification. In: WACV
24. Zhao R, Ouyang W, Wang X (2014) Learning mid-level filters for person reidentification. In: IEEE conference on computer vision and pattern recognition
25. Singh S, Gupta A, Efros AA (2013) Representing videos using mid-level discriminative patches. In: CVPR
26. Boureau Y-L, Bach F, LeCun Y, Ponce J (2010) Learning mid-level features for recognition. In: CVPR
27. Quattoni A, Torralba A (2009) Recognizing indoor scenes. In: CVPR
28. Li L.-J, Su H, Fei-Fei L, Xing EP (2010) Object bank: a high-level image representation for scene classification and semantic feature sparsification. In: NIPS
29. Singh S, Gupta A, Efros AA (2012) Unsupervised discovery of mid-level discriminative patches. In: ECCV

30. Lin D, Lu C, Liao R, Jia J (2014) Learning important spatial pooling regions for scene classification. In: IEEE conference on computer vision and pattern recognition
31. Cheng G, Han J, Guo L, Liu T (2015) Learning coarse-to-fine sparselets for efficient object detection and scene classification. In: CVPR
32. Sivic J, Zisserman A (2003) Video Google: a text retrieval approach to object matching in videos. In: ICCV
33. Kim G, Torralba A (2009) Unsupervised detection of regions of interest using iterative link analysis. In: NIPS

An Overview on Data Deduplication Techniques

Xuecheng Zhang and Mingzhu Deng

Abstract The massive data puts forward higher requirements on the capacity of storage devices, but from a practical point of view, the increasement of capacity is far more behind the growth of data. Deduplication technique, for its high efficiency, few resource consumption and extensive application scope, comes to the fore among various data reduction techniques. The so-called data deduplication refers to find and eliminate redundant data among the storage system. For local storage system, the only one data object is needed to store to save limited storage space; for network system, not only storage space can be saved, but also transmission bandwidth can be reduced to increase the transmission rate. It is a compromise to achieve the purpose of efficient storage at cost of computational overhead. This article will introduce data deduplication techniques, describe basic principles and processes, summarize the main technique of the current study and provide recommendations for future development.

Keywords Data deduplication · Chunking · Optimization · Reliability · Scalability · Load balance

1 Introduction

The arrival of the era of big data, we need to store massive amounts of data safely and effectively, the capacity of the storage system needs to be increased. However, the increasement of storage capacity is far behind the growth of data, the storage system is facing serious challenges. Deduplication technique is to deal with the status of the data explosion and provide an effective means of data storage.

X. Zhang (✉) · M. Deng
National University of Defense Technology, Changsha, China
e-mail: zhang.xuecheng@foxmail.com

M. Deng
e-mail: dk_nudt@126.com

V.E. Balas et al. (eds.), *Information Technology and Intelligent Transportation Systems*, Advances in Intelligent Systems and Computing 455,
DOI 10.1007/978-3-319-38771-0_35

This paper focuses on data deduplication techniques and conducts a comprehensive study of the system, which starts with the basic principles and sets out the specific process of deduplication in detail, then given several classifications of deduplication, and analyzes the related techniques about deduplication combined with existing instances from five aspects, then points out the problems and challenges facing in the future based on these techniques. At last, we conclude and present our current work.

2 Basics

Deduplication is to detect and remove redundant data for an efficient storage. The system includes the following sections: access control protocols, file services, content analysis, chunks filter, chunks storage and indexing. The access control protocols layer supports all types of file access and receive requests of the upper application. The file services layer analyses requests into actual operation, then the file system respond to requests. Traditional storage systems usually sent the operations to the general block layer directly. However, with data deduplication technique, there is content analysis layer after file services layer, which analyzes the content of files or data stream and divides them into a series of chunks according to the presupposed strategy. Chunks are compared and judged whether is a duplication at the filter layer, and the duplication will be "filtered" out. Finally, nonredundant chunks will be stored and the index will be updated. The structure shown in Fig. 1 [29].

2.1 *Process*

Generally speaking, the basic flow of the entire deduplication include the following steps which is shown in Fig. 2 [29].

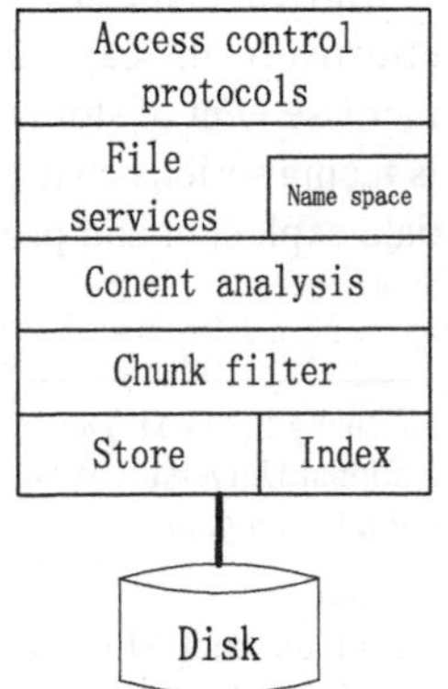

Fig. 1 Data deduplication system

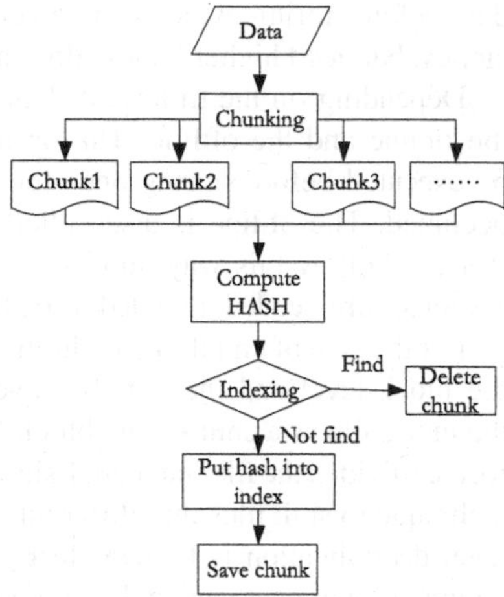

Fig. 2 Algorithm chart of data deduplication system

Chunking: Depending on the strategies such as whole file chunking, fixed size chunking and content defined chunking, files or data stream are divided into chunks. It should be noted that the chunking will have a direct impact on the subsequent steps. If the size of chunks are larger, the computational overhead is smaller, but the effect of deduplication is not obvious. Conversely, the reverse.

Compute: Compute eigenvalue of each chunk, the feature will be used as uniquely identification of the chunk, and as the basis of next step to determine whether is duplicated. So encryption algorithms, like MD5, SHA-1 and so on, are commonly used which have an ability of anti-collision.

Lookup: lookup the index table and compare the eigenvalue with the existed ones to determine whether it is a duplicate chunk. The index table will be expanded along with the growth of data, so the huge amount of data can degrade system performance.

Delete: According to the results above, if the chunk is duplicated, it can be discarded directly. But the metadata information pointed to the only one chunk needs to record in case that the chunk will be accessed.

Store: After deduplication, the unique chunks will be stored. The eigenvalue as a new entry will be added to the index table, and the unique chunks will be assigned the appropriate address to store.

2.2 Classification

According to different principles, there are mainly three methods of classification.

According to the position of deduplication, there are the source and the destination. The source deduplication deletes duplicate data before transmitting which reduces the amount of data and improves bandwidth utilization, but it takes up the limited resources and affect performance. The destination deduplication deletes duplicate data before storing, which can detect data from multiple clients and get higher efficiency, but need higher bandwidth and longer time.

Depending on the timing of deduplication, deduplication can be classified into the online and the offline. The online is a real-time mode, so deduplication will be executed before storing once the data generates, and less storage resources are occupied. The offline is also called as post-process, which delete duplicate after storing. Follow this way, more spaces are occupied to store data but less system resources are needed, and deduplication can be done whenever the system is idle.

For the size of chunking, deduplication can be classified into the whole file level, the block level and the byte/bit level. The whole file level deduplication regards the entire file as a chunk. The block level includes fixed-size and variable-size. The former divides the file into equal-sized chunks and the latter uses the sliding window technique to split files into different size chunks based on the contents. The byte/bit level deduplication is the smallest granularity, which need much more additional computational overhead and storage space.

3 Recent Work

3.1 Data Chunking

The fundamental purpose of deduplication is to reduce the amount of data and only nonredundant information is stored. Good chunking mechanism can detect the maximum repetition rate of the data and is the premise of improving deduplication efficiency.

The whole file chunking (WFC) is the easiest way as shown in Fig. 3 [15]. This method calculates and compares the entire file as a chunk. Therefore it is convenient to implement and the overhead is minimum, some products such as Centera [4] and Windows 2000 [3] adopt this method. But the size is too large to detect internal

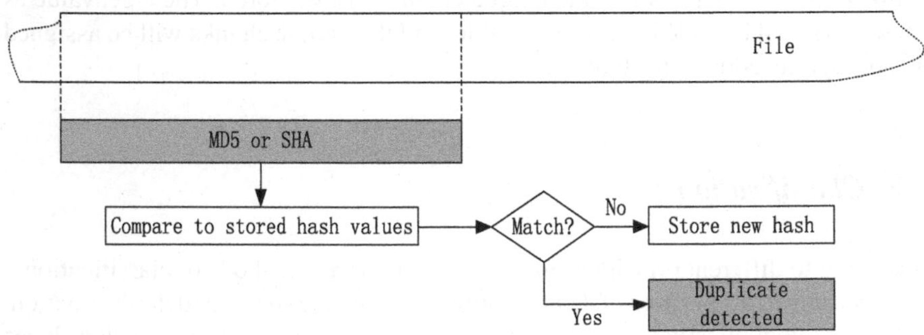

Fig. 3 Whole file chunking

duplicate data of files and is extremely sensitive to minor changes to files, so it is very limited. Generally, it is fit for scenes with many small files.

Denehy [7] summarized three methods of chunking: fixed size chunking (FSC), content defined chunking (CDC) and fixed-size sliding window chunking (FSWC). The division of FSC is shown in Fig. 4 [7]. Every object will be divided into non-overlapping fixed-size chunks, and the hash values and corresponding addresses of each chunk are stored in the index table. Venti [24], Oceanstore [14] and the traditional snapshot technology have adopted this approach. This division method can eliminate the redundancy between files and inside file. However, this division was unable to detect the insertion and deletion. When inserting or deleting at a point, all the chunks following the point will be affected, but it cannot be detected. Under the worst condition, if the insertion or deletion takes place in the beginning of the file, then the entire file will need to restore.

In order to solve the problems above, Muthitacharoen [23] divided the files into variable-size chunks based on the contents in their low-bandwidth network file system (LBFS), as shown in Fig. 5 [7]. For variable-size chunks, Rabin fingerprint algorithm is used to calculate the sliding window boundaries to determine the chunk size. This

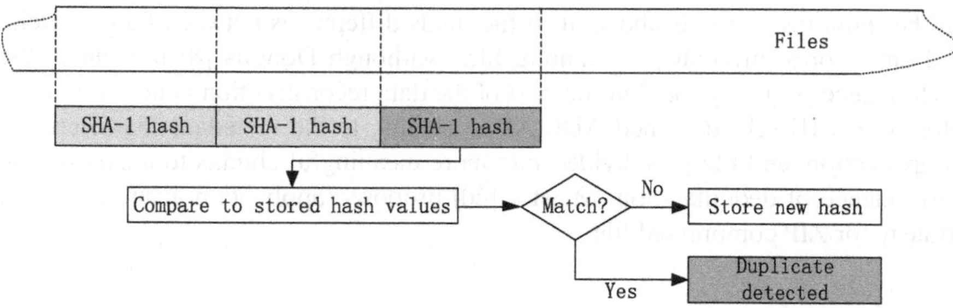

Fig. 4 Fixed size chunking

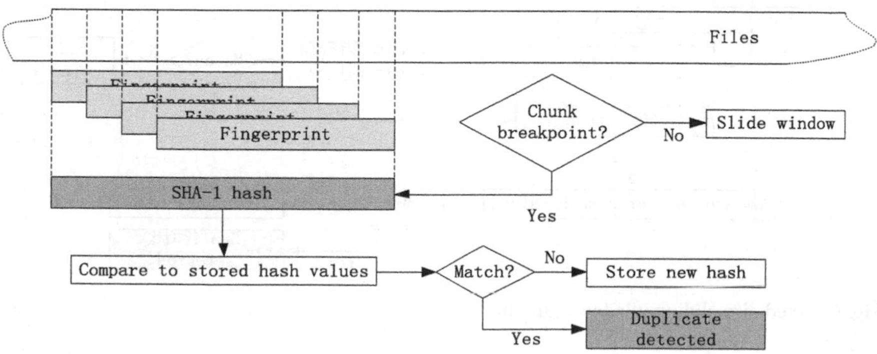

Fig. 5 Content defined chunking

method can deal with the case of insertion or deletion and fit for data updating more frequently. Therefore, it is usually have a higher deduplication rate than FSC, and it has been widely used in the backup system Pastiche [5], the archiving system Deep store [31] and so on. However, this method requires a lot of computing for fingerprint values and extra storage space to store the metadata, and deduplication efficiency is limited by the maximum window of the chunk. Lawrence [30] had analyzed CDC with experiments and shown that if the window is too small, such as less than 256 bytes, it would result in an increasing amount of computational burden and storage space. Kruus [13] put forward Bimodal CDC technique which maintained two sizes of chunking strategies, then chose between the two sizes dynamically according to data querying.

The CDC mentioned above solves the problem of insertion and deletion, but also brings new problems. For storage systems, storing variable-size chunks is much more complex than fixed-size, so the two methods can be combined to fixed-size sliding window, as shown in Fig. 6 [7]. This method uses rsync algorithm and the fixed-size sliding window to calculate each chunk's checksum, then compares values to determine whether is repeated or not. Jain made use of such fixed-size sliding window to design TAPER [12].

In addition, we can also use the Delta to realize byte/bit-level deduplication. Based on the chunking methods above, it further finds differences between files by Delta and only stores different parts among files. Although Douglis [8] thought of the performance is quite good, but the cost of the data reconstruction is not optimistic. Moreover, LIU [18] designed ADMAD according to the metadata characteristics of application, and file was divided into more meaningful chunks to maximize the performance of deduplication. Meister [20] proposed application-aware chunking strategy for ZIP compressed file.

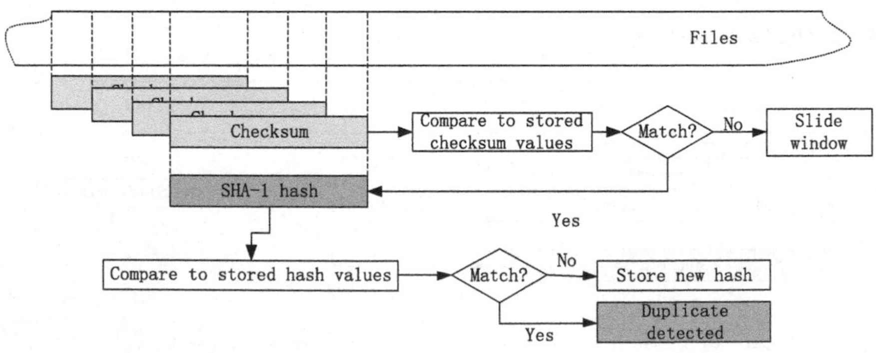

Fig. 6 Fixed-size sliding window chunking

3.2 *Performance Optimization*

Deduplication is a typical compromise to achieve the purpose of efficient storage at cost of computational overhead, so performance optimization should be considered. Youjip [26] proposed an increment of K-modulo algorithm to save chunking time. Jaehong [22] proposed an approach of integrating the two chunking methods based on the previous incremental model. By aware of the contents of the file, it would determine to use fixed-size or variable-size method. P-Dedupe [27] system took full advantages of idle resources with multi-core and uses multiple threads to chunking and calculating in parallelism.

With the increasing amount of data, the index table will be too large to reside in memory. Therefore, performance optimization is more important for index table and disk I/O. D2D [16] used a sparse indexing technique by sampling and the data locality to reduce the capacity of the index table. Extreme Binning [2] made chunks both in memory and external storage simultaneously and calculated the minimum file eigenvalues which used as identification of different files. It reduced the number of queries to disk and improved I/O performance. Benjamin used three techniques in DDFS [34] to optimize index query: (1) Bloom filter was used to detect whether the data segment was new to avoid the overhead of a query which was not in the index table; (2) Storing the data segment and its fingerprint according to the same order of files, then the spatial locality can be used; (3) Base on the previous locality, fingerprint values will be prefetched to the cache. DDFS prevented about 99 % of the disk I/O query by the ways above. However, I/O performance of traditional disk is limited by its mechanical properties severely. Because the disk query latency is too high, even if it takes only 0.1 ms but for the main memory is 1,000 times slower. With the development of industry, a large number of new storage medium appear in people's field of vision, such as flash memory, phase-change memory, etc. These medium use new technology to improve disk performance. So Meister designed dedupv1 [21] system based on SSD and Biplob [6] used flash to optimizing index query.

3.3 *Reliability*

While deduplication saves storage space, it is also at the expense of reliability.

One case is that a valid chunk is missed when calculating eigenvalues. Hash values cannot guarantee the uniqueness of the data detecting, especially when the amount of data increases rapidly and hash values may cause conflict [11]. One of the most direct and effective ways is to determine whether the chunk is the same with further comparison by-byte. Another way is to use of stronger anti-collision hash algorithm or to combine with different hash algorithms.

The more common is because multiple duplicate files share only one copy, then all of these files will be affected and cannot be recovered once the copy is damaged. So the direct way is to retain partial redundant at the cost of storage space to obtain reliability. But the number of copies should be reasonable, which shouldn't cause greater impact on deduplication efficiency. Deepavali [1] calculated a copy standard of each chunk according to the dependence and importance, and retained different numbers of redundant copy selectively. Zhou [32] assigned an optimal repeat degree according to each individual chunks universality in the data set and kept a minimum number of copies. In addition to retaining copies, the traditional data protection mechanisms can use less storage overhead to achieve better reliability. D2D [16] and DDFS [34] used RAID-6 error correction technology to ensure reliability which can recover the entire disk. R-ADMAD [19] packed the variable-size chunks in the form of packets into fixed-size object, then the object is recoded by the ECC and store on the redundancy nodes in distribution. Cluster deduplication systems HYDRAstor [9] stored the debris of the chunks on the distributed node evenly and used erasure codes to protect chunks.

3.4 Scalability

Although deduplication can reduce the amount of duplicate data, but with the accumulation of time, the amount of data which is not repeated will also increase, so the deduplication system faces expansion problems.

Firstly, increasing the capacity of a single node needs to improve the industrial technology. Even if the larger capacity storage device can be created, but the increasing amount of data will result in the bigger index table and the performance will be affected. Thus a disguised way is to reduce the space occupied by the index table and increase the amount of data that is indicated by the index of per unit. ChunkStash [6] stored the index information on the Flash, and only compressed signature keys are saved in memory rather than the hash of entire block. Guo [10] improved the sampling which sampling less and less gradually, thereby reducing the space occupied by the index. In addition, he also saved the index table in SSDs and only the key information is stored in the memory. Compared to the amount of index, the data occupies larger space, so more studies are aimed at reducing the data. Typically the compression and the deduplication technology are used in combination, just as Migratory Compression [17] did.

Since a single node capacity is so limited, using cluster system by adding the number of nodes is an effective way. Deepavali [2] built a cluster deduplication systems and designed Extreme Binning strategy which migrated the metadata and data of similar files and stored on the same node for enhancing scalability. HYDRAstor [9] identified the prefix of fingerprint and assigned hash value to each node, then specified different nodes are responsible for different areas so that deduplication of each node can execute independently.

3.5 *Load Balance*

Data is stored on the device ultimately, so how to design a reasonable layout of data in order to achieve load balance is also needed to consider.

Tan found the data placement in non-linear would break the space locality and had an impact on performance especially the throughput and the read performance. So De-Frag [25] is designed to reduce the nonlinear data placement, which picked some redundant data written to disk selectively, so the rate of deduplication was sacrificed to enhance space locality, but the performance was improved.

Zhou [33] studied on the distributed environment of data deduplication system layout in depth. In order to improve the utilization of the storage node, he designed and implemented the data placement strategy based on capacity-aware, which optimized data placement to achieve load balance between each node.

Xu [28] found that the imbalance caused by the placement between nodes that was the main factor for lower read performance. Under the load balance, he explored reliable distributed deduplication storage system and designed a EDP algorithm which is not only to ensure storage efficiency but also to ensure data reliability.

4 Issues and Challenges

With the environments of big data, the significant position of deduplication is increasing. However, deduplication as an essential technique in today's storage system, is also facing a variety of challenges. According to the analysis above, the problems and challenges mainly focus on the following aspects:

Rationality of the chunk division. In general, the bigger the granularity is and the lower the deduplication rate is, but the achievement is easier. Whereas the smaller the granularity is, the higher the deduplication rate is, but the cost is greater. Therefore the analysis should be based on the specific cases, or different division methods are combined to use together. For different conditions, different division methods are taken to improve the deduplication rate;

Performance optimization. Deduplication is bound to save space at the expense of performance for the price, include computational overhead and query cost. How to reduce the loss of performance as much as possible is needed to consider. The size of chunks affects the performance, so appropriate method need to be selected. In addition, optimization of the index table is also worth exploring, especially the bottlenecks of disk I/O.

Ensure data reliability. Deduplication only retains thc unique copy and all files share the copy by metadata pointer, thus the reliability reduces greatly. Appropriate redundancy and other data protection mechanisms should be used to ensure the reliability of data.

Scalability of system. Single node is unable to meet the growing requirement for the amount of data, so deduplication and data compression are combined to reduce the amount of data greatly. Another more efficient way is to use deduplication in a multi-node cluster system to achieve better effects.

Data placement of deduplication. After deduplication, due to the reduction of efficient data, the read performance will be decreased. So a reasonable way of data placement will have a direct impact on system throughput and read performance. In particular, each node can work independently in cluster deduplication system, so the parallelism between the nodes can be made use of to achieve load balance.

Deduplication has been used in backup and archiving systems and occupied an important position, which has a great significance to save storage space and is an important approach to achieve Green IT. Compared to the existing storage systems, the deduplication system added computational overhead and extra storage overhead, so the research on deduplication focuses on improving deduplication efficiency, optimizing system performance, enhancing data reliability and enhancing scalability. However, the exploration for load balance of data placement after deduplication is imperfect. So our work will focus on optimizing deduplication system for load balancing and design a file-aware deduplication system to take advantages of the parallelism between nodes to improve throughout and read performance.

References

1. Bhagwat D, Pollack K, Long DD, Schwarz T, Miller EL, Paris JF (2006) Providing high reliability in a minimum redundancy archival storage system. In: 14th IEEE international symposium on modeling, analysis, and simulation of computer and telecommunication systems, MASCOTS. IEEE, pp 413–421
2. Bhagwat D, Eshghi K, Long DDE, Lillibridge M (2009) Extreme binning: scalable, parallel deduplication for chunk-based file backup. Modeling analysis and simulation of computer and telecommunication systems MASCOTS, pp 1–9
3. Bolosky WJ, Corbin S, Goebel D, Douceur JR (2000) Single instance storage in windows. In: Proceedings of the 4th USENIX windows systems symposium, pp 13–24. Seattle, WA
4. Centera E (2004) Content addressed storage system
5. Cox LP, Murray CD, Noble BD (2002) Pastiche: making backup cheap and easy. ACM SIGOPS Oper Syst Rev 36(SI):285–298
6. Debnath BK, Sengupta S, Li J (2010) Chunkstash: speeding up inline storage deduplication using flash memory. In: USENIX annual technical conference
7. Denehy TE, Hsu WW (2003) Duplicate management for reference data. Technical report, Research Report RJ10305, IBM
8. Douglis F, Iyengar A (2003) Application-specific delta-encoding via resemblance detection. In: USENIX annual technical conference, general track, pp 113–126
9. Dubnicki C, Gryz L, Heldt L, Kaczmarczyk M, Kilian W, Strzelczak P, Szczepkowski J, Ungureanu C, Welnicki M (2009) Hydrastor: a scalable secondary storage. In: FAST, vol 9, pp 197–210
10. Guo F, Efstathopoulos P (2011) Building a high-performance deduplication system. In: USENIX annual technical conference
11. Henson V (2003) An analysis of compare-by-hash. In: HotOS, pp 13–18
12. Jain N, Dahlin M, Tewari R (2005) Taper: tiered approach for eliminating redundancy in replica synchronization. In: Proceedings of the 4th conference on USENIX conference on file and storage technologies, vol 4, pp 21–21. USENIX Association

13. Kruus E, Ungureanu C, Dubnicki C (2010) Bimodal content defined chunking for backup streams. In: FAST, pp 239–252
14. Kubiatowicz J, Bindel D, Chen Y, Czerwinski S, Eaton P, Geels D, Gummadi R, Rhea S, Weatherspoon H, Weimer W et al (2000) Oceanstore: an architecture for global-scale persistent storage. ACM SIGPLAN Not 35(11):190–201
15. Li AO, Shu JW, Ming-Qiang LI (2010) Data deduplication techniques. J Softw 1(21):430–433
16. Lillibridge M, Eshghi K, Bhagwat D, Deolalikar V, Trezis G, Camble P (2009) Sparse indexing: large scale, inline deduplication using sampling and locality. In: Fast, vol 9, pp 111–123
17. Lin X, Lu G, Douglis F, Shilane P, Wallace G (2014) Migratory compression: coarse-grained data reordering to improve compressibility. In: FAST, pp 257–271
18. Liu C, Lu Y, Shi C, Lu G, Du DH, Wang DS (2008) Admad: application-driven metadata aware de-duplication archival storage system. In: Fifth IEEE international workshop on storage network architecture and parallel I/Os, SNAPI'08. IEEE, pp 29–35
19. Liu C, Gu Y, Sun L, Yan B, Wang D (2009) R-admad: high reliability provision for large-scale de-duplication archival storage systems. In: Proceedings of the 23rd international conference on supercomputing. ACM, pp 370–379
20. Meister D, Brinkmann A (2009) Multi-level comparison of data deduplication in a backup scenario. In: Proceedings of SYSTOR 2009: the Israeli experimental systems conference. ACM, p 8
21. Meister D, Brinkmann A (2010) dedupv1: improving deduplication throughput using solid state drives (SSD). In: IEEE 26th symposium on mass storage systems and technologies (MSST). IEEE, pp 1–6
22. Min J, Yoon D, Won Y (2011) Efficient deduplication techniques for modern backup operation. IEEE Trans Comput 60(6):824–840
23. Muthitacharoen A, Chen B, Mazieres D (2001) A low-bandwidth network file system. In: ACM SIGOPS operating systems review, vol 35. ACM, pp 174–187
24. Quinlan S, Dorward S (2002) Venti: a new approach to archival storage. In: FAST, vol 2, pp 89–101
25. Tan Y, Yan Z, Feng D, He X, Zou Q, Yang L (2015) De-frag: an efficient scheme to improve deduplication performance via reducing data placement de-linearization. Clust Comput 18(1):79–92
26. Won Y, Kim R, Ban J, Hur J, Oh S, Lee J (2008) Prun: eliminating information redundancy for large scale data backup system. In: International conference on computational sciences and its applications, ICCSA'08. IEEE, pp 139–144
27. Xia W, Jiang H, Feng D, Tian L, Fu M, Wang Z (2012) P-dedupe: exploiting parallelism in data deduplication system. In: IEEE 7th international conference on networking, architecture and storage (NAS). IEEE, pp 338–347
28. Xu M, Zhu Y, Lee PP, Xu Y, Even data placement for load balance in reliable distributed deduplication storage systems
29. Yinjin F, Nong X, Fang L (2012) Research and development on key techniques of data deduplication [j]. J Comput Res Dev 1:002
30. You L, Karamanolis CT (2004) Evaluation of efficient archival storage techniques. In: MSST, pp 227–232. Citeseer
31. You LL, Pollack KT, Long DD (2005) Deep store: an archival storage system architecture. In: Proceedings of the 21st international conference on data engineering, ICDE. IEEE, pp 804–815
32. Zhengda Z, Jingli Z (2010) A novel data redundancy scheme for de-duplication storage system. In: 3rd international symposium on knowledge acquisition and modeling (KAM). IEEE, pp 293–296
33. Zhou Z, Zhou J (2012) High availability replication strategy for deduplication storage system. Adv Inf Sci Serv Sci 4(8):115
34. Zhu B, Li K, Patterson RH (2008) Avoiding the disk bottleneck in the data domain deduplication file system. In: Fast, vol 8, pp 1–14

Classification of Foreign Language Mobile Learning Strategy Based on Principal Component Analysis and Support Vector Machine

Shuai Hu, Yan Gu and Yingxin Cheng

Abstract To improve the classification accuracy of foreign language mobile learning (m-learning) strategies applied by college students, an evaluation model based on principal component analysis (PCA) and support vector machine (SVM) is proposed. PCA was first employed to reduce the dimensionality of an evaluation system of foreign language m-learning strategies and the correlation between the indices in the system was eliminated. The first 5 principal components were extracted and a classification model based on SVM was established by taking the extracted principal components as its inputs. Gaussian radial basis function was adopted as the kernel function and the optimal SVM model was realized by adjusting the parameters C and g. The classification result was compared with those produced by a BP neural network model and a single SVM model. The simulation results prove that the PCA-SVM model has a simpler algorithm, faster calculating speed, higher classification accuracy and better generalization ability.

Keywords Principal component analysis · Support vector machine · Mobile learning strategy · Classification

1 Introduction

The classification of foreign language mobile learning (m-learning) strategies is in essence the classification of multi-dimensional vectors [1, 2]. In recent years, artificial neural networks, which have been applied in various areas, could possibly

S. Hu (✉) · Y. Gu · Y. Cheng
Teaching and Research Institute of Foreign Languages,
Bohai University, Jinzhou 121013, Liaoning, China
e-mail: hushuai6@163.com

Y. Gu
e-mail: waiyubugy@163.com

Y. Cheng
e-mail: 17435303@qq.com

V.E. Balas et al. (eds.), *Information Technology and Intelligent Transportation Systems*, Advances in Intelligent Systems and Computing 455,
DOI 10.1007/978-3-319-38771-0_36

provide a solution to the above-mentioned classification problem. Among all types of neural networks, the most widely used one is BPNN, however, when it comes to the classification of foreign language m-learning strategies, BPNN has drawbacks such as poor computation accuracy caused by inadequacy of training samples, slow convergence speed and over-fitting due to the complicated network structure. While a support vector machine (SVM) overcomes shortcomings of neural networks such as complex structures and propensity for local minima. The sound theoretical foundation and good generalization performance of a support vector machine makes it very popular in nonlinear, small-sampled, high-dimensional pattern classification and regression estimation [3, 4]. Currently, a large amount of researches manage to improve the approximation performance of BPNN by improving its algorithm, whereas its classification accuracy is still weakened by heavy information overlapping between evaluation indices. Due to the high correlation between the indices in the evaluation system of college student foreign language m-learning strategies, a direct input of all the index values into a SVM is bound to affect the final accuracy. To solve this problem, a classification model based on principal component analysis (PCA) and least square SVM is proposed in this paper. The indices in the evaluation system were first analyzed to reduce the datum dimensionality by PCA. Next, a classification model based on SVM was established. Its classification performance was tested against those produced by a BPNN and a SVM without PCA processing.

2 Theoretical Basis

2.1 PCA Algorithmic Principle

PCA is a statistical method to reduce datum dimensionality by replacing a large quantity of correlated variables with a rather small amount of independent and integrated variables, while retaining most of the information of the original variables [5]. Its algorithm is as follows:

Let X be a data set with h samples. Each sample has g evaluation indices. X is standardized by mean standard deviation method to obtain the standardized sample matrix X', as shown in Eq. (1).

$$X' = \begin{pmatrix} x'_{11} & x'_{12} & x'_{13} & \cdots x'_{1g} \\ x'_{21} & x'_{22} & x'_{23} & \cdots x'_{2g} \\ \vdots & \vdots & \vdots & \cdots \vdots \\ x'_{h1} & x'_{h2} & x'_{h3} & \cdots x'_{hg} \end{pmatrix}_{h \times g} \tag{1}$$

Compute the correlation matrix R according to Eq. (2).

$$R = \begin{pmatrix} r_{11} & r_{12} & r_{13} \cdots & r_{1g} \\ r_{21} & r_{22} & r_{23} & \cdots r_{2g} \\ \vdots & \vdots & \vdots & \cdots & \vdots \\ r_{g1} & r_{g2} & r_{g3} & \cdots r_{gg} \end{pmatrix}_{g\times g} \quad (2)$$

The characteristic equation of R is computed using Jacobi method and g non-negative eigenvalues can be obtained: $\lambda_i, i = (1, 2, \ldots, g)$. The contribution rate of the ith principal component reflects the information retention rate of the original matrix, as is shown in Eq. (3).

$$G_i = \lambda_i \Big/ \sum_{i=1}^{g} \lambda_i \quad (i < g) \quad (3)$$

The accumulative contribution rate of the first i principal components is shown in Eq. (4).

$$G_{1-i} = \sum_{j=1}^{g} \lambda_j \Big/ \sum_{k=1}^{g} \lambda_k \quad (j < k) \quad (4)$$

The amount of principal components depends on their cumulative variance contribution rate. The common practice is to select the components whose cumulative contribution rates are above 85 % [6]. The eigenvector of λ_i is shown in Eq. (5).

$$T = \left(T_1^{(i)}, T_2^{(i)}, \ldots, T_g^{(i)}\right)_{g\times i} \quad (5)$$

The new sample matrix after datum deduction can be computed according to Eq. (6).

$$Y = X' \cdot T \quad (6)$$

2.2 SVM Algorithmic Principle

The idea of SVM is to improve the generalization ability by widening classification intervals by a discriminant function. For all linear classification problems, let the training set be $\{x_k, y_k\}, (k = 1, 2, \ldots, n)$, the mathematical expression of the optimal hyper plane is Eq. (7), in which b is the threshold, ω is the weight vector. The discriminant function is shown in Eq. (8), where a_k is the Lagrange multiplier, b is the threshold determined by the training samples, x_k and y_k are the support vectors among any two classes and $r(x_k, x)$ is the kernel function [7, 8].

$$f(x) = \omega \cdot \phi(x) + b \quad (7)$$

$$f(x) = \operatorname{sgn}\left(\sum_{k=1}^{n} y_k \cdot a_k \cdot r(x_k, x) + b\right) \quad (8)$$

In this paper, libsvm-3.20 toolbox of MATLAB (R2013a) is employed to create an SVM classification model, which includes library functions such as parameter optimization, and it supports multi-classification.

3 Establishment of a PCA-SVM Classification Model

The classification method based on PCA and SVM is to first transform samples by PCA rather than directly create a sample hype-plane. Then SVM is used to train the transformed samples to obtain the optimal hyper-plane to classify the samples.

3.1 *Establishment of m-Learning Strategy Evaluation Indices and Sample Gathering*

The original data of college student foreign language m-learning strategies are obtained through questionnaire. The questionnaires combine questions from the Oxfords language learning strategy questionnaire and the online learning version of Learning and Study Strategies Inventory (LASSI). There are 5 first grade indices and 22 s grade indices reflecting college student m-learning strategies in areas like cognition, meta-cognition, emotion, resource management and information processing. Feedbacks of 60 sophomore students of grade 2013 to the questionnaires were gathered and analyzed by 5 experts, with the target classification number being 5.

3.2 *PCA Modeling*

The original data were normalized before PCA processing. The normalized data of college student m-learning strategies are shown in Table 1.

Figures of the visualized datum distribution were made to observe the data more clearly. Figure 1 is a scatter diagram of each index showing that there are large differences in the range of the original data, which indicates that datum normalization is necessary. Figure 2 is a box plot of the original data, which shows that a few off-group points can be found in indices X_1, X_7, X_8, X_{10}, X_{15}, X_{16}, X_{17} and X_{18}. The shorter box-whiskers of most indices indicate that their property values are relatively concentrated, meaning information overlap is heavy, which in turn affects the classification accuracy.

Table 1 Normalized data of college student m-learning strategies sample

Sample code	X_1	X_2	X_3	Index ...	X_{20}	X_{21}	X_{22}	Target class
1	0.8230	0.9093	1.0000	...	1.0000	0.8889	0.9450	1
2	0.8850	0.9070	0.9763	...	0.9825	0.9231	0.9445	1
3	0.0553	0.2884	0.1293	...	0.4614	0.4103	0.3486	1
4	0.2434	0.1744	0.2322	...	0.3912	0.4274	0.4312	1
5	0.1881	0.1047	0.1953	...	0.0579	0.2359	0.1339	2
...	...	...	...	...	...		...	... ...
56	0.6881	0.5581	0.6570	...	0.4579	0.3590	0.3394	5
57	0.5863	0.3837	0.6702	...	0.4737	0.2308	0.3853	5
58	0.5376	0.4721	0.6966	...	0.3912	0.3675	0.3945	5
59	0.4978	0.5233	0.6570	...	0.4140	0.3590	0.3725	5
60	0.5376	0.5395	0.3668	...	0.4789	0.4205	0.3028	5

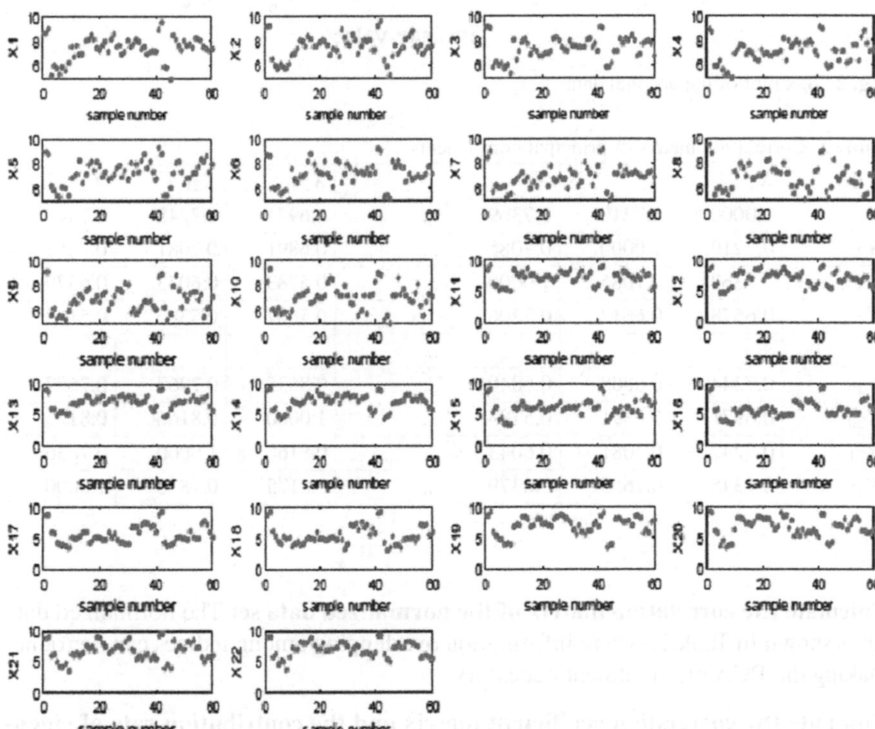

Fig. 1 Scatter diagram of each index

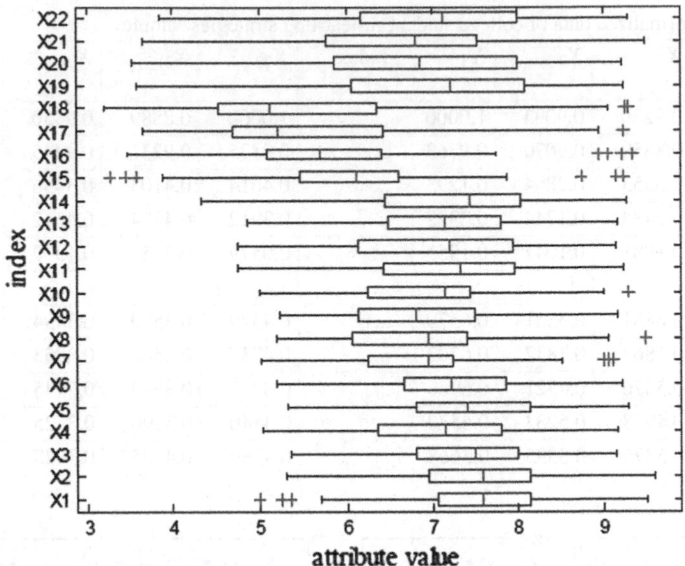

Fig. 2 Box plot of the original data

Table 2 Correlation matrix of principal components

Index	X_1	X_2	X_3	...	X_{20}	X_{21}	X_{22}
X_1	1.0000	0.7210	0.7369	...	0.6971	0.7242	0.7345
X_2	0.7210	1.0000	0.7085	...	0.6891	0.7081	0.7655
X_3	0.7369	0.7085	1.0000	...	0.5783	0.6043	0.6379
X_4	0.6578	0.6612	0.7206	...	0.5607	0.5386	0.5192
..	...	...	...	...	...	...	...
X_{19}	0.7344	0.6808	0.6840	...	0.8635	0.7967	0.7657
X_{20}	0.6971	0.6891	0.5783	...	1.0000	0.8166	0.8175
X_{21}	0.7242	0.7081	0.6043	...	0.8166	1.0000	0.7836
X_{22}	0.7345	0.7655	0.6379	...	0.8175	0.7836	1.0000

Calculate the correlation matrix of the normalized data set The normalized data set is shown in Table 2, where information overlapping among indices can be found, making the PCA pre-treatment necessary.

Compute the correlation coefficient matrix and the contribution rate of eigenvalues The computed eigenvalue contribution rates and the cumulative contribution rates of the principal components are shown in Fig. 3.

It can be seen that the first 5 eigenvalues are 14.0287, 2.83917, 0.953432, 0.553761 and 0.501261, and their contribution rates are 63.76, 12.90, 4.33, 2.56 and 2.24 % respectively. The cumulative contribution rate of the first 5 principal components

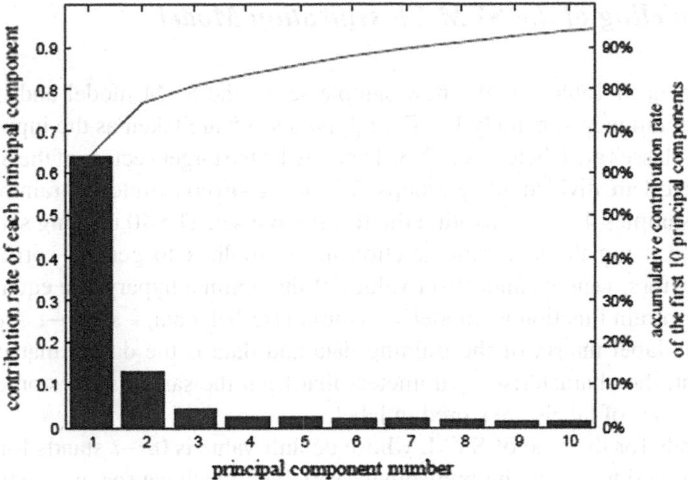

Fig. 3 The contribution rates and cumulative contribution rates of the first 10 principal components

Table 3 The new sample set

Sample code	F_1	F_2	New data set F_3	F_4	F_5	Target class
1	0.8842	0.7178	0.4022	0.6199	0.9475	1
2	0.8044	0.7624	0.3645	0.6364	0.9832	1
3	0.7263	0.7118	0.6608	0.5335	0.1992	1
4	0.7685	0.7889	0.6446	0.5024	0.2253	1
5	0.5205	0.4436	0.3665	0.4535	0.0604	2
…	…	…	…	…	…	…
6	0.5051	0.3517	0.8586	0.8597	0.5294	5
57	0.5754	0.2516	0.9788	0.9964	0.4742	5
58	0.4717	0.3638	0.8255	0.9016	0.5022	5
59	0.3297	0.3130	0.1634	0.5933	0.4144	5
60	0.5558	0.1342	0.1990	0.5160	0.3897	5

is up to 85.99 %, which indicates that they can explain most of the variables in the characteristic parameter matrix.

Extract the principal component eigenvectors Extract the principal component vectors. Multiply the normalized 6022 dimensional sample matrix with the 225 dimensional eigenvector matrix of the first 5 principal components. A new 605 dimensional sample set of the PCA-SVM classification model is created as shown in Table 3.

3.3 Modeling of the SVM Classification Model

Take the data in Table 3 as the new sample set of the SVM model and the first 5 principal components, namely F1, F2, F3, F4 and F5 are taken as the input vectors. The target classes are labeled as 1, 2, 3, 4 and 5 to be the target vectors of the SVM. The data in Table 3 are divided into 2 groups. Samples 1–40 constitute the training sample set while samples 41–60 constitute the test sample set. The 40 training samples are first trained using the svmtrain function in the toolbox to get the corresponding support vectors, weights and offset values of the optimal hyperplane equation. The form of svmtrain function is: model = svmtrain (la-bel, data, $-s$ $-t$ $-c$ $-g$), where label is the label matrix of the training data and data is the datum matrix. In this experiment, the characteristic parameters that train the samples are stored in data, while the types of labels are stored in label.

$-s$ stands for the type of SVM, whose default value is 0. $-t$ stands for the type of kernel function. $-c$ is the punishment variable, which causes over-learning if it gets too big, and lowers the classification accuracy if it gets too small. $-g$ refers to the setting of the *gamma* function, which reflects the features of the training sample set. What is vital for the creation of a SVM classification model is choosing a proper kernel function.

As a nonlinear function, RBF can effectively simplify the computation in sample training process, so RBF is taken as the kernel function. The punishment variable C and gamma function g of the SVM classification model determine the learning ability and classification accuracy. In this paper, optimization of parameters C and g is realized by particle swarm optimization (PSO). The evolution frequency is set to be 100. The optimal values of c and g can be obtained through *psoSVMcgFor*-Class function, and they are 19.4675 and 1.2133 respectively. Finally the mobile learning strategy classification model based on PCA-SVM is established.

4 Result and Discussion

To prove the effectiveness of the PCA-SVM classification model, a single SVM model and a standard BPNN model are established to compare with it.

Divide the data in Table 1 that are not treated via PCA into two groups. Samples 1–40 are taken as the training sample set, while samples 41–60 are taken as the test sample set. Input the training sample set to the SVM model and optimize the parameters using PSO. The parameter C at the highest classification accuracy is 9.9458. The optimal g is 7.3483. The biggest iteration frequency is 200. The population amount is 20. Then a single SVM classification model is created.

To maximize the training speed of BPNN, *mapminmax* function is used to normalize the data of the training sample set. The number of output layer neurons of the BPNN equals to the number of the target classes. The number of output layer neurons is 5, with (10000) representing class 1, (01000) representing class 2, (00100) representing class 3, (00010) representing class 4 and (00001) representing class 5.

The generalization performances of the SVM classification model and the BPNN classification model are tested, with sample 41 to sample 60 in Table 1 being the test sample set. And samples 41–60 in Table 3 are taken as the test sample set to test the generalization performance of the PCA-SVM model. The generalization performances of the three models are shown in Table 4.

The classification results of the PCA-SVM model and the single SVM model are shown in Fig. 4. As for the single SVM model, 4 samples are classified incorrectly by it, leaving the total accuracy being just 80 %. While as for the PCA-SVM model, only 1 sample is classified with error. It is obvious that the PCA-SVM model is superior in terms of classification accuracy, which means the employment of PCA in this case raises the classification accuracy from 80 to 95 %. So PCA processing is necessary. It can be seen from Table 4 that not only is the classification accuracy of the PCA-SVM model significantly higher than those of the single SVM model and the standard BPNN model, but the classification time is also much shorter. This indicates that computation work is largely reduced by replacing the original evaluation indices with the data after dimension reduction by PCA.

Table 4 Test of the generalization performances of the three models

Network type	Number of test samples	Average accuracy/%	Classification time/(s)
PCA-SVM	20	95.00	0.003589
SVM	20	80.00	0.044131
Standard BP	20	75.00	0.212988

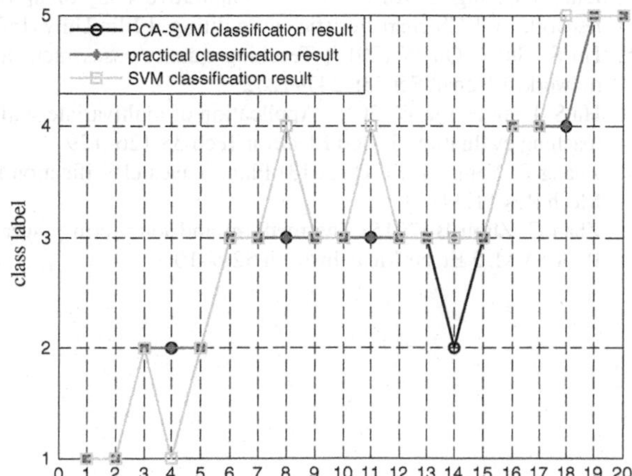

Fig. 4 Comparison between the classification results of the PCA-SVM model and the SVM model

5 Conclusion

This paper proposes a method based on PCA and SVM for college students foreign language m-learning strategy classification. To reduce datum dimensionality, PCA was used to transform the original data from 22 indices to 5 integrated independent ones. Then the PCA-processed data were input to the SVM model and parameter optimization of the established SVM classifier was done using PSO algorithm. The experiment results indicate that the proposed method improves the computation speed, classification accuracy and generalization performance of the single SVM model. Acknowledgements. The research work was supported by Social Science Foundation of Liaoning Province under Grant No. L14CYY022 and Grant No. L15AYY001. It was also supported by Liaoning Provincial Department of Education under Grant No. W2015015.

Acknowledgments The research work was supported by Social Science Foundation of Liaoning Province under Grant No. L14CYY022 and Grant No. L15AYY001. It was also supported by Liaoning Provincial Department of Education under Grant No. W2015015.

References

1. Liu Y (2012) An empirical study on mobile learning of college English B: design and development based on Kooles FRAME model. J Open Educ Res 18:76–81
2. Mao Y, Wei Y (2013) The empirical analysis of university students mobile learning needs. J Libr Inf Serv 57:82–90
3. Ding S, Wu Q (2012) Performance comparison of function approximation based on improved BP neural network. J Comput Mod 11:10–13
4. Ding S, Chang X, Wu Q (2013) Comparative study on application of LMBP and RBF neural networks in ECS characteristic curve fitting. J Jilin Univ (Inf Sci Ed) 31:203–209
5. Hu S, Gu Y, Qu W (2015) Teaching quality assessment model based on PCA-LVQ neural network. J Henan Sci 33:1247–1252
6. Hu S, Jiang H, Qu W (2015) Application of multivariate statistical analysis in foreign language teaching evaluation. J Mod Electron Tech 38:126–129
7. Zhang Z, Yang M, He D (2013) Plant leaves classification based on PCA and SVM. J Agric Mech Res 11:34–38
8. Zhan C, Zhou B (2015) The medium and long term power load forecasting model based on PCA-SVM. J Electr Meas Instrum 52:6–10

A Fuzzy Least Square Support Tensor Machines Based on Support Vector Data Description

Ruiting Zhang and Yuting Kang

Abstract Most of the traditional machine learning algorithms are based on the vector, but in tensor space, Tensor learning is useful to overcome the over fitting problem in vector-based learning. In the meanwhile, tensor-based algorithm requires a smaller set of decision variables as compared to vector-based approaches. We also would require that the meaningful tensor training points must be classified correctly and would not care about some training points like noises whether or not they are classified correctly. To utilize the structural information present in high dimensional features of an object and fuzzy membership, this paper presents a novel fuzzy classifier for Image processing based on support vector data description (SVDD), termed as Fuzzy Least Squares support tensor machine (FLSSTM), where the classifier is obtained by solving a system of linear equations rather than a quadratic programming problem at each iteration of FLSSTM algorithm as compared to STM algorithm. This in turn provides a significant reduction in the computation time, as well as comparable classification accuracy. The efficacy of the proposed method has been demonstrated in ORL database and Yale database.

Keywords Support tensor machines · Support vector data description · Tensor learning · Fuzzy least square support tensor machines

1 Introduction

Nowadays, a tremendous amount of data have continuously flooded into our society. The explosive growth of data has generated an urgent need for new techniques and new skills in the data mining areas. Machine learning is an effective method and an important branch of data mining. One of the core tasks of machine learning is how to

R. Zhang (✉) · Y. Kang
Canvard College, Beijing Technology and Business University, Beijing 101118, China
e-mail: ruitingzh@163.com

Y. Kang
e-mail: 897245070@qq.com

V.E. Balas et al. (eds.), *Information Technology and Intelligent Transportation Systems*, Advances in Intelligent Systems and Computing 455,
DOI 10.1007/978-3-319-38771-0_37

represent of data. High dimensional data with many attributes are often encountered in the real applications, Thus, how to efficiently represent image data has a fundamental effect on the classification. Most of the traditional learning algorithms are based on the vector space model, such as SVM and LSSVM.

However, in practice, a lot of objects need to be expressed in tensor, such as the gray images and the gray images sequence. For example, the gray image (Fig. 1) can be represented by the second order tensor. In the most machines learning algorithms, firstly the tensor data is converted into the vector data, it may destroy the data structural information or lead to high dimensional vectors. In recent years, because of the drawbacks, the machine learning based on tensor space has attracted significant interest from the research community. A set of algorithms have been extended to deal with tensors, for example, support tensor machine (STM) [1–6]. With utilizing the tensor representation, the number of parameters estimated by the tensor-based learning can be greatly reduced. Therefore, the tensor-based learning algorithms are especially suitable for solving the small-sample-size (S3) problem, which the number of samples available for training is small and the number of input features used to represent the data is large.

Meanwhile, in many practical engineering applications community, the training data is often corrupted by those abnormal outliers and noise. Since SVM is particularly sensitive to outliers. Fuzzy support vector machine (FSVM) [7–9] provides an effective method to deal with this problem. In FSVM, each training sampling is associated with a fuzzy membership and different memberships represent different contributions to the learning of decision surface. It can reduce the effects of outliers by fuzzy membership functions.

Fig. 1 A gray image can be represented by a matrix

In this paper, we propose a novel method called Fuzzy Least Square Support Tensor Machine (FLSSTM), which is a tensor version of LSSVM, or fuzzy version of LSSTM. LSSTM is based on the tensor space, which directly accepts order-2 tensors as inputs. To obtain a classifier in the tensor space not only retains the data structure information, but also helps overcome the over-fitting problem encountered mostly in vector-based learning. In comparison to solving a QPP at every iteration of the STM algorithm, FLSSTM solves a system of linear equations in an iterative fashion, which eventually converges to an optimal solution after a few iterations.

The rest of the paper is organized as follows. Section 2 provides an overview of Fuzzy membership and Support vector data description (SVDD). In Sect. 3 we present a novel FLSSTM based on SVDD algorithm. Section 4 demonstrates experimental results.

2 Fuzzy Membership and Support Vector Data Description

2.1 *Fuzzy Membership*

In many real-word application community, due to over fitting in SVMs, the training process is particularly sensitive to those abnormal outliers in the training dataset which are far away from their own class. A key difficulty with real dataset is that parts of abnormal outliers are noise, which tends to corrupt the samples. In order to decrease the effect of those outliers or noises, we assign each data point in the training dataset with a membership. Samples with a higher membership value can be thought of as more representative of that class, while those with a lower membership value should be given less importance, so that those abnormal data with a low membership contribute to total error term decreases. In fact, this fuzzy membership value determines how important it is to classify a data sample correctly. One of the important things for machine learning is to choose the appropriate fuzzy memberships. At present, most of people define a fuzzy membership basing on the distance between each point and its class center [10, 11].

2.2 *Support Vector Data Description*

In this section, we briefly review the basic idea of SVDD algorithm. It is developed by Tax and Duin to solve one-class classification problems. Given a set of training samples $\{x_i, i = 1, 2, \dots l\}$ belonging to a given class. The goal idea of SVDD is to find a hypersphere of radius $R > 0$ and center a with a minimum volume containing most the target class samples. Then we can solve the following optimization problem:

$$\min \ R^2 + C\sum_{i=1}^{l} \xi_i$$
$$\text{s.t.} \ \ \|\Phi(x_i) - a\|^2 \le R^2 + \xi_i, \xi_i \ge 0, i = 1, 2, \ldots, l \tag{1}$$

Where ξ_i is a slack variable, C is a regularization parameter controls the tradeoff between the volume of the hypersphere and the errors, $\Phi(x)$ is a nonlinear function which maps x_i to a high dimensional feature space, By introducing Lagrange multiples and Applying the KKT conditions we solve the following dual problem:

$$\max \ \sum_{i=1}^{l} \alpha_i K(x_i, x_i) - \sum_{i=1}^{l}\sum_{j=1}^{l} \alpha_i\alpha_j K(x_i, x_j)$$
$$\text{s.t.} \ \sum_{i=1}^{l} \alpha_i = 1, 0 \le \alpha_i \le C, i = 1, 2, \ldots, l \tag{2}$$

Then we can compute R by any Support vector x_i with $\alpha_i > 0$:

$$R^2 = K(x_i, x_i) - 2\sum_{j=1}^{l} \alpha_j K(x_i, x_j) - \sum_{i=1}^{l}\sum_{j=1}^{l} \alpha_i\alpha_j K(x_i, x_j) \tag{3}$$

3 Fuzzy Least Squares Support Tensor Machines

Given the training samples: $\mathcal{T} = \{(X_1, y_1, s_1), \ldots, (X_l, y_l, s_l)\} \in (^{n_1} \otimes ^{n_2} \times \mathcal{Y} \times s)^l$, where $X_i \in {}^{n_1} \otimes {}^{n_2}$ is a training sample, y_i is the class label of X_i and $s_i \in [0, 1]$ is the fuzzy membership degree of X_i belonging to y_i. The classification problem is modeled by the following programming:

$$\min_{u,v,b,\xi} \ \frac{1}{2}\left\|uv^{\mathrm{T}}\right\|_{\mathrm{F}}^2 + \frac{C}{2}\sum_{i=1}^{l} \xi_i^2,$$
$$\text{s.t.} \ y_i(u^{\mathrm{T}}X_i v + b) = 1 - \xi_i, i = 1, \ldots, l. \tag{4}$$

where $C > 0$ is a regularization parameter and ξ_i is a slack variable, is the membership generalized by some outlier-detecting methods. For solving the optimization problem (4), firstly we construct the Lagrange function as follows:

$$L(u, v, b, \xi, \alpha) = \frac{1}{2}\left\|uv^{\mathrm{T}}\right\|_{\mathrm{F}}^2 + \frac{C}{2}\sum_{i=1}^{l} s_i\xi_i^2 - \sum_{i=1}^{l} \alpha_i\left[y_i\left(u^{\mathrm{T}}X_i v + b\right) - 1 + \xi_i\right] \tag{5}$$

The KKT necessary conditions of the Lagrange function for the optimality are:

$$u = \frac{\sum_{i=1}^{l} \alpha_i y_i X_i v}{v^{\mathrm{T}} v} \tag{6}$$

$$v = \frac{\sum_{i=1}^{l} \alpha_i y_i u^T X_i}{u^{\mathrm{T}} u} \tag{7}$$

$$\sum_{i=1}^{l} \alpha_i y_i = 0 \tag{8}$$

$$\xi_i = \alpha_i / C s_i, i = 1, \ldots, l \tag{9}$$

From Eqs. (6) and (7), it is obviously that and rely on each other, and cannot be solved with traditional methods. Like STM, we use alternating iterative algorithm [3, 4]. We first fix , for example, we let $\beta_1 = \|u\|^2$ and $x_i = X_i^T u$, Let D be a diagonal matrix where $D_{ii} = y_i$. The optimization problem (6) can be reduced to the following QPP:

$$\begin{aligned} \min_{v,b,\xi} \quad & \frac{1}{2}\beta_1 v^T v + \frac{1}{2} C \xi^T S \xi \\ \text{s.t.} \quad & D(Xv + eb) + \xi = e \end{aligned} \tag{10}$$

where $X = \left(x_1^T, x_2^T, \ldots, x_m^T\right)^T$, and e is a vector of ones of appropriate dimension. It can be seen that (10) is similar in structure to LSSVM. For solving (10), we consider its Lagrange function, then applying the KKT necessary and sufficient optimization conditions, we obtain the following:

$$\beta_1 v = X^T D \alpha \tag{11}$$

$$e^T D \alpha = 0 \tag{12}$$

$$C S \xi = \alpha \tag{13}$$

Substituting the values of v, b and ξ form (11), (12) and (13) into the equality contraints of (10) yields the following system of linear equations for obtaining α:

$$\begin{pmatrix} O & -Y^T \\ Y & \frac{DXX^T D}{\beta_1} + \frac{S^{-1}}{C} \end{pmatrix} \begin{pmatrix} b \\ \alpha \end{pmatrix} = \begin{pmatrix} 0 \\ e \end{pmatrix} \tag{14}$$

where S^{-1} is a diagonal matrix where $S_{ii}^{-1} = s_i^{-1}$. Then a, v, b can be computed using (11), (12) and (13) respectively. Once v is obtained, Let $\beta_1 = \|v\|^2$ and $\tilde{x}_i = X_i v$, On the similar lines, u can be computed by the following QPP:

$$\min_{u,b,\xi} \ \frac{1}{2}\beta_2 u^T u^2 + \frac{1}{2} C\xi^T S\xi \tag{15}$$
$$\text{s.t.} \ \ D(\tilde{X}v + eb) + \xi = e$$

Thus, u and v can be obtained by iteratively solving the optimization problems (10) and (15).

4 Experiments

In this section, in order to verify the effectiveness of the proposed FLSSTM, We will compare the results of FLSSTM with the vector-based classification method FLSSVM and tensor-based classification method LSSTM. All the algorithms have been implemented in MATLAB7.8 (R2009a) on Ubuntu running on a PC with system configuration Intel Core5 Duo (2.60GHz) with 2 GB of RAM. The data sets to be used are taken from the UCI Repository. We randomly select our training images from the entire set and repeat the experiment 10 times.

The ORL database of the face images has been provided by AT&T laboratories from Cambridge. It contains 400 images of 40 individuals, with varying lighting, facial expression and facial details. All images were normalized to a resolution of 3232 pixels with (1024) gray levels. Since we consider the binary problem of learning a classifier, we randomly chose two classes images to distinguish. For each binary classification experiment, we consider a subset of ten images of both the subjects for training (training ratio is 10–50 %), while the rest is considered for testing. Table 1 shows the mean recognition rates and standard deviations of all algorithms in our experiments with different ratio of training sets and test sets on ORL database. From the Table 1, it can be seen that when training ratio is small, FLSSTM outperforms FLSSVM and LSSTM, and the advantage of FLSSTM gradually reduced when training set becomes larger. When training ratio is 10 %, the maximum difference

Table 1 The Comparison between LSSTM, FLSSVM and FLSSTM

Training ratio (%)	Recognition rates and standard deviations		
	LSSTM	FLSSVM	FLSSTM
10	89.702.95	87.9012.1	**91.982.54**
20	93.823.60	94.4415.3	**94.513.28**
30	95.792.37	**96.519.82**	95.711.56
40	**99.351.76**	98.0610.1	96.672.89

of the accuracy between FLSSTM and FLSSVM is 3.08 %. However, the maximum difference of the accuracy between FLSSTM and LSSTM is only 2.28 %.

The Yale database of the face images has been provided by Yale University. It contains 165 images about 15 individuals, where each person has 11 images. These images with varying lighting condition (left-light, center-light) and facial expression (happy, sad, normal, sleepy, surprised, wink). All images were normalized to a resolution of 100100 pixels with (10000) gray levels. One object from Yale database is displayed in Fig. 2. We consider eleven particular examples of binary classification in Yale database. It consists of three subject pairs with similar facial features (smile, beard, glasses), and three subject pairs with distinct facial features.

Table 2 shows the mean recognition rates of FLSSTMLSSTM and FLSSVM algorithms in our experiments with different size of training sets and test sets on Yale database. It is evident from the Table 2 that FLSSTM roughly outperforms LSSTM in most of the cases. From Table 2, it can be seen obviously that the maximum difference of the accuracy between FLSSTM and LSSTM is 5.06 % while the minimum difference of the accuracy is only 0.83 % for the different subject pairs when the training size k = 1.

From Table 2, it can be seen obviously that the introduction of fuzzy membership improves the classification ability, the results have been provided on the small training size but not the large training size. The main reason lies in less the information for classifying planes when the training size is small, the fuzzy membership can increase the test accuracy when building a classifier. The superiority of fuzzy membership gradually reduced when training sets become larger.

Fig. 2 Eleven facial samples from a subject within the Yale database

Table 2 Mean Recognition rates (%) on Yale Database

Training Size k	Methods	Recognition rates with dissimilar facial features			Recognition rates with similar facial features		
		(1, 4)	(5, 11)	(12, 15)	(2, 7)	(6,14)	(8, 13)
k = 1	LSSTM	82.44	94.54	72.17	81.89	71.89	79.00
	FLSSVM	82.28	96.50	66.89	78.50	71.00	78.33
	FLSSTM	**87.50**	**97.06**	**73.17**	**83.78**	**75.72**	**79.83**
k = 2	LSSTM	87.59	99.20	77.72	89.89	81.79	89.26
	FLSSVM	89.14	**99.57**	76.36	90.56	80.12	**94.38**
	FLSSTM	**89.63**	96.36	**78.89**	**90.80**	77.78	90.56
k = 3	LSSTM	**91.67**	97.64	**83.61**	95.63	85.49	91.87
	FLSSVM	89.31	**99.86**	80.69	**95.67**	82.85	**96.88**
	FLSSTM	89.86	95.49	81.25	85.83	73.12	83.75

5 Conclusions

This paper firstly consider the fuzzy membership in the training dataset in tensor space, and propose an improved tensor-based method FLSSTM algorithm to learn better from datasets in the presence of outliers or noises. For solving the small-sample-size (S3) problems, the tensor representation always performs better than the vector representation. The above several numerical experiments show that the tensor-based methods have more advantages than vector-based methods for small-sample-size (S3) problems. The formulation of FLSSTM requires solving a system of linear equations at every step of an iterative algorithm, using the alternating projection method similar to STM, in contrast to solving a QPP. This makes FLSSTM a fast tensor-based linear classifier.We will continue to research the method of selecting better fuzzy membership function.

References

1. Kotsia I, Guo WW, Patras I (2012) Pattern Recognit 45:4192–4203
2. Kotsia I, Patras I (2011) Support tucker machines. In: Proceedings of IEEE Conference on computer vision and pattern recognition, Colorado, USA, pp 633–640
3. Cai D, He XF, Han JW (2006) Learning with tensor representation. Department of Computer Science Technical Report No.2716, University of Illinois at Urbana-Champaign (UIUCDCS-R-2006-2716)
4. Cai D, He XF, Wen JR, Han J, Ma WY (2006) Support tensor machines for text cate-gorization. Department of Computer Science Technical Report No.2714, University of Illinois at Urbana-Champaign (UIUCDCS-R-2006-2714)
5. Tao D, Li X, Hu W, Maybank SJ, Wu X (2005) Supervised tensor learning. In: ICDM 2005: proceedings of the 5th IEEE international conference on data mining, p 450C457
6. Khemchandani R, Karpatne A, Chandra S (2012) Proximal support tensor machines. In-t J Mach Learn Cybern. online
7. Lin Chunfu, Wang Shengde (2002) Fuzzy support vector machines. IEEE Trans. Neural Netw 13(2):464–471
8. Inoue T, Abe S (2001) Fuzzy support vector machines for pattern classification. Proceedings of IJCNN01, Washington DC, vol 2, pp 1449–1454
9. Huang H-P, Liu Y-H (2002) Fuzzy support vector machine for pattern recognition and data mining. Int J Fuzzy Syst 4(3):826–835
10. Fan XH, He ZG (2010) A Fuzzy support vector machine for Imbalanced Data classifica-tion. IEEE ICPOIP,2010 International conference on optoelectronics and image processing, pp 11–14
11. An W, Liang M (2013) Fuzzy support vector machine based on within-class scatter for classification problems with outliers or noises. Neurocomputing 110:101–110
12. Tax D, Duin RP (1999) Support vector domain description. Pattern Recognit Lett 20: (11C13) 1191C1199

On the Detection Algorithm for Faster-than-Nyquist Signaling

Biao Cai, Xiaohu Liang, Qingshuang Zhang and Yuyang Zhang

Abstract FTN has been proposed for several decades, and it is attractive for the reason that it can boost the symbol rate without changing the power spectral density. Actually, FTN has been used in some 5G standards. FTN signaling detection is an important element in FTN signaling scheme. In this paper, we shall introduce some FTN signaling detection algorithms that were proposed in literatures.

Keywords FTN signaling · VA · BCJR · GMP · Frequency domain equalization · Turbo equalization

1 Introduction

As the spectrum resources become more and more scarce, the techniques to increase the spectral efficiency are extremely attractive in recent years. One of them is Faster-than-Nyquist signaling. The concept of Faster-than-Nyquist signaling was first proposed by Mazo in 1975 [1]. In classic band-limited scenarios, T-orthogonal pulse, mostly being a rRC pulse, is used to form a intersymbol interference free system, assuming the symbol interval is given by T. With a match filter, the receiver can detect the signal symbol-by-symbol in some simple algorithms. By contrast, in FTN signaling system, the symbol interval is set to be smaller than T, which could lead to higher symbol rate than conventional system.

B. Cai (✉) · X. Liang · Q. Zhang · Y. Zhang
PLA University of Science and Technology, Nanjing 210007, China
e-mail: 553897138@qq.com

X. Liang
e-mail: liangxiaohu688@163.com

Q. Zhang
e-mail: zhangqspku@163.com

Y. Zhang
e-mail: yuyangzhang1991@126.com

V.E. Balas et al. (eds.), *Information Technology and Intelligent Transportation Systems*, Advances in Intelligent Systems and Computing 455,
DOI 10.1007/978-3-319-38771-0_38

Easy to see that using the lower symbol interval without changing pulse shape will cause the unavoidable ISI which needs a high detection complexity in order to eliminate the ISI. Owing to the development of semi-conductor, the complexity of FTN signaling detection becomes bearable. At present, a great deal of detection algorithms have been proposed in literatures.

This paper is organized as follows. In Sect. 2 we give a FTN signaling model and in Sect. 3 we introduce the FTN detection algorithms that have been well established by now. In Sect. 4, the analysis of those algorithms are presented. And the Sect. 5 is the conclusion.

2 FTN Signaling Model

We consider a baseband binary PSK system model using a T-orthogonal pulse $\varphi(t)$ in the AWGN channel. The signal transmitted in the channel is written as follow

$$S(t) = \sum_{n=-\infty}^{+\infty} x[n]\varphi(t - n\tau T) \tag{1}$$

where $x[n]$ are independent identically distributed data symbols that $\in \{+1, -1\}$ and τT is the symbol interval in which τ is a symbol's packing ratio ranging from 0 to 1. For $\tau = 1$, the model becomes a conventional communication model. For $\tau < 1$, the symbols are sent faster than nyquist criterion, we say it is FTN signaling. In [2], it has been proved that for MLSE there exist constant C_1 and C_2 so that the probability of symbol error P_s can be bounded by

$$C_1 Q\left(\sqrt{d_{min}^2 \frac{E_b}{N_0}}\right) \le P_s \le C_2 Q\left(\sqrt{d_{min}^2 \frac{E_b}{N_0}}\right) \tag{2}$$

where d_{min} is the minimum Euclidean distance of the signaling. According to Mazo's paper, the τ for binary sinc pulse can be reduced to 0.802 without suffering any distance loss, which means that the symbols can be sent faster almost without any performance loss. With the $\tau < 1$, there is no difference between FTN signaling with T-orthogonal signaling in the power spectral density. In this way, FTN signaling is a promising information transmission method to increase the spectral efficiency.

The notations used in this paper are as follows. Lower case letters denote scalars, bold lower case letters denote column vectors, and bold upper case letters denote matrices.

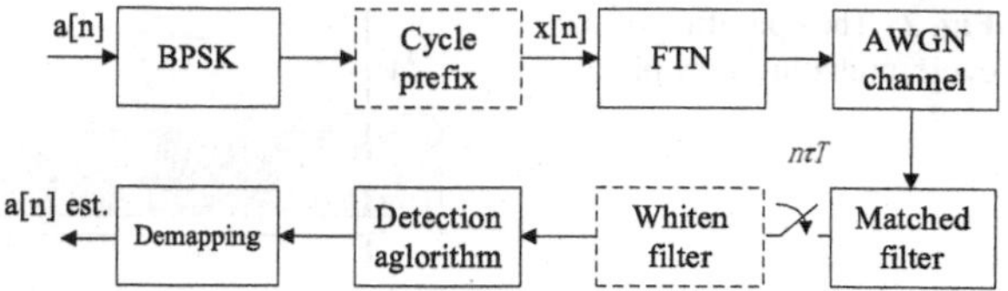

Fig. 1 The general structure of FTN system model. The Data symbols a[n] that $\in \{1, 0\}$ are spaced every τT

3 FTN Detection Algorithm

The general FTN system model is shown in Fig. 1. The modules in dashed boxes are not necessary for some detection algorithms in this paper. And the specific module needed in the certain algorithm is shown in the table below in which WF stands for whiten filter and CP stands for cycle-prefix.

VA	FDE	BCJR	GMP
WF	CP	WF	

3.1 VA

Viterbi algorithm (VA) [3] is a common tool in communication. VA is widely used in demodulation, decoding, equalization, etc. Since the FTN signaling introduces the ISI, we can apply the VA to make an equalization. The VA can maximize the probability $P(\mathbf{r}|\mathbf{x})$ in which $\mathbf{x}$ is the symbol vector, and $\mathbf{r}$ is the received signal vector.

Signal is sampled at $n\tau T$ after the match filter. And the samples are in maximum SNR because of the existence of the match filter. It is well known that the noise in the samples is not white which is not meeting the condition to use VA. Thus, a whiten filter is needed before the detection.

3.2 FDE

In the frequency domain equalization (FDE) detection [4], we assume the ISI length is finite. Actually, in practical system, such as RC pulse, the ISI attenuate quickly that leads the assumption reasonable.

In the transmitter, we partition the data into block that the length is N. Under the assumption that the ISI length is v, we add a $2v$-length cycle-prefix to the tail of the block. Thus, the cycle-prefix module is needed in the transmitter. Figure 2 illustrates the operation.

Then, we delete the first and the last v symbols to form a N-length block. The set of operations make we can detect the FTN signaling block-by-block. The finally received symbols $\mathbf{r} = \left[r_0, \ldots r_{N-1}\right]^T$ can be expressed by

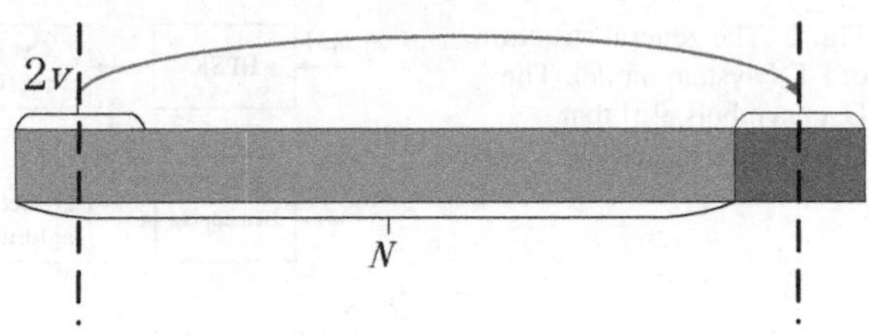

Fig. 2 The operation of the cycle-prefix in every block

$$\mathbf{r} = \mathbf{G}\mathbf{x} + \mathbf{n} \tag{3}$$

where $\mathbf{G}$ is a $N \times N$ size circular matrix, $\mathbf{x} = \left[x_0, \ldots x_{N-1}\right]^T$ represents the symbol block transmitted, and $\mathbf{n}$ is the noise. Since the $\mathbf{G}$ is circular matrix, we can decompose it by

$$\mathbf{G} = \mathbf{Q}^T \boldsymbol{\Lambda} \mathbf{Q}^* \tag{4}$$

$\mathbf{Q}^*$ is a FFT factor matrix, which means the left multiplication of $\mathbf{Q}^*$ is equivalent to FFT. Thus, we have

$$\mathbf{r}_f = \mathbf{Q}^* \mathbf{r} \tag{5}$$

$$= \boldsymbol{\Lambda} \mathbf{Q}^* \mathbf{x} + \mathbf{Q}^* \mathbf{n} \tag{6}$$

$$= \boldsymbol{\Lambda} \mathbf{x}_f + \mathbf{n}_f \tag{7}$$

Applying the MMSE criterion, we obtain the time domain estimation defined by

$$\hat{\mathbf{x}} = \mathbf{Q}^T \mathbf{M} \mathbf{r}_f \tag{8}$$

$\mathbf{M}$ is the equalization diagonal matrix, calculated by

$$m_{(i,i)} = \lambda^*_{(i,i)} / \left(\lambda^2_{(i,i)} + N_0\right) \tag{9}$$

where $m_{(i,i)}$ and $\lambda_{(i,i)}$ represent respectively the ith-row and ith-column element of the $\mathbf{M}$ and $\boldsymbol{\Lambda}$. N_0 is the single side-band power spectrum density of the noise.

3.3 BCJR

The BCJR algorithm was proposed by L.R. Bahl et al. in 1974 [5], in order to minimize the symbol error rate. The BCJR can be used to solve the detection problem while the source is Markov. The key to BCJR is to iteratively calculate the LLR defined by

$$LLR\left(x\left[n\right]\right) = \ln \frac{P\left(x\left[n\right] = +1 \mid \mathbf{y}\right)}{P\left(x\left[n\right] = -1 \mid \mathbf{y}\right)} \tag{10}$$

in which $\mathbf{y}$ is the received symbols vector.

The FTN signaling can be treated as a real valued convolutional code, meeting the requirement of Markov. However, the specialty different from the convolutional code is that the noises in the samples at $n\tau T$ have correlation. The pre-condition to use the BCJR algorithm is that the noise is white. Thus, we need to add the whiten filter after the matched filter in the receiver.

Actually, the structure can not be directly used in the practical scenario. The FTN signaling has the long ISI taps, leading to a high detection complexity. There are many modified-BCJR algorithms in the literature to reduce the states. In order to use those algorithm, we make some adjustments on the structure of the receiver. Let whiten filter be an allpass whiten filter which creates the max phase outputs[1]. Then, through a block reverse module, we obtain the min phase beneficial to use the modified BCJR such as Offset-State-BCJR [6] and M-BCJR [7].

3.4 GMP

Gaussian Message Passing (GMP) is a graph technique to perform LMMSE estimation for a Gaussian linear system. The enhanced message passing rules are detailed in [8].

In the FTN signaling, we can send the symbols frame-by-frame, which means we can model the system by (3). We rewritten the Eq. (3) as

$$\mathbf{r} = \mathbf{g}_j x_j + \boldsymbol{\xi}_j \tag{11}$$

where x_j is the jth coded bit, $\mathbf{g}_j$ is the jth column of $\mathbf{G}$, and the remaining part defined by

$$\boldsymbol{\xi}_j = \sum_{i \neq j} \mathbf{g}_i x_i + \mathbf{n} \tag{12}$$

First, we discuss the joint Gaussian approach, in which the vector $\mathbf{n}$ is assumed to be white and the $\boldsymbol{\xi}_j$ are joint Gaussian. Then, we have

$$\mathbf{m}_{\xi_j} = \mathbf{G}\mathbf{m}_{\mathbf{x}}^{prio} - \mathbf{g}_j m_{x_j}^{prio} \tag{13}$$

$$\mathbf{V}_{\xi_j} = \mathbf{G}\mathbf{V}_{\mathbf{x}_{(j)}}\mathbf{G}^T + \sigma^2\mathbf{I} \tag{14}$$

where m and V donate the means and the variances of subscript. $\mathbf{I}$ is the identity matrix with proper size. The $\mathbf{m}_{\mathbf{x}}^{prio}$ and the $\mathbf{V}_{\mathbf{x}_{(j)}}$ can be easily calculated under the assumption that data symbol is equiprobable.

And applying the proposed GMP based on the factor graph in [9], the $m_{x_j}^{post}$ and the $v_{x_j}^{post}$ can be obtained by the forward recursion and the backward recursion. The

[1]The min phase is theoretically feasible, but unstable.

$m_{x_j}^{post}$ is used to make the decision of $x[n]$. It is worth noting that the $m_{x_j}^{post}$ can be complex valued, which means that with the increasing of constellation size, the complexity of the algorithm is the same.

However, the algorithm above is under the assumption that the vector $\mathbf{n}$ is white. The modification work is showed in [10]. First, we use a p-order AR process to describe the colored noise $\mathbf{n}$,

$$n_j = \sum_{i=1}^{p} a_i n_{j-i} + n_j^{'} \tag{15}$$

where n_j is the ith sample of the noise, a_i is the coefficient and $n_j^{'}$ is the white innovation term. Let the vector $\mathbf{g} = \left[g_L, g_{L-1}, \ldots, g_{-L}\right]$ donates the ISI coefficient vector and the $\mathbf{A} = \left[a_p, a_{p-1}, \ldots a_1\right]$ donates the AR process parameter. The jth element of $\mathbf{r}$ is defined by

$$r_j = \mathbf{g}\left[x_{j-L}, x_{\mathrm{j-L+1}}, \ldots, x_{\mathrm{j+L}}\right]^T + n_j \tag{16}$$

where

$$n_j = \mathbf{A}\left[n_{j-p}, n_{j-p+1}, \ldots, n_{j-1}\right]^T + n_j^{'} \tag{17}$$

so

$$r_j = [\mathbf{g}\ \mathbf{A}][x_{j-L}, \ldots, x_{j+L}, n_{j-p}, \ldots, n_{j-1}]^T \tag{18}$$

Now, the model becomes the same as it in [9], and the similar GMP can be used to calculate the $m_{x_j}^{post}$ and the $v_{x_j}^{post}$.

3.5 Algorithm Modification in Turbo Equalization

The turbo equalization [11] was proposed in 1995 by C. Douillard et al. In the turbo equalization, the ISI is treated as the outer convolutional code that the code rate is 1. It has been proved that the turbo code can approximately arrive at the Shannon Limit. The iterative structure like turbo code shall have a good performance. The pulses shaping, channel, and the matched filter are equivalent to the discrete time ISI channel. In order to make use of turbo equalization, we add the convolutional encode module and interleaver before the BPSK modulation to complete the structure of SCCC. The structure of the turbo equalization is shown in Fig. 3. The decoder in the turbo equalization applies the maximum posterior probability (MAP) algorithm such as BCJR.

Actually, all the algorithms in Sect. 3 are able to be used in the turbo equalization with some modification.

For the VA, the original VA do not produce the soft information which is necessary in the turbo equalization. So, in turbo equalization, we apply the SOVA [12]. The

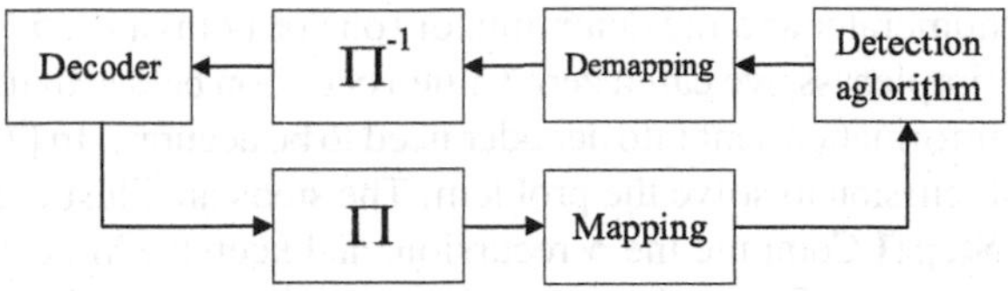

Fig. 3 The structure of turbo equalization. Π represents the interleaver

SOVA calculate the branch metric in the same way as VA, and it also calculate the erroneous decision probability p_j for jth bit. Then the LLR of the jth is expressed as $LLR(x_j) = \ln \frac{1-p_j}{p_j}$.

For the FDE, the modified work is detailed in [13], and we make a simplified statement below. Assuming that the $\tilde{\mathbf{x}} = [\tilde{x}_0, \ldots, \tilde{x}_{N-1}]$ are the soft symbols generated by the *prior* information offered by the decoder, we modify the received vector by

$$\tilde{\mathbf{r}} = \mathbf{r} - \mathbf{G}\tilde{\mathbf{x}} \tag{19}$$

$$= \mathbf{G}(\mathbf{x} - \tilde{\mathbf{x}}) + \mathbf{n} \tag{20}$$

By take the FFT operation on the both side of (20), we obtain

$$\tilde{\mathbf{r}}_f = \mathbf{Q}^* \tilde{\mathbf{r}}_f \tag{21}$$

$$= \mathbf{\Lambda} \mathbf{Q}^* (\mathbf{x} - \tilde{\mathbf{x}}) + \mathbf{n}_f \tag{22}$$

Applying the MMSE criterion, the time domain estimate vector $\hat{\mathbf{x}}$ is given by

$$\hat{\mathbf{x}} = \mathbf{Q}^T \mathbf{D} \tilde{\mathbf{r}}_f \tag{23}$$

in which $\mathbf{D}$ is a diagonal matrix. The ith-row and the ith-column of $\mathbf{D}$ is

$$d_{(i,i)} = \frac{\lambda^*_{(i,i)}}{-|\lambda_{(i,i)}|^2 \sum_{j=0}^{N-1} |\tilde{\mathbf{x}}_j|^2 / N + N_0} \tag{24}$$

Finally, we arrive at the extrinsic information defined by

$$L_{ext}(\mathbf{x}) = \frac{\alpha \tilde{\mathbf{x}} + \hat{\mathbf{x}}}{1 + \alpha (1 + \sum_{j=0}^{N-1} |\tilde{x}_j|^2)} \tag{25}$$

where $\alpha = trace(\mathbf{\Lambda D})/N$.

For the BCJR, it can be directly used in the turbo equalization. But, with the ISI length increases, the complexity increases quickly. One of the modification widely used is M-BCJR. Since it only keeps M paths in every stage, the occasion that the

numerator and the denominator (one or both) are zero may happens. For expression simpleness, we call it zero valued occasion below. In turbo equalization, the extrinsic information sent into decoder need to be accurate. In [7], the author introduced a third recursion to solve the problem. The steps are illustrated below.
Step:1 Compute the α recursion, and keep the M best paths in each stage.
Step:2 Compute the β recursion, and keep the M best paths in each stage.
Step:3 Compute the LLR in (10), and check the whether the zero valued occasion exists. If, take step 4.
Step:4 Decide the symbols from the respective zero valued occasion stage to the stage-ISI length to obtain the state of this stage. And trace forward a few stages to produce a few of α, The αs obtained from the one branch form the probability of the one symbol outcomes, and the αs obtained from the another branch form the probability of the another symbol outcomes. Then, we can calculate LLR.

For the GMP, the extrinsic information is defined by

$$L_{ext}\left(x_j\right) = ln\frac{p(\mathbf{r}|x_j = +1)}{p(\mathbf{r}|x_j = -1)} \tag{26}$$

$$= ln\frac{\exp\left[-\frac{1}{2}\left(\mathbf{r} - \mathbf{g}_j - \mathbf{m}_{\xi_j}\right)^T \mathbf{V}_{\xi_j}^{-1}\left(\mathbf{r} - \mathbf{g}_j - \mathbf{m}_{\xi_j}\right)\right]}{\exp\left[-\frac{1}{2}\left(\mathbf{r} + \mathbf{g}_j - \mathbf{m}_{\xi_j}\right)^T \mathbf{V}_{\xi_j}^{-1}\left(\mathbf{r} + \mathbf{g}_j - \mathbf{m}_{\xi_j}\right)\right]} \tag{27}$$

Equation (27) contains the matrix inversion. The complexity can be high if the size of **G** is large. In [9], it has been proved that the Eq. (27) is equal to

$$L_{ext}(x_j) = 2\left(\frac{m_{x_j}^{post}}{v_{x_j}^{post}} - \frac{m_{x_j}^{prio}}{v_{x_j}^{prio}}\right) \tag{28}$$

in which $m_{x_j}^{post}$ and the $v_{x_j}^{post}$ can be easily carried out in GMP recursion.

4 Performance Analysis

The performance and the complexity of the algorithms above are obviously different. The FTN signaling is detected symbol-by-symbol in frequency domain equalization. The complexity is low, and performance is not good under the low SNR. With the SNR increasing, the BER decreases quickly, making the frequency domain equalization detection useful under the high SNR. And the introduction of FFT and IFFT makes it operated in frequency domain. So, the long taps FTN scenario such as low roll-off factor RC can be detected without leading to any complexity increasing.

The performance of BCJR and the VA are similar, and both are sequence estimate algorithms. The MLSE makes full use of the constraint in the FTN wave form as a convolutional code. But the BCJR and the VA are in high complexity increasing exponentially with constellation size and the number of ISI taps.

And the GMP has a lower complexity than BCJR and VA, owing to the fact that only a small amount matrix inversion need to be done, but higher than FDE. The more attractive advantage is that the complexity has nothing to do with constellation size.

The complexity of turbo equalization with the algorithms above is high caused by the iteration. And the performance of those algorithms in turbo equalization are similar. The BER curve can approximately reach the theoretical performance of the convolutional code when the SNR reaches over 4–5 dB, which means that the ISI is almost eliminated.

5 Conclusion

Owing to the studies of many scholars, a lot of practical algorithms have been proposed. And the detection of FTN signaling is more and more mature. However, there are still a lot of things shall be done. The algorithm and the method to deal with ISI is much more that those in this paper, we may find the lower complexity algorithm which can be used in turbo equalization. We may even change the iterative structure of turbo equalization, such as the introducing of the ADMM. And the challenge ahead is that the wireless channel is complex, the FTN detection algorithm in fading channel need to be studied. We hope the theory about FTN will be more and more thorough.

References

1. Mazo JE (1975) Faster-than-Nyquist signalling. Bell Syst Tech J 54:1451–1462
2. Forney GD Jr (1972) Maximum likelihood sequence estimation of digital sequences in the presence of intersymbol interference. IEEE Trans Inform Theory 18:363–378
3. Forney GD (1973) The Viterbi algorithm. Proc IEEE 61(3):268–278
4. Sugiura S (2013) Frequence-domain equalization of faster-than-Nyquist signaling. IEEE Wirel Commun Lett 2:555–558
5. Bahl LR, Cocke J, Jelinek F, Raviv J (1974) Optimal decoding of linear codes for minimize the symbol error rate. IEEE Trans Inform Theory 20:284–287
6. Anderson JB, Prlja A, Rusek F (2009) New reduced state space BCJR algorithms for the ISI channel. In: ISIT 2009. Seoul, Korea
7. Prlja A, Anderson JB (2012) Reduced-complexity receivers for strongly narrowband intersymbol interference introduced by faster-than-Nyquist signaling. IEEE Trans Commun 60:2591–2601
8. Loeliger HA, Hu J, Korl S, Guo Q, Li P (2006) Gaussian message passing on linear models: an update. In: Internatioanl symposium on turbo codes and related topics
9. Guo Q, Ping L (2008) Lmmse turbo equalization based on factor graphs. IEEE J Sel Areas Commun 26(2):311–319
10. Şen P, Aktaş T, Yılmaz AO (2014) A low-complexity graph-based LMMSE receiver designed for colored noise induced by FTN-signaling. In: IEEE WCNC, pp 642–647
11. Douillard C, Jézéquel M, Berrou C (1995) Iterative correction of intersymbol interference: turbo-equalization. Eur Trans Commun 6:507–511

12. Hagenauera J, Hoeher P (1989) A Viterbi algortihm with soft-decision outputs and its applications. In: IEEE GOLBECOM 1989, Dallas, vol 3, pp 1680–1686
13. Sugiura S, Hanzo L (2015) Frequency-domain equalization aided iterative detection of faster-than-Nyquist signaling. IEEE Trans Veh Technol 64:2122–2128

A Novel Audio Segmentation for Audio Diarization

Xuehan Ma

Abstract The Speaker change detection task usually contains two passes: potential change detection and verification. BIC criterion is often utilized to measure the dissimilarity. However, insufficient data may lead to modeling error, which cannot represent one speaker correctly. In this paper, we propose combining prosodic feature with LSP (Linear Spectrum Pair) feature to detect change points. Prosodic feature can contribute to eliminate false change points locally. The experiments show an improvement compared with the traditional speaker change point mechanism.

Keywords Speaker change detection · Prosodic feature · BIC verification

1 Introduction

Speaker Diarization is increasingly on the spot among the researchers, however the diversity of audio types may result in the relatively high Diarization Error Rate. How to detect the change points including both the audio types and different speakers efficiently is still a tough task. Supposing we have already correctly segment the input audio into speech, music, environment sounds. The next processing is to segment the speech segments into speaker-homogeneous regions, that is, each segment just be-longs to only one speaker. As a result, speaker segmentation, or speaker tracking indexes the speech stream by speaker identities, in which there is no prior knowledge about the number and identities of speakers, particularly in continuous speech stream.

The task can be viewed as a combination of two parts, one is to select representative features, and the other is to choose rational similarity measures. As to the first one, it can be stated as grasping personal traits, including the vocal tract and phonation characters. Both long-term and short-term features can be extracted, reflecting prosodic information and voicing parameters respectively. A lot of study has shown

X. Ma (✉)
School of Information and Communication Engineering, Beijing University of Posts and Telecommunications, Beijing, China
e-mail: maxh1205@163.com

V.E. Balas et al. (eds.), *Information Technology and Intelligent Transportation Systems*, Advances in Intelligent Systems and Computing 455,
DOI 10.1007/978-3-319-38771-0_39

that short-term features, such as Mel frequency cepstral coefficients (MFCCs) and linear spectral pair (LSP), occupy the dominance position in speaker recognition. However, according to [1], a systematic investigation of speaker discriminability of 70 long-term features was conducted to get the top-10 features, when combined with the short-term features to show a consistent improvement. As depicted in [2], spectrogram is formulated as a matrix to extract the features to distinguish different audio types in a supervised manner. Similarity measures are implemented by calculating the distance between two models to quantify the homogeneity. The more similar, the smaller the distance. Different criteria may result in results. In the work of [2], six kinds of criteria are provided to test the classification procedure. When the speaker turn is frequent, rational model should be considered before applying distance criteria. In [3], different criteria are chosen according to the duration of segments, compensating the insufficient data.

In this paper, we present a high-accuracy algorithm for speech segmentation. The speech is discriminated in 3-s window sliding by 1-s, as depicted in [4]. A novel work is the combination of prosodic features with frame-based LSP features to achieve low false alarm rate, which leads to 8 % relative improvement.

The rest of the paper is organized as follows. Section 2 discuss the structure of the proposed segmentation system is introduced. In Sect. 3, the proposed algorithm is presented. In Sect. 4, the way to calculate the thresholds is introduced. In Sect. 5, some experimental results are presented. Finally, a conclusion is drawn.

2 Related Works

As already explained in Sect. I, the goal of segmentation is to answer the question, Who spoke when? [5]. No prior knowledge about the identity or number of speakers in the recording is given. Either the scene occasion or the speaking speed is given, thus we need to tackle different scenarios at either slow or fast speed.

Although filler and overlap problems are very common in spontaneous speech and has been studied thoroughly in [6, 7], in our context, we assume that people do not speak simultaneously and the filler is neglected. Filler parts take the form of an interjection, like /uh/ and /um/ in English, which have flat pitch and energy, while the overlapping is that there are more than two speakers talking simultaneously.

Feature selection and extraction is the front-end of Diarization. Researchers devoted to seek different features for characterizing personal traits. In [8], jitter and shimmer are proposed in speaking verification, which reflect the variability of pitch (F0). In [3], PMVDR, SZCR, FBLC are combined and then analyzed using principal component analysis (PCA). In this paper, we adopt F0_median feature described in [1].

Distance curve reflects the variation of distance in time, previous work has been done through different criteria. DISTBIC [9] applies well to slow speaker change speeches, while encountered problems in case of short segments and require a high computation cost. T2-BIC improve the computation cost, while depends on many empiric parameters which is not robust. Finally, DIST-T2-BIC is a hybrid algorithm which improve the detection of speaker turns close to each other with a low cost.

These above two-pass algorithm is summed up as follows: a first pass is based on peak-picking in the distance curves to find out potential change points, and a second pass is to validate these candidates by applying Bayesian Information Criterion (BIC). Because only when segments are large enough, BIC criteria gives better result since model estimation becomes more accurate [10], that is to say, we must guarantee the length of segments in second pass no shorter than 2-s.

Recently, other frameworks use kernel methods in high-dimension feature space and information-theoretic methods based on entropy [11]. In [12], 1-class SVM segmentation method and exponential family model is proposed to tackle the problem in just one segmentation step to avoid the model selection procedure. In [11], a framework of information geometry for exponential families is presented to detect changes by monitoring the information rate of the signals. In this paper, the DIST-BIC is adopted combining with discriminative prosodic features as stated in [1].

3 Audio Segmentation

In extracting feature procedure, all input audio stream are down sampled into 8 KHz sample rate and subsequently segmented by 3-s window, with a step size of 1-s. Each segment is preprocessed by removing silence and unvoiced frames. Each segment is processed to find speaker change points by distance curve and local F0_median value. Finally, the candidates are verified using model selection criteria–BIC.

3.1 Features

In our experiments, we used four features: zero cross rate (ZCR), short time energy (STE), F0_median, linear spectral pair (LSP). Below, we will mainly introduce LSP and F0_median.

• LSPs are derived from linear predictive coefficients (LPCs), reflects the spectrum envelop. In each 1-s window, the audio clip is further divided into 25-ms non-overlapping frames, on which LSP is extracted applying a 15 Hz band-width.

• The F0 and its statistical values are influenced by the length and mass of the vocal folds in the larynx. The main difference between speakers with respect to F0 is related to gender and age. Furthermore, F0 also varies depending on the speaker mood and speech content. Hence, F0 can be regarded as a capable speaker discriminative feature.

3.2 Preprocessing

LSP is the most important feature used to model each segment. STE and ZCR features are used to discriminate silence frames and unvoiced frames, which will be excluded when estimating speaker model. The threshold of STE and ZCR are set by the minimum mean of GMM model, which contains three components, respectively.

3.3 Potential Change Points Detection

At this step, speaker model is formulated using voiced segments in each window. Due to the insufficient data in each window, the model is in diagonal covariance. LSP divergence distance is measured between each two adjacent window [4]. In this paper, we apply distance measure between the current window and the after next window, thus the difference is more prominent. Consequently, a series of change points will be gotten.

The distance criteria encompass Kullback–Leibler Distance (KL), Generalized Likelihood Ratio (GLR), Hotelling T2-statistic, BIC and so on. In this paper, we choose KL criteria excluding the second parts which contain both covariance and mean. According to [4], the dissimilarity measure between two models can be defined as

$$D = \frac{1}{2}tr[(C_{SP} - C_{LSP})(C_{SP}^{-1} - C_{LSP}^{-1})] \tag{1}$$

where C_{SP} and C_{LSP} are the estimated LSP covariance matrices, respectively.

This distance measure is quite effective to discriminate different speakers. Thus, a local peak represents a speaker change point. Adaptive and rule-based threshold is adopted in [4, 9]. Here, we employ peaks finding method to find local peaks to make sure there is no true change points missed.

3.4 Prosodic Verification

From the section C, numerous candidates can be detected, which include many false change points. Hence, we need to eliminate these false points locally to make sure more data is used to model one speaker the next step.

As well known, the pitch reflects different people traits. In [7], F0_median is ranked top 1 through Fisher calculation as the most discriminative feature supplementary to MFCC, and the result is improved in contrast to the base system ICSI. Here, we further validate the discriminant ability of F0_median using one-way ANOVA (one-way Analysis of Variance). From Fig. 1, we can see the discrimination of F0_median clearly. In this paper, F0 is extracted from Praat software.

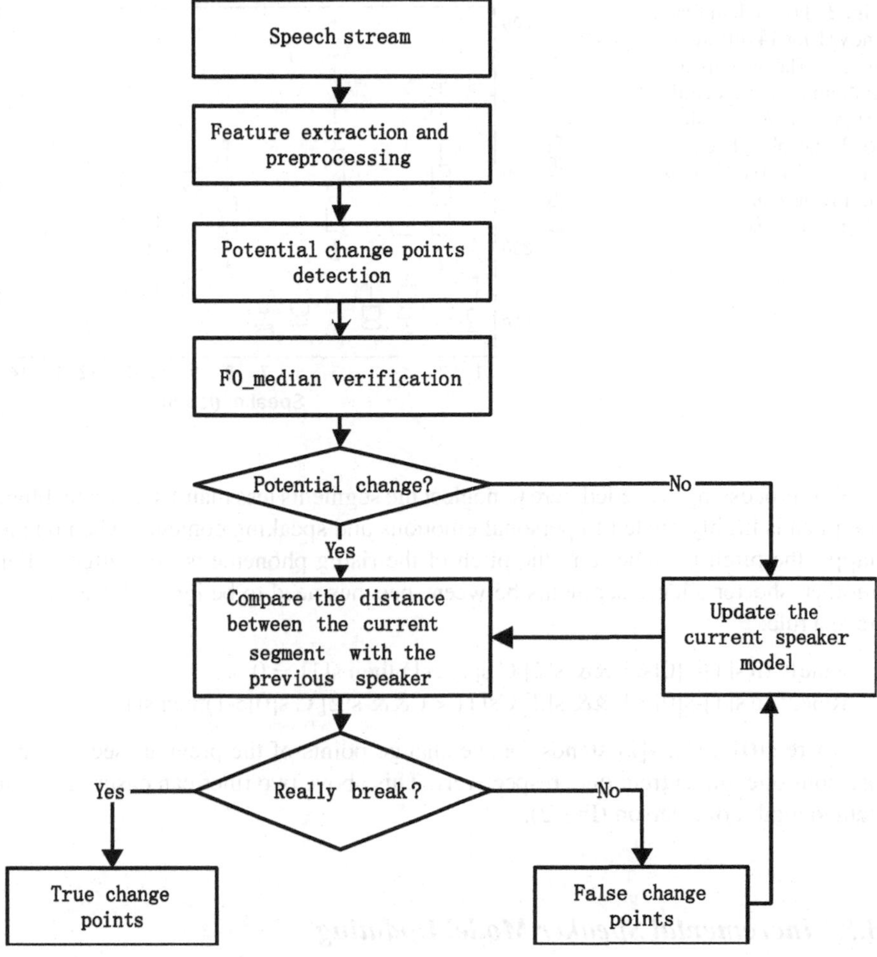

Fig. 1 Audio segmentation system flowchart

Through comparing the pitch distance between two adjacent change points, we can get a series of distance. Let $D(i, j)$ denote the distance between the ith and jth speech segment, if $D(i, j)$ is greater than TH_i, then its a true change point. Here, the is automatically set according to the previous distances, i.e.

$$TH_i = \frac{1}{N}\sum_{j=0}^{N} D(i-j, i-j-1) \tag{2}$$

where N is set 3 or 5 empirically in our algorithm. The threshold determined in this way works satisfactory in different situations, even though different persons have different pitch values and fluctuation ranges.

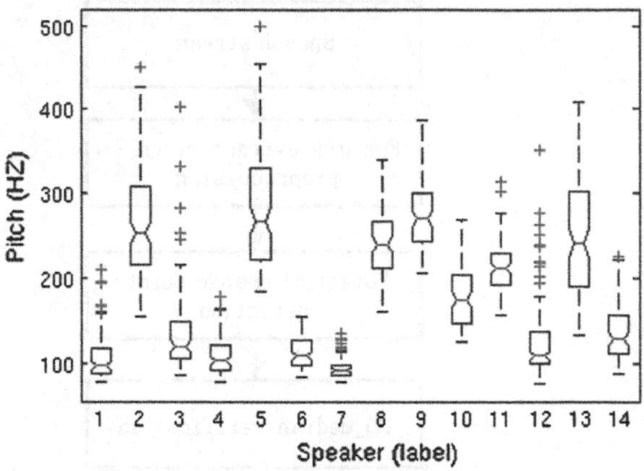

Fig. 2 F0_median features anoval for 14 different people. The median and its 25th and 75th percentile of F0_median feature shown for 14 peoples. The p value smaller than 0.001 shows the significance of the F0_median feature

Post processing is needed here to neglect the segments less than 1 s. For one thing, the pitch is highly related to personal emotions and speaking contents, when one is happy, the pitch is higher and the pitch of the rising phoneme is often higher. For another, shorter silence segments between speeches need to be ignored. Hence, we have a rule:

Rule1: If(s[1]-s[0]<1 && s[2] C s[1] >1) then s[1] = 0
Rule2: If(s[1]-s[0]<1 && s[2] Cs[1] <1 && s[2] C s[0]>1)then s[1] = 0

where $s[0]$, $s[1]$, $s[2]$ stands for the change points of the previous second one, previous one and current one, respectively. This above two rules can ensure enough data to model one person (Fig. 2).

3.5 Incremental Speaker Model Updating

More data is needed to model one speaker more accurately. If there is no change point, the next segment is assumed as the same speaker as previous one. Thus, we use this new data to update the current speaker model. Here, we employ a quasi-Gaussian mixture model instead of original GMM using EM algorithm [4].

3.6 Candidates Verification

Even though prosodic verification can reduce the number of change points, there still exists false candidates. This refinement algorithm is based on the current segment and the previous speaker quasi-Gaussian model at each potential change point. In this step, Bayesian information criteria(BIC) is used to measure the model dissimilarity. Suppose two Gaussian model are $N(\mu_1, C_1)$ and $N(\mu_2, C_2)$, the number of feature vectors are and; when one Gaussian model is used to estimate these two speech segments, the model is $N(\mu, C)$. The BIC difference between the two models is:

$$BIC(C_1, C_2) = \frac{1}{2}((N_1 + N_2)\log|C| - N_1 \log|C_1| - N_2 \log|C_2|) - \frac{1}{2}\lambda(d + d)\log(N_1 + N_2) \quad (3)$$

where λ is a penalty factor for compensated for small size cases, and is the feature dimension. Generally, $\lambda = 1$.

Suppose at each candidate, the model of previous speaker is $N(\mu, C_i)$, and the model of current segment is $N(\mu, C)$. Then the distance is estimated as weighted sum of the distance between $N(\mu, C)$ and each $N(\mu_i, C_i)$, that is, the distance does not consider the $GMM - s$ as an integral one, but as several independent components. According to BIC theory, if the result is positive, the change point is discarded; otherwise, its a true point.

4 Experiment Results

4.1 Data

The testing materials used for speaker segmentation evaluations include broadcast news, talk shows and a few meeting videos. In total, they are about 2 h. All the audio tracks are down sampled to 8 KHz in mono-channel before processing.

4.2 Potential Change Points

The performance evaluations of speaker change detection are described with FAR (False Alarm Rate) and MDR (Missed Detection Rate) [9]. Because false alarms are more tolerant than the missed ones, we assign higher cost of missed alarms.

At this step, we test our algorithm step by step, including the potential change detection and prosodic verification. In each step, we must guarantee the lower MDR and longer duration as well.

From the evaluation test described in Table 1, we can see even through higher FAR is derived, the MDR is near to zero. Peak finding method can ensure most of the change points can be detected, even though the peak is not prominent or the true change points is near to each other. The Detected in the table represents the number of change points that are correctly detected in the true points.

Table 1 Potential change points on development data

Collections	Accuracy (100 %)				
	Detected	Miss	False	FAR	MDR
CCTV	188	4	120	38.96	2.08
Talk-show	196	8	123	38.56	3.92
Meeting	186	5	116	38.41	2.61

Table 2 F0_median verification

Collections	Accuracy (100 %)				
	Detected	Miss	False	FAR	MDR
CCTV	188	10	48	14.29	5.05
Talk-show	196	11	50	20.33	5.31
Meeting	186	9	47	20.17	4.61

Table 3 BIC Verification

Collections	Accuracy (100 %)				
	Detected	Miss	False	FAR	MDR
CCTV	188	15	24	11.32	7.39
Talk-show	196	13	26	11.71	6.22
Meeting	186	12	23	11	6.06

Higher FAR means insufficient data available for speaker modeling in the next procedure. Hence, we employ F0_median to reduce FAR and guarantee lower MDR as well. From Table 2, a conclusion can be derived that F0_median can eliminate the false change points greatly, which can be seen from the decrease in FAR. It implies the significance of F0_median in discriminating different speakers.

However, a little increase in MDR will arise. Because F0_median can be influenced by speakers mode and speaking content, and people of the same sex may have similar pitch characteristics. The point is just what is needed to modify in the future.

In the step of rule based method, long enough duration is pledged for the next procedure, thus the BIC model selection criterion can give better performance [10].

From Table 3, we can see that BIC verification is quite effective in decreasing the FAR. It is found that, in BIC verification phase, the factor of BIC formula needs to be adjusted according to different audio types. Here, we usually set it as 0.8 empirically. In the experiment, we found that if the length of one speaker is shorter than 1 s, it will be neglected. Our algorithm is limited to find out those points effectively, so it needs to change the sliding window mechanism or new algorithm to improve the result.

5 Conclusions

In this paper, the proposed algorithm can guarantee lower MDR and improve the FAR. Further, the system can run with little time delay, we have tested the computational complexity of our algorithm in term of CPU time. Due to the diagonal modeling and low distance calculation cost, the algorithm can run about 20 % time of the real time. However, the robustness and stability still needs to be improved in the future, especially in different surroundings and inverse condition [12].

References

1. Friedland G, Vinyals O, Huang Y, Muller C (2009) Prosodic and other long-term features for speaker diarization. IEEE Trans Audio Speech Lang Process 17(5):985–993
2. Ghoraani B, Krishnan S (2011) Time frequency matrix feature extraction and classification of environmental audio signals. IEEE Trans Audio Speech Lang Process 19(7):2197–2209
3. Huang R, Hansen JHL (2006) Advances in unsupervised audio classification and segmentation for the broadcast news and NGSW corpora. IEEE Trans Audio, Speech, Lang Process 14(3):907–919
4. Lu L, Zhang H, Jiang H (2002) Content analysis for audio classification and segmentation. IEEE Trans Speech Audio Process 10(7):504–516
5. Tranter S, Reynolds D, Member S (2006) An overview of automatic speaker diarisation systems. IEEE Trans Speech, Audio Lang Process 14:1557–1565
6. Li Y, He Q, Li T (2008) A novel detection method of filled pause in mandarin spontaneous speech. In: Proceedings of the seventh IEEE/ACIS international conference on computer and information science (icis 2008) ieee computer society, pp 217–222
7. Yella SH, Bourlard H (2014) Overlapping speech detection using long-term conversational features for speaker diarization in meeting room conversations. IEEE/ACM Trans Audio Speech Lang Process 22(12):1688–1700
8. Hernando J (2009) Using Jitter and Shimmer in speaker verification. Signal Process 3(4):247–257
9. Delacourt P, Wellekens CJ (2000) DISTBIC: a speaker-based segmentation for audio data in-dexing. Speech Commun 32:111–126
10. Chen S, Gopalakrishnan P (1998) Speaker, environment and channel change detection and clustering via the bayesian information criterion. In: DARPA speech recognition work-shop
11. Dessein A, Cont A (2013) An information-geometric approach to real-time audio segmentation. Signal Process Lett IEEE 20(4):331–334
12. Hachem K, Davy M, Ellouze N (2010) Robust unsupervised speaker segmentation for audio diarization. In: Signal processing, pp 307–320

Research on LogGP Based Parallel Computing Model for CPU/GPU Cluster

Yongwen Wu, Junqiang Song, Kaijun Ren and Xiaoyong Li

Abstract CPU/GPU heterogeneous computing has become a tendency in scientific and engineering computing. The level of heterogeneity in modern computing systems gradually rises, and CPU/GPU Heterogeneous system contains three levels of heterogeneity. Conventional parallel computation models cannot be used to estimate the running time under the CPU/GPU heterogeneous computing environment. In this paper, a new model named VLogGP is proposed, and the communication and memory access characteristics are both abstracted based on CPU/GPU heterogeneous system. We map the model to TH-1A platform, and measure all model parameters for this kind of platforms. The model can be used to study the behavior of parallel applications, estimate the execution time and guide the optimization of parallel programs.

Keywords Heterogeneous system · Parallel computing model · Communication · LogGP model

Y. Wu (✉)
College of Computer, National University of Defense Technology, Changsha, China
e-mail: wuyongwen@nudt.edu.cn

J. Song · K. Ren · X. Li
Academy of Ocean Science and Engineering, National University of Defense Technology, Changsha, China
e-mail: junqiang@nudt.edu.cn

K. Ren
e-mail: renkaijun@nudt.edu.cn

X. Li
e-mail: sayingxmu@nudt.edu.cn

V.E. Balas et al. (eds.), *Information Technology and Intelligent Transportation Systems*, Advances in Intelligent Systems and Computing 455,
DOI 10.1007/978-3-319-38771-0_40

1 Introduction

Traditional developed high performance computers using general CPUs encountered significant challenges in system power, cooling and cost. GPUs are attractive as accelerators for better computing ability, high memory bandwidth, low power and programmability which make them an alternative in heterogeneous computer systems design. CPU/GPU heterogeneous computing has become a tendency in scientific and engineering computing. Parallel computing generally adopt universal models, as for emerging CPU/GPU heterogeneous system, due to their three levels of heterogeneous characteristics, existing models are unable to accurately describe their structural features and performance factors. There is an urgent need to research on parallel computing model for heterogeneous clusters, thereby providing support for the current and future development of parallel applications based on heterogeneous platforms.

Parallel computing model is used to abstract/reflect the underlying characteristics of parallel computer system with a small number of parameters, which builds a bridge between parallel programming model and parallel computer systems, and of significance to algorithm designers. The traditional parallel computing models mainly contain the synchronous processors PRAM model [7], Postal model [1], the bulk synchronous parallel model (BSP) [12] and the LogP model [5] which can predict the communication time for short messages; in a heterogeneous parallel computer system, there are Cluster-M [6], HCGM [11], HBSPk [13], MMGP [3], HlogGP [4], mPlogP [10] model, they are correspond to different special architectures based on traditional models. We analyze MMGP, mPlogP, HlogGP models in the following which are closely related to what we researched. MMGP and mPlogP are used in heterogeneous architectures describe lower level data communication via high speed interconnection network to transfer data between different processor/memory module. HlogGP not only simulates parallel computing system in a variety of communication operations (storage and network communications), it can also simulate data copying operation in memory spaces. However, the MMGP model does not distinguish the communication overhead between HPU (Host Processing Units) and HPU (Accelerator Processing Units), Unity with the formula: $C_{i,j} = o_i + L + o_j$. Secondly, MMGP model is idealized, ignore the sub-task communication overhead, assuming that in addition to the main overhead of assigning tasks at the beginning and result processing at the final stage, the other overheads can be ignored or hidden in the calculation. It is worth mention that the mPlogP model considers computing core and accelerate core that are located on the same chip, with different usability, there is a great difference with the hybrid architecture such as CPU/GPU; the mPlogP model takes less consideration of new communication technology, in particular virtual address and Direct V2.0 technology, etc. These can make mPlogP model's prediction time for the CPU/GPU heterogeneous cluster program not accurate. HlogGP mainly considers loosely coupled heterogeneous cluster communication operations, there are two kinds of inadequacy: on the one hand, this model has $(2M^2 + 4M)$ parameter, it is difficult to accurately mapped to specific large heterogeneous clusters;

on the other hand, it does not comprehensively consider node heterogeneity within the construction as well as its storage traffic overhead, and thus can't effectively describe the behavior of the current GPU clusters' communication behavior.

Conventional parallel computation models cannot be used to estimate the running time under the CPU/GPU heterogeneous computing environment. In this paper, a new model named VLogGP is proposed, the communication and memory access are both abstracted based on the characteristic of CPU/GPU heterogeneous system. The rest of the paper is organized as follows. Section 2 provides an overview of the CPU/GPU communication technology. In Sect. 3 we extend the LogGP model and put forward the VLogGP model which is particularly for CPU/GPU clusters. Section 4 shows measurement of the parameters. Section 5 provides a summary of the implementation and initial results, and conclusions are presented in Sect. 6.

2 CPU/GPU Communication Technology

The storage and communication overhead are important factors in the overall performance of large scale GPU parallel programs. Academia and industry research centers have conducted extensive studies to reduce network traffic overhead, as to how to reduce the communication overhead of applications on GPU clusters. Figure 1 shows three different ways of inter-GPUs data transmission, as early as 2010, before the release of GPUDirect technology, data communication between GPUs undergo the following five steps: (1) GPU writes data to a special space in the main memory which can't be accessed by other third party software; (2) CPU copy the data to another storage area of the main memory, this space may be accessed by third party software, such as InfiniBand; (3) Sending the data to a remote node by means of network devices; (4) Do copy operation like step 2 in CPU main memory; (5) Finally, send the data to the target GPU memory.

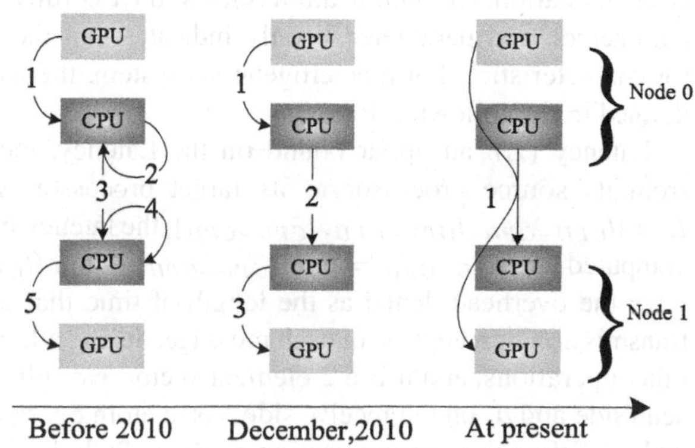

Fig. 1 Inter nodes CPU/GPU data transmission methods

CPUs are involved in the process of data communication between different GPU nodes, and memory copy operations relate to step 2 and 4 are performance bottlenecks of the entire communication process. For this reason, at the end of 2010, NVIDIA and Mellanox jointly launched GPUDirect technology which enables network equipment and GPU share the same system memory, eliminating the need for additional copy operations, and there is no need for CPU participating in the communication process, which can reduce communication latency, and greatly improves the efficiency of communication. While the existing means of communication is much better than no use of GPUDirect technology, the whole process still contains three steps. The release of Kepler GK110 supports GPUDirect RDMA, their purpose is to allow third party devices, such as IB adapters, NIC, and SSD, to direct access to GPU memory. Under CUDA5.0 version, GPUDirect RDMA provides the following important features: no data buffering on the CPU side; direct memory access between the NIC and GPU (DMA); significantly improved MPISend/MPIRecv efficiency between the GPUs and other nodes; eliminate CPU bandwidth and latency bottlenecks; work with a variety of third-party network and storage devices. CUDA5.0 integrates all the communication process as a step, making communication delay and bandwidth between CPU and GPU no longer performance bottlenecks.

3 The Parallel Computing Model VlogGP

To apply extended LogGP model to CPU/GPU heterogeneous cluster. We still use five parameters L, o, g, G, P. However, given the connection between CPU and GPU within a node is PCIe, whereas the connection between nodes is through dedicated network, the delays will be different, in addition, different transport modes result in a corresponding different bandwidths, especially there are differences between CPU and GPU, point to point communication bandwidth will be affected by communication direction too. The proposed model emphasizes the importance of storage communication. Communication relates to CPU fully reflected in the L, o, g, G, P parameters, and parameter V only indicates that the model parameters have vector characteristics. For a heterogeneous system, the five parameters of VlogGP are defined in the following forms:

Latency (L): an upper bound on the Latency, incurred in sending a message from its source processor to its target processor, which is 3-element vectors. $L = \{l_{CPU-GPU}, l_{GPU-CPU}, l_{CPU-CPU}\}$, the latency between GPU and GPU can be computed as: $l_{GPU-GPU} = \min(l_{CPU-GPU}) + \min(l_{CPU-CPU}) + \min(l_{GPU-CPU})$.

o: the overhead dened as the length of time that a processor is engaged in the transmission or reception of each message, during which the processor can't perform other operations, and it is a 2-element vector, we split the overhead o into o_s on the send side and o_r on the receive side. As to atom operators, it's hard to distinguish o_s and o_r. When the communication media is PCIe bus, consider that GPU and CPU have different frequencies, and GPU is in a subordinate position, we only consider overhead related to CPU.

g: the gap between messages, dened as the minimum time interval between consecutive message transmissions or receptions at a processor.

G: the gap per byte for long messages, dened as the time per byte for a long message. The reciprocal of G characterizes the available processor communication bandwidth for long messages, which is a 3-element vector. $G = \{G_{GPU-CPU}, G_{CPU-CPU}, G_{CPU-GPU}\}$.

P: the number of processor/memory modules, which can represents computing capacity of different nodes and is a 2-element vector, $P = \{P_C, P_G\}$.

4 Determination of the Model Parameters

The model includes 10 major uncertain parameters. We first introduce how to deduce these parameters, and then apply the model to the TH-1A system. Commonly, we use *RTT* (Round Trip Time) to evaluate the performance of a network. Many network testing procedures [2] using ping-pong communication mode to obtain *RTT* as show in Fig. 2a, the client starts the communication operator and measures *RTT* time, the server will forward each received data packet to the client. Figure 2b shows another strategy of getting *RTT*, namely the ping-ping program: The client consecutively sends n messages, after the server receives all the messages, the server only forwards the last received data to the client. From the client's perspective, the entire ping-ping procedure involves n transmitting operations and only one receiving operation. Hoefler [9] proposed an accurate measurement of LogGP's parameters.

The notion of the parametrized round trip time (*PRTT*) is defined as a function of other specific parameters. The parameters are the number of ping-ping packets (n), the delay between each packet (d) and the message size (s). A specific combination of n, d and s is denoted as *PRTT(n,d,s)*. The following subsections show that the notion of *PRTT(n,d,s)* is sufficient to assess all model parameters related to CPUs accurately without network flooding or unnecessary contention. The *PRTT* for a single ping-ping message without delay can be expressed as follows:

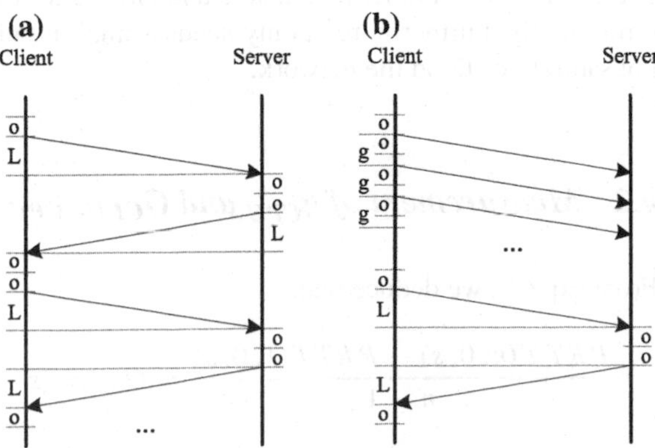

Fig. 2 Ping-pong and ping-ping tests. **a** Ping-pong. **b** ping-ping

$$PRTT(1,0,s) = 2 \times \left(o_r + L_{CPU-CPU} + (s-1)G_{CPU-CPU} + o_s\right) \quad (1)$$

The cumulative hardware gap defined as: $G_{all} = g_{cpu} + (s-1)G_{CPU-CPU}$, then n ping-ping messages can be modeled as:

$$PRTT(n,0,s) = PRTT(1,0,s) + (n-1)G_{all} \quad (2)$$

If we add a delay d variable to Eq. (2), then it can be written as:

$$PRTT(n,d,s) = PRTT(1,0,s) + (n-1) \times \max\{o_s + d, G_{all}\} \quad (3)$$

Parameters of the model can be inferred from Eq. (1)–(3).

4.1 Measurement of o_s and o_r

Overhead would vary with message size, and we now show how to account over-head for different message sizes. First we rewrite Eq. (3):

$$\frac{PRTT(n,d,s) - PRTT(1,0,s)}{n-1} = \max\{o_s + d, G_{all}\} \quad (4)$$

Then choose a suitable d_G which meets $d_G \geq G_{all}$, we can derive o_s:

$$\frac{PRTT(n,d,s) - PRTT(1,0,s)}{n-1} - d_G = o_s \quad (5)$$

From Eq. (5), o_s can be measured from $PRTT(n, d_G, s)$ and $PRTT(1,0,s)$. d_G can be denoted by $PRTT(1,0,s)$, which guarantees $d_G \geq G_{all}$. If a network with a very low latency L and a very high gap g, we choose $d_G = PRTT(2,0,s)$ to guarantee $d_G \geq G_{all}$, which would unnecessary prolong the measurement time. The above assessment method of have two kinds of advantages: on the one side, o_s doesn't depend on other parameters and only relies on $PRTT$ which would reduce error greatly. Furthermore, it only needs a small number of messages and thus does not saturate or flood the network.

4.2 Measurement of g_{CPU} and $G_{CPU-CPU}$

From Eq. (2), we deduce that:

$$\frac{PRTT(n,0,s) - PRTT(1,0,s)}{n-1} = G_{all} = (s-1)G_{CPU-CPU} + g_{CPU} \quad (6)$$

We can get a linear function of $G_{CPU-CPU}$ and g_{CPU}, namely $f(s) = (s-1)G_{CPU-CPU} + g_{CPU}$. For two differents, we can obtain the values of $G_{CPU-CPU}$ and g_{CPU} from the known $PRTT(n, 0, s)$ and $PRTT(1, 0, s)$. However, different networks would lead deviations for different data sizes. It's hard to detect the impaction of protocol changes in low levels when using this method. We chose many different s, and fit a linear function to the values (such as using least square method). The slope of (6) represents $G_{CPU-CPU}$, and $g_{CPU} = f(1)$. The above assessment of $G_{CPU-CPU}$ and g_{CPU} have two kinds of advantages: We only use a small n to benchmark each single message, thus, it wouldn't arose network flooding or additional overload and our results are not influenced by anomalies for specific message size. Further-more, we are able to detect changes of the underlying communication protocol.

4.3 Measurement of $l_{CPU-CPU}$

In order to minimize the RTT, on modern networks, it's difficult to accurately access $l_{CPU-CPU}$, We use $PRTT(1, 0, 1)/2$ to represent $l_{CPU-CPU}$.

4.4 Assessment of Parameters Relate to GPU

Five parameters relate to the GPU part, which are $l_{CPU-GPU}$, $l_{GPU-CPU}$, $G_{GPU-CPU}$, $G_{CPU-GPU}$ and P_G. P_G represents the computing capacity of GPU; as to $G_{GPU-CPU}$ and $G_{CPU-GPU}$, we use $\frac{1}{G_i}$ to describe the bandwidth of different communication media. The reciprocal of the bandwidth can be used to denote G; as to $l_{CPU-GPU}$, $l_{GPU-CPU}$, they can be directly assessed like the measurement of $l_{CPU-CPU}$.

5 Results

5.1 The Architecture of TH-1A System

We adopt the TH-1A system [14] as our experimental platform. The compute subsystem is constructed with 7168 compute nodes, each of which includes two Intel six-core X5670 CPUs and one Tesla M2050 GPU. The communication subsystem contains two types of networks, namely, the proprietary high speed interconnection net-work and GigaEthernet. The former is responsible for the data exchange between all nodes and has the characteristics of low latency (1.57us) and high bandwidth, while the latter collects the status of each component and assists in monitoring the whole system.

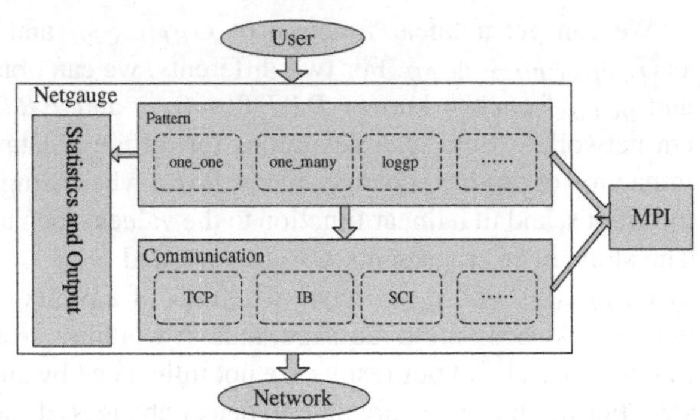

Fig. 3 Netgauge's patterns and modules

5.2 Testing of Parameters Relate to CPU Communication

Netgauge [8] is a high-precision network parameter measurement tool. It supports benchmarking of many different network protocols and communication patterns. The general framework is shown in Fig. 3. The model defines many communication methods, such as one_one, one_many, loggp etc., which are based on TCP, IB and SCI network protocols. We make some modification for Netgauge tool, CPU communication parameters can be obtained accurately through the method mentioned on the above section, several tests are launched for different message lengths, transmission intervals, and RTT of vary number of messages transmission, that are derived from the formula mentioned above.

In the process of measuring the parameters, the processes are divided into two sets (S_{client} and S_{server}, where $S_{client} \bigcap S_{server} = \emptyset$), for each process $P_c \in S_{client}$ has a unique corresponding process $P_s \in S_{server}$. As the client process P_c is concerned, it first calls the high precision time measurement function to get the timestamp t_1, then P_c sends n times d_siz bytes message to P_s and receive a feedback message from the P_s, at last, P_c calls the high precision time measurement function to get timestamp $t2$ and the whole time overhead of communication operation. Correspondingly, after received all the messages successfully, P_s sends a d_siz bytes messages to P_c. All operations are launched at the client end throughout the testing process.

Table 1 lists the real value of the parameters under different message length, we use $\frac{PRTT(1,0,1)}{2}$ to approximate $l_{CPU-CPU}$, which doesn't depend on message size. The received overhead is defined as the time of handling MPI_Recv (), which is of low accuracy. When the message size is less than 8 KB, the inter-node latency $l_{CPU-CPU}$ between CPUs is relatively small, and increase slowly, when the message size is greater than 8 KB, the latency $l_{CPU-CPU}$ grows rapidly.

As to the parameter g, its value is far less than o_r and o_s under most message sizes, when a message gets ready, the last message has been sent to the network, so there is no conflict if messages sent consecutively. The parameter $1/G$ characterizes the main communication network bandwidth. It can be drawn from the Table 1 when

Table 1 Inter nodes CPU communication parameter values

Size	$l_{CPU-CPU}$(us)	Os(us)	Or(us)	G(us)	H2H (MB/s)
1 B	1.27	0.05	0.16	nan	4.77
1 KB	1.51	0.15	0.30	0.17	2400.69
2 KB	1.73	0.26	0.41	0.18	2516.31
3 KB	1.89	0.31	0.53	0.18	2734.07
4 KB	2.00	0.30	0.64	0.19	3076.45
5 KB	2.17	0.35	0.76	0.20	2597.44
6 KB	2.35	0.41	0.88	0.21	2875.54
7 KB	2.53	0.47	0.99	0.22	3215.26
8 KB	2.71	0.52	1.13	0.23	3608.96
16 KB	6.86	2.04	3.76	0.43	3963.60
32 KB	12.20	3.41	7.73	0.18	4228.27
64 KB	16.38	6.52	8.80	0.19	4507.06
128 KB	27.16	10.99	15.11	0.18	4538.72
256 KB	63.47	25.38	37.03	0.19	4502.77
512 KB	139.04	60.47	77.51	0.19	4531.63
1 MB	252.56	113.10	138.40	0.20	4554.48
2 MB	587.66	276.68	309.92	0.17	4496.73

the sending message is greater than 64 KB, the network basically reaches its full bandwidth state. The peak bandwidth is approximately 5 GB/s.

5.3 *Testing of Latency and Bandwidth Relate to GPU*

We choose ping-pong strategy in the latency test. The sender sends a message of certain size to the receiver, and waits for a response from the receiver. After the receiver receives the message, it sends a similar size data to the sender. Repeated the ping-pong test many times, and calculate the average one-way latency. Some blocking function (MPI_Send, MPI_Recv) was used for testing. In bandwidth test, the sender sends a fixed number of (equal to the window size) back-to-back two-way messages to the receiver, and waits for response from the receiver. The receiver sends a response message to the sender until receipt all messages. Repeated the test many times, the bandwidth depending on the time overhead and the number of bytes sent, the purpose is to determine what sustained data transmission rate achieved. Therefore, non-blocking MPI function (MPI_Isend, MPI_Irecv) is adopted.

Since the parameters in this section relate to GPU operations, so the corresponding atomic communication (cudaMcpyDTH or cudaMcpyHTD) is occurs within a single node and the entire communication process is controlled by the originator.

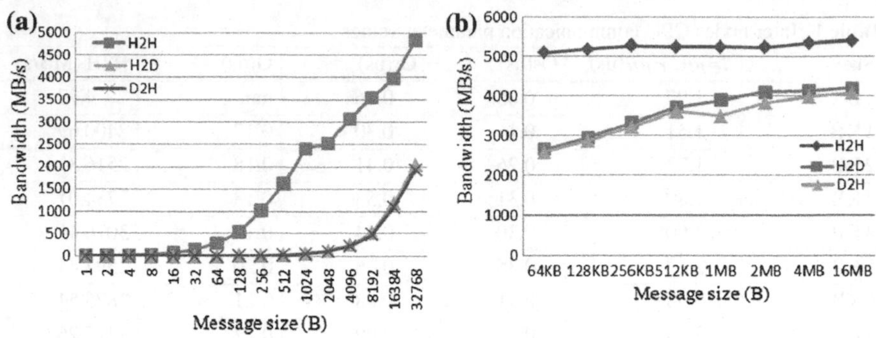

Fig. 4 Bandwidth values for different message lengths. **a** Short messages. **b** Long messages

For D2H (*D* represents device and *H* represents host) bandwidth test, at first *n* times cudaMcpyDTH operation was performed, and then perform a cudaMcpyHTD operation; accordingly, the H2D bandwidth test is constructed by *n* times cudaMcpyHTD operations and one cudaMcpyDTH operation. Figure 4 shows results of message bandwidth test (H2H means host to host, H2D indicates host to device), (a) results for short messages show that the overall efficiency of H2H bandwidth is higher than H2D and D2H bandwidth; for GPU data transmission, the reached bandwidth of D2H is higher than H2D bandwidth; for small messages, the bandwidth of devices is mainly limited by the overhead of starting devices. (b) When the message size is greater than 32 KB, the gained stable bandwidth basically reaches 5 GB/s or so, as the GPU is concerned, the maximum bandwidth can only be gained when the message size is larger than 2 MB, so when it comes to frequent short message communications, an efficient approach is to aggregate multiple small messages into a big one, thus reduce communication overhead.

Table 2 shows the results of latency test, there is a great difference of data communication latency between intra-node and inter-nodes. for GPU data transmission, the latency of D2H data transmission is lower than H2D data transmission; intra-node latency is one order of magnitude lower than inter-nodes latency; when message size is small, CPU to CPU data communication latency is around 2 s, but it is greater than 20 s for CPU to GPU data transmission, the reason is that the overhead of starting GPU device is greater than CPU. The results is consistent with the above mentioned bandwidth test results.

So far, we have mapped the model to the TH-1A system, and get the value of 10 abstracted uncertain parameters that better characterize the existed GPU clusters, after a deliberate analysis of these parameters, the model can be used to effectively guide algorithm design for CPU/GPU heterogeneous clusters.

Table 2 Latency values under different message length

Message(B)	CPU–CPU (1Node)(us)	CPU–CPU (2Nodes)(us)	CPU–GPU (1Node)(us)	GPU–CPU (1Node)(us)
1 B	0.2	1.95	21.43	19.85
8 B	0.22	2.01	21.49	21.44
64 B	0.24	2.87	22.74	21.97
512 B	0.31	4.42	23.56	23.63
4 KB	0.8	8.33	30.24	30.99
16 KB	2.93	17.85	35.68	34.87
64 KB	8.27	57.62	58.76	58.01
256 KB	31.03	150.46	81.26	79.6
512 KB	59.93	308.51	142.65	134.79
1 MB	116.68	589.68	397.23	385.42
2 MB	239.5	1025.73	720.62	701.84
4 MB	512.28	2137.81	1526.38	1427.32

6 Conclusion

This paper put forward a parallel computing model-VlogGP to simulate communication operation and computing capacity of CPU/GPU cluster. Compared with the existing models MMGP, HlogGP, mPlogP, the proposed VlogGP model can accurately portray the characteristics of inter and intra node communication and node computing capacity for CPU/GPU heterogeneous cluster. We deduce how to obtain the real value of each parameter. The model is based on the current CPU/GPU heterogeneous architecture, communication between CPU and GPU must interact through PCIe bus. Our future work is to map the proposed model to architecture like AMD fusion processor. Furthermore, we will further study the complicated storage hierarchy and the parameter configuration of hybrid programming model for CPU/GPU heterogeneous cluster.

Acknowledgments This work is supported by National Natural Science Foundation of China (Grant No. 61572510, 61502511) and China National Special Fund for Public Welfare (Grant No. GYHY201306003).

References

1. Bar-Noy A, Kipnis S (1992) Designing broadcasting algorithms in the postal model for message-passing systems. In: Proceedings of the fourth annual ACM symposium on parallel algorithms and architectures. ACM, pp 13–22
2. Benchmarks-PMB P.M. (2000) Part mpi-1. Pallas GmbH

3. Blagojevic F, Feng X, Cameron KW, Nikolopoulos DS (2008) Modeling multigrain parallelism on heterogeneous multi-core processors: a case study of the cell be. High performance embedded architectures and compilers. Springer, Berlin, pp 38–52
4. Bosque JL, Perez LP (2004) Hloggp: a new parallel computational model for heterogeneous clusters. In: IEEE international symposium on cluster computing and the grid, 2004. CCGrid 2004. IEEE, pp 403–410
5. Culler D, Karp R, Patterson D, Sahay A, Schauser KE, Santos E, Subramonian R, Von Eicken T (1993) LogP: towards a realistic model of parallel com-putation, vol 28. ACM
6. Eshaghian MM, Shaaban ME (1994) Cluster-m parallel programming paradigm. Int J High Speed Comput 6(02):287–309
7. Fortune S, Wyllie J (1978) Parallelism in random access machines. In: Proceedings of the tenth annual ACM symposium on theory of computing. ACM, pp 114–118
8. Hoefler T, Mehlan T, Lumsdaine A, Rehm W (2007) Netgauge: a network performance measurement framework. In: HPCC, vol 7. Springer, Berlin, pp 659–671
9. Hoefler T, Schneider T, Lumsdaine A (2009) Loggp in theory and practice-an in-depth analysis of modern interconnection networks and benchmarking methods for collective operations. Simul Model Prac Theory 17(9):1511–1521
10. Li L, Zhang X, Feng J, Dong X (2010) mplogp: a parallel computation model for heterogeneous multi-core computer. In: 2010 10th IEEE/ACM international conference on cluster, cloud and grid computing (CCGrid). IEEE, pp 679–684
11. Morin P (1998) Coarse grained parallel computing on heterogene-ous systems. In: Proceedings of the 1998 ACM symposium on applied computing. ACM, pp 628–634
12. Valiant LG (1990) A bridging model for parallel computation. Commun ACM 33(8):103–111
13. Williams TL (2000) A general-purpose model for heterogeneous computation. Ph.D. thesis, Citeseer
14. Xie M, Lu Y, Liu L, Cao H, Yang X (2011) Implementation and evaluation of network interface and message passing services for tianhe-1a supercom-puter. In: 2011 IEEE 19th annual symposium on high performance interconnects (HOTI). IEEE, pp 78–86

Unmanned Surveillance System Based on Radar and Vision Fusion

Chaoqi Ma, Wanzeng Cai and Yafei Liu

Abstract Unmanned surveillance system is mainly composed of objects detection and classification tasks. Traditional methods which are mostly based on vision cannot obtain the velocity and distance information of the objects. So they suffer from high false alarm rate and miss alarm rate. In this paper, a new kind of unmanned surveillance system based on radar and vision fusion is designed and realized. The main two contributions of the paper are: (1) We first introduce the objects' area which is calculated by fusing the radar and vision data as a robust feature for objects detection and classification; (2) We proposed a new approach to detect human of different postures away from other objects. Experiments show the effectiveness and robustness of the proposed system.

Keywords Fusion · Moving objects detection · Objects classification

1 Introduction

It's of extremely demanding of security at key places such as airports, warehouses, reservoirs, nuclear power plants, borders lines, etc. The area of these places tend to be very broad. It's not possible for human on duty at 24 h for the reason of high cost and the variety of human factors (e.g. the staff's responsibility), which may greatly decrease the reliability of the system's security. So Unmanned Surveillance System based on various sensing devices, which can offer 24 h monitoring work and of low false alarm rate, is quite suit for ensuring the security of these senses.

C. Ma (✉) · W. Cai · Y. Liu
National University of Defense Technology, Changsha 410073, China
e-mail: machaoqi1990@hotmail.com

W. Cai
e-mail: zengzeng2016@sina.com

Y. Liu
e-mail: lyfustb@163.com

V.E. Balas et al. (eds.), *Information Technology and Intelligent Transportation Systems*, Advances in Intelligent Systems and Computing 455,
DOI 10.1007/978-3-319-38771-0_41

The current mainstream unmanned surveillance systems are mostly based on one of infrared [3], fiber optic vibration [7], radar [4] and video analysis [6]. So there are some basic technical defects for these systems. Infrared, fiber optic vibration and radar systems will cause miscarriage of justice because they cannot identify and classify the object, and affect the user experience. Leaves jitter, and other small animals straying into these systems are the reason of miscarriage of justice. Unmanned surveillance system based on video analysis is difficult to adapt to harsh environments due to its own visible defects. Therefore, unmanned surveillance system based on multiple sensor fusion become the hot topic in the research.

Aiming at overcoming the drawback of the misjudgment or poor environmental adaptability of single sensor systems, this paper researches unmanned surveillance systems based on radar and video fusion. The working process is described as follows: (1) radar sensor detects intrusion objects and calculate the objects' range and azimuth information, while video analysis subsystem analysis the invasion objective type and alarms based on some predefined rules. The system combines the capability of all-weather radar's high sensitivity for moving object detection with comprehensive object identification and classification capabilities of video analysis, to guarantee the system at a very low false alarm rate, greatly improving the system performance.

The paper is organized as follows: The second chapter briefly describes the structure of the proposed system; the third chapter elaborates how to integrate the object range of radar measurements with calculation process in the object area of video analysis; the fourth chapter describes the object classification algorithm based on radar and video features; experimental procedures and results analysis are presented in Chap. 4; conclusion is drawn at last.

2 Structure of the System

The main function of the systems is to alarming when detecting a suspicious object (typically human) appearing in the scenes at 24 h. The potential objects usually contain human, dog, cat and others. We need to detect human with various postures (e.g. standing, squatting) away from dogs, cats and other animals. The system consists of three functional modules, namely objects detection, objects identification and alarming module, as shown in Fig. 1.

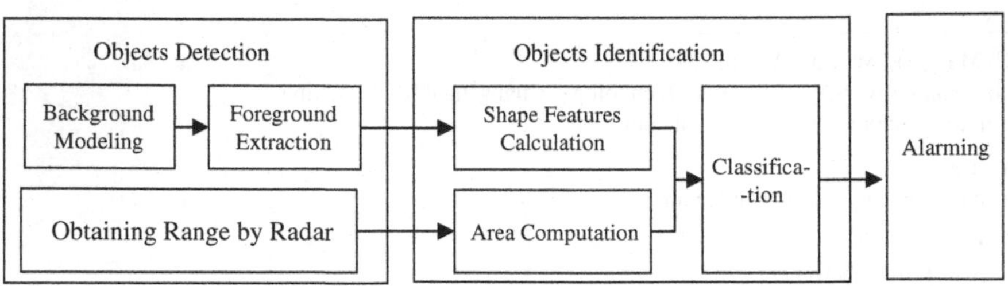

Fig. 1 Structure of the system

Objects detection module monitors whether there is any object appears in the scene and gets the location information of the object, which comes from video analysis and radar measurement. We use Gaussian Mixture Model to model the video's back-ground. And then the moving object's appearance is obtained by subtracting the current frame with the background model. Meanwhile, the paper also uses a millimetre-wave radar to obtain the object range and azimuth.

Objects identification module identifies the object's category based on its appearance and its distance information, which includes human, cats, dogs, etc. Particularly, the module calculates the shape features of the detected object, while computing its actual area based on the distance information measured by radar and its pixels, to generate the final feature vector. SVM classifier is used to classify the feature vector to obtain the precise category.

Alarming module gives an alarm when abnormalities is detected.

3 Object's Area Calculation by Imaging Principle and Radar Ranging

The goal of radar ranging is to get the objects' distance accurately. The objects' actual area is calculated by the imaging principle. In this manner the area that affected only by internal parameters of cameras and radar. But after calibration, the parameters' influence by be eliminated, which will lead to a better environmental adaptability.

As shown in Fig. 2, the relationship between the actual area S_T and after imaging area S_I of the object is:

$$S_T = \frac{u^2}{v^2} S_I \tag{1}$$

where u is the object's distance, is the camera focal length. Next, we give a brief proof to this relationship, and show why it's not influenced by the camera's internal parameters. The actual area of the object is

$$S_T = \int_0^{H_T} W_T\left(h_T\right)\, dh_T \tag{2}$$

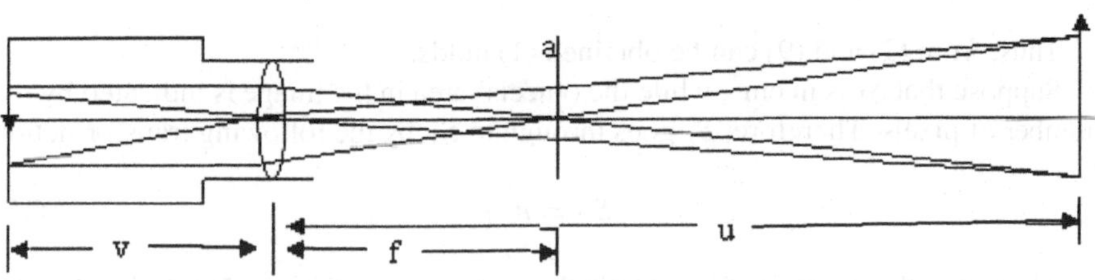

Fig. 2 The imaging principle of camera

where H_T is the actual height of the object, and $W_T\ (h_T)$ is the actual width at the height h_T of the object.

Area of the object in the image is

$$S_I = \int_0^{H_I} W_I\ (h_I)\ dh_I \tag{3}$$

where H_I is the height of the object in the image, and $W_I\ (h_I)$ is the in the actual width at the height h_I of the object.

According to the imaging principle, we have:

$$W_T\ (h_T) = W_T\left(\frac{u}{v}h_I\right) = \frac{u}{v}W_p\ (h_I) \tag{4}$$

$$H_I = \frac{v}{u}H_T \tag{5}$$

where h_I is corresponding to the height h_T, which is the actual height of the object in the image. Thus, the following is derived:

$$S_T = \int_0^{H_T} W_T\ (h_T)\ dh_T \tag{6}$$

where $h_T = \frac{u}{v}h_I$. Then we have

$$S_T = \int_0^{\frac{v}{u}H_T} W_T\left(\frac{u}{v}h_I\right)\ d\frac{u}{v}h_I \tag{7}$$

According to (3) and (7), the following can be derived

$$S_T = \int_0^{H_I} \frac{u}{v}W_I\ (h_I)\ d\frac{u}{v}h_I \tag{8}$$

and

$$S_T = \frac{u^2}{v^2}\int_0^{H_I} W_I\ (h_I)\ dh_I \tag{9}$$

Thus, from (2) and (9) can be obtained (1) holds.

Suppose that S_I is in cm^2, while the objects' area in the image is indicated by the number of pixels. Therefore, S_I goes through to S_P by the following transformation

$$S_P = nS_I \tag{10}$$

where n is the number of image pixels per square centimetre. It follows that the conversion relationship between the actual area and the pixel area of object

$$S_T = cu^2 S_P \tag{11}$$

where u is the distance from the object to the camera given by the radar, and $c = \frac{n}{v^2}$ is determined by the camera's internal parameters, which can be obtained after calibration. Thus, the whole transformation process is not influenced by external factors, making the measurement process of the object's area be of great environmental adaptability.

4 Object Classification by Novel Features

Object classification is a complex nonlinear problems. The methods using a single feature or single rule will suffer from great limitation. Currently most common object classification methods consist of feature extraction, and feature classification.

Pedestrian detection by histogram of gradient (HOG) feature and SVM classifier is one of the most popular algorithms [1], which achieves a very high detection accuracy. However, this algorithm is of high computational complexity, which make it hard to realize real-time performance. So the video's resolution has certain limitation. In addition, the SVM method based HOG lacks of better adaptability for deformation caused by non-rigid motion. Human and animals will show certain similarity with some postures and cannot be distinguished either. So single HOG+SVM can't complete the classification task.

Another method is based on the Haar feature, which uses cascade classifier to classify the objects. It got good results in detecting human organs [9]. Liu et al. applied this method to the traffic scene [5]. They extracted shape and motion characteristics of the objects, followed by a plurality of weak classifiers. The weak classifiers became a strong classifier and was used to perform object classification. However, this method is only effective when the object's feature does not change along with motion, or when lateral movement occurs. If there is longitudinal movement of the objects, the shape and size of the objects change a lot, resulting in the different features. So the classifier can't accurately classify the objects.

The problems have two-fold: on one hand, shape change caused by the non-rigid movement of humans and animals, makes the shape feature has a large range of variation. In some cases the shape of human beings and animals is even similar. On the other hand, longitudinal moving will leads to changing of the size of objects. So is difficult to compare and distinguish a feature vector of different objects. How to deal with these two main issues is the key to classify objects accurately. For the first problem, this article deals human of different posture as objects of different categories. For the second problem, we use radar and video fusion to get object's accurate area. Meanwhile, some features that are of translation invariance and scale invariance characteristics are used, such as aspect ratio, tightness, duty-factor, and so on. Classifiers based on these features have better adaptability for the change of environment, the object's longitudinal movement, etc.

Shape features of objects used in this paper include:

1. Aspect ratio. $\rho = \frac{h}{w}$, where h and w are the object's height and width of the circumscribed rectangle.
2. Duty-factor. $\delta = \frac{S_0}{S_r}$, where S_0 is the object occupying area, S_r is the object area of the circumscribed rectangle.
3. Tightness. $\varphi = \frac{c^2}{S_0}$, where c is the object's contour length. Tightness of object used to describe the complexity of the object's profile. In general, human and animals' tightness is higher than vehicles'.
4. The head and body width ratio $w = \frac{W_h}{W_b}$. The median width of upper $\frac{1}{7}$ of the object is considered as the head width W_h, while the median width of upper $\frac{1}{4}$ to $\frac{1}{2}$ of the object is considered as the body width W_b.
5. Invariant moments ϕ_t. Invariant moments is derived from the second and third order normalized center distance, which is of translation invariance, rotational in-variance and scaling invariance properties. It is a typical feature to describe the shape of the object. It is defined as follows:

$$\phi_1 = \eta_{20} + \eta_{02} \tag{12}$$

$$\phi_2 = (\eta_{20} - \eta_{02})^2 + 4\eta_{11}^2 \tag{13}$$

$$\phi_3 = (\eta_{30} - 3\eta_{12})^2 + (3\eta_{21} - \eta_{03})^2 \tag{14}$$

$$\phi_4 = (\eta_{30} + \eta_{12})^2 + (\eta_{21} + \eta_{03})^2 \tag{15}$$

$$\begin{aligned}\phi_5 = {} & (\eta_{30} - 3\eta_{12})(\eta_{30} + \eta_{12})\left[(\eta_{30} + \eta_{12})^2 - 3(\eta_{21} + \eta_{03})^2\right] \\ & +3(\eta_{21} - \eta_{03})(\eta_{21} + \eta_{03})\left[3(\eta_{30} + \eta_{12})^2 - (\eta_{21} + \eta_{03})^2\right]\end{aligned} \tag{16}$$

$$\begin{aligned}\phi_6 = {} & (\eta_{20} - \eta_{02})\left[(\eta_{30} + \eta_{12})^2 - (\eta_{21} + \eta_{03})^2\right] \\ & +4\eta_{11}(\eta_{30} + \eta_{12})(\eta_{21} + \eta_{03})\end{aligned} \tag{17}$$

$$\begin{aligned}\phi_7 = {} & (3\eta_{21} - \eta_{03})(\eta_{30} + \eta_{12})\left[(\eta_{30} + \eta_{21})^2 - 3(\eta_{21} + \eta_{03})\right] \\ & +(3\eta_{12} - \eta_{30})(\eta_{21} + \eta_{30})\left[3(\eta_{30} + \eta_{12})^2 - (\eta_{21} + \eta_{03})^2\right]\end{aligned} \tag{18}$$

where η_{ij} is $i + j$ order of the normalized center distance, which is defined as follows:

$$\eta_{ij} = \frac{\mu_{ij}}{\mu_{00}^r} \tag{19}$$

where $r = \frac{i+j}{2} + 1$, $i + j = 2, 3, \ldots$, and μ_{ij} is the center distance of the image $f(x, y)$:

$$\mu_{ij} = \sum\nolimits_x \sum\nolimits_y (x - \bar{x})^i (y - \bar{y})^j f(x, y) \tag{20}$$

6. Profile projection $[TH_k, TV_k, TH_{max}, TV_{max}]$. TH_k are ten value uniformly taken from the horizontal projection left to right, while TV_k are ten value uniformly taken from vertical projection top to down. In order to different object shape feature parameters with comparability, TH_k and TV_k need to be normalized. TH_{max} and TV_{max} is the subscripts of maximum value of TH_k and TV_k. TH_k and TV_k describe the shape of the top contour and the right contour, TH_{max} and TV_{max} describe the peak position of the top contour and the right contour [2].

The shape features described above and object area S determined previously are combined to form the feature vector:

$$X = [S, \sigma, \delta, \varphi, w, \phi_t, TH_k, TV_k, TH_{max}, TV_{max}] \tag{21}$$

According to analysis of the feature vector X above, samples of the same type are generally clustered in the feature space. So we use SVM classifier to learn the feature vectors. SVM is the statistical learning theory proposed by Vapink in 1990s [8], which is based on structural risk minimization principle, and is able to find an optimal hyperplane between the two categories of high-dimensional feature space.

But objects' non-rigid movement and its different location/orientation may cause dramatic changes in their features. The samples with different postures are aggregated not in one area of the feature space but multiple areas. Standing and squatting human are the most obvious. Their aspect ratio, profile projectors and other features show a big difference. Differences of these features will make the same type of samples in the feature space distributed in multiple areas. Therefore, simply using SVM to strike human, animal classification may cause larger error classification.

Based on the above analysis, in order to be able to accurately identify human under various postures, we propose to broken down human into two sub-category, that is human in standing and human in squatting. Both are trained by the classifier. The results show that with this method, the classification accuracy can be greatly improved.

5 Experiments and Analysis

In this section, we will first show the effectiveness of our method in actual area calculation. Then we will show the great improvement of classification accuracy by applying the area features and adopting the method by treating human with different gestures as different categories.

First, we uses a camera whose focal length is 4 mm, to measure pixel area of a white-board with the size 90 cm × 60 cm at different distances. Measured data is used to fit the Eq. (11). $S_T = 0.54\,\text{m}^2$ is the exact area value of the white-board area. The fitted curve is shown in Fig. 3. The results of the coefficients is $c_{4\,\text{mm}} = 1147500^{-1}$.

As can be seen from Fig. 3, except the case where the distance from the camera is too close, the fitting curve can perfectly reflect the relationship between the object

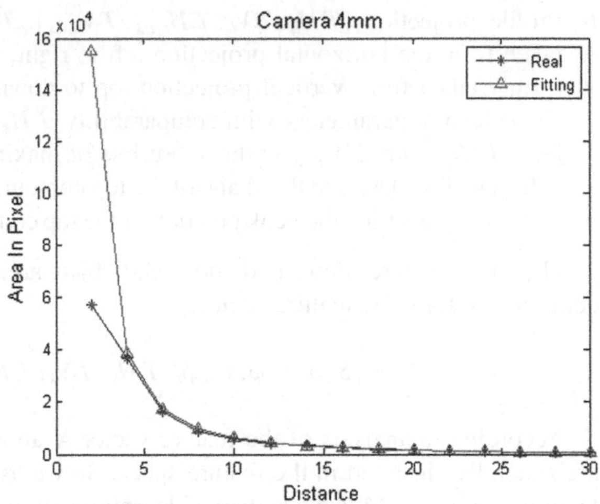

Fig. 3 Fitting results of the 4 mm camera ("Real" is the curve of actual pixel area about the distance, "Fitting" is the curve of the fitting)

distance and the pixel area. Thus, Eq. (11) can calculate the actual area of the object accurately from the pixel area.

For object classification, we first create a dataset by ourselves. Totally 2283 samples are collocated from the camera and radar, in which 1052 of them are human in standing, 295 of them are human in squatting, and the remain 936 samples are doges. 80 % of the total sample as training set, and the remaining 20 % as a test set for cross validation. The ground truth is manually segmented by ourselves. Table 1 shows our dataset's composition. Figures 4, 5 and 6 show parts of the training samples.

In order to verify the impact of area information for object classification, this paper in the following two feature space X and $X' = X - \{S\}$ is to train and test the samples. X include the actual area of objects, while X' not include. We don't subdivide people according to posture. We just divide the samples into two categories: human and dogs.

Table 1 Composition of our dataset

Category	Human in standing	Human in squatting	Doges	Total
Number	1052	295	936	2283
Percentage	46.08 %	12.92 %	41.00 %	100 %

Fig. 4 Parts of the training sample of standing people

Fig. 5 Parts of the training sample of squatting people

Fig. 6 Parts of the training sample of dog

Table 2 The results without the area size

Category	Human	Dog	Total
Accuracy	89.54 % ± 3.46 %	75.78 % ± 11.24 %	83.92 % ± 9.19 %

Table 3 The results without the area size

Category	Human	Dog	Total
Accuracy	82.72 % ± 10.35 %	90.40 % ± 5.24 %	85.80 % ± 7.29 %

First we use X' as feature vector, and the results shown in Table 2.

Then we use X as feature vector, and the results shown in Table 3.

As can be seen from the two comparative experiments above, after the addition of area characteristics, classification accuracy has been improved. The dog's classification accuracy was dramatically improved from 75.78 to 90.40 %. Because that area feature increases discrimination between squatting-human and dog, so the case that the dog is misclassified human is reduced.

In order to show the advantage of adopting the method by treating human with different gestures as different categories, two experiments are employed based on feature vector X. The first experiment breaks human into two sub-categories, while the second not. Note that in the first experiment, the results from human in standing and human in squatting will be combined as one category. Experiment results are

Table 4 Results considering human with different postures as different sub-categories

Class of samples	Human		Dog	Total
	Standing	Squatting		
Accuracy	76.24 % ± 13.45 %	43.31 % ± 18.64 %	96.08 % ± 3.29 %	79.95 % ± 5.66 %
Combination	98.35 % ± 0.53 %		96.08 % ± 3.29 %	97.52 % ± 1.68 %

shown in Table 4. We can see that, although individual classification accuracy rate of the different attitude of people is not high, but after ignoring the misclassification between different posture of people, the classification accuracy rate of humans and dogs has increased dramatically than Table 3.

The experiment results above are based on single frame analysis. In practice, the objects usually stay for some time within the monitoring range. So if we use several successive frames for detection and classification task, and combine the results in a statistical way, the system can further reduce the false alarm rate and improve the system's performance and robustness.

6 Summary

The paper designs a unmanned surveillance system based on radar and video analysis fusion. The radar and video cameras capture moving objects within the scope of monitoring, and alarm the abnormal situation after identifying judgment.

In the classification process, the paper uses measured object distance by radar as an additional input to accurately calculate the actual object area. The area feature is efficiently combined with shape features, and used to the classification task. This feature can deal with the object's longitudinal motion problem and non-rigid motion problem. Moreover, by treating human with different postures as different sub-categories, the system gets dramatically performance increase, and show great robustness. The proposed unmanned surveillance system can accurately capture anomalies within the scope of monitoring.

References

1. Dalal N, Triggs B (2005) Histograms of oriented gradients for human detection. In: IEEE computer society conference on computer vision and pattern recognition, 2005. CVPR 2005, vol 1. IEEE, pp 886–893
2. Hu Q (2008) Technique of motion detection and objects decision in visual surveillance. Master's thesis, National University of Defense Technology
3. Li CY (2014) The design and research of home security system based on gprs and infrared detection technology. Master's thesis, Hunan University

4. Li HM, Wang Y, Sun Jl (2012) Design and relization of remote monitoring system based on radar network. Mordern Electron Tech 35(1):31–33
5. Liu B, Zhou XH, Zhou HQ (2004) Vehicle detection and recognition in multi-traffic scenes. J Univ Sci Technol China 5:013
6. Ma Jf (2012) Research and implementation of security system with the remote video surveillance. Master's thesis, Chongqing University
7. Mahmoud SS, Katsifolis J (2009) Elimination of rain-induced nuisance alarms in distributed fiber optic perimeter intrusion detection systems. In: SPIE defense, security, and sensing. International society for optics and photonics, pp 731604–731604
8. Vapnik V (2013) The nature of statistical learning theory. Springer Science & Business Media
9. Viola P, Jones MJ (2004) Robust real-time face detection. Int J Comput Vis 57(2):137–154

Attribute Extracting from Wikipedia Pages in Domain Automatically

Fenglong Su, Chuanzhen Rong, Qingquan Huang, Jiyuan Qiu, Xinhong Shao, Zhenjun Yue and Qinghua Xie

Abstract In the age of Big Data, input determines output. There is a large amount of data on the internet, but little knowledge. So researchers develop different kinds of methods to automatically extract knowledge from different data platforms. The traditional methods of supervised learning cost more time and labor, which are willing to be gradually replaced by the semi-supervised and unsupervised learning methods. In this paper we proposed a new semi-supervised method to complete this task, which costs just little, called TSVM (Transductive Support Vector Machine). In order to improve the accuracy and the intelligent level, we also add the Word Embeddings to the semi-supervised method. The AP (Affinity Propagation) algorithm makes a contribution to the word clustering automatically. Experimental results demonstrate a better performance to extract the attribute information in the military transportation domain from the Wikipedia compared with the traditional supervised leaning method.

F. Su · C. Rong · Q. Huang · J. Qiu · X. Shao · Z. Yue
Institute of Communication Engineering, PLA University of Science and Technology, Nanjing, China
e-mail: xuexi2016@qq.com

C. Rong
e-mail: rcz@foxmail.com

Q. Huang
e-mail: 554834712@qq.com

J. Qiu
e-mail: jyqiu@qq.com

X. Shao
e-mail: 353335378@qq.com

Z. Yue
e-mail: zhenjuny@qq.com

Q. Xie (✉)
Institute of National Defense Engineering, PLA University of Science and Technology, Nanjing, China
e-mail: tougao2015@qq.com

V.E. Balas et al. (eds.), *Information Technology and Intelligent Transportation Systems*, Advances in Intelligent Systems and Computing 455,
DOI 10.1007/978-3-319-38771-0_42

Keywords Attribute extraction · Word embedding · Semi-supervised learning · AP algorithm · Information extraction · Word clustering

1 Introduction

Attribute is the natural properties of the entity. In domain it can describe the entity in detail. In the era of information explosion, we urgently need to understand an entity from all aspects. Such as a celebrity [1], his name, schools, birthplace, spouse, religion and so on. If we want to get useful information of an entity, attribute extraction is needed. There are a lot of scholars studying this issue. Such as, extracting school attributes, extracting product attributes, extracting character attributes and so on. They also have been making their research through different data platforms. Such as, business website [2], Wikipedia pages, twitters [3] and so on.

1.1 Information Extraction and Entity Attribute

IE (Information Extraction) is a base task in the NLP (Nature Language Processing) field which is a basic work serving for the high level tasks, such as Attribute Extraction, Relation Extraction, Event Extraction, Topic Tracking, Sentiment Analysis, and Intelligent Question-Answering. Attribute extraction contains several tasks in this work: named entity recognition, attribute recognition, and to determine whether they are related. According to the ACE (Automatic Content Extraction) and MUC (Message Understanding Conference), Attribute Extraction means that: extracting useful information from the structured or unstructured natural language textual carrier, consist of well-known entities, the relevant attributes and attribute values, and representing them normatively. It belongs to the shallow understanding of the text.

1.2 Word Embedding and Language Model

Deep Learning is a hotspot of research in recent years, which has a breakthrough in many fields: Image Classification, Speech Recognition and Text Mining. Word Embedding is a kind of unsupervised learning methods, which is based on deep learning framework. It has the advantages of low dimension, quick and accurate, when compared with the traditional VSM (Vector Space Model). In this paper, we make use of the RNNLM (Recurrent Neural Network Language Model) to train the word embeddings, which is faster and better than the Bengio's NNLM (Neural Network Language Model).

1.3 Affinity Propagation and Word Clustering

Affinity Propagation (AP) as a novel clustering algorithm was put forward in the journal of Science in 2007 [4]. It is based on the similarity between the N data points to cluster them. AP algorithm makes all data points passing two types of messages (Responsibility and Availability) to choose the data clustering center. When we get the word embeddings (vectors) from the language model in the large corpus, they can be clustered by the AP algorithm easily. AP algorithm continuously updates responsibility and availability values of each point in iterative process, until it generates some high-quality exemplars. Some variables have different meanings.

Exemplar: it refers to the clustering centers.

Similarity: **S** (i, j), the similarity of data points. In this paper we use Euclidean distance to calculate the similarity.

Preference: P (i) or **S** (i, i), the measure of the possibility of the point i become a clustering center. In this paper we use median similarity.

Responsibility: **R** (i, k), the measure of the suitability when point k choose point i to become its clustering center.

Availability: **A** (i, k), the measure of the suitability when point i choose point k to become its clustering center (Fig. 1).

1.4 Transductive Support Vector Machine and Semi-supervised Learning

Support Vector machine (SVM) is currently a popular research direction in machine learning, which has been proved to be a powerful learning machine model in many fields [5]. The traditional support vector machine is mainly used for supervised learning problems, but there are many semi-supervised learning problems in the reality. Transductive Support Vector Machine (TSVM) is a semi-supervised learning

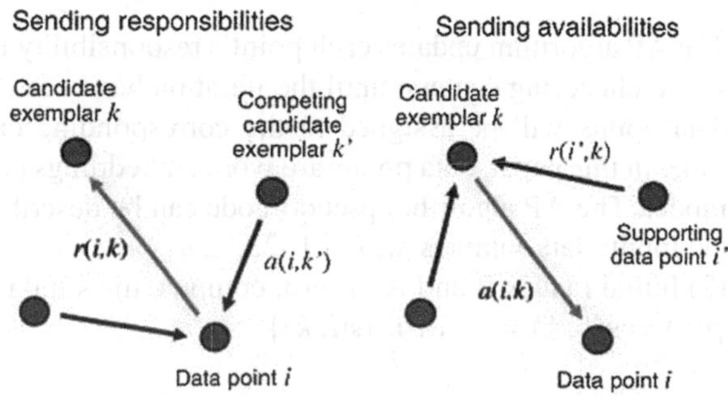

Fig. 1 Responsibility and availability of AP algorithm

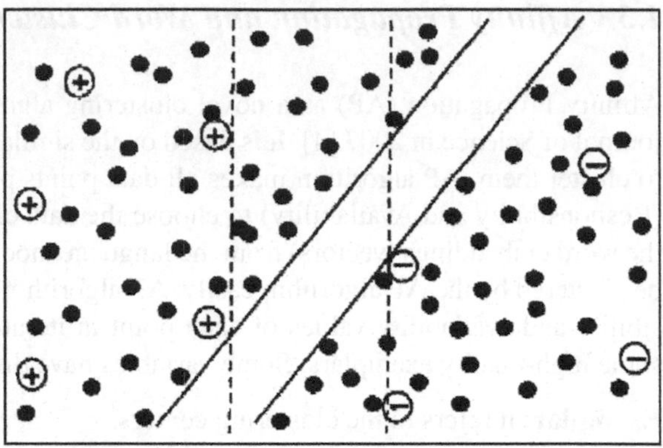

Fig. 2 Classification boundaries of TSVM and SVM

model which can infer unknown samples according to the known samples [6]. In this method we can generalize the classifier and reduce the cost. In this way, TSVM is just an extension of standard SVM algorithm to predict the unlabeled samples. Solving binary classification problems is to use the identified data in the sample space to find an optimal hyper-plane between the two classes of samples in the traditional SVM. But TSVM can use the labeled samples and unlabeled samples to find the optimal classification hyper-plane at the same time, which will maximize the space between the original labeled samples and unlabeled samples. The new found optimal classification boundary should meet the smallest generalization error for the classification of the original unlabeled samples. In Fig. 2, the dotted lines are the classification of the boundaries after standard SVM learning from the original labeled samples. The solid lines are the classification of the boundaries after TSVM learning from the original labeled samples and unlabeled samples. "+" and "−"dots: positive and negative labeled samples; black dots: unlabeled samples.

2 Algorithm and Model

2.1 AP Algorithm

The AP algorithm updates each point's responsibility R and availability A to choose stable clustering centers, until the iteration has been convergenced. The non-center data points will be assigned to the corresponding clustering centers at the same time. In this paper, data points are word embeddings (vectors) trained from language model. The AP algorithm pseudo-code can be described as these steps:

(1) Input: data samples $\mathbf{x}_i$, $i = 1, 2, \ldots, n$

(2) Initial matrix R and A to zero, compute the similarity matrix S

$$p(k) = \mathbf{s}(k, k) = \underset{i \neq k}{median}\{\mathbf{s}(i, k)\}$$

(3) Update matrix R, by rules:
$\mathbf{r}(i,k) \leftarrow \mathbf{s}(i,k) - \max_{k^* \neq k}\{\mathbf{a}(i,k^*) + \mathbf{s}(i,k^*)\}$
$\mathbf{r}(k,k) = p(k) - \max_{j \neq k}\{\mathbf{a}(k,j) + \mathbf{s}(k,j)\}$
(4) Update matrix A, by rules:
$\mathbf{a}(i,k) \leftarrow \min\{0, \mathbf{r}(k,k) + \sum_{i^* \notin \{i,k\}} \max[0, \mathbf{r}(i^*,k)]\}$
$\mathbf{a}(k,k) \leftarrow \sum_{i^* \neq k} \max(0, \mathbf{r}(i^*,k))$
(5) Do step (2) and (3), until $\mathbf{r}\,(k,k) + \mathbf{a}\,(k,k) > 0$, loop
(6) Put the data point i into the corresponding clustering center k
(7) Output: data clustering centers M_i, $i = 1, 2, \ldots, k$.

2.2 TSVM Classifier

It is crucial for machine learning methods to select features. We choose Word features, POS features, Entity category features, Ordinal features and Distance features in this paper. Relevant feature templates are in Table 1. Then we use the TSVM classifier which needs these features to build our model.
The TSVM classifier can be roughly divided into the following steps:
(1) Input: labeled samples: $(\mathbf{x}_1, y_1) \ldots (\mathbf{x}_n, y_n)$ unlabeled the samples: $\mathbf{x}_1^*, \ldots \mathbf{x}_k^*$
(2) Specify the parameters C and C^*, train an initial SVM classifier with labeled samples;
(3) Use the model to mark the unlabeled samples in step (2);
(4) Select some highest value of $(\mathbf{w} \cdot \mathbf{x}_j^* + b)$ from unlabeled samples and mark them "+1", the others mark "−1", and specify a temporary impact factor C_{temp}^*;
(5) Retrain the SVM classifier with labeled samples and unlabeled ones;
(6) Exchange a pair of marked values of different test samples, Loop, until it not make the value decrease in maximum in Formula (1);
(7) Increase the value of C_{temp}^* gradually and go back to step (6), until $C_{\text{temp}}^* > C^*$, Loop;
(8) Output: the values of unlabeled samples after an assignment: $y_1^*, y_2^* \ldots y_k^*$
Some parameters are explained in 2.3 Our Model

2.3 Our Model

We use the word embedding to cluster the attribute words firstly. Then the PC chooses some better results. In the next step, better results will be put into the TSVM classifier. In our work, these word pais which are called samples are the entities and their relevant candidate attribute keywords which both appear in the same sentences. In this paper our model can be explained:

Table 1 Feature set in this paper

Feature set number	Feature selection
$F1$	Word features
$F2$	POS features
$F3$	Entity category features
$F4$	Ordinal features
$F5$	Distance features

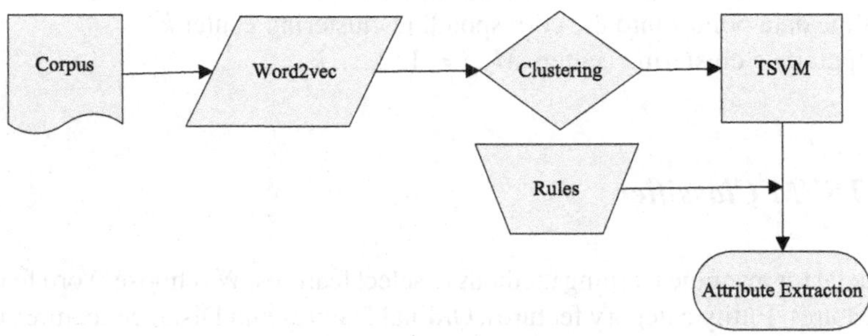

Fig. 3 Model flowchart

Given a set of training word pairs which are independent and identically distributed with labels $(\mathbf{x}_1, y_1) \ldots (\mathbf{x}_n, y_n)$, $\mathbf{x}_n \in \mathbf{R}^m$, $y_n \in \{-1, +1\}$ and another group from the same distribution of sample word pairs without labels: $\mathbf{x}_1^*, \mathbf{x}_2^*, \ldots \mathbf{x}_k^*$;

In linear inseparable condition generally, TSVM can be described as the following optimization problem:

$$\left.\begin{array}{ll} \text{Minimize over } (y_1^*, \ldots y_k^*, \mathbf{w}, b, \xi_1, \ldots \xi_n, \xi_1^*, \ldots \xi_k^*) : & \frac{1}{2}\|\mathbf{w}\|^2 + C\sum_{i=1}^{n}\xi_i + C^*\sum_{j=1}^{k}\xi_j^* \\ \text{Subject to :} & \forall_{i=1}^{n} : y_i(\mathbf{w}\cdot\mathbf{x}_i + b) \ge 1 - \xi_i, \xi_i \ge 0 \\ & \forall_{j=1}^{k} : y_j(\mathbf{w}\cdot\mathbf{x}_j^* + b) \ge 1 - \xi_j^*, \xi_j^* \ge 0 \end{array}\right\}, \quad (1)$$

C and C*: the parameters for the user to specify and adjust; ξ^*: impact factor in the process of training on unlabeled sample word pairs;*: to be labeled. Model flowchart is given in Fig. 3 simply.

3 Experiments and Discussion

We crawled some military transportation equipment articles from Wiki Encyclopedia pages and made them preprocessed simply. For example, Word Segmentation, Part-of-Speech Tagging, Phrasing and Named Entity Recognition. We use these tools:

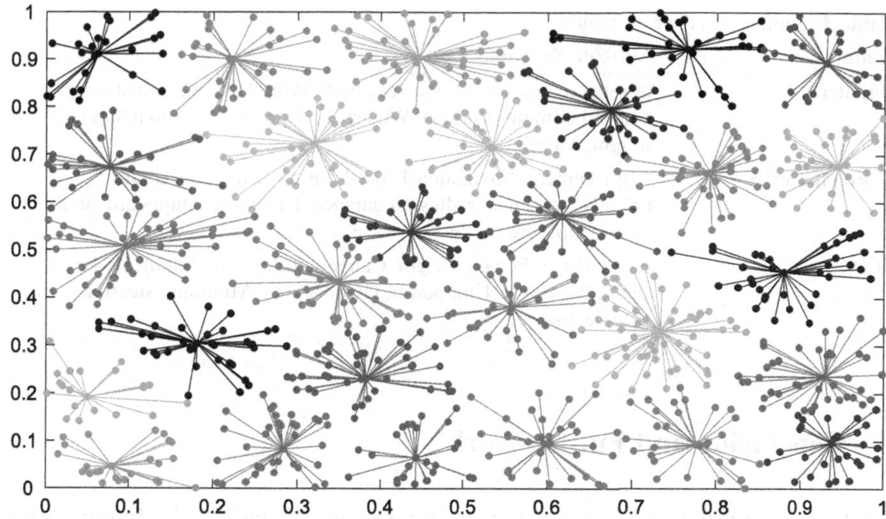

Fig. 4 Some words clustering by AP

Table 2 *F*-values of two algorithms in different numbers of samples (%)

Algorithm	Samples						
	20	80	200	500	1000	1500	2000
TSVM	63.4	54.3	61.8	64.5	63.8	65.4	60.2
SVM	0	15.3	28.8	35.4	43.1	65.1	64.0

Stanford NLP Toolkit, CRF++-0.58 Toolkit and SVM-lightV3.5. And the evaluation criteria is the F measure formula: $F = \frac{2RP}{R+P} \times 100\%$, ($R$- recall rate, P- precision rate).

Different dimensions of word vectors and different window sizes both have an effect on the attribute clustering. After several test, we decided to use 500 dimensions and 10 windows in the word2vec toolkit (in this work we use the Python gensim package). In the AP clustering process, we also find some related words. They are just related with the entities but not be the attribute keywords. The reason may be caused by the corpus and trained word vectors. Some words clustering by AP are show in Fig. 4. And TSVM has a better performance than SVM in a small number of data sets. When we give them 20 pairs of words, TSVM can achieve a beautiful result (F-value is 63.4). The experimental results are in Table 2.

After the above steps we filter and classify them with some rules. Part of the test results are shown in Table 3.

Table 3 Entities and their attributes

Entities	Attribute Keywords
Fighter	Combat radius, Flight, Rate of climb, Wing loading, Thrust-weight ratio, Maximum overload, Wingspan, Bare weight, Maximum take-off weight, Missile…
Aircraft Carrier	Displacement, Navigational speed, Firing range, Firing rate, Effectively combat radius, Endurance, Passenger, Shipboard aircraft, Ship beam, Power, Nuclear power…
Tank	Firing range, Power, Weight, Caliber artillery, Maximum stroke, Dip angle, Passenger, Unit power, Grade ability, Minimum steering radius, Vehicle armor…

4 Conclusion and Future Work

On the basis of previous work, we made some improvement simply. In this paper, Word Embedding combining with the AP algorithm are introduced into the words clustering study. After clustering we use expert dictionary to fitter them. Then we use TSVM method to predict the more sentences in the semi-supervised learning Framework. The experimental result is satisfactory. But there is also some shortage. High-quality and large-scale corpus is needed to further study; Clustering attribute words by AP is not accurate enough and we should extend and fitter them; TSVM model can be further optimized. In the future study, linguistic features and sentence structure will be taken into consideration. At the same time we will make use of newly unsupervised learning method in deep learning framework.

References

1. Zhang K, Wang M, Cong X (2014) personal attributes extraction based on the combination of trigger words, dictionary and rules. In: Proceedings of the third CIPS-SIGHAN joint conference on Chinese language processing, pp 114–119
2. Mauge K, Rohanimanesh K, Ruvini J-D (2012) Structuring e-commerce inventory. In: proceedings of the 50th annual meeting of the association for computational linguistics, pp 805–814
3. Li J, Ritter A, Hovy E (2014) Weakly supervised user profile extraction from twitter. In: proceedings of the 52nd annual meeting of the association for computational linguistics, pp 165–174
4. Frey BJ, Dueck D (2007) Clustering by passing messages between data points. Science 315:972–976
5. Vapnik V (1999) The nature of statistical learning theory. Springer, New York
6. Joachims T (1999) Transductive inference for text classification using support vectror machine. In: Proceedings of the sixteenth international conference on machine learning, Morgan Kaufmann, pp 148–156

Oscillation Source Detection for Large-Scale Chemical Process with Interpretative Structural Model

Yaozong Wang, Xiaorong Hu, Sun Zhou and Guoli Ji

Abstract In large-scale chemical processes involving a number of control loops, oscillations propagate to many units through control loops and physical connections between units. The propagation may result in plant-wide oscillation that is closely related to product quality, costs and accident risk. In order to detect the root cause of the plant-wide oscillation, this paper presents a new method that is based on interpretative structural model (ISM) that is established according to process topology. Firstly, a topological graph is obtained by considering process prior knowledge of the flowchart. Then according to the graph the adjacency matrix and reachability matrix are calculated. After that a multilayered structure of all control loops is obtained by establishing ISM. Thus the root causes of the oscillations are determined by propagation analysis. The superiority of this method is that the oscillation source can be effectively determined and clearly interpreted by the ISM. An application to a typical process from Eastman Chemical Company plant is provided to illustrate the methodology.

Keywords Plant-wide oscillation · Root cause · Fault diagnosis · Interpretative structural model

---

Y. Wang · X. Hu · S. Zhou (✉) · G. Ji
Department of Automation, Xiamen University, Xiamen 361005, People's Republic of China
e-mail: zhousun@xmu.edu.cn

Y. Wang
e-mail: ssyzwang@gmail.com

X. Hu
e-mail: 632662110@qq.com

G. Ji
e-mail: glji@xmu.edu.cn

V.E. Balas et al. (eds.), *Information Technology and Intelligent Transportation Systems*, Advances in Intelligent Systems and Computing 455,
DOI 10.1007/978-3-319-38771-0_43

1 Introduction

It is an important issue to quickly and effectively diagnose the faults taking place in industrial system, which are closely related to product quality, economic benefit and accident risk. Oscillations are a commonplace type of disturbance in chemical process. Ones occur at local area of the plant and diffusely propagate to the whole plant through control loops and physical connections between equipment. Most of process variables are caught in interminable oscillating disturbance. Therefore it is imperative to develop root cause diagnosis for the plant-wide oscillation.

Some detection method of oscillations is developed for large-scale control loops. Hagglund [1] first gave the concept of oscillation detection and a real-time detection approach using the integrated absolute deviation (IAE) was proposed. Jiang et al. [2] modified and extended spectral envelope to detect single or multiple oscillations from the view point of frequency domain. However, root cause diagnosis for oscillations is also crucial. Several possible reasons give rise to the oscillation, including valve stiction, nonlinearity of process, poor control system, ill-advised controller parameters and external oscillating disturbance [1]. Among these, the first one is often the main cause. Horch et al. [3] proposed a new method based on the cross-correlation between controller input and process output to differentiate whether oscillations stem from valve stiction. Jiang et al. [2] provided the oscillation contribution index (OCI) to evaluate the influence of process variables on oscillations. And an automated qualitative shape analysis (QSA) [4] was developed to locate root causes of oscillations, which classifies output signals according to wave information.

However, all the above methods are data-driven without utilization of prior knowledge, and most of ones are proposed for some specific causes. Furthermore, with an inherent limitation, it is hard to diagnose real causes employing pure data-driven methods when abnormalities contain unknown or multiple faults. For the past few years, graph-based methods for fault diagnosis have drawn extensive attention and achieved rapid development. Maurya [5] contributed an overview in regard to the signed digraph (SDG), and also discussed some potential applications of graph-based methods about fault diagnosis and process safety analysis. Chiang [6] incorporated cause-and-effect diagram and multivariate statistics to improve detection and diagnosis accuracy. A novel method called dynamic uncertain causality graph (DUCG) [7] was proposed to online diagnosis causes utilizing dynamic evidences. In addition, Thornhill [8] gave a comprehensive overview about oscillations detection and diagnosis and listed a few existing methods.

As a graph-based method, interpretative structural model (ISM) was first come up with by Warfield [9]. And it has gradually become a popular approach for analysis of relationships between variables within complex system. One is already applied into many fields such as business decision making [10] and strategy formulation for city development [11]. It is worth saying that ISM is knowledge-based with avoiding complicated calculation, and it is particularly suitable to diagnose root causes of oscillations in large-scale control loops. Therefore this paper adopts ISM to handle that problem. A main novelty is that we give a clearer visualized multilayered structure to describe propagative routes of oscillations using graph-based method.

2 Oscillation Source Diagnosis with ISM

In this section, the process topology and ISM are used to diagnose the plant-wide oscillation in chemical process. At first, prior process knowledge are considered to establish process topology, and the chemical process will be simplified as adjacency matrix and reachability matrix based on the process topology. And these two matrices contain the information of loops and interactions between these loops. Next, based on reachability matrix, a multilayered structure is acquired using ISM to discuss and diagnose root causes and propagations of oscillations.

2.1 Topological Graph of Process Flowchart

A complex system is consisted of a number of entities that can be physical components, devices, events or subsystems. Besides, relationships between entities are the part of complex system as well, including cause-and-effect, ordinal or correlational relationship etc. As a type of complex system, chemical process often involves large-scale control loops, and the plant-wide oscillation phenomenon always occurs. The main reason is that oscillations are propagated through control loops and physical connecting devices. To describe the chemical process and to obtain propagation information of oscillations, process topology is employed.

For a specific target chemical process, we need to establish process topology for the after oscillation diagnosis. First of all, we investigate process background information and determine system elements. Specifically, choose controllers as the element of chemical process. Then define each controller as node $s_i (i = 1, 2, \ldots, N)$ that also means control loop too. And analyze whether there has direct interaction between two adjacent controllers according to prior information. If the controller output (OP) of controller s_i can directly affect the process variable (PV) of s_j, a directed line segment is added to indicate this relationship. And, a diagraph of control loops, or called process topology, will be obtained that contains the information about all controllers and their interplays. A specific analysis for practice process will be introduced later.

2.2 Adjacency Matrix and Reachability Matrix

Besides using relational graph to describe complex system, relational matrix corresponding to the relational graph also can be used. In order to obtain the mathematical form of process topology, adjacency matrix is employed to describe interplays between every two controllers. Therefore, the $N \times N$ adjacency matrix based on process topology is given as

$$A = \begin{bmatrix} a_{1,1} & a_{1,2} & \dots & a_{1,N} \\ a_{2,1} & a_{2,2} & \dots & a_{2,N} \\ \dots & \dots & \dots & \dots \\ a_{N,1} & a_{N,2} & \dots & a_{N,N} \end{bmatrix} \quad (1)$$

where N denotes the number of controllers. And if here exists a direct interaction from controller s_i to s_j, the (i, j)th element of adjacency matrix A is assigned with a value of "1". If not, a value of "0" is set. And the information of all adjacent relationships are incorporated into A.

In order to acquire more relational information, we consider another relational matrix that is reachability matrix. Then we first define any matrix X satisfies the following Boolean equivalent [12]

$$X^{\#}(i, j) = \begin{cases} 0, & X(i, j) = 0 \\ 1, & X(i, j) \neq 0 \end{cases} \quad (2)$$

So according to (2), we have

$$(A + I)^2 = (I + A + A^2)^{\#} \quad (3)$$

and then $(A + I)^3, (A + I)^4, \dots, (A + I)^k$ are calculated as well. In the same way, one can be proved as

$$(A + I)^k = (I + A + A^2 + \dots + A^k)^{\#} \quad (4)$$

If there exists a constant k and the matrix A satisfies

$$(A + I)^{k-1} \neq (A + I)^k = (A + I)^{k+1} = \dots = (A + I)^N = R \quad (5)$$

the result R would be defined as reachability matrix. Therefore the reachability matrix that is also square matrix can be calculated by

$$R = (A + I)^N = (I + A + A^2 + \dots + A^N)^{\#} \quad (6)$$

and if the (i, j)th element of R has the value of "1", that means here exists a reachable route from controller s_i to s_j, or else here does not. For the large-scale control system with N loops, the maximum length of route is no more than $N - 1$. And if the values of all elements in reachability matrix are "1", it would indict that the OP of any controller can affect all PVs of the others. To sum up, the reachability matrix incorporates with both of direct and indirect relationships of controllers.

2.3 ISM for Large-Scale Control Loops

ISM is a structural modeling technology for analysis of entities and their relationships in system. By using prior knowledge of system structure, ISM can separate system into subsystems or system elements, and then reconstructs original system to multilayered description. More concretely, the relationships between entities are resolved into new hierarchical geometric model on the basis of system topology and matrix theory. Come to the concerned chemical process involving large-scale loops, ISM can be used to assign and classify control loops, and provide a final hierarchical structure that is helpful for a better understanding of loops interplay and oscillation propagation. ISM includes the following steps:

Step (a) Definition for sets. According to R, we define the reachable set and the antecedent set as

$$\begin{cases} P(s_i) = \{s_j | s_j \in S, r_{i,j} = 1, j = 1, 2, \ldots, N\} & i = 1, 2, \ldots, N \\ Q(s_i) = \{s_j | s_j \in S, r_{j,i} = 1, j = 1, 2, \ldots, N\} & i = 1, 2, \ldots, N \end{cases} \tag{7}$$

where the former includes the controllers that have reachable route from s_i. And the controllers in the latter have reachable route to s_i, and S is assumed as the set containing all controllers. After that, we also define a intersection as

$$T = P(s_i) \cap Q(s_i), i = 1, 2, \ldots, N \tag{8}$$

Step (b) Domain decomposition. Decompose R into many domains denoted as $\{D_1, D_2, \ldots\}$. Assume u and v are any two controllers, and u and v are in the same domain when $T(u) \cap T(v) \neq \varnothing$, or else they are in different domain when $T(u) \cap T(v) = \varnothing$.

Step (c) Interstage decomposition. This step is done in each foregoing domain. The core equation is $P(s_i) = P(s_i) \cap Q(s_i) = T, i = 1, 2, \ldots, N$, and a iteration is given as

$$\begin{cases} L_j = \{s_j \in D - L_1 - L_2 - \cdots L_{j-1} | P_{j-1}(s_i) \cap Q_{j-1}(s_i) = P_{j-1}(s_i)\} \\ P_{j-1}(s_i) = \{s_i \in D - L_0 - L_1 - L_2 - \cdots L_{j-1} | r_{i,j} = 1\} \\ Q_{j-1}(s_i) = \{s_i \in D - L_0 - L_1 - L_2 - \cdots L_{j-1} | r_{j,i} = 1\} \end{cases} \tag{9}$$

where L_j denotes the set containing all controllers at ith level, and $j \geq 2$. Repeat the interstage decomposition until $D - L_1 - L_2 - \cdots - L_{j-1} = \varnothing$, or else set $j = j + 1$.

Step (d) Establish multilayered structure. Resort all elements of R in line with the order $L_1, L_2, \ldots$ to obtain a sorted reachability matrix R. According to the sorted one, directed line segment between every two controllers at different level can be draw. And the final hierarchical structure is the needful description.

For all levels, the lower level can influence the higher one, but not vice versa. Put attention to oscillations, they can propagate from the lower level to the higher one through the one-way routes between every two levels, which leads to the plant-wide oscillation. And thats the core in this paper. An application example to practice process will be introduced later.

3 Application to Industrial Process

In this section, a practice industrial process is applied to demonstrate the effectiveness of the proposed method. A control loop diagraph is provided utilizing prior process knowledge, and then the adjacency matrix and reachability matrix are calculated. In the end, a true root cause of oscillations is diagnosed using the given hierarchical structure based on ISM.

3.1 Control Loop Diagraph of the Process

A typical process from Eastman Chemical Company plant [13] is considered and its schematic is shown in Fig. 1. The process has three distillation columns, two decanters and some recycle streams. And there are 14 PID controllers that have been numbered

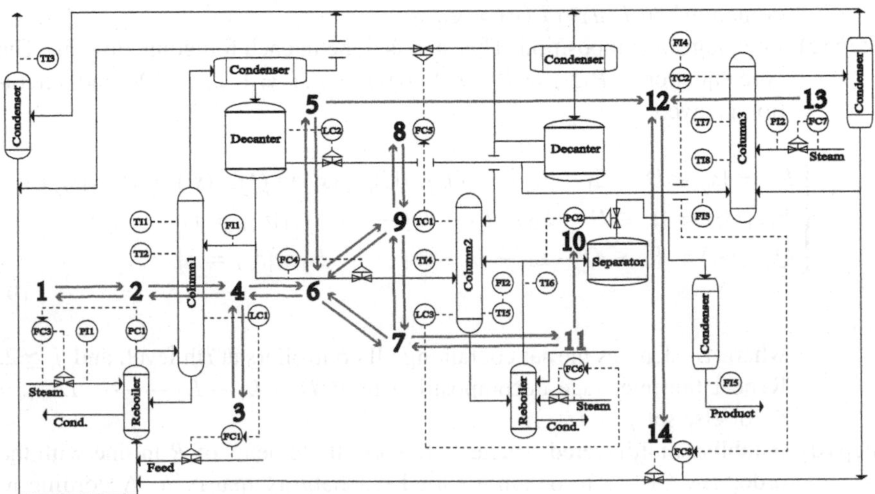

Fig. 1 Control loop digraph of the process from Eastman Chemical Company plant

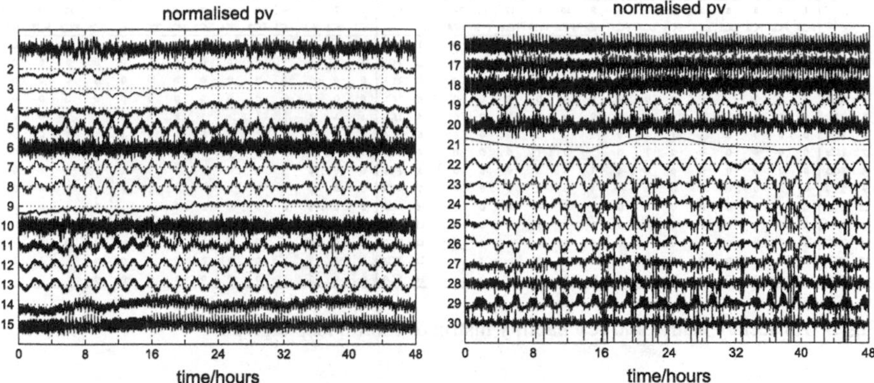

Fig. 2 Time series of the plant-wide disturbances in the process from Eastman Chemical Company plant [13]

from 1 to 14 on the schematic. Among these controllers, TC, LC, FC, PC and AC represent temperature, level, flow, pressure and composition controllers, respectively. In the same way, the indicators of temperature, level, flow, pressure and rotor speed are denoted as TI, LI, FI, PI and SI as well. In addition, the plant-wide oscillation was detected as shown in Fig. 2, and a data were from a data set provided by [13]. The industrial data tags 1–30 denote the PVs of controllers and indicators. Obviously, the most of PVs are periodical changed with short spiky deviations. Therefore the root cause diagnosis of oscillations is imperative to be developed.

The interactions between every two adjacent controllers of the process are analyzed using loop information on schematic. For instant, FC6 controls steam flowrate, and the steam runs from reboiler to separator so that the pressure of latter and PC2 will be influenced. Meanwhile, the steam leaves reboiler and then flows into distillation column 2 to produce liquid, and the leftover comes back to reboiler. Therefore there is certain interplay between LC3 and FC6. Figure 1 also gives control loop diagraph including both of unidirectional and bidirectional relationships. From another perspective, the diagraph also shows all potential propagative routes of oscillations. We do use the one to seek root causes of oscillations. And then adjacency matrix and ISM are employed to provide a clearer description for control loops.

3.2 Adjacency Matrix and Reachability Matrix

With an investigation of system elements and their relationships, the calculation of adjacency matrix is introduced. We collect, from Fig. 1, correlations between every two controllers into A. According to the aforementioned introduction, the value of "1" is assigned to the (i, j)th element of A when here exists a one-way interaction from s_i to s_j. If not, set the value as "0". Besides, all diagonal elements of A are

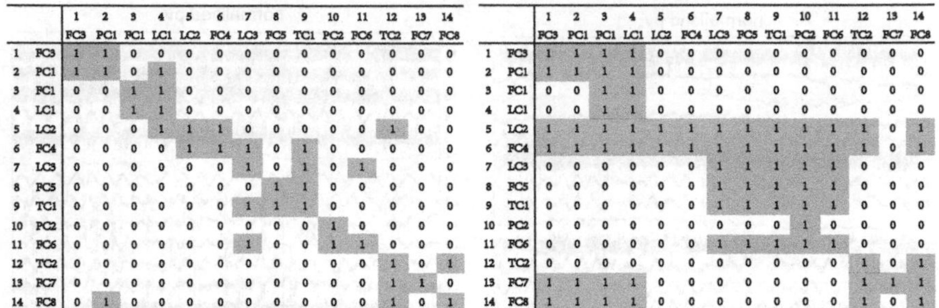

		1 FC3	2 PC1	3 FC1	4 LC1	5 LC2	6 FC4	7 LC3	8 FC5	9 TC1	10 PC2	11 FC6	12 TC2	13 FC7	14 FC8
1	FC3	1	1	0	0	0	0	0	0	0	0	0	0	0	0
2	PC1	1	1	0	1	0	0	0	0	0	0	0	0	0	0
3	FC1	0	0	1	1	0	0	0	0	0	0	0	0	0	0
4	LC1	0	0	1	1	0	0	0	0	0	0	0	0	0	0
5	LC2	0	0	0	1	1	1	0	0	0	0	0	1	0	0
6	FC4	0	0	0	0	1	1	1	0	1	0	0	0	0	0
7	LC3	0	0	0	0	0	0	1	0	1	0	1	0	0	0
8	FC5	0	0	0	0	0	0	0	1	1	0	0	0	0	0
9	TC1	0	0	0	0	0	0	1	1	1	0	0	0	0	0
10	PC2	0	0	0	0	0	0	0	0	0	1	0	0	0	0
11	FC6	0	0	0	0	0	0	1	0	0	1	1	0	0	0
12	TC2	0	0	0	0	0	0	0	0	0	0	0	1	0	1
13	FC7	0	0	0	0	0	0	0	0	0	0	0	1	1	0
14	FC8	0	1	0	0	0	0	0	0	0	0	0	1	0	1

		1 FC3	2 PC1	3 FC1	4 LC1	5 LC2	6 FC4	7 LC3	8 FC5	9 TC1	10 PC2	11 FC6	12 TC2	13 FC7	14 FC8
1	FC3	1	1	1	1	0	0	0	0	0	0	0	0	0	0
2	PC1	1	1	1	1	0	0	0	0	0	0	0	0	0	0
3	FC1	0	0	1	1	0	0	0	0	0	0	0	0	0	0
4	LC1	0	0	1	1	0	0	0	0	0	0	0	0	0	0
5	LC2	1	1	1	1	1	1	1	1	1	1	1	1	0	1
6	FC4	1	1	1	1	1	1	1	1	1	1	1	1	0	1
7	LC3	0	0	0	0	0	0	1	1	1	1	1	0	0	0
8	FC5	0	0	0	0	0	0	1	1	1	1	1	0	0	0
9	TC1	0	0	0	0	0	0	1	1	1	1	1	0	0	0
10	PC2	0	0	0	0	0	0	0	0	0	1	0	0	0	0
11	FC6	0	0	0	0	0	0	1	1	1	1	1	0	0	0
12	TC2	0	0	0	0	0	0	0	0	0	0	0	1	0	1
13	FC7	1	1	1	1	0	0	0	0	0	0	0	1	1	1
14	FC8	1	1	1	1	0	0	0	0	0	0	0	1	0	1

Fig. 3 Adjacency matrix and reachability matrix of control loop digraph

set the value of "1", which indicates that the controller OP can also affect its own PV. Adjacency matrix is given in Fig. 3 (left-hand side)and the highlight data is for a clearer look. Next, reachability matrix is calculated by (2) and (3). Figure 3 (right-hand side) also shows the one. From the adjacency matrix we can see that there are no reachable route from s_{10} to any other controllers. On the contrary, there always have reachable routes from s_5 or s_6 to others but s_{13}.

3.3 Multilayered Structure of Control Loops

After obtaining of reachability matrix, the step of ISM for all controllers is given. Figure 4 displays the specific procedure through four times repetition of step (c) in Sect. 2.3. The controllers that belong to level 1 are highlighted in the first subgraph, and the controllers in the second subgraph belong to level 2 and the like. Obviously, all controllers are assigned to four levels. In addition, the level of first decomposition is defined as the highest level so that level 4 is the lowest. Namely, controllers 5, 6

Level 1

	P{Si}	Q{Si}	P{Si}∩Q{Si}
1	1,2,3,4	1,2,5,6,12,13,14	1,2
2	1,2,3,4	1,2,5,6,12,13,14	1,2
3	3,4	1,⋯,6,12,13,14	3,4
4	3,4	1,⋯,6,12,13,14	3,4
5	1,⋯,12,14	5,6	5,6
6	1,⋯,12,14	5,6	5,6
7	7,8,9,10,11	5,⋯,9,11	7,8,9,11
8	7,8,9,10,11	5,⋯,9,11	7,8,9,11
9	7,8,9,10,11	5,⋯,9,11	7,8,9,11
10	10	5,⋯,11	10
11	7,8,9,10,11	5,⋯,9,11	7,8,9,11
12	1,2,3,4,12,14	5,6,12,13,14	12,14
13	1,2,3,4,12,13,14	13	13
14	1,2,3,4,12,14	5,6,12,13,14	12,14

Level 2

	P{Si}	Q{Si}	P{Si}∩Q{Si}
1	1,2	1,2,5,6,12,13,14	1,2
2	1,2	1,2,5,6,12,13,14	1,2
5	1,2,5,⋯,9, 11,12,14	5,6	5,6
6	1,2,5,⋯,9, 11,12,14	5,6	5,6
7	7,8,9,11	5,⋯,9,11	7,8,9,11
8	7,8,9,11	5,⋯,9,11	7,8,9,11
9	7,8,9,11	5,⋯,9,11	7,8,9,11
11	7,8,9,11	5,⋯,9,11	7,8,9,11
12	1,2,12,14	5,6,12,13,14	12,14
13	1,2,12,13,14	13	13
14	1,2,12,14	5,6,12,13,14	12,14

Level 3

	P{Si}	Q{Si}	P{Si}∩Q{Si}
5	5,6,12,14	5,6	5,6
6	5,6,12,14	5,6	5,6
12	12,14	5,6,12,13,14	12,14
13	12,13,14	13	13
14	12,14	5,6,12,13,14	12,14

Level 4

	P{Si}	Q{Si}	P{Si}∩Q{Si}
5	5,6	5,6	5,6
6	5,6	5,6	5,6
13	13	13	13

Fig. 4 Domain decomposition and interstage decomposition for control loop

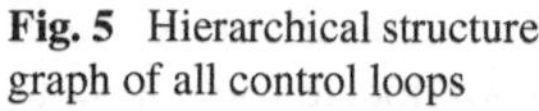

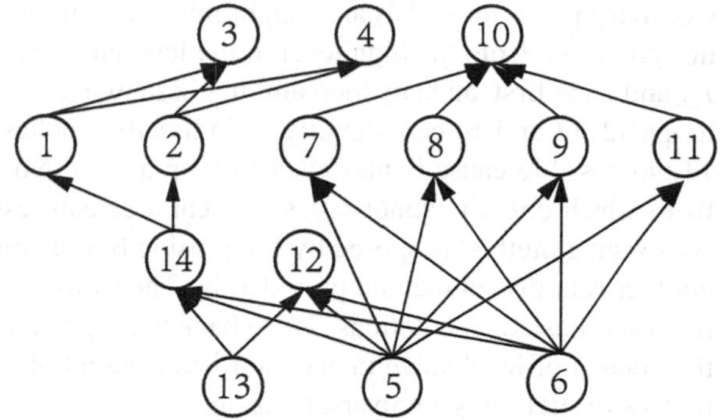

Fig. 5 Hierarchical structure graph of all control loops

and 13, or called control loops, at level 4 are truly underlying loops in the chemical process.

Then obtain resorted reachability matrix R in line with the order $L_1, L_2, \ldots,$ and draw directed line segment for every two controllers at different level. Note that the following two rules should be followed. First one is that there is no directed line segment if two controllers belong to same level such as controllers 5 and 6. The another one suggest that if there, for the controller at current level, is no interaction with any controller at next level, then go to the level after next to find. For a simple example, controllers 14 and 12 that are non-connective to controller 7 represent all of the controllers in L_3 so that controller 7 only builds relationship with controllers 5 and 6 at level 4. A final description of multilayered structure is shown in Fig. 5. We can see that the process based on large-scale control loops is consisted of four-level hierarchical structure, and filtered relationships of loops are distinctly displayed. With the one-way interaction from the low level to the high one, the controllers in L_4 impact the controllers in L_1 through such reachable routes, but not vice versa. Put attention to root causes diagnosis, oscillations also propagate to every level through these routes.

Compare this hierarchical structure with Fig. 1 by the way, we have some interesting finds. Loops 5 and 6 can reach all loops but 13 with the most widely influence, which is may well be the reason that these two are assigned to underlying level. But, there is no any route to loop 13 and that also may be the real reason why loop 13 is at the level 4. On the contrary, loop 10 deserves the highest level because there is no route from itself to other loop.

3.4 Root Cause Determination

Based on previous parts, now discussion and diagnosis of oscillation root causes are given. For example, if only control loops 1, 2, 3 and 4 have oscillations, loop 1 or 2 would be the cause with analyzation of hierarchical structure, and we can get the same result by foregoing schematic. In the process from Eastman Chemical

Company plant, most PVs are caught into oscillations, and these oscillations would not propagate from the high level to the low one. Therefore we locate root cause to L_3 and L_4 at first, and any loop among these two levels is possible. However, any of loops 12, 13 and 14 cant directly or indirectly reaches any of loops 7, 8, 9, 10 and 11, so possible cause is narrowed to loop 5 or 6. No matter oscillations produced from which one, the abnormities in local area can result in plant-wide oscillation. An existing method proposed by Jiang using bi-coherence and PV-OP plot [14] and another data-driven method provided by Thornhill [13] all ensure that loop 5 is the real root cause of oscillations. Trace back to loop 5 in practical process and find that the stiction of level valve in decanter is the essential cause. That also demonstrates that valve stiction is a universal reason.

4 Conclusion

This paper investigates oscillation propagation problem in large-scale control loops. A topology-based method is used with utilization of target process information and design core of control system. In order to diagnose root cause of oscillations quickly and effectively, ISM is developed. Adjacency matrix and reachability matrix are calculated in proper order based on the diagraph. And then all controllers are assigned to specific level. As a final result, a hierarchical structure description containing all loops and their interactions is given. The main advantage of the proposed graph-based method is that the cause diagnosis of oscillations does not involve complicated calculation only using prior process knowledge. In addition, it is novel that the proposed method provides a clearer hierarchical structure for control loops. A process from Eastman Chemical Company plant is developed to demonstrate the effectiveness of the aforementioned method.

Acknowledgments This work was supported by the National Natural Science Foundation of China (Nos. 61304141, 61573296), Fujian Province Natural Science Foundation (No. 2014J01252), the Specialized Research Fund for the Doctoral Program of Higher Education of China (No. 20130121130004), the Fundamental Research Funds for the Central Universities in China (Xiamen University: Nos. 201412G009, 2014X0217, 201410384090, 2015Y1115) and the China Scholarship Council award.

References

1. Hagglund T (1995) A control-loop performance monitor. Control Eng Prac 3(11):1543–1551
2. Jiang HL, Choudhury MAAS, Shah SL (2007) Detection and diagnosis of plant-wide oscillations from industrial data using the spectral envelope method. J Process Contr 17(2):143–155
3. Horch A, Isaksson AJ, Forsman K (2002) Diagnosis and characterization of oscillations in process control loops. Pulp Pap-Canada 103(3):19–23
4. Rengaswamy R, Hagglund T, Venkatasubramanian V (2001) A qualitative shape analysis formalism for monitoring control loop performance. Eng Appl Artif Intel 14(1):23–33

5. Maurya MR, Rengaswamy R, Venkatasubramanian V (2003) A systematic framework for the development and analysis of signed digraphs for chemical processes. 1. Algorithms and analysis. Ind Eng Chem Res 42(20):4789–4810
6. Chiang LH, Braatz RD (2003) Process monitoring using causal map and multivariate statistics: fault detection and identification. Chemometr Intell Lab 65(2):159–178
7. Zhang Q, Geng SC (2015) Dynamic uncertain causality graph applied to dynamic fault diagnoses of large and complex systems. ieee t reliab 64(3):910–927
8. Thornhill NF, Horch A (2007) Advances and new directions in plant-wide disturbance detection and diagnosis. Control Eng Prac 15(10):1196–1206
9. Warfield JN (1974) Developing subsystem matrices in structural modeling. IEEE Trans Syst Man Cybern smc 4(1):74–80
10. Xiao DD, Zhu GL, Xu Z (2009) Analysis of influence factors of corporate technology innovation based on interpretative structural model. In C Ind Eng Eng Man, pp 1943–1947
11. Liu JY, Zhu MZ (2011) Analysis of the factors influence on urban economic development based on interpretative structural model. Adv Intel Soft Compu 106:347–351
12. Mah RSH (1982) Chemical process structures and information flows. Chem Eng Prog 78(7):84–89
13. Thornhill NF, Cox JW, Paulonis MA (2003) Diagnosis of plant-wide oscillation through data-driven analysis and process understanding. Control Eng Prac 11(12):1481–1490
14. Jiang HL, Patwardhan R, Shah SL (2009) Root cause diagnosis of plant-wide oscillations using the concept of adjacency matrix. J Process Contr 19(8):1347–1354

Disease Candidate Gene Identification and Gene Regulatory Network Building Through Medical Literature Mining

Yong Wang, Chenyang Jiang, Jinbiao Cheng and Xiaoqun Wang

Abstract Finding key genes associated with diseases is an essential problem of disease diagnosis and treatment, and drug design. Bioinformatics takes advantage of computer technology to analyze biomedical data to help finding the information about these genes. Biomedical literatures, which consists of original experimental data and results, are attracting more attention from bio-informatics researchers because literature mining technology can extract knowledge more efficiently. This paper designs an algorithm to estimate the association degree between genes according to their co-citations in biomedical literatures from PubMed database, and to further predict the causative genes associated with a disease. The paper also uses hierarchical clustering algorithm to build a specific genes regulation network. Experiments on uterine cancer shows that the proposed algorithm can identify pathogenic genes of uterine cancer accurately and rapidly.

Keywords Literature mining · Co-citation · Pathogenic gene · Gene regulatory network · Hierarchical clustering

Y. Wang (✉) · C. Jiang · J. Cheng
School of Software, Beijing Institute of Technology, Beijing 100081, China
e-mail: wangyong@bit.edu.cn

C. Jiang
e-mail: jiangchenyang1@qq.com

J. Cheng
e-mail: 554091788@qq.com

X. Wang
Institute of Biophysics, Chinese Academy of Sciences, Beijing 100101, China
e-mail: xiaoqun.wang@gmail.com

V.E. Balas et al. (eds.), *Information Technology and Intelligent Transportation Systems*, Advances in Intelligent Systems and Computing 455,
DOI 10.1007/978-3-319-38771-0_44

1 Introduction

Finding genes critical to the formation and development of a disease from disease candidate genes are of significance to diagnosis and treatment of diseases, which is an important goal of bioinformatics [1]. Bioinformatics technologies are used to manage and analyze mass data generated by biomedical experiments, and to give some predictions or directive conclusions for researchers further works. In recent years, a new branch of bioinformatics, literature mining [2] is more and more used to partially replace biomedical experiments, and finally to accelerate the whole research process. Literature is the most important way to publish experimental data, experimental results and experimental conclusion, so it contains a lot of original knowledge that has been proved in experiments. Literature mining can automatically analyze the huge amount of biomedical literature, if combined with artificial verification, then the researchers can not only save a lot of manpower by greatly narrowing down the scope of artificial reading, but also can extract and discover knowledge more efficiently and can give reliable guide to biomedical experiments.

AI. Mubaid et al. [3] explore the relationship between genes and diseases by studying cooccurrence of genes and they find 6 candidate genes, all of which can be verified by PubMed literatures. Chun et al. [4] developed a named entity recognition and mining system based on maximum entropy for automatic identification of prostate cancer related genes, then apply it to Medline database with the accuracy of 92.1 %. Some researchers [5] take advantage of Protein–Protein Interaction (PPI) networks to predict disease related genes. So there are many PPI databases, such as HPRD (Human Protein Reference Database) [6], DIP (the Database of Interacting Proteins) [7] and OPHID (Online Predicted Human Interaction Database) [8]. Some researchers don't use these well-built databases, rather build their own gene networks. Ozgur et al. built a prostate cancer related gene network based on some seed genes through text mining, and only one candidate gene cannot be supported by literatures [9]. Many different algorithms have been proposed to evaluate gene networks, such as shortest path [10, 11], similarity assisted [12, 13] and centrality based [9, 14].

In this paper, we design an algorithm to predicate disease candidate genes by doing text mining on the PubMed biomedical literature database combined with human gene ontology database. We first compute the degree of the association between genes, and then predict the causative genes associated with diseases. Further, we use hierarchical clustering algorithm to build specific genes regulation network and analyze the network by taking the results of text mining into consideration. The experimental results show that the proposed algorithm can identify pathogenic genes of uterine cancer accurately and rapidly. Those less referenced genes (low-frequency genes) should get more attention in future research. This work can provide a good guide for the study in the field of biomedical area, and show the broad application prospects of text mining in the biomedical field as well.

2 System Design and Implementation

2.1 Getting Literature Data Source

First, we download the data of more than 20,000 genes encoding proteins in the human genome (exons) from NCBI Gene database (ftp://ftp.ncbi.nih.gov/gene/). The data contains the ontology of each gene and the corresponding gene ID, saved as CSV files. Then write a Java program to read the CSV file for each gene ontology symbol, and set the searching parameters through URL, and then search the literature which relate to genes ontology in the database through the E-utilities API, and finally retrieve XML flies from search results. For each gene, parse the XML files, retrieve the PMID of top 100 most related papers. Remove duplicate PMID and then add the PMIDs to a set. For each PMID, search and retrieve the XML files of corresponding literature in PubMed, parse these files to get the titles, abstracts and keywords. Finally we store them into a local MySQL database, and take each literature's PMID as the primary key. The article's title, abstract and keywords are data source for the following text mining.

2.2 Processing Literature Data

Identifying the gene ontologies. We use word segmentation method to retrieve gene ontologies in data source. There are many ways to represent gene ontologies in these literatures and some representations in early papers are nonstandard, so we need a collection of standard gene ontologies for reference before segmentation. In order to identify gene ontologies more easily, we make the following special preprocessing to the data source before segmenting it. First, read CSV files of all IDs of gene ontologies that describe human genome, then read CSV files of the corresponding gene ontologies and genes alias from NCBI GENE database. Store these files into memory, and add these files to a hash table named HashMap<String,String> sym2id, the key can be gene ontology or gene's alias, and the value is a standard gene ontology. This hash table will be referenced in following steps. When segment the data source, we first replace special symbols, such as '(', ')', '+', ',', '.', '[', ']', '<', '>', '?', '/', '\', '&' with space. Then process some fuzzy words, for example, spaces before and after was, an, and met should be removed. Remove "T cells" or "T cell" to prevent confusion in segmentation. For each segmented word, lookup in the hash table sym2id to determine whether the word is a certain gene, namely whether it is a gene ontology symbol (a standard gene ontology symbol or alias). If it is, it will be the only standard symbol of the gene ontology mapping for the gene, and will be stored in hash table, waiting for further processing. To evaluate the above identification algorithm, we select 100 papers randomly from the local database and identify gene ontologies in them, then evaluate the precision rate by manual verification. The accuracy (F1-Measure) of algorithm reaches more than 97 % as shown in Table 1. At the aspect

Table 1 Gene Ontology Identification Evaluation Results

No.	The number of correct gene ontology identification	The total number of identified gene ontologies	The total number of all gene ontologies in selected papers	Precision rate (PR)	Recall rate (RR)	F1-measure
1	698	707	731	0.987	0.955	0.971
2	1002	1007	1025	0.995	0.977	0.986
3	423	423	442	1	0.957	0.978
4	486	486	499	1	0.974	0.987
5	769	774	786	0.994	0.978	0.986
6	553	559	571	0.989	0.968	0.979
7	673	673	695	1	0.968	0.984
8	733	733	741	1	0.989	0.995
9	529	533	552	0.992	0.958	0.975
10	638	638	653	1	0.977	0.989

of performance, the whole process of identify gene ontologies in 20,000 papers only takes about 5 s.

In which, PR, RR and F1-Measure (the harmonic mean of PR and RR) are computed as in (1), (2) and (3) respectively.

$$PR = \frac{the\, number\, of\, correct\, gene\, ontology\, identification}{the\, total\, number\, of\, identified\, gene\, ontologies} \tag{1}$$

$$RR = \frac{the\, number\, of\, correct\, gene\, ontology\, identification}{the\, total\, number\, of\, gene\, ontologies\, in\, sample\, papers} \tag{2}$$

$$0 < F1 - Measure = \frac{PR + RR}{2} < 1 \tag{3}$$

Calculating the correlation degree between genes. Both identifying candidate genes of diseases and building gene regulation networks need analyze specific genes correlation, so we design an algorithm based on co-occurrence of two genes in the same literatures to calculate quantitative correlation between genes.

First, calculate the eigenvalue of the nth gene ontology in the mth paper, i.e. $TFIDF_{mn}$, which represents the importance of each gene in each literature and the focus of a paper as well. $TFIDF_{mn}$ can be calculated as in (4), (5) and be normalized as in (6). In which, IDF (Inverse Document Frequency) represents the distribution of a gene ontology in a document, TF (Term Frequency) represents the frequency of a gene ontology in a literature. S_m represents the total number of gene ontologies in first m papers. The closer W_{mn} to 1, the more influence the gene will have on this paper.

$$IDF = \lg\left(\frac{the\,number\,of\,correct\,gene\,ontology\,identification}{the\,total\,number\,of\,identified\,gene\,ontologies}\right) \quad (4)$$

$$W_{mn} = TF_{mn} \times IDF_n \quad (5)$$

$$TFIDF_{mn} = \frac{TF_{mn} \times IDF_n}{\sqrt{\sum_{n=1}^{S_m}(TF_{mn} \times IDF_n)^2}} \quad (6)$$

If two genes both appear in a literature and their normalized eigenvalues are similar, then the two genes are likely to have similar influence in the literature. So we can find genes closely related to other genes by evaluate the similarity between the genes' normalized eigenvalues in all the documents.

After calculating all the normalized eigenvalues, we can get an $n \times m$ matrix **P**, where n represents the total number of gene ontologies in human genome, m represents the total number of papers associated with these gene ontology. If a gene ontology doesn't appear in a literature, the corresponding W_{mn} is set to 0. Then the ith row of **P** is the eigenvalue vector of the ith gene, denoted as $\boldsymbol{A}_i = [W_{1i}, W_{2i}, \ldots, W_{mi}]$. For the ith gene and jth gene, we can use *cosine* similarity of vector $\boldsymbol{A}_i$ and vector $\boldsymbol{A}_j$ to evaluate their similarity in literatures.

Predicting disease candidate genes. First, we predict candidate genes of disease based on some already known causal genes. For each virulence gene of a disease, select the most relevant genes (whose eigenvalue vector similarity is larger than a threshold) from biomedical literature corpus using the algorithm described in the above section. After the process, there will be some collections of genes associated with virulence genes, then the elements in the intersection of these collections are the candidate genes of the disease.

When we know nothing about a disease's virulence genes, we cannot use genes' association for prediction directly. But a similar gene ontology mining method can still be used to do prediction. Taking the disease terms (such as "`lung cancer`") in Medical Subject Headings (`MeSH`) and the gene ontologies in human genome as a special dictionary, we do text mining in biomedical literatures to calculate the terms' TF, TF-IDF and normalized eigenvalues, and finally look for gene ontologies with high enough co-occurrences rate with the disease terms in the literatures, which would be the disease's candidate genes.

Using hierarchical clustering algorithm to build gene regulation network. We combine the association degrees between genes and hierarchical clustering algorithm to construct gene regulation network. In general, we find out 50 most relevant genes to the target gene through mining the medical literatures. According to the normalized eigenvalues matrix **P**, we establish a matrix **P1** of the `cosine distances` between all the 51 genes, and then use a hierarchical clustering algorithm to build a binary tree based on **P1**. The process consists of the following steps:

1. Initialization: Each gene is a cluster, so there are 51 clusters at beginning.
2. Clustering: Select two clusters A and B with the nearest cosine distance in matrix **P1**, **A** and **B** compose a new cluster **C**. Calculate the relevance (cosine distance) between **C** and other clusters except **A** and **B**.
3. Updating cosine distance matrix **P1**: Delete the rows and columns that contain cluster **A** and **B** from matrix **P1**, insert a new row and column for cluster **C**, and update all elements in matrix **P1**.
4. Repeat step 2 and 3 until there is no new clusters inserted.

The above process of clustering can be recorded to construct gene regulation networks. So we save the information of cluster **A**, cluster **B** and cluster **C** in each execution of step 2, and all the information can be used to build a regulation network for a specific gene.

3 Evaluation and Results Analysis

We design some experiments to evaluate the disease candidate gene identification algorithms and the gene regulation network building method described in Sect. 2.

3.1 Evaluations of the Prediction Results of Candidate Gene

Evaluating candidate genes predication based on some known causal genes. The candidate genes for 7 common diseases based on some known causal genes, that are grouped by their frequencies are shown in Table 2.

According to the frequency of genes appeared in the literatures, genes can be divided into four categories: ultrahigh frequency genes, high frequency genes, medium frequency genes and low frequency genes. It can be seen that most candidate genes are ultrahigh frequency genes or high frequency genes, which are the current research focus. Those medium frequency genes and low frequency genes are less studied by researchers, so they would become new research fields in future.

Evaluating candidate genes predication without known causal genes. We find out the most relevant 20 genes associated with lung cancer as shown in Table 3.

3.2 Evaluation of the Gene Regulation Network Construction Algorithm

We construct the gene regulation network for COMT as shown in Fig. 1. COMTs GRN is similar to a biology evolutionary tree, which reflects the genes correlations in such a way that the branches of higher relevant genes meet earlier. There are 50

Table 2 Disease Candidate Genes

Disease	Known causal genes	Candidate genes			
		Ultrahigh frequency genes	High frequency genes	Medium frequency genes	Low frequency genes
Lung cancer	EGFR, NRAS, HRAS, KRAS, TP53, ALK	CDKN2A, PIK3CA, BRAF			
Uterine cancer	PGR, ESR1	CYP1A1	CYP1A2, GSTM1, NQO1, CYP19A1, CYP17A1	GSTM2, NUDT1, GSTT2	ADH1B
Neuroma	NF2, SEMA3F	BRAF, PTEN, AKT1	SHH, SHANK3	ATRX, NF1, PRKAR1A	IL17RA
Lung carcinoma	EGFR, NRAS, HRAS, KRAS, ALK	CDKN2A, TP53, BRAF, PIK3CA			
Goiter	TG, TPO, TRH	MBP, BDNF			
Pre-eclampsia	FLT1, PGF	EN2, MBP, KIT, PTEN, STAT3, ANG, EGFR	PAX3, CXCR4, NOTCH4	TGFA, ADM	BVES
Microcephaly	MCPH1, ASPM, CDK5RAP2		CDK6	STIL, WDR62	ZNF335, CENPJ, CASC5, CEP135

Table 3 Lung Cancer Candidate Genes

TRIM21	CRP	CHI3L1	ADAMTS8	IL6
0.0920	0.0725	0.1542	0.0598	0.1342
IRAK3	LYZ	VWF	MIB1	LPIN2
0.0659	0.0819	0.0903	0.1960	0.085
TANK	IL1RN	CCL21	TLR2	MMP13
0.1253	0.0613	0.1052	0.0712	0.0818
PDGFRB	CILP	ALG1	SETD1A	TIMP2
0.0873	0.1465	0.0725	0.0794	0.0810

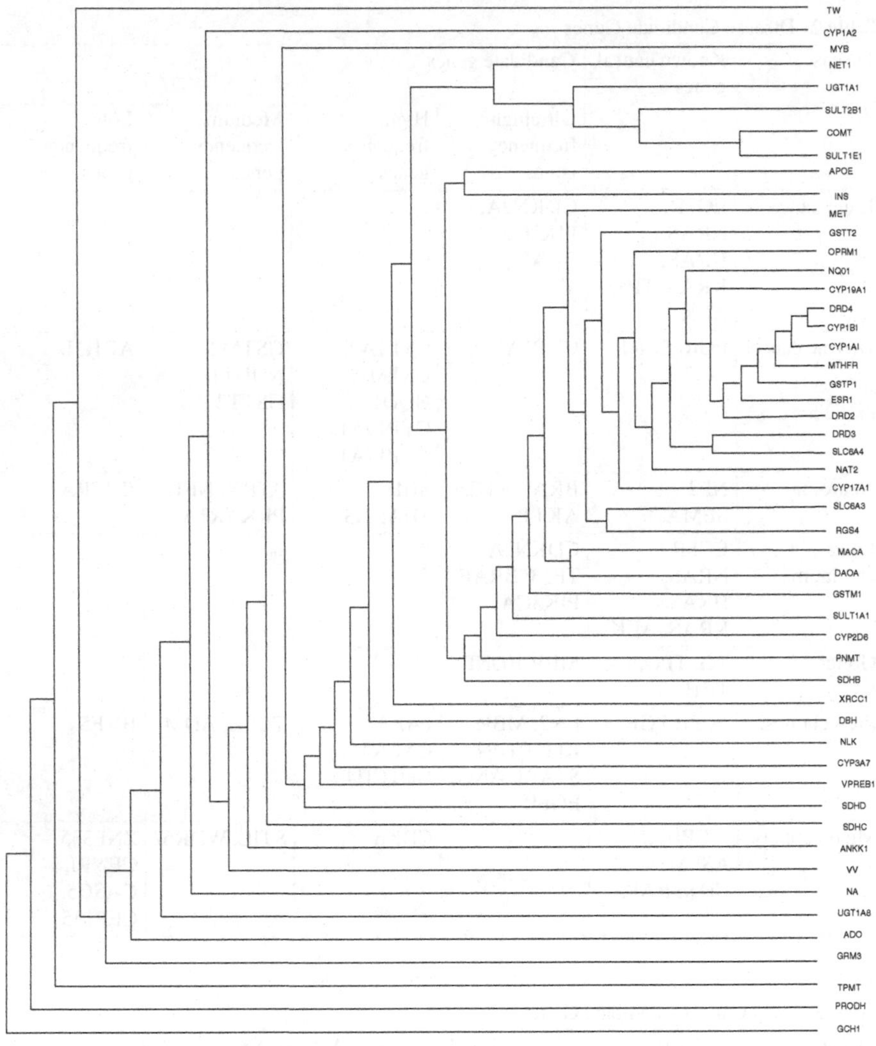

Fig. 1 Simple gene regulation network for COMT

most relevant genes of COMT in the network. In general, the high frequency genes such as SULT1E1, SULT2B1 converge with other genes most early, while the low frequency genes such as TW, GCH1 and PRODH meet with others in the end.

The goal of our regulation networks for specific genes is help researchers reduce their workload of building GRN. Since we dont involve any biological semantic during the literature mining process, we cannot determine whether one genes expression will influence other genes expression, so we cannot build a complete Boolean network. But our network does provide reasonable structure of GRN for researchers, so that they can determine the gene ontology and relevance between genes in GRN more easily.

4 Conclusion

In this paper, we studied the approaches to evaluate the correlation between genes through mining the huge amount of biomedical literatures. Based on the relationship information, we further design algorithms to predicate candidate virulence genes of diseases, and to construct simple gene regulation network for specific genes. The experiments on PubMed biomedical literature database show that our algorithms can help researchers determine their goal genes for future biomedical research efficiently.

Our future work includes improving the efficiency of retrieving papers from PubMed database and integrating semantic into gene regulation network.

References

1. Lander ES, Weinberg RA (2000) Genomics: journey to the center of biology. Science, U.S. 287, pp 1777–1782
2. Jensen LJ, Saric J, Bork P (2006) Literature mining for the biologist: from information retrieval to biological discovery. Nat Rev Gen Lond 7:119–129
3. AI-Mubaid H, Singh RK (2005) A new text mining approach for finding protein-to-disease associations. Am J Biochem Biotechnol 1(3):145–152
4. Chun HW, Tsuruoka Y, Kim JD et al (2006) Automatic recognition of topic-classified relations between prostate cancer and genes using MEDLINE abstracts. BMC Bioinform 7(1):1–8
5. Chen JY, Shen C, Sivachenko AY (2006) Mining Alzheimer disease relevant proteins from integrated protein interactome data. Pac Symp Biocomput 11:367–378
6. Human Protein Reference Database (2009). http://www.hprd.org/
7. Database of Interacting Proteins (2014). http://dip.doe-mbi.ucla.edu/dip/Main.cgi
8. Interologous Interaction Database (2015). http://ophid.utoronto.ca/ophidv2.204/
9. Ozgur A, Vu TG, Radev DR (2008) Identifying gene-disease associations using centrality on a literature mined gene-interaction network. Bioinformatics 24(13):i277–i285
10. Liu B, Jiang T, Ma S et al (2006) Exploring candidate genes for human brain diseases from a brain-specific gene network. Biochem Biophys Res Commun 349(4):1308C–1314
11. Radivojac P, Peng K, Clark WT, Peters BJ et al (2008) An integrated approach to inferring gene-disease associations in humans. Proteins 72(3):1030–1037
12. Wu X, Liu Q, Jiang R (2009) Align human interaetome with phenome to identify causative genes and networks underlying disease families. Bioinformatics 25(1):98–104
13. Miozzi L, Piro RM, Rosa F, Ala U, Silengo L et al (2008) Functionnl annotation and identification of candidate disease genes by computational analysis of normal tissue gene expression data. PLoS One 3(6):24–39
14. Ortutay Y, Vihinen M (2009) Identification of candidate disease genes by integrating Gene Ontologies and protein-interaction networks: case study of primary immunodeficiencies. Nucleic Acids Res 37(2):622–628

CP-ABE Based Access Control for Cloud Storage

Yong Wang, Longxing Wei, Xuemin Tong, Xiaolin Zhao and Ming Li

Abstract CP-ABE (Cipher-text Policy Attribute Based Encryption) can help providing reliable, fine-grained access control in untrusted cloud storage environment, since users can access to data files only if their attributes satisfy the access policies associated with the files. However, CP-ABE has two main drawbacks: its policies are not expressed using standard languages and it can't support non-monotonic policies. So we extended CP-ABE to support XACML (eXtensible Access Control Markup Language) based policy transformation and to support logical NOT in policies through De Morgan's Laws. And then we applied it to a secure overlay cloud storage system called FADE to deploy access control for Amazon S3 cloud storage service. The simulation results show that our proposal is practical and time efficient.

Keywords Cloud storage · Access control · CP-ABE · Rich policy · FADE

Y. Wang (✉) · L. Wei · X. Tong · X. Zhao
School of Software, Beijing Institute of Technology, Beijing 100081, China
e-mail: wangyong@bit.edu.cn

L. Wei
e-mail: longxingwei@126.com

X. Tong
e-mail: aietuzi@vip.qq.com

X. Zhao
e-mail: zhaoxl@bit.edu.cn

M. Li
MPS Information Classified Security Protection Evaluation Center,
Third Research Institute of Ministry of Public Security, Shanghai 200031, China
e-mail: li.ming@cspec.org.cn

V.E. Balas et al. (eds.), *Information Technology and Intelligent Transportation Systems*, Advances in Intelligent Systems and Computing 455,
DOI 10.1007/978-3-319-38771-0_45

1 Introduction

As the rapid development of cloud storage in the past few years, data security in cloud storage issues become increasingly critical. Because cloud storage providers (CSPs) are untrusted, we can neither rely on the CSPs to guarantee data' security, nor entrust them with traditional server-based access control enforcement. At present, the most practical solution is encrypting data before upload them to the cloud and using cryptographically enforced access control. Many researchers proposed ABE (Attribute Based Encryption) [1] based solutions, in which users' attributes are embedded into access policies and then used for encrypting data. Such solutions can provide fine-grained access control without the CSPs getting involved.

CP-ABE [1] is one form of ABE, where cipher-texts are associated with access policies defined by data owners to control which kinds of users can decrypt cipher-texts. CP-ABE is the most widely used ABE technology in cloud storage, since it can fit the scenarios very well. However, the CP-ABE policies must be monotonic, that is users can not use any policies with logical NOT. In addition, policies are not expressed in a standard format, which would limit the possible interoperation with other systems in the future. In this paper, we extended CP-ABE from two aspects: using XACML [2] to describe access control policies; supporting any policies with logical **NOT**.

We implemented the extended CP-ABE by modifying the open source cpabe toolkit [3], and particularly modified the client part and the key manager to support policy format transformation. The extended CP-ABE is then applied to an access control system for a secure overlay Amazon S3 cloud storage called `FADE` [4, 5] through modifying the interface to CP-ABE.

The rest of the paper is organized as follows. Section 2 introduces XACML policy syntax and then describes the transformation between XACML policy and CP-ABE policy. Section 3 presents how to extend CP-ABE to support rich policy with logical NOT. Section 4 shows the design and implementation of the cloud storage access control system. Section 5 evaluates the system. In the last section we draw the conclusions.

2 Transformation Between XACML Policy and CP-ABE Policy

CP-ABE uses strings to describe policies, which results in poor platform independency and weak interoperability with other systems. So we proposed an approach to transform CP-ABE policies to standard format through XACML. XACML is an XML based language widely used to describe policies for attribute based access control. We first introduced how to generate XACML policy from existing CP-ABE policy, and then provided an interface to transform XACML policy to CP-ABE policy.

2.1 Generating XACML Policy from CP-ABE Policy

A rich CP-ABE policy is a logical expression like "**NOT** ($A_1 == value1$) **AND** (($A_2 == value2$) **OR** ($A_3 == value3$))". We used eclipse plugin `ALFA` to transform CP-ABE policy to XACML format. Because the function "**n-of**" and the function "**not**" in XACML can't be used in Target elements, we embedded the logical **NOT** part in XACML conditions, which means putting logic inside rules.

Any value in XACML must correspond to an attribute, while values without corresponding attribute names can exist in CP-ABE policy. In order to support values without attribute names, the system associated such values with a specific attribute named "$unclassified$".

2.2 Transformation from XACML Policy to CP-ABE Policy

An XACML policy can be expressed as a tree T_x, where x is the root node. The transformation algorithm starts from the root node and proceeds in a recursive way. If the node type is XML_TEXT_NODE, then return the value of the node combined with the attribute. If the node type is $XML\_ELEMENT\_NODE$, then recursively process its' child nodes and combine the results returned by the child nodes with the operator (**AND**, **OR**, **NOT** or **N-OF**) of current node. In particular, **N-OF** should be replaced by "**n-of**" which is a function of XACML. We implemented the above transformation algorithm using `Libxml2` [6].

3 CP-ABE Extension to Support Rich Policy

3.1 CP-ABE

User's private key is associated with any number of attributes in CP-ABE and ciphertext is associated with policy. If and only if the recipient's private key satisfies the policy used to encrypt data, they can decrypt the encrypted data. CP-ABE provides four basic algorithms: `Setup`, `Encrypt`, `KeyGen` and `Decrypt`. John Bethencourt, et al. developed the `cpabe` toolkit [3], which provides four corresponding command line tools: `cpabe-setup`, `cpabe-keygen`, `cpabe-enc` and `cpabe-dec`.

CP-ABE uses an access tree to express a policy. Each non-leaf node represents a threshold gate described by its children and a threshold value. Let num_x represent the number of children of node x and k_x is its threshold value. When $k_x = 1$, the threshold gate is an OR gate, when $k_x = num_x$, it is an AND gate and each leaf node x is described by an attribute and it's threshold value $k_x = 1$.

In order to extend CP-ABE to support non-monotonic policy, we modified CP-ABE in two steps:

1. Transformed policy access trees based on De Morgan's laws so that the "**NOT**" operators only appear in leaf nodes.
2. Modified CP-ABE to support logical "**NOT**" operations.

3.2 Policy Transformation Based on de Morgan's Laws

De Morgan's laws are a pair of transformation rules that allow the expression of conjunctions and disjunctions purely in terms of each other via negation in propositional logic and Boolean algebra [7]. The rules can be expressed as: "**NOT** (A **AND** B)" is equivalent to "**NOT** A **OR NOT** B"; "**NOT** (A **OR** B)" is equivalent to "**NOT** A **AND NOT** B". Similarly, the threshold gate "**NOT** (***k*-of** $(A_1, A_2, \ldots, A_n)$)" is equivalent to "**(*k* + 1)-of** (**NOT** A_1, **NOT** $A_2, \ldots,$ **NOT** A_n)".

Let T express the original access tree, we transform T based on De Morgan's laws to access tree T′, where "**NOT**" operator only appears in leaf nodes. The transformation starts from the root node and proceeds recursively, which is transforming the sub-tree if the current root node is non-leaf node and has "**NOT**" operator. Firstly, delete the "**NOT**" operator. Secondly, transform from "**AND**" to "**OR**", or from "**OR**" to "**AND**", or from "**k-of**" to "**(k+1)-of**". Finally, add "**NOT**" operator for all child nodes of current node. Repeat this process for all child nodes. If the current node doesn't have any "**NOT**" operator, then process all child nodes recursively. If it is leaf node, just return. After the process, "**NOT**" operators only appear in leaf nodes.

3.3 Support of Logical NOT in CP-ABE

CP-ABE policies support 7 types of expressions, all of which can be transformed so that the logical **NOT** operator only exists in leaf nodes. As shown in Table 1, a policy except the expression No. 1, can be converted to another one when **NOT** works on it. For example, "**NOT** ($TAG == value$)" can be converted to "$TAG! = value$".

In order to support logical **NOT** in CP-ABE, we transformed **NOT** expression to logical **OR** expressions as follows:

1. Preprocess all attributes and classify them. Link all attributes that belong to the same classification in a link.
2. Transform policy using the algorithm shown in the above section to make sure that logical **NOT** only exists in leaf nodes.
3. Replace leaf nodes with **NOT** operators. Let x represents a leaf node with negative attributes and attr(x) represent its attributes. First, find the links (created in preprocessing) which involves attr(x). Second, transform node x to non-leaf node

Table 1 Expressions of CP-ABE policy

No.	Expression
1	$value$
2	$TAG == value$
3	$TAG > value$
4	$TAG < value$
5	$TAG <= value$
6	$TAG >= value$
7	$TAG! = value$

by setting $k_x = 1$ and $num_x = n - 1$, where n represents the number of attributes in the link that attr(x) belong to. Finally, create $n - 1$ leaf nodes (These nodes are child nodes of node x.) using attributes in the link except attr(x).

We implemented the algorithm on `cpabe` toolkit, by modifying `policy_lang.y` to parse new rich policies and adding components to transform policies and evaluate access trees.

4 Design and Implementation of CP-ABE Based Access Control System For Cloud Storage

Our access control system improves `FADE` [5] from the following three aspects: XACML is used to express policies; the extended CP-ABE is used; client and key manager is updated to support transformation of policy files and attribute files.

4.1 Overview

Architecture. The system consists of three parts as same as FADE:

1. Cloud: Cloud maintained by a third-party provider is used to provide cloud storage service. It only needs to provide basic operations such as file uploading and file downloading through APIs.
2. Key Manager: Each key manager is an independent entity. It is used to maintain policies and keys for access control.
3. Client: Client is an interface between data owners or data users and the cloud. Client encrypts data before it is uploaded and decrypts data after it is downloaded. Client also interacts with key managers to perform policy interaction and key operation.

Table 2 Notations used in this paper

Notation	Description
F	Data file from client, which will be uploaded
K	Data key used to encrypt data file F
P_i	Policy name with index i
PF_i	Policy file associated with P_i
A_i	Attribute set associated with P_i
p_{ij}, q_{ij}	RSA prime numbers for policy P_i on the jth key manager
n_{ij}	$n_{ij} = p_{ij} * q_{ij}$
(e_{ij}, d_{ij})	RSA control key pair for policy P_i on the jth key manager
S_i	Secret key associated with P_i
$\{data\}_{key}$	Encryption of data with the symmetric key "key"
$[data]_{abe}$	Encryption of data under policy "abe" using CP-ABE
R	A random number used for blind

Security assumptions. Users can access data only if they satisfy the policies associated with the encrypted data, so our system has two security assumptions: 1. All encryption mechanisms are safe; 2. Only fail-stop failures may happen in key managers.

The notations used in the description of our system are shown in Table 2.

4.2 Access Control for File Access

File uploading. The file uploading process is illustrated in Fig. 1.

1. The client sends PF_i, A_i to N Key Managers.
2. All the N Key Managers create control key pairs. The private control keys (n_{ij}, d_{ij}) $(0 < i <= N)$ are maintained by the Key Managers. They send the public control keys (n_{ij}, d_{ij}) $(0 < i <= N)$ to the client.
3. The client generates random key K and S_i associated with P_i. Client splits S_i to N sub-fragments according to Shamir's Secret Sharing Schema [8, 9] and the result sub-fragments are denoted as S_{i1}, $S_{i2}, \ldots, S_{iN}$. Client encrypts F using symmetric encryption with key K as $\{F\}_K$. Then it computes the HMAC signature for $\{F\}_K$ and appends the signature to the encrypted data. Then client encrypts K using symmetric encryption with key S_i as $\{K\}_{S_i}$, encrypts sub-fragments of S_i using RSA as $S_{ij}^{e_{ij}}(\text{mod} n_{ij})$. Finally, client sends encrypted data $\{K\}_{S_i}$, $S_{i1}^{e_{i1}}$, $S_{i2}^{e_{i2}}$, $\ldots$, $S_{iN}^{e_{iN}}$, $\{F\}_K$ and the policy file name P_i to the cloud, then deletes keys $K, S_i, S_{i1}, S_{i2}, \ldots, S_{iN}$.

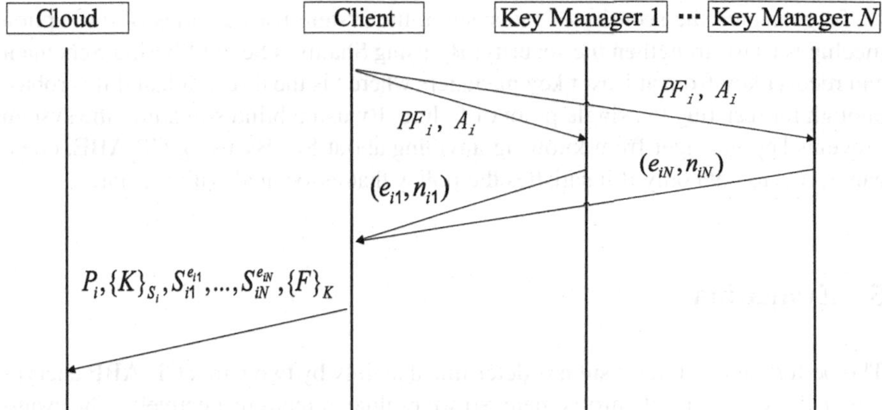

Fig. 1 File uploading process

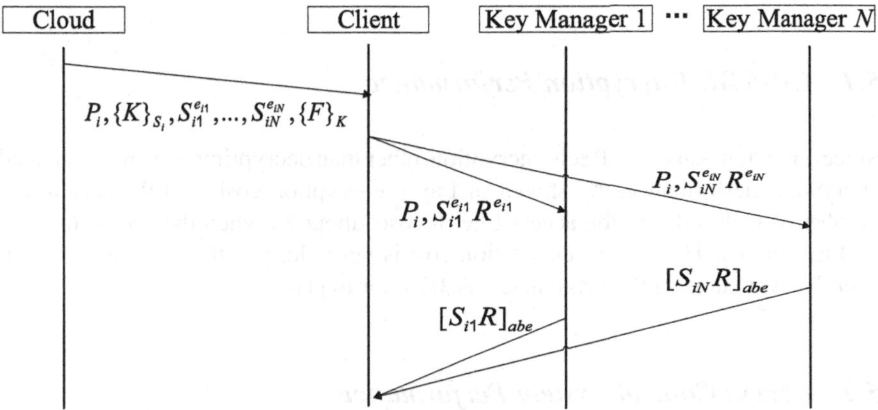

Fig. 2 File downloading process

File downloading. The file downloading process is illustrated in Fig. 2.

1. The client downloads related data from the cloud and checks HMAC signature of the encrypted data file $\{F\}_K$, then generates a random number R and sends $S_{i1}^{e_{i1}} R$, $S_{i2}^{e_{i2}} R$, …, $S_{iN}^{e_{iN}} R$ to related Key Managers respectively.
2. The Key Managers decrypt $S_{ij}^{e_{ij}} R^{e_{ij}}$ and get $S_{ij} R$ using private control key (n_{ij}, d_{ij}), and send $S_{ij} R$ back after encrypting it with policy file PF_i using CP-ABE.
3. Client decrypts $\left[S_{ij} R\right]_{abe}$ if it satisfies policy file PF_i, then unblinds $S_{ij} R$ and gets S_{ij} [10]. And it recovers S_i through Shamir's Secret Sharing Schema. Since S_i is recovered, $\{K\}_{S_i}$ can be decrypted, so does $\{F\}_K$. Finally, client can recover the original data file F.

To guarantee the security of the system, the system uses a series of encryption mechanisms to strengthen the security: By using Shamir's Secret Sharing Schema it can recover key from at least t key managers where t is the threshold, and it is robust enough for resisting the single point of failure. By using blind signature, the system prevents key manager from knowing anything about S_{ij}. By using CP-ABE, client can access to data only if it satisfies the policy that associated with the data.

5 Evaluation

The performance of the system is determined mainly by two parts: CP-ABE encryption and access control enforcement, so we evaluated them respectively. The evaluation was done on a virtual machine with single-core processor and 512 M memory with Ubuntu 10.04. The key managers were deployed in local computers.

5.1 *CP-ABE Encryption Performance*

Since our extension only affects encryption, other than decryption, we only evaluated encryption performance. As shown in Fig. 3, encryption cost is still linear to the number of leaf nodes in the access tree. It costs about 2 s when the access tree has 100 leaf nodes. The overall encryption cost is very close to the original CP-ABE (For the evaluation of the original CP-ABE, refer to [1]).

5.2 *Access Control System Performance*

We evaluated time performance of the cloud storage access control system.

Figure 4 shows the time cost of file uploading at the aspects of total time, file uploading time, metadata uploading time, attribute file and policy file transferring time. It's can be seen that file uploading costs most of the time, because the size

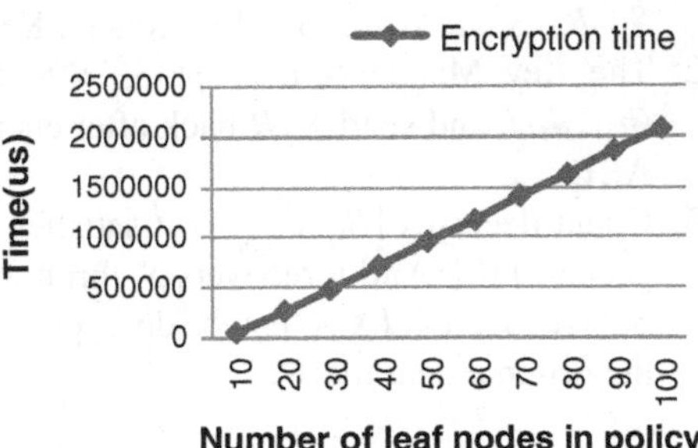

Fig. 3 Time cost of encryption

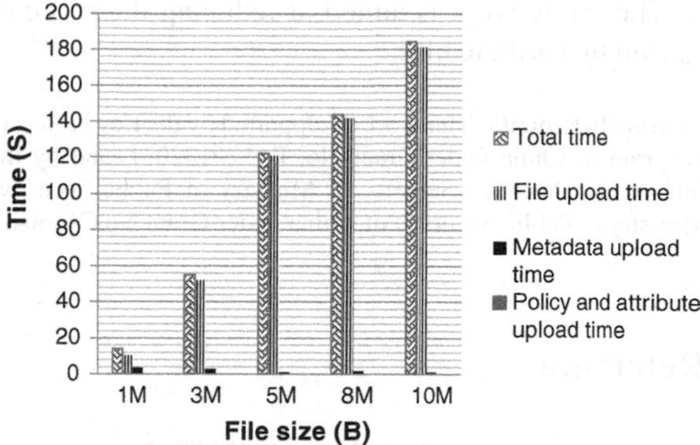

Fig. 4 Time cost of file uploading

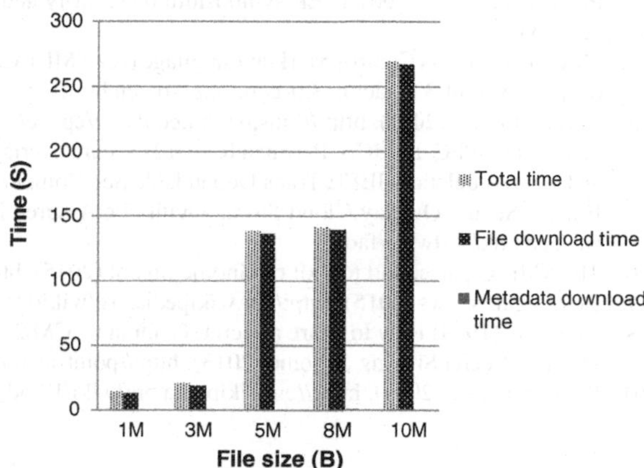

Fig. 5 Time cost of file downloading

of metadata, policy file and attribute file is no more than a few KB, which is much smaller than data file. Figure 5 shows the time cost of downloading, which also consists of total time, file downloading time and metadata downloading time. The most time consuming part is the file downloading. It's worth noting that the network bandwidth on campus is very low, only 30 k/s, so it takes a long time while upload and download files of any size. All transfer time will be reduced obviously if the bandwidth is larger.

6 Summary

The main contributions of this work are: XACML is used to describe policy to make it platform-independent and interoperable; CP-ABE is extended to support rich policy without much higher time cost; Based on extended CP-ABE and XACML policy, we implement an access control system for cloud storage based on FADE.

The future work is aimed at reducing the space cost of the extended CP-ABE caused by attribute files.

Acknowledgments This work is supported by the Key Project of National Defense Basic Research Program of China under Grant No. B1120132031, and by the Opening Project of Key Lab of Information Network Security of Ministry of Public Security (The Third Research Institute of Ministry of Public Security) of China under Grant No.C10604.

References

1. Bethencourt J, Sahai A, Waters B (2007) Ciphertext-policy attribute-based encryption. In: Proceedings of the 2007 IEEE symposium on security and privacy ieee computer society, pp 321–334
2. eXtensible Access Control Markup Language (XACML) Version 3.0 (2013). http://docs.oasis-open.org/xacml/3.0/xacml-3.0-core-spec-os-en.html
3. CPABE toolkit (2011). http://hms.isi.jhu.edu/acsc/cpabe/
4. Tang Y, Lee PPC, Lui JCS, Perlman R (2012) Secure overlay cloud storage with access control and assured deletion. IEEE Trans Dependable Sec Comput 9(6):903–916
5. FADE - Secure Overlay Cloud Storage with File Assured Deletion (2011). http://ansrlab.cse.cuhk.edu.hk/software/fade/
6. The XML C parser and toolkit of Gnome libxml (2015). http://xmlsoft.org/
7. De Morgan's laws (2015). http://en.wikipedia.org/wiki/De_Morgans_laws
8. Shamir A (1979) How to share a secret. Commun ACM 22(11):612–613
9. Shamir's Secret Sharing Scheme (2013). http://point-at-infinity.org/ssss/
10. Blind signature (2014). http://en.wikipedia.org/wiki/Blind_signature

Bayesian Approach to Fault Diagnosis with Ambiguous Historical Modes

Sun Zhou, Zhongyuan Cai and Guoli Ji

Abstract Fault diagnosis plays an important role in diverse engineering areas including industrial control loop systems, mechanical systems, etc. Bayesian methods are a class of data-driven fault diagnosis methods developed in recent years. However, one difficulty with Bayesian methods is that they do not deal with the case that there is uncertainty about the underlying mode in the historical data. For this problem, a new approach is proposed in this paper, through which the ambiguous modes are softly classified by combing historical data, current evidence and the prior knowledge under a Bayesian framework. In addition, weighted kernel density estimation instead of classic histogram method is used for likelihood estimation to enhance diagnosis. The proposed Bayesian approach is tested on the fault diagnosis of Tennessee Eastman (TE) process using benchmark data and the proposed approach performs better in comparison with typical previous methods.

Keywords Fault diagnosis · Ambiguous mode · Bayesian approach

1 Introduction

Fault diagnosis has been an increasingly significant issue to improve the efficiency, stability and security for diverse processes. Owing to the complexity of industry processes, it is complicated to model a real industrial system. An alternative is to use data-driven methods such as Bayesian methods. It provides a reliable way for dealing with uncertainty conditions and is prove to be effective in engineering diagnosis.

S. Zhou (✉) · Z. Cai · G. Ji
Department of Automation, Xiamen University,
Xiamen 361005, People's Republic of China
e-mail: zhousun@xmu.edu.cn

Z. Cai
e-mail: zyycai@foxmail.com

G. Ji
e-mail: glji@xmu.edu.cn

V.E. Balas et al. (eds.), *Information Technology and Intelligent Transportation Systems*, Advances in Intelligent Systems and Computing 455,
DOI 10.1007/978-3-319-38771-0_46

In the presence of disturbances, Bayesian inference provides a fit way to solve the industrial process monitoring and diagnosis. It is one of the most widely used methods in faults diagnosis. Bayesian diagnosis method is capable of incorporating historical data and prior knowledge for fault diagnosis and is proved to perform well in industrial applications.

A main branch of fault detection and diagnosis is control-loop performance diagnosis methods that was first proposed by Harris [1], and was further developed in [2]. Huang and Shan [3] put forward a linear quadratic regulator (LQG) benchmark. Then further development was reported on evaluating performance of other components such as instruments [4, 5] and valves [6, 7], etc. In these works different monitoring algorithms are used to detect corresponding single problem and. However, a change in operating mode may influence more than one instrument, resulting in the possibility of wrong diagnosis. While Bayesian methods have no such limitation and have the superiority to diagnose unspecific faults. Nevertheless, one difficulty with Bayesian methods is that there must be exact knowledge of the underlying mode in the historical data.

To deal with such difficulty where there is uncertainty about the underlying mode, some approaches [8, 9] were proposed including incomplete data method, direct probability approach, second-order approximation method, etc. The incomplete data method just simply ignores the ambiguous data and the remaining data is utilized to apply Bayesian diagnosis method. Thus the diagnosis performance is affected because of too few data is used for training. When the proportion of ambiguous data is bigger, the diagnosis result will become more unreliable. Direct probability approach probability boundaries are too large to make a diagnosis. Regarding the second-order approximation method, it approximates the likelihood through Taylor series expansion to yield better estimation, but there is still useful information that is not taken into consideration by this method. In this paper, a new approach is proposed, through which the ambiguous modes are softly classified by combing historical data, current evidence and the prior knowledge under a Bayesian framework.

2 Bayesian Approach for Diagnosis with Ambiguous Modes

2.1 Problem Formulation

Let us take control loop diagnosis as example. A typical control loop consists of the following components: controller, actuator, process and sensor. Each component may subject to some abnormalities which may influence control loop performance. We assume that all or some monitors are designed for the components in the control loop. However, each of monitors is possible to be affected by turbulence. In addition, any component may influence the monitors designed for other components. An allocation of operating state of all the components in the control loop is called a mode denoted as m, m can take different values and its specific value is denoted by m_k. The allocation of monitor readings is called evidence, and is denoted as $e = (\pi_1, \pi_2, \ldots, \pi_S)$, where π_i is the ith monitors reading, and S is the sum of monitor. Historical data are obtained from the past data record where the mode of control loop and the monitor readings are also recorded. Each sample d^t at time t in the historical data set D consists of the evidence e^t and the underlying mode m^t, which can be denoted as $d^t = \{e^t, m^t\}$.

Given an evidence e, the historical data set D, posterior probability can be expressed as

$$p(m|e, D) = \frac{p(e|m, D)p(m|D)}{p(e|D)} \tag{1}$$

where $p(m|e, D)$ is posterior probability, which is what we want to obtain under the condition of knowing current evidence e and historical data D. $p(m|D)$ is prior probability of mode m. $p(e|D)$ is a constant, and can be calculated as $p(e|D) = \sum_m p(e|m, D)p(m|D)$.

2.2 Bayesian Approach

Here we present a Bayesian approach so that the ambiguous modes can be softly classified. By marginalization over all possible likelihood parameters, the likelihood

$$p(e|m = m_k, D) = \int_{\Omega} p(e|\Theta_{m_k}, m = m_k, D) f(\Theta_{m_k}|m = m_k, D) d\Theta_{m_k} \tag{2}$$

The set $\Theta_{m_k} = (\theta_{e_1,m_k}, \theta_{e_2,m_k}, \ldots, \theta_{e_N,m_k})$ and $\theta_{e_i,m_k} = p(e = e_i|m = m_k)$ are parameters for all available evidence e. N is the total number of the evidence. Ω contains all parameters under all possible modes. $f(\Theta_{m_k}|m = m_k, D)$ can be calculated through the method of Bayesian rules:

$$f(\Theta_{m_k}|m = m_k, D) = \frac{p(D|\Theta_{m_k}, m = m_k) f(\Theta_{m_k}|m = m_k)}{p(D|m = m_k)} \tag{3}$$

Then we assume that $f(\Theta_{m_k}|m = m_k)$ subjects to Dirichlet distribution, we can obtain the following:

$$f(\Theta_{m_k}|m = m_k) = \frac{\Gamma(\sum_{q=1}^{N} \alpha_q^{m_k})}{\prod_{i=1}^{N} \Gamma(\alpha_q^{m_k})} \prod_{q=1}^{N} \theta_{q,m_k}^{\alpha_q^{m_k}-1}, \alpha_q^{m_k} > 0 \tag{4}$$

$\alpha_q^{m_k}$ is the parameter of Dirichlet distribution called prior simple about mode m_k. $\Gamma(x)$ is the gamma function.

The probability $p(D|\Theta_{m_k}, m = m_k) = \prod_{t=1}^{N_{m_k}} p(d_{m_k}^t|\Theta_{m_k}, m = m_k)$, where $d_{m_k}^t$ is the data point at time t with mode m_k. When $e^t = e_i$

$$p(d^t_{m_k}|\Theta_{m_k}, m = m_k) = \theta_{i|m_k} \tag{5}$$

The conclusion $p(d^t_{m_k}|\Theta_{m_k}, m = m_k) = \prod_{i=1}^{N} \theta_{i|m_k}^{n_{i|m_k}}$ is made. Then we can calculate the value of the likelihood:

$$p(e = e_i|m = m_k, D) = \frac{n_{i|m_k} + \alpha_{i|m_k}}{N_{m_k} + A_{m_k}} \tag{6}$$

$n_{i|m_k}$is the number of historical data point when $e = e_i$ and $m = m_k$. $\alpha_{e_i|m_j}$ is the amount of prior samples that is distributed to evidence e_i with mode m_j; $N_{m_j} = \sum_i n_{e_i|m_j}$ and $A_{m_j} = \sum_i \alpha_{e_i|m_j}$.

Assume there is an ambiguous mode M_K. Before we calculate the proportion parameter, we should first figure out the following parameters through (6):

$$\hat{\theta}_{d_i}\{\frac{m_j}{M_k}\} = \frac{(n_{e_i|m_j} + \alpha_{e_i|m_j})/(N_{m_j} + A_{m_j})}{\sum_{m_k \in m_Q}(n_{e_i|m_k} + \alpha_{e_i|m_k})/(N_{m_k} + A_{m_k})} \tag{7}$$

where d_i denotes a data point in historical data set and $d_i = (e_i, M_K)$, $n_{e_i|m_j}$ is the amount of historical samples when the evidence $e = e_i$, and mode $m = m_j$; $\alpha_{e_i|m_j}$ is the amount of prior samples that is distributed to evidence e_i with mode m_j; m_Q is a set of all available modes. In addition, where $N_{m_j} = \sum_i n_{e_i|m_j}$ and $A_{m_j} = \sum_i \alpha_{e_i|m_j}$.

Then normalize $\hat{\theta}_{d_i}\{\frac{m_j}{M_K}\}$ to obtain proportion parameters:

$$\theta_{d_i}\{\frac{m_j}{M_K}\} = \frac{\hat{\theta}_{d_i}\{\frac{m_j}{M_K}\}}{\sum_{m \in M_K}\hat{\theta}_{d_i}\{\frac{m_j}{M_K}\}} \tag{8}$$

Summary the formulas (6) and (8), likelihood can be estimated as

$$p(e_i|m) = \frac{n(e_i|m) + \sum_{m \in M_k}(\theta_{d_i}\{\frac{m}{M_k}\}n(e_i|M_k)) + \alpha_{e_i|m}}{n(m) + \sum_{d_i \in D}\sum_{m \in M_k}(\theta_{d_i}\{\frac{m}{M_k}\}n(e_i|M_k)) + A_m} \tag{9}$$

Then we can obtain the final posterior probability through Bayesian method.

3 Enhancing Diagnosis Through Weighted Kernel Density Estimation

In addition, to enhance diagnosis we propose weighted kernel density estimation instead of classic histogram method for likelihood estimation.

The advantage of kernel density estimation is that it reflects the true shape of the data and can fully reflects the distributions regardless of the data as shown in Fig. 1.

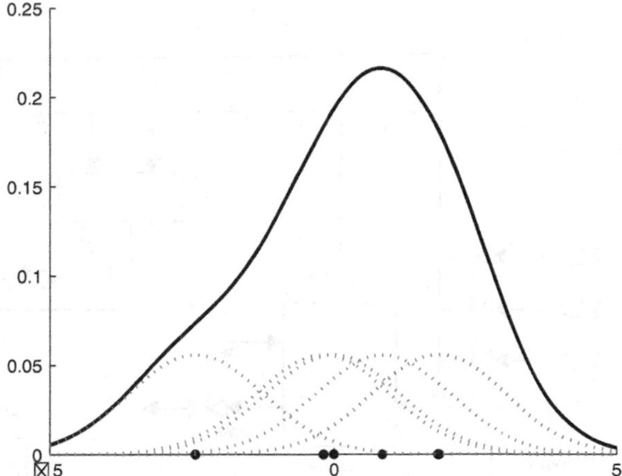

Fig. 1 Kernels summing to a kernel density estimate [10]

In addition, a kernel density estimate can be obtained in a simple step. So kernel density estimates can yield the same result even with a single data set. However, it is in strong connection with the discrete method, the kernel density method subjects to the issue of dimensionality. When in high dimension, performance will depress.

Kernel density estimation [10] performs precisely instead of histogram density estimation. Diverse modes correspond to diverse weights. Assume the weight of mode m is θ_m, the kernel density estimation function will be:

$$\hat{p}(x) = \frac{1}{Nh} \sum_{i=1}^{N} \theta_m K(\frac{x - x_i}{h}) \tag{10}$$

where N is the total number of evidence, and h is a kernel density constant, $K(\cdot)$ is kernel function. x_i denotes the current evidence. $\hat{p}(x)$ is the estimation function.

4 Application to TE Process

To verify the capability of the proposed method, it is applied to diagnose of Tennessee Eastman problem. The Tennessee Eastman Plant-wide Industrial Process Control Problem is called as TE problem [12], which has been viewed as a criterion to assess the performances of various diagnostic methods. Figure 2 shows a procedure of the TE process. The process comes into being two products from four reactants. Also there is a byproduct and an inert. The reactions are:

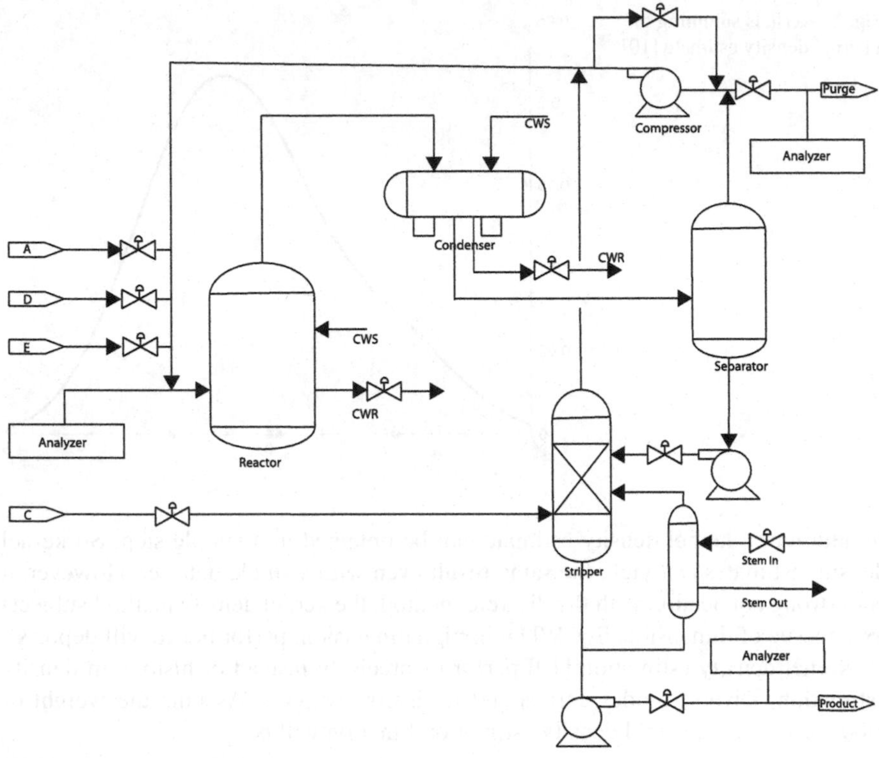

Fig. 2 The Tennessee Eastman process [11]

$$
\begin{aligned}
A(g) + C(g) + D(g) &\rightarrow G(liq), Product1 \\
A(g) + C(g) + E(g) &\rightarrow H(liq), Product2 \\
A(g) + E(g) &\rightarrow F(liq), Byproduct \\
3D(g) &\rightarrow 2F(liq), Byproduct
\end{aligned}
\tag{11}
$$

The gaseous reactants are sent to the reactor, in which the reactants reactor to come into being liquid products. The products run out of the reactor in the form of vapor as well as the unreacted reactants, and enter a cooler to be condensed. Then they go through a vapor/liquid separator. The components that are not condensed return to the reactor by a compressor. The condensed components are sent to the stripper to remove the rest reactants by reacting with reactant C. The products export the stripper at the bottom of the system. The inert and byproduct are purged from the system in the form of vapor.

In this paper we implement our method with the data set used by Russell et al. [13](available at http://web.mit.edu/braatzgroup), where the TE process is controlled by the measure proposed by Lyman and Georgakis [14]. In the paper, three representative modes (IDV2:B composition, A/C ratio constant; IDV6:A feed loss;IDV8:A,B,C feed composition) were selected, two measurement variables (XMEAS(4), XMEAS(8)) were used.

In this simulation, since both training data and test data have exact information about evidence and mode, we shall mask some of the data as ambiguous according to resemblance of data toward other modes. During the masking course, a parameter called likelihood ratio threshold is established, as a result, if another mode is similar enough, the data point will be classified as ambiguous. For example, assume the data comes from mode m_1. If there is a data point d^k, the likelihood ratio R between mode m_k and mode m_1 is large enough, then the mode m_k was added. As a result, the mode with d^k is assumed to be ambiguous. By adjusting the value of the threshold, certain amounts of data can be classified as ambiguous. After that, the aforementioned data sets are utilized in the proposed method.

$$R = \frac{p(d^k|m_k)}{p(d^k|m_1)} > Threshold \tag{12}$$

Here we apply the incomplete method and the proposed method to the fault diagnosis problem of TE problem. The final posterior probability calculated by each of the methods is shown in Fig. 3. Horizontal coordinates of every graph in Fig. 3

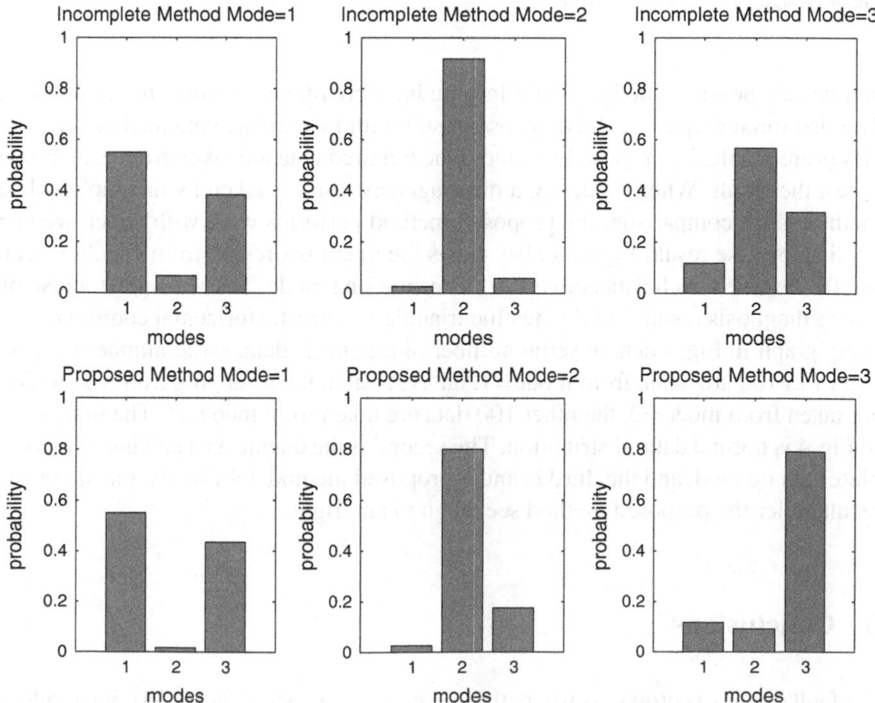

Fig. 3 Posterior probability comparison

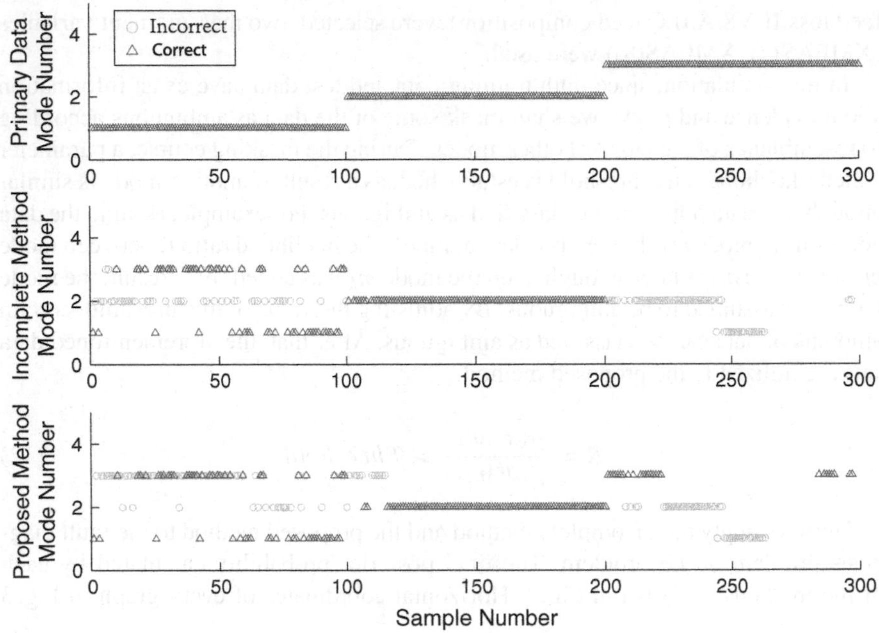

Fig. 4 Diagnosis result comparison

denotes all possible modes, while longitudinal coordinate denotes the possibility. The first three graphs are the diagnosis result with incomplete data method, and the others are results by proposed method. One hundred data are taken from each mode to test the result. When mode=3, a misdiagnosis result is taken by incomplete data method. As a comparison, the proposed method performs well, with which we can obtain a precise result. Figure 4 also shows the diagnosis result, from which we can briefly observe which data belongs to corresponding mode. The red circle represents wrong diagnosis results, while the blue triangle is correct. Horizontal coordinates of every graph in Fig. 4 denote serial number of diagnosis data, serial number ranging from 1 to 100 are taken from mode=1, datas serial number ranging from 101 to 200 are taken from mode=2, the other 100 data are taken from mode=3. The first graph of Fig. 4 is normal datas distribution. The second is the diagnosis result under incomplete data method, and the third is under proposed method. Obviously, the diagnosis result under the proposed method seems good and right.

5 Conclusions

For fault diagnosis problems where there is uncertainty about the underlying mode, a new approach is put forward under Bayesian framework to combine historical data, current evidence and prior knowledge together and to softly classify the data under ambiguous mode. Furthermore, weighted kernel density estimation instead of classic histogram method is proposed for likelihood estimation to enhance diagnosis. The proposed Bayesian approach is tested on the faults diagnosis problem of Tennessee Eastman (TE) process and is proved to perform better than typical previous methods.

Acknowledgments This work was supported by the National Natural Science Foundation of China (Nos. 61304141, 61573296), Fujian Province Natural Science Foundation (No. 2014J01252), the Specialized Research Fund for the Doctoral Program of Higher Education of China (No. 20130121130004), the Fundamental Research Funds for the Central Universities in China (Xiamen University: Nos. 201412G009, 2014X0217, and 201410384090) and the China Scholarship Council award.

References

1. Harris TJ (1989) Assessment of control loop performance. Can J Chem Eng 67(5):856–861
2. Huang B, Shah S, Kwok K (1995) Am Control Conf 2:1250–1254
3. Huang B, Shah S (1999) Performance Assessment of Control Loops. Springer, Berlin
4. Gonzalez R, Huang B, Xu F, Espejo A (2012) Dynamic Bayesian approach to gross error de-tection and compensation with application toward an oil sands process. Chem Eng Sci 67(1):44–56
5. Qin SJ, Li W (2001) Detection and identification of faulty sensors in dynamic processes. AIChE J 47(7):1581–1593
6. Choudhury M, Jain M, Shah S (2008) Stiction C definition, modeling, detection and quantification. J Process Control 18(3–4):232–243
7. He Q, Wang J, Pottmann M, Qin S (2007) A curve fitting method for detecting valve stiction in oscillation control loops. Ind Eng Chem Res 46(13):4549–4560
8. Gonzalez R, Huang B (2013) Control loop diagnosis with ambiguous historical operating modes: Part 1. A proportional parameterization approach. J Process Control 23:585–597
9. Gonzalez R, Huang B (2013) Control loop diagnosis with ambiguous historical operating modes: Part 2, information synthesis based on proportional parameterization Ruben. J Process Control 23:1441–1454
10. Gonzalez R, Huang B (2014) Control loop diagnosis using continuous evidence through kernel density estimation. J Process Control, 24:640–651
11. Ricker N (1996) Decentralized control of the Tennessee Eastman challenge process. J Process Control 6(4):205–221
12. Downs JJ, Vogel EF (1993) A plant-wide industrial process control problem. Comput Chem Eng 17:245–255
13. Russell EL, Chiang LH, Braatz RD (2000) Data-Driven Methods for Fault Detection and Diagnosis in Chemical Processes. Springer, Berlin
14. Lyman PR, Georgakis C (1995) Comput Chem Eng 19:321–331

Analyzing Grammatical Evolution and πGrammatical Evolution with Grammar Model

Pei He, Zelin Deng, Chongzhi Gao, Liang Chang and Achun Hu

Abstract Grammatical evolution (GE) is an important automatic programming technique developed on the basis of genetic algorithm and context-free grammar. Making changes with either its chromosome structure or decoding method, we will obtain a great many GE variants such as πGE, model-based GE, etc. In the present paper, we will examine the performances, on some previous experimental results, of GE and πGE with model techniques successfully applied in delineating relationships of production rules of context-free grammars. Research indicates modeling technology suits not only for GE constructions, but also for the analysis of GE performance.

Keywords Genetic programming · Grammatical evolution · Finite state transition system · Model

P. He (✉) · C. Z. Gao · A. C. Hu
School of Computer Science and Educational Software,
Guangzhou University, Guangzhou 510006, China
e-mail: bk_he@126.com

C. Z. Gao
e-mail: gaochz@hotmail.com

A. C. Hu
e-mail: 1065168399@qq.com

P. He · Z. Deng
School of Computer and Communication Engineering,
Changsha University of Science and Technology, Changsha 410114, China
e-mail: zl_deng@sina.com

L. Chang
Guangxi Key Laboratory of Trusted Software,
Guilin University of Electronic Technology, Guilin 541004, China
e-mail: kxrj@guet.edu.cn

V.E. Balas et al. (eds.), *Information Technology and Intelligent Transportation Systems*, Advances in Intelligent Systems and Computing 455,
DOI 10.1007/978-3-319-38771-0_47

1 Introduction

Genetic programming (GP) is an important automatic programming approach developed by Koza [1] on the framework of genetic algorithm (GA) [2]. Since its advent in 1992, many progresses have been made in this area. Not only does it emerge into a big family in the last decades, including a great many of GP variants [3–8] such as Grammatical Evolution (GE), Gene Expression Programming (GEP), Multi-Expression Programming (MEP), Hoare logic-based Genetic Programming (HGP), etc., finding wide applications [5, 9–12] in mathematical modeling, software engineering, financial prediction, and circuit design, but also its complex tree-based genotypic representation is changed into some kinds of linear structures. This change makes GP much more easy to use, putting it into practice as a result.

GE is a typical example of linear GP comprised of GA and context-free grammar. It differs from classical GP in ways of representations and decoding method. In GE, chromosomes are usually represented by strings of codons, and decoded into phenotypes in terms of a special genotype-to-phenotype mapping. Varying either the codon or the mapping, we can get different GP variants. So, performance comparisons among GPs become an open issue [13]. In this paper, we will analyze the results of GE and πGE [6] through using model framework of [13, 14]. This approach can help answering part of the concerned problems to some degree.

2 Background and Motivation

2.1 Grammatical Evolution

Grammatical evolution is an automatic programming technique pioneered by Ryan, Collins and ONeill in 1998 [15]. It represents programs by sequences of codons (each codon consists of 8bits), and decodes them into grammatical derivations of a BNF (Backus Naur Form) grammar using the following rule.

Rule = (codon value) mod (number of rules available for the current non-terminal)

Given a grammar as shown in Quartic Symbolic Regression problem of Sect. 4.1, an example of decoding process is given in Fig. 1.

Fig. 1 Decoding of individual of quartic symbolic regression problem

Expression	Con	Num	R=Con mod Num	Rule
<expr>	8	2	0 (the firstrule)	<expr>::= <expr><op><expr>
<expr><op><expr>	11	2	1 (the secondrule)	<expr>::=<var>
<var><op><expr>	11	1	0 (the firstrule)	<var>::= x
x<op><expr>	10	4	2 (the thirdrule)	<op>::=*
x*<expr>	11	2	1 (the second rule)	<expr>::= <var>
x*<var>	10	1	0 (the first rule)	<var>::=x
x*x				

2.2 πGrammatical Evolution

As we can see from Sect. 2.1, GE is essentially a leftmost derivation system of a BNF grammar. If any non-terminal of a sentential form can be chosen as the *current non-terminal* during rule selections of GE, a novel GE variant called πGE will be obtained [15, 16]. In view of this, a πGE codon should be defined as a pair (NT, rule), where NT is used for choosing of the non-terminal from the sentential form as the current non-terminal, and the value of rule is applied for the production selection. Henceforth, a chromosome of πGE consists of a vector of these pairs. Derivations may take place anywhere non-terminals occur.

2.3 Fitness Measure

One of the major steps for GP as well as its variants concerns the fitness measure. It is a primary mechanism defining what a good solution would be, and guiding the GP system to synthesizing the desired result. In this paper, we will evaluate fitness by running the evolved programs on sample data set. For the examples [6] discussed in the following sections, the fitness on cases of Quartic Symbolic Regression $f(x) = x + x^2 + x^3 + x^4$, Santa Fe Ant Trail are given by the reciprocal of the sum, taken over 100 cases, of the absolute value of the error between the evolved and target functions and finding all 89 pieces of food located on a trail within a specified number of time steps, respectively.

3 Grammar Model

Model-based grammatical evolution (MGE) [13, 14] is a relatively new GE variant developed over its grammar model. As we know, any codon of a GE chromosome has no definite meaning, therefore is difficult to understand. To this end, we have investigated relationships among production rules, delineating them with finite state transition system. This system is the so called grammar model. Given a context-free grammar $G = (V_N, V_T, S, P)$, where V_N stands for a set of non-terminals, V_T a set of terminals, $S(inVN)$ the start symbol, and P the set of productions, we can construct its leftmost grammar model as follows [13, 14].

(1) Draw two vertices S_N and S_N^α for each production rule $N \to \alpha \in P$ (when N has many alternatives, we should treat them separately as different production rules). Where $S_N = \{\alpha \| \alpha$ is a sentence of G or a sentential form of Gwhose leftmost non-terminal symbol is $N\}$, $S_N^\alpha = \{\beta_N^\alpha | \beta \in S_N$ and β_N^α is the string obtained from substitution of α for the leftmost non-terminal symbol N in $\beta\}$.
(2) Draw an ϵ arrow from V to V for each vertex V of step 1;

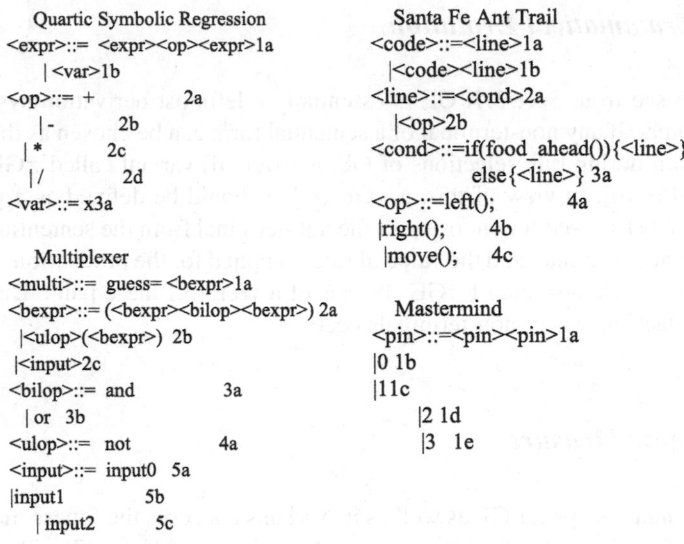
Quartic Symbolic Regression
<expr>::= <expr><op><expr>1a
| <var>1b
<op>::= + 2a
| - 2b
| * 2c
| / 2d
<var>::= x3a
Multiplexer
<multi>::= guess= <bexpr>1a
<bexpr>::= (<bexpr><bilop><bexpr>) 2a
|<ulop>(<bexpr>) 2b
|<input>2c
<bilop>::= and 3a
| or 3b
<ulop>::= not 4a
<input>::= input0 5a
|input1 5b
| input2 5c
Santa Fe Ant Trail
<code>::=<line>1a
|<code><line>1b
<line>::=<cond>2a
|<op>2b
<cond>::=if(food_ahead()){<line>}
else{<line>} 3a
<op>::=left(); 4a
|right(); 4b
|move(); 4c
Mastermind
<pin>::=<pin><pin>1a
|0 1b
|11c
|2 1d
|3 1e

Fig. 2 Grammars of four examples

(3) Draw an arrow from S_N to S_N^α for vertices S_N, S_N^α of step 2 if $N \to \alpha \in P$; naming the arrow either with the production rule or the rule name.
(4) Draw an arrow from S_M^α to S_N for each pair of vertices (S_M^α, S_N) in $\{(S_M^\alpha, S_N)|S_M^\alpha$, S_N are vertices of $LMGM(G)$ in which there exist sentential forms taking non-terminal symbol N as their leftmost non-terminal symbol$\}$.

The transition diagram obtained above reflects the leftmost derivation relation. We also call it the leftmost grammar model, denoted LMGM(G). Similarly, we can define a grammar model for sentential forms $\{\alpha|S \overset{*}{\Rightarrow} \alpha, \alpha \text{ is a sentential form of } G\}$. This is an extension to LMGM(G), denoted L1GM(G). It covers all valid grammatical derivations of the concerned grammar. Figure 6 shown in Sect. 4 is the leftmost grammar model of Santa Fe ant trail problem. And Fig. 4 through Fig. 5 are the L1GM(G)s of it and Quartic Symbolic Regression problem.

Fig. 3 A comparison of the results obtained for GE and πGE access all the problems analyzed

	Mean Best Fitness (Std. Dev.)	Mean Average Fitness (Std. Dev.)	Successful Runs
Santa Fe Ant			
πGE	47.47 (10.98)	12.14 (0.44)	1
GE	**53.37 (20.68)**	**12.57 (0.68)**	**4**
Symbolic Regression			
πGE	**0.665 (0.42)**	**0.123 (0.07)**	**18**
GE	0.444 (0.4)	0.76 (0.05)	10
Mastermind			
πGE	**0.905 (0.04)**	0.89 (0.001)	**14**
GE	0.897 (0.02)	0.89 (0.003)	7
Multiplexer			
πGE	**0.958 (0.06)**	**0.861 (0.04)**	**20**
GE	0.904 (0.05)	0.828 (0.05)	7

4 Experiments and Analysis

4.1 Experiments

In [6] ONeill etc. proposed a πGE and examined its effectiveness based on such specific problems as Santa Fe Ant Trailand Quartic Symbolic Regression problem, etc. The Grammars used in these problems are shown in Fig. 2. After setting the following parameters, they got Fig. 3 which demonstrated that πGE outperforms GE in solving Quartic Symbolic Regression problem, but scores poorly on the former case.

Part of the major parameters is: Number of Codons in a chromosome is in the range of [1–20], mutation (probability of 0.01 per bit) and one point crossover (probability of 0.9) are adopted.

4.2 Analysis

Before answering why πGE outperforms GE in solving Quartic Symbolic Regression problem, but scores poorly on the case of Santa Fe Ant Trail, we will construct their grammar models as shown in Fig. 4 through 5, and Fig. 6. Owing to the limited space, only parts of them are given here. Now lets paraphrase the figures. According to the modeling algorithm (see Sect. 4), each production rule $P \rightarrow \alpha$ contributes two states $S_P, S_P{}^{\alpha}$, we deduce, for the grammar of the Quartic Symbolic Regression problem, the maximal set of states contains 10 states, i.e. $\{S_{<expr>}, S_{<op>}, S_{<var>}, S_{<expr>}^{<expr><op><expr>}, S_{<expr>}^{<var>}, \ldots, S_{<var>}^{x}\}$. So, denoting these states as 1a, 2a, 3a, …, 3aR, L1LM(Quartic Symbolic Regression) as shown in Fig. 4. is obtained.

Clearly, πGE, GE correspond to L1GM(G) (Fig. 4 or Fig. 5) and LMGM(G) respectively. It is easy to see that the grammar models represented by Fig. 4 and Fig. 5 are always more complex than those by Fig. 6. In the case of Fig. 4, a program can

T	1a	2a	3a	1aR	1bR	2aR	2bR	2cR	2dR	3aR
1a				1a	1b					
2a						2a	2b	2c	2d	
3a										3a
1aR										
1bR										
2aR										
2bR										
2cR										
2dR										
3aR										

The transition matrix T(x,y) is defined as follows:

$$T(x,y) = \begin{cases} z & \text{exisitingarrowslabeledwith z leadingfromstate x to y} \\ \varepsilon & \text{all valuesnot givenherefor thetablecellsare } \varepsilon\text{s, which mean thereexisit } \varepsilon \text{ arrowsleadingfromstate x to y} \end{cases}$$

Fig. 4 The transition matrix (L1GM) T for quartic symbolic regression

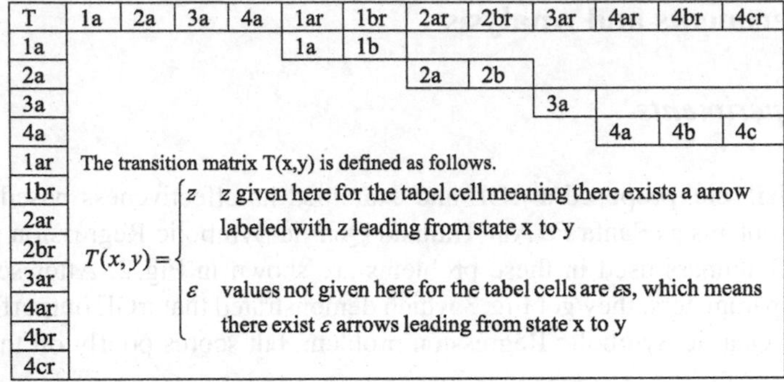

T
1a
2a
3a
4a
1ar
1br
2ar
2br
3ar
4ar
4br
4cr
1a
1b
2a
2b
3a
4a
4b
4c
The transition matrix T(x,y) is defined as follows.
T(x, y) =
z z given here for the tabel cell meaning there exists a arrow labeled with z leading from state x to y
ε values not given here for the tabel cells are εs, which means there exist ε arrows leading from state x to y

Fig. 5 The transition matrix (L1GM) T for Santa Fe Ant Trail

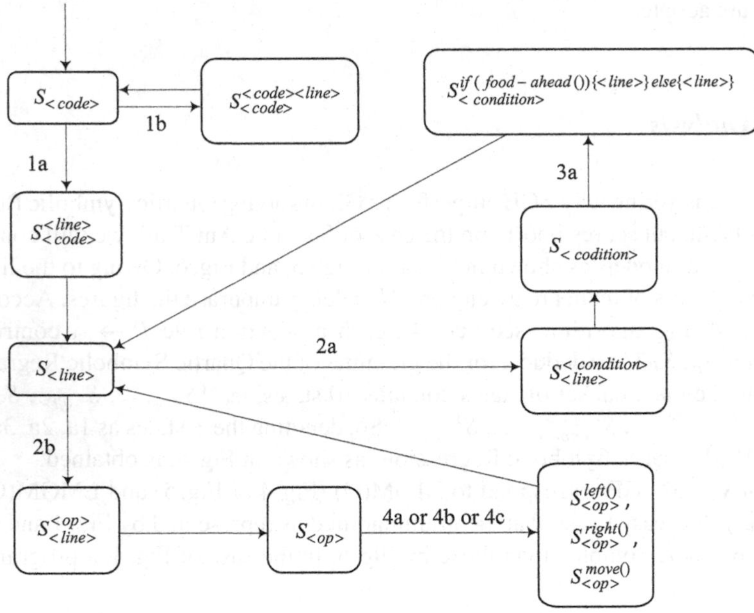

Fig. 6 The LMGM transition diagram for Santa Fe ant trail. Note that arrows without names are ε arrows. $S_{<code>}$ is the start state. Besides, ε arrows over all states are omitted here

be developed in a great number of ways. However, in the case of Santa Fe Ant Trail, each grammatical derivation of a sentence in Fig. 5, for the sake of unambiguousness of the grammar, corresponds to only one sequence of derivations in Fig. 6. This provides a new explanation on the experimental results. The analysis for the other two cases are similar.

5 Conclusion

Up to now, we have introduced grammar model as well as its applications in performance analysis. The method could be summarized as modeling the concerned context-free grammars first, and then analyzing the complexities of them. Our future work will focus on the analysis of their structures. It will play a critical role in design of effective GE variants such as HGP, MGE, etc.

Acknowledgments This work was supported by the National Natural Science Foundation of China (Grant Nos.61170199, 61363030), the Natural Science Foundation of Guangdong Province, China (Grant No.2015A030313501), the Scientific Research Fund of Education Department of Hunan Province, China (Grant No.11A004), and the Open Fund of Guangxi Key Laboratory of Trusted Software (Guilin University of Electronic Technology) under Grant No. kx201208.

References

1. Koza JR (1992) Genetic programming: on the programming of computers by means of natural selection. MIT Press, Cambridge
2. Mitchell M (1996) An Introduction to Genetic Algorithm. MIT Press, Cambridge
3. ONeill M, Ryan C (2001) Grammatical evolution. IEEE Trans Evol Comput 5(4):349–358
4. Ferreira C (2001) Gene expression programming: a new adaptive algorithm for solving problems. Complex Syst 13(2):87–129
5. Oltean M, Grosan C, Diosan L, Mihaila C (2009) Genetic programming with linear representation: a survey. Int J Artif Intell Tools 19(2):197–239
6. ONeill M, Brabzaon A, Nicolau M, Mc Garraghy S, Keenan P (2004) Grammatical evolution. In: Deb Ed K (ed) Proceedings of GECCO. LNCS vol 3103, pp 617–629
7. He P, Kang LS, Fu M (2008) Formality based genetic programming. IEEE congress on evolutionary computation, Hong Kong
8. He P, Kang LS, Johnson CG, Ying S (2011) Hoare logic-based genetic programming. Sci China Inf Sci 54(3):623–637
9. Langdon WB, Harman M (2015) Optimizing existing software with genetic programming. IEEE Trans Evol Comput 19(1):118–135
10. Burbidge R, Wilson MS (2014) Vector-valued function estimation by grammatical evolution. Inf Sci 258:182–199
11. Alfonseca M, Gil FJS (2013) Evolving an ecology of mathematical expressions with grammatical evolution. BioSystems 111:111–119
12. Risco-Martin JL, Colmenar JM, Hidalgo JI (2014) A methodology to automatically optimize dynamic memory managers applying grammatical evolution. J Syst Softw 91:109–123
13. He P, Johnson CG, Wang HF (2011) Modeling grammatical evolution by automaton. Sci China Inf Sci 54(12):2544–2553
14. He P, Deng ZL, Wang HF, Liu ZS (2015) Model approach to grammatical evolution: theory and case study, Soft Computing
15. Ryan C, Collins J, ONeill M (1998) Grammatical evolution: evolving programs for an arbitrary language. In: Banzhaf W, Poli R, Schoenauer M, Fogarty T (eds) Proceedings of the first European workshop on genetic programming (EuroGP98). LNCS, vol 1391. Springer, Berlin, pp 83–96
16. Fagan D, Hemberg E, ONeill M, McGarraghy S (2013) Understanding expansion order and phenotypic connectivity in GE, EuroGP 2013. LNCS, vol 7831, pp 33–48

A Comparison of Assessment Methods for Muscle Fatigue in Muscle Fatigue Contraction

Xinyu Huang and Qingsong Ai

Abstract Muscle fatigue often occurs in daily life and work. The assessment method of muscle fatigue based on the surface electromyography (sEMG) has been well reported in some comprehensive reviews. Recently, the time domain parameters, frequency domain parameters, discrete wavelet transform and non-linear methods have been widely used for the evaluation of muscle fatigue. In this paper, one fatigue indice was proposed: area ratio modified index (ARMI) was compared to seven assessment indices: root mean square (RMS), integrated electromyography (IEMG), mean power frequency (MPF), median frequency (MF), wavelet energy (WE), wavelet entropy (WEn) and Lemple–Ziv complexity (LZC). Ten healthy participants completed a fatigue experiment with isometric contraction of biceps brachii. It was showed that there was a significant positive rate of change of RMS, IEMG, WE and ARMI of sEMG, while a negative rate of change of MPF, MF and LZC of sEMG when fatigue occurred, and WEn of sEMG basically had no change. Meanwhile, the percentage deviation in ARMI was greater than the other seven indexes. It was proved that ARMI can be used as effective indicator and was more sensitive than the other seven indexes to evaluate the muscle fatigue.

Keywords sEMG · Muscle fatigue assessment · Assessment indice · ARMI

1 Introduction

Muscle fatigue is a common experience in daily life and work. For people engaged in manual labor, it's easy to cause damage to their joints and muscles due to muscle fatigue by the repetitive stresses. How to effectively evaluate muscle fatigue is of great significance for the prevention of muscle damage.

X. Huang (✉) · Q. Ai
School of Information Engineering, Wuhan University of Technology, Wuhan, Hubei, China
e-mail: xinyuhuangwhut@163.com

Q. Ai
e-mail: qingsongai@whut.edu.cn

V.E. Balas et al. (eds.), *Information Technology and Intelligent Transportation Systems*, Advances in Intelligent Systems and Computing 455,
DOI 10.1007/978-3-319-38771-0_48

The surface electromyography signal (sEMG) is a one-dimensional non-stationary time series recorded from the muscle surface through electrode collection and amplification [1]. It explains activities of neuromuscular system. It's widely used for the evaluation of muscle activity because it is non-invasive, real-time, and easy to operate. In recent years, the research on sEMGs becomes one hotspot in the field of clinical diagnosis, rehabilitation medicine and sports medicine [2]. The change of sEMG signal reacts muscle functional status to a certain extent so that muscle fatigue can be evaluated by analyzing its characteristics [3].

The sEMG-based approaches to muscle fatigue assessment were well reported in a comprehensive review [4], which provides a summarization of linear and non-linear methods for the evaluation of muscle fatigue. The traditional time domain and frequency domain methods are widely used to the evaluation of muscle fatigue. During fatigue contraction, the sEMG amplitude was found to increase [5, 6], at the same time, there was an decrease in frequency parameters of the sEMG [7–9]. Chowdhury SK et al. [10] used discrete wavelet transform analysis to conduct quantitative assessment of neck and shoulder muscle fatigue during dynamic contraction. It was found that power of wavelet in lower frequency showed the higher sensitivity to fatigue caused by the repetitive stresses. Balasubramanian V et al. [11] investigated the muscle activity status based on different bicycle designs. The changes in RMS and MPF of the sEMG signal were used to evaluate muscle fatigue. Hussain M et al. [12] presented that the wavelet transform could be used as a effective method for the understanding of muscle fatigue during walk. sEMG is non-stationary signal. Currently more and more non-linear methods were used to evaluate muscle fatigue. Rogers D R et al. [13] developed the first component of Principal component analysis (PCA) and the generalized mapping index (GMI) to assess muscle fatigue. Daniuseviciute L et al. [14] used Shannon wavelet entropy to detect changes of fatigued muscle complexity. Talebinejad M et al. [15] presented a Lemple–Ziv complexity method to estimate muscle fatigue.

The specific objective of this paper is to propose one fatigue indice to quantitatively evaluate the muscle fatigue and compare with other different fatigue assessment indexes, namely the traditional time domain parameters (RMS and IEMG), frequency domain parameters (MPF and MF), WE, WEn and LZC during fatigue contraction of biceps brachii. The rest of this paper is organized as follows: Sect. 2 describes methods and materials. Results of experiments are showed in Sect. 3. Section 4 presents discussion and future work. In the last section, the conclusion are given.

2 Methods and Materials

2.1 Participants

Ten male volunteers participated in this study after they were briefed extensively objectives of this study. Participants had the following physical characteristics (mean

± SD); age = 24.5 ± 1.3 yrs, body mass = 63.5 ± 8.4 kg, body height = 172.4 ± 4.3 cm. These volunteers were asked to take a complete rest for 24 h before the exercise.

2.2 *Electromyography (EMG) System*

A multi-channel sEMG acquisition system (Biometrics Ltd, UK) was used to collect sEMG data. The performance parameters of the instrument were as follows: 8, 16, 24 independent programmable channels, a maximum data acquisition rate of each channel is 20 kHZ, the maximum total acquisition rate of the wireless data acquisition is 160 KHz. The frequency range of the band-pass filter was 10–500 Hz, and the frequency of EMG data collection was set at 1000 Hz.

2.3 *Data Acquisition and Analysis*

Before the experiment alcohol and cotton balls were used to wipe right arms of the subjects biceps brachii surface, and remove grease on the skin surface. sEMG electrodes were attached to the biceps brachii through a perforated medical tape, and the reference electrode was placed on the elbow joint. In the process of fatigue contraction, the subjects kept the following state: right arms natural prolapsed, upper arm and forearm kept at a 90° angle, right hand held a 5 kg dumbbell, held the palm up, the elbow closed to the body, sustained and repeated isometric contraction of biceps brachii until the subjects felt intense fatigue. sEMG of the subjects was acquired by the sEMG acquisition system in whole experiment. Besides, at the end of experiment, the participants were asked to report discomfort in the biceps brachii using Borgs subjective rating scale [16].

2.3.1 Time Domain Parameters

The RMS and IEMG are defined as follows

$$RMS = \sqrt{\frac{1}{n}\sum_{n} x_n^2} \quad (1)$$

$$IEMG = \frac{1}{n}\sum_{n} |x_n| \quad (2)$$

where x_n is the value of the sEMG signal at a specific sampling point, and n is the number of samples.

2.3.2 Frequency Domain Parameters

The MPF is defined as follows

$$MPF = \frac{\int_{f1}^{f2} f \cdot PS(f) \cdot df}{\int_{f1}^{f2} PS(f) \cdot df} \tag{3}$$

where $PS(f)$ is the spectrum of sEMG power, $f1$ is lowest frequency of the bandwidth of sEMG and $f2$ is highest frequency of the bandwidth of sEMG.

The MF is defined as follows

$$\int_{f1}^{MF} PS(f) \cdot df = \int_{MF}^{f2} PS(f) \cdot df \tag{4}$$

where $PS(f)$ is the spectrum of sEMG power, $f1$ is lowest frequency of the bandwidth of sEMG and $f2$ is highest frequency of the bandwidth of sEMG.

2.3.3 Wavelet Energy

At each discrete wavelet transform decomposition, sEMG signal x is filtered using a series of low pass filters and high pass filters to analyze the low frequencies and high frequencies. The output of the low pass filters and high pass filters represent the approximate coefficients (CA) and detail coefficients (CD), respectively [17]. The sEMG signal is decomposed into five levels of wavelet transform using the sym2 wavelet (Fig. 1). The energy at the jth level can be calculated by integrating the squares of wavelet coefficient values.

$$E_j = |C_j|^2 \tag{5}$$

2.3.4 Wavelet Entropy

By discrete wavelet transform the total and relative wavelet energy can be obtained in the followings:

$$E_{tot} = \sum_j E_j \tag{6}$$

$$p_j = \frac{E_j}{E_{tot}} \tag{7}$$

where E_j is the wavelet energy at resolution level j, E_{tot} is the sum of the wavelet energy at different level and p_j is the probability density distribution of wavelet energy of different level.

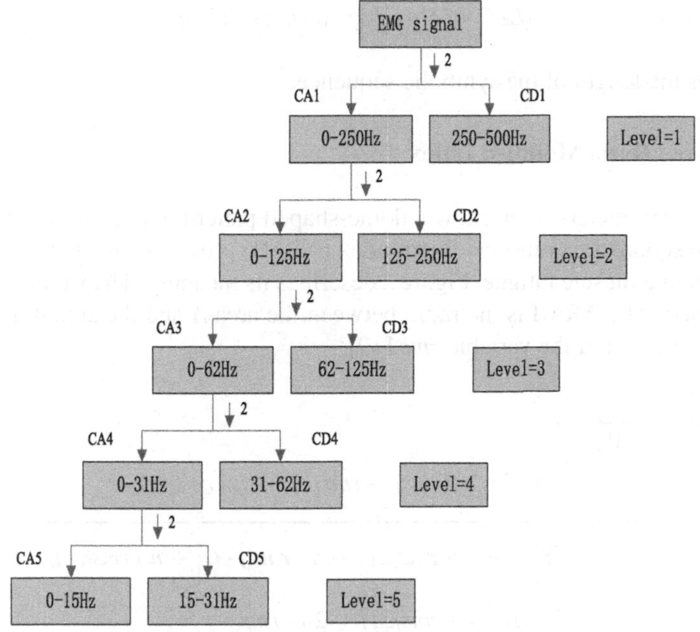

Fig. 1 Five level of decomposition algorithm of DWT used in this study

The wavelet entropy is defined as follows

$$WEn = -\sum_{j} p_j \cdot \log_2(p_j) \tag{8}$$

where p_j is the probability density distribution of wavelet energy of different level.

2.3.5 Lempel–Ziv Complexity

The signal $x(i)$ is converted into a binary sequence $s(i)$ by comparing the signal with a threshold value S_d. When the signal $x(i)$ is smaller than S_d, one maps the signal to 0; otherwise, to 1. This can be defined as follows.

$$s(i) = \begin{cases} 0, & \text{if } x(i) < S_d \\ 1, & \text{otherwise} \end{cases} \tag{9}$$

One good choice of Sd is the median of the signal. Several parsing methods have been proposed. The parsing method put forward by the original authors [18] of LZ complexity was used in this study. A detailed description of this algorithm is reported in [18]. $C(n)$ is normalized for obtaining complexity measurement independent of sequence length.

$$LZC = c(n)\left[\log_2\{c(n)\} + 1\right]/n \tag{10}$$

where n is the length of the symbolic sequence.

2.3.6 Area Ratio Modified Index

Amplitude parameters often show a dome-shaped pattern with respect to time. The root mean square area ratio modified index (ARMI) provides a quantitative method to evaluate the muscle fatigue. Figure 2 describes its meaning. The variations of *rms* is quantified. The ARMI is the radio between the area A and the area $A + B$ which is under the curve of the variable *rms* [19].

$$\begin{aligned} ARMI &= \frac{A}{|A| + B} \\ &= \frac{(t_c/2)\sum_{i=1}^{n}(rms_i + rms_{i-1}) - n \cdot rms_0 \cdot t_c}{\left|(t_c/2)\sum_{i=1}^{n}(rms_i + rms_{i-1}) - n \cdot rms_0 \cdot t_c\right| + n \cdot rms_0 \cdot t_c} \\ &= \frac{\sum_{i=1}^{n}(rms_i + rms_{i-1}) - 2n \cdot rms_0}{\left|\sum_{i=1}^{n}(rms_i + rms_{i-1}) - 2n \cdot rms_0\right| + 2n \cdot rms_0} (i = 1, 2, \ldots) \end{aligned} \tag{11}$$

where t_c is the duration of the entire contraction and *rms*0 is the reference value (the first *rms* value). A positive ARMI (0–1) expresses an increase when *rms* value increase with time; on the other hand, the ARMI value becomes negative (−0.5–0) when *rms* value decreases with time [19].

The slope of characteristic is calculated and analysed to evaluate fatigue. It is defined as follows

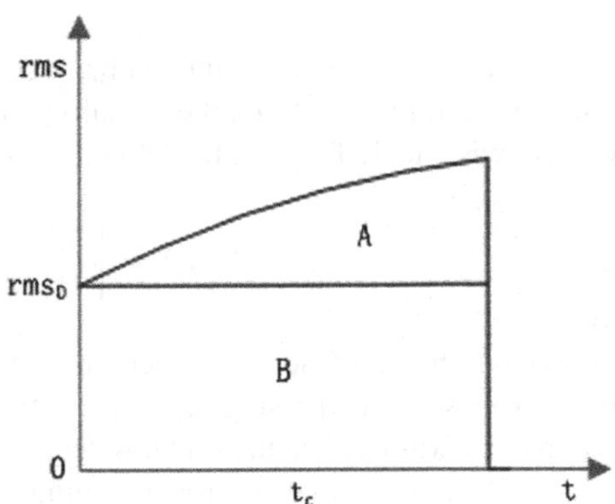

Fig. 2 Definition of the root mean square area ratio modified index

$$\%Deviation\ in\ ch = \frac{ch(after) - ch(before)}{ch(before)} \tag{12}$$

where ch is the characteristic value.

All original sEMG signals were sampled by the frequency of 1000 Hz and processed through a band-pass filter (20–500 Hz) and a band-stop filter (49–51 Hz). A characteristic value of sEMG was calculated every 5 s. These data were processed and analyzed in MATLAB. The first characteristic value was used as characteristic value before fatigue and the last characteristic value was used as characteristic value after fatigue except for ARMI. The first characteristic of ARMI is zero, so the second characteristic of ARMI was used as characteristic value before fatigue.

2.4 Statistical Analysis

Statistical software (SPSS19.0) was used to determine the number of subjects in this study. The analysis used the mean and SD values of the extracted features with an alpha level of 0.05. The effect of time was set at two levels: T1 (before fatigue) and T2 (after fatigue). The effect of time and the difference among deviation of different characteristics were evaluated by using pairwise t-test. The significance level was set to $p < 0.05$ for all analysis.

3 Results

In the process of muscle fatigue contraction, the degree of fatigue of subjects increased and subjects felt intense fatigue eventually. Average time duration for static was 68 ± 14 s. A mean discomfort of 7 (=strong discomfort) in the biceps brachii was reported by the subjects at the end of experiment.

Different characteristic value of sEMG from T1 to T2 was shown in Fig. 3. The average of RMS value and IEMG value increased from T1 to T2. Meanwhile, the average of MPF value, MF value and LZC value decreased from T1 to T2. In addition, the average of WE value and ARMI value at T1 was larger than these at T2. The average of WEn at T1 was basically equal to that at T2. Pairwise t tests showed a distinct pattern of statistical significance between the time instances for different indexes. There was statistical significance on the eight indexes except for WEn from T1 to T2 (Table 1).

Figure 4 showed the percentage deviation of eight indexes. There was a significant positive rate of change of RMS, IEMG, WE and ARMI with a negative rate of change of MPF, MF and LZC, WEn basically had no change. The percentage deviation in ARMI was largest among eight indexes, percentage deviation in WE was second largest, percentage deviation in WEn was lowest.

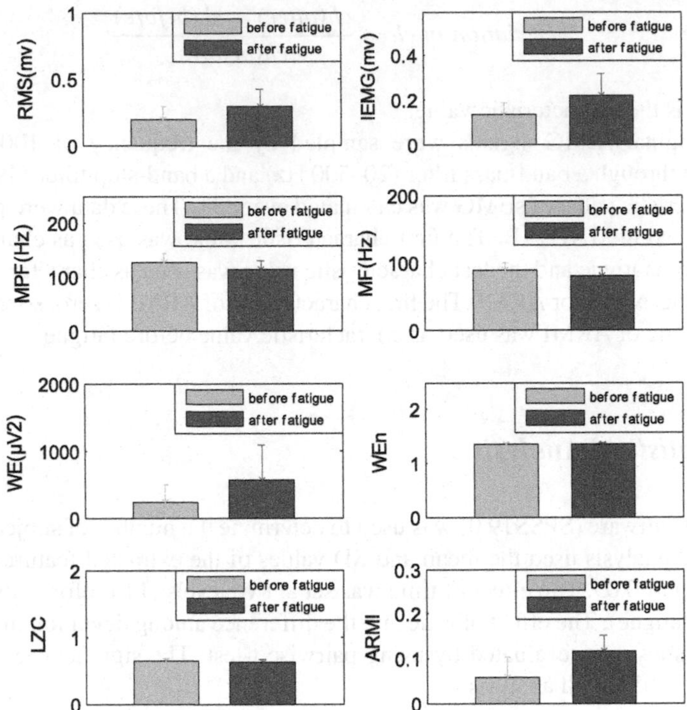

Fig. 3 Different mean characteristic value of sEMG before and after fatigue

Table 1 P-values of the pairwise t-test for the effect of time on the characteristics

	RMS	IEMG	MPF	MF	WE	WEn	LZC	ARMI
T1 versus T2	0.001*	0.003*	0.009*	0.021*	0.008*	0.528	0.001*	0.004*

T1 and T2 refer to before fatigue and after fatigue, respectively. Statistically significant values are marked with asterisks (*)

The p-values of the pairwise t-test for percentage deviation in different indexes were given in Table 2. It showed that there was no difference among RMS, IEMG and ARMI and the percentage deviation in MPF, MF and LZC showed no difference. Meanwhile, Statistical significance was observed between ARMI and the others except for RMS, IEMG and WE.

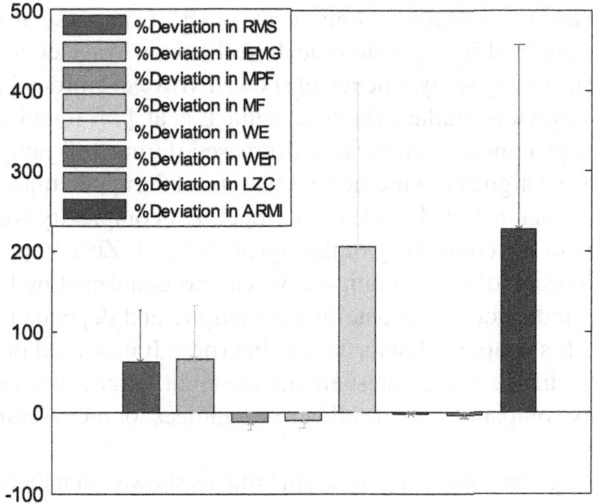

Fig. 4 Different percentage deviation in eight indexes

Table 2 P-values of the pairwise t-test for percentage deviation in different characteristics

	IEMG	MPF	MF	WE	WEn	LZC	ARMI
RMS	0.78	0.001*	0.001*	0.005*	0.002*	0.001*	0.079
IEMG		0.001*	0.001*	0.008*	0.002*	0.001*	0.091
MPF			0.605	0.002*	0.031*	0.085	0.028*
MF				0.003*	0.033*	0.118	0.029*
WE					0.004*	0.003*	0.786
WEn						0.001*	0.032*
LZC							0.030*

Statistically significant values are marked with asterisks (*)

4 Discussion

The purpose of this paper was to compare the performance of eight fatigue assessment indexes. Ten subjects took part in isometric contraction experiments of biceps brachii. Continuously performed repetitive contraction increased the subjective discomfort in the biceps brachii, confirming the fatigue in that muscle. The change of characteristics of sEMG was used to reflect the muscle fatigue. In the past, a significant change of the spectrum and amplitude of sEMG signal due to muscle fatigue has been demonstrated. In this study, the increase of RMS and IEMG value and the decrease of MPF and MF value due to fatigue was consistent with the experimental results of other researchers [20, 21]. There was a significant increase in the WE during muscle fatigue contraction, which was also reported by Chowdhury SK et al. [10].

The entropy is a non-linear method to measure the complexity of the signal. Due to the physiological mechanisms of muscle fatigue, high frequency components of sEMG decrease and low frequency components of sEMG increase. It appears that the entropy may be affected by a physiological mechanism. Wavelet entropy was used to evaluate fatigue in this study. The result showed Wavelet entropy basically had no change. In a few previous studies, Daniuseviciute L et al. [14] found wavelet entropy of sEMG of biceps femoris significantly decreased during 100 jumps. But entropy was not affected by fatigue in isometric contractions of the upper trapezius muscle that was reported by Troiano A et al. [22]. The Lempel–Ziv complexity is another method for measurement of the complexity of the signal. Lempel–Ziv complexity decreased during the progression of muscle fatigue, which was confirmed by Liu J et al. [23]. ARMI is a new indicator to evaluate muscle fatigue and depends on a reasonable reference value. It's simple, reliable, easy to interpret. It increased in muscle fatigue contraction. The change was greatest among the eight fatigue indices and showed a higher sensitivity compared with the other seven indices for the assessment of muscle fatigue.

The preliminary test comparing the eight indexes showed that there was no difference among RMS, IEMG and ARMI, the main reason was that all of them reflected the change of amplitude of sEMG. Similarly, MPF and MF reflected the change of frequency spectrum of sEMG with no difference. WEn was the combination of time-frequency domain analysis and complexity. LZC reflected the complexity of the signal. In the future work, the eight indexes mentioned by this study will be used to evaluate muscle fatigue in dynamic fatigue tests and the effect of assessment remains to be further validation.

5 Conclusion

By comparing eight fatigue indexes, it was found that RMS, IEMG, WE and ARMI of sEMG increased with the decrease of MPF, MF, LZC of sEMG in muscle fatigue contraction, and WEn basically had no change. Meanwhile, ARMI showed the highest sensitivity among eight indexes. Through the statistical analysis of percentage deviation in eight indexes, it was found that there was significant difference between ARMI and the others except for RMS, IEMG and WE. Therefore, ARMI can be used as an effective indicator for the evaluation of muscle fatigue.

Acknowledgments This work was supported by the National Natural Science Foundation of China (Grant No. 51475342).

References

1. Phinyomark A, Phothisonothai M, Suklaead P et al (2011) Fractal analysis of surface electromyography (EMG) signal for identify hand movements using critical exponent analysis. Software engineering and computer systems. Springer, Berlin, pp 703–713
2. Merletti R, Farina D (2009) Analysis of intramuscular electromyogram signals. Philos Trans R Soc A Math Phys Eng Sci 367(1887):357–368

3. Cifrek M, Medved V, Tonković S et al (2009) Surface EMG based muscle fatigue evaluation in biomechanics. Clin Biomech 24(4):327–340
4. González-Izal M, Malanda A, Gorostiaga E et al (2012) Electromyographic models to assess muscle fatigue. J Electromyogr Kinesiol 22(4):501–512
5. Ferguson SA, Marras WS, Allread WG et al (2012) Musculoskeletal disorder risk during automotive assembly: current vs. seated. Appl Ergon 43(4):671–678
6. Staudenmann D, van Dieën JH, Stegeman DF et al (2014) Increase in heterogeneity of biceps brachii activation during isometric submaximal fatiguing contractions: a multichannel surface EMG study. J Neurophysiol 111(5):984–990
7. Li X, Shin H, Zhou P et al (2014) Power spectral analysis of surface electromyography (EMG) at matched contraction levels of the first dorsal interosseous muscle in stroke survivors[J]. Clin Neurophysiol 125(5):988–994
8. Halim I, Omar AR, Saman AM et al (2012) Assessment of muscle fatigue associated with prolonged standing in the workplace. Saf. Health at Work 3(1):31–42
9. Venugopal G, Navaneethakrishna M, Ramakrishnan S (2014) Extraction and analysis of multiple time window features associated with muscle fatigue conditions using sEMG signals. Expert Syst Appl 41(6):2652–2659
10. Chowdhury SK, Nimbarte AD, Jaridi M et al (2013) Discrete wavelet transform analysis of surface electromyography for the fatigue assessment of neck and shoulder muscles. J Electromyogr Kinesiol 23(5):995–1003
11. Balasubramanian V, Jagannath M, Adalarasu K (2014) Muscle fatigue based evaluation of bicycle design. Appl Ergon 45(2):339–345
12. Hussain M, Mamun M (2012) Effectiveness of the wavelet transform on the surface EMG to understand the muscle fatigue during walk. Meas Sci Rev 12(1):28–33
13. Rogers DR, MacIsaac DT (2013) A comparison of EMG-based muscle fatigue assessments during dynamic contractions. J Electromyogr Kinesiol 23(5):1004–1011
14. Daniuseviciute L, Pukenas K, Brazaitis M, et al (2010) Wavelet-based entropy analysis of electromyography during 100 jump. Electron Electr Eng–Kaunas: Technologija (8):104
15. Talebinejad M, Chan ADC, Miri A (2011) A Lempel-Ziv complexity measure for muscle fatigue estimation. J Electromyogr Kinesiol 21(2):236–241
16. Borg G (1998) Borg's perceived exertion and pain scales. Human Kinetics Publishers, Champaign
17. Kumar DK, Pah ND, Bradley A (2003) Wavelet analysis of surface electromyography. IEEE Trans Neural Syst Rehabil Eng 11(4):400–406
18. Lempel A, Ziv J (1976) On the complexity of finite sequences. IEEE Trans Inf Theory 22(1):75–81
19. Olmo G, Laterza F, Presti LL (2000) Matched wavelet approach in stretching analysis of electrically evoked surface EMG signal. Signal Process 80(4):671–684
20. Ming D, Wang X, Xu R et al (2014) sEMG feature analysis on forearm muscle fatigue during isometric contractions. Trans Tianjin Univ 20:139–143
21. Zwambag DP, Brown SHM (2014) The effect of contralateral submaximal contraction on the development of biceps brachii muscle fatigue. In: human factors: the journal of the human factors and ergonomics society, p 0018720814550034
22. Troiano A, Naddeo F, Sosso E et al (2008) Assessment of force and fatigue in isometric contractions of the upper trapezius muscle by surface EMG signal and perceived exertion scale. Gait Posture 28(2):179–186
23. Liu J, Wang J, Luo X (2003) sEMG signal complexity and entropy changes during exercise-induced local muscle fratigue. Shengwu Wuli Xuebao 20(3):198–202

Design of Products Non-contact Detection System

Juan Zhu, Nan Li and Wang Yulan

Abstract Qualified for the detection of cases of industrial products, the proposed non-contact detection system based on the MFC. CCD based imaging and product distribution geometry and boundary conditions, combined with digital image processing technology for industrial product testing, given the overall design concept detection system, complete MFC detection system based on PC software, use the button controls to achieve the start-stop control of the process, with the sound of an alarm signal is given substandard products, software for image preprocessing first, then the outline of the product for testing, with the realization of testing products based matching approach. Experimental results show that the system can be very good to complete product testing capabilities, the system is robust and real-time, accurate detection rate is higher than 998, the detection time of less than 10 ms, fully meet the requirements of industrial inspection.

Keywords Non-contact · Product testing · Image processing · Border measure

1 Introduction

The final process of product is detection in industrial production to see whether the product is qualified. Unqualified products go into the market will greatly reduce the company's reputation, and affect their development prospects. Therefore, efficient

J. Zhu (✉)
Electronic Information College of Changchun Guanghua University,
Changchun 130031, China
e-mail: zhuj_guanghua@qq.com

N. Li
Electronic Engineering Department of Armored
Force Technology College, Changchun 130031, China
e-mail: 572210786@qq.com

W. Yulan
Scientific Research Department of Changchun University, Changchun 130031, China
e-mail: 14525451@qq.com

V.E. Balas et al. (eds.), *Information Technology and Intelligent Transportation Systems*, Advances in Intelligent Systems and Computing 455,
DOI 10.1007/978-3-319-38771-0_49

and reliable product inspection system has broad market prospect. With the rapid development of computers, image processing technology has become increasingly mature. In order to improve the efficiency and effectiveness of product detection, this paper presents a product detection system based on image processing technology. Detect the appearance of the product is most important. Use camera to obtain the product image, the image data were analyzed using the algorithm determines whether the product is qualified. This system is simple without artificial measurement, and has high visibility, which can greatly reduce the detection time for the company.

2 Design of the Detection System

After the product is completed, a detection is required after passing the factory. To achieve automatic detection of industrial products, in the product passing through the position fixed camera, use the camera to capture images of products. The data of product images is transmits to up computer. Embed image processing algorithm to up computer, using algorithms to detect cases of qualified products, the result of detection by serial transmission to the next crew of the CPU to control whether the disposal stage of the product to make culling. The entire inspection system block diagram shown in Fig. 1 to achieve. Wherein the dashed box for PC image processing algorithms need to write a program, dotted line box to the hardware part of the system, where the system chosen economical microcontroller to complete the hardware control functions.

Figure dashed box contents is the contents of the PC, the PC used in Microsoft Foundation Classes Acronym MFC development platform, its rich programming interface controls can make visibility, simple operation, using MFC to do software platform, the detection algorithm can be updated as needed, in order to improve the

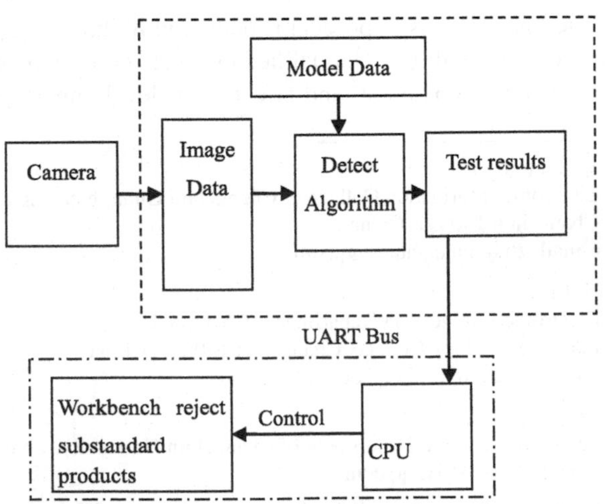

Fig. 1 Detection system block diagram

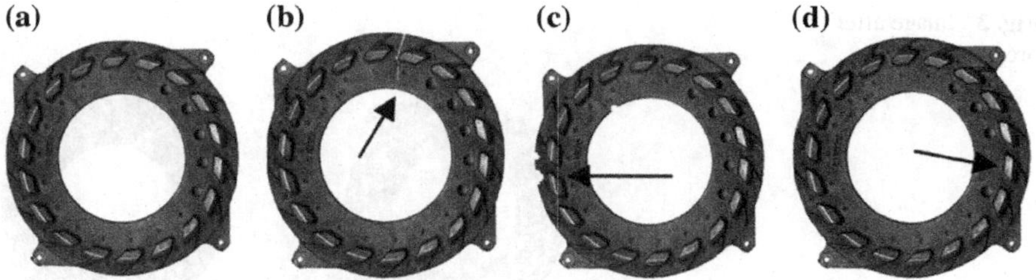

Fig. 2 Qualified parts and defective parts

accuracy of test results, and the algorithm is simple porting the software has turned video, and other interactive detection start button, and the display window to display products.

3 Detection Method of Products

The system of the host computer in the MFC environment to achieve camera using C++ programming language to read the image acquisition, display, testing and other functions. Because industrial products qualified or extremely strict requirements to Fig. 2a parts, for example, only in full compliance with the parts shown only called qualified products. When the deviation reaches a certain level are considered to be substandard products, as shown in Fig. 2b–d.

4 Design of the Detection System

The arrows indicate the tip position of the defective parts, the part in which a figure cracks, b in figure edge parts defective, missing a recess on part c in figure. These parts should be deemed to have failed parts. In response to this demand for such parts testing, design the corresponding image processing detection algorithm.

4.1 Image Data Preprocessing

When industrial product testing, using the camera to obtain image data part, data obtained at this time will carry a lot of noise, the other parts of the image will not only data, but also inevitably carry background data, so image preprocessing to be from the above analysis, the part may contain details of real data, therefore we must ensure that the process of de-noising keep the details of the image, the paper kind median filtering way to smooth the noise, in order to target and background

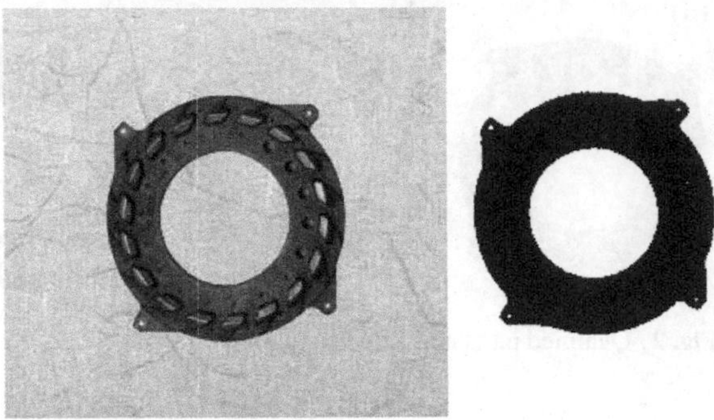

Fig. 3 Image after preprocessing

have sufficient reparability, using the log-likelihood function to extract the target, removing the background area. In the center of the image (target area) and image periphery (background area) were sampled to give the target and the background histogram $H_{object}(i)$ and $H_{back}(i)$, n_{object} and n_{back}, respectively, with the number of target and back-ground sampling point, the target and background normalized histogram are:

$$H'_{obj}(i) = \frac{H_{obj}(i)}{n_{obj}} \tag{1}$$

$$H'_{back}(i) = \frac{H_{back}(i)}{n_{back}} \tag{2}$$

To establish the number of each pixel likelihood expression:

$$L(i) = \log \frac{\max(H'_{obj}(i), \delta)}{\max(H'_{back}(i), \delta)} \tag{3}$$

Here, δ is very small, 0.0000001, It is to prevent the expression does not make sense. When $L(i)$ is greater than the threshold value, the corresponding pixel gray-point i is considered to be the target area, otherwise considered to be the background, the target area is obtained The white, black for the background area, showing that the method can distinguish between good target and background (Fig. 3).

4.2 Parts Inspection

The template is stored in the memory part, the part to be detected in real time to read, since there will be detected angle between the part and the template parts rotate, difference, change the light of changes in illumination, etc. can be overcome through the front of the front background extraction mode, then Part detection algorithm is

Fig. 4 Code of serial port

```
HANDLE hCom;
DWORD dwError;
hCon = CreateFile ("COM1",
GENERIC_READ| GENERIC_WRITE, 0,
NULL, OPEN_EXISTING, 0, NULL);
if(hCom==(HANDLE)0xFFFFFFFF)
{
dwError= GetLastError();
MessageBox(dwError);
}
```

mainly to solve the problem of parts rotation and scaling, in order to solve these two problems, with the SIFT algorithm calculates a target angle and zoom level selection.

Firstly SIFT operator match the characteristics of the target, since the parts most of the questions are part of the part to be detected to calculate the rotation angle, then the parts to be tested the same angle reverse rotation process, so that the angle to be detected is consistent parts and template part, and then to the target template edge detection, de-noising, etc., and performs a full match. Edge detection target image and the template to calculate the similarity. Error parts below the threshold considered acceptable parts, otherwise considered to be substandard parts.

4.3 Output the Detection Result

The test results sent to the CPU via the serial port, serial port API functions directly transferred in MFC: CreateFile. Open the serial code is shown in Fig. 4.

5 Experimental Results

To test the effectiveness of the present system of four different parts were selected to the 1000 test samples for testing, where each part has about 90 % and 10 % of qualified products of substandard products, test results is shown in Table 1.

From the experimental results, the detection accuracy of the system up to 998, and the detection time of less than 10 ms, fully able to meet the requirements of industrial processes.

Table 1 Detect results

Sample	Pass/Fail	Detect result pass/fail	Error rate ‰	Detection time (ms)
1	900/100	899/101	1	6
2	903/97	903/97	0	4
3	901/99	899/101	2	8
4	898/12	899/11	1	5

6 Summary

This paper proposes a detection system for industrial products, given the system's overall design philosophy and detection algorithms, from the experimental data show that the system can fully meet the requirements of industrial production, auxiliary plant for testing of parts, greatly improving plant work efficiency and reduce overhead, the system is only suitable for the detection plane parts, parts for all aspects of three-dimensional detection system can not be achieved, which is to study the contents of the future.

References

1. Meili MA, Yizhong MA, Yongzhong Z (2010) Image processing technology in industrial products measurement electronic design engineering (3)
2. Xianjun Y (2007) Product surface image defects automatic detection method microcomputer information (18)
3. Gong M, Tang Y, Qicai L (1995) Egg quality and achieve automatic detection method automation and instrumentation (05)
4. Hong L, Jin T, Shu-zhi L (2014) Image processing platform designed with matlab GUI. J Jiangxi Univ Sci Technol 35(3):79–84
5. Rafael C, GonzalecRichard E (2011) Woods digital image processing. Electronic Industry Press, Beijing
6. Haibo LIU, Jing SHEN, Song GUO (2010) Visual c++, a digital image processing technology. Mechanical Industry Press, Beijing
7. Xueli B (2014) An open image processing platform based on.NET virtual machine. Comput Meas Control 22(7):2174–2177
8. Shaosheng D, Yue Z (2005) A design method of universal image processing plat-form. Semicond Ptoelectronics 26(1):66–68

The Application of SPA in Pattern Recognition

Jian Shi and Jianzhong Jiang

Abstract Set Pair Analysis (SPA) theory is a relatively new approach to deal with uncertainty, and has been successfully applied to areas including decision making, data fusion, product design, etc. The set pair is defined as a pair that consists of two interrelated sets, SPA's main idea is considering the relation of certainties and uncertainties of the set pair, then analyzing and processing the relation. In this study, we employ SPA theory in pattern recognition to enhance accuracy and speed. Class $c_j (j = 1, 2, ..., n)$ is divided into several subclasses which are represented as binary connection number, the value of coefficient i is optimized by genetic algorithm. To validate the usefulness of our method, experiments were carried out and the results indicated that accuracy and speed may be improved significantly by using our method.

Keywords Set pair analysis · Connection number · Genetic algorithm · Pattern recognition

1 Introduction

For the purpose of understanding and manipulating uncertainty knowledge, researchers have proposed many methods, such as probability theory, fuzzy set theory and rough set theory. Set Pair Analysis (SPA, also named Connection Mathematics) is another method and overlaps with the others, proposed by Keqin Zhao in 1989 [1].

In SPA theory, depicting an object needs two sets (namely set pair, SP). The underlying idea of SPA is analyzing certain and uncertain relations of SP, then computing connection number $\mu = a + bi + cj$, where a and c indicate certainty, b indicates uncertainty. The coefficient j is specified as -1, the coefficient i represents

J. Shi (✉) · J. Jiang
National Digital Switching System Engineering and Technological R&D Center, Zhengzhou, People's Republic of China
e-mail: shijianloko@163.com

J. Jiang
e-mail: 1228244885@qq.com

V.E. Balas et al. (eds.), *Information Technology and Intelligent Transportation Systems*, Advances in Intelligent Systems and Computing 455,
DOI 10.1007/978-3-319-38771-0_50

uncertainty in micro-level (b represents uncertainty in macro-level). SPA describes uncertainty from different perspectives and different levels, thus portrays uncertainty with a more realistic and practical approach, and has been successfully applied to areas including data fusion [2], decision making [3], product design [4] and network planning [5, 6], etc. Binary Connection Number (BCN) is a special case of connection number, defined as $\mu = A + Bi$. We mainly study BCN instead of connection number in this paper, and have put it into pattern recognition. Traditional method in pattern recognition, Nearest Neighbor (NN) for example, is inefficient. In this paper, each class $c_j (j = 1, 2, ..., n)$ is classified into $k_j (j = 1, 2, ..., n)$ subsets which are presented as $\mu_{j_h} = A_{jh} + B_{jh} i_{jh} (j = 1, 2, ..., n \, and \, h = 1, 2, ..., k_j)$ by clustering algorithm k-means. By SPA theory, the coefficient i_{jh} is an uncertain value between -1 and 1, i.e. $i_{jh} \in [-1, 1]$, and the value of i_{jh} can be different in various circumstances. There have been many studies on exploring how to ascertain the value of i_{jh} [7–10]. In this paper, coefficient i_{jh} is optimized by Genetic Algorithm (GA).

The rest of the paper is organized as follows. Section 2 describes basic concepts of SPA amply, and Sect. 3 proposes our method about the application of SPA in pattern recognition. In the next section the explanation for the research design and experiments are represented, and in the final section, the conclusions of the study are represented.

2 Set Pair Analysis Theory

SPA theory, proposed by Keqin Zhao in 1989, is a new mathematical tool to analyze and process uncertainty, and has been proved a more reasonable uncertain theory. The basic principles of SPA consist of four aspects [11]:

(1) Anything in the world are common contacts;
(2) Uncertainty principle;
(3) Things or concepts are exist in pairs;
(4) All things are interrelated and interact with each other.

The most significant concepts of SPA are set pair and connection number.

2.1 Set Pair

Set pair is a concept derived from the thought of Russells paradox. To depict an objective target, two sets are needed simultaneously. For example, while talking about positive number, we compare it with negative number unconsciously, thus the positive and negative number may compose a set pair.

Definition 1 Set Pair (SP) is the combination of two sets which are needed when representing an objective target.

Given objective target O, set A_0 and set B_0, SP is expressed as $H_0 = (A_0, B_0)$.

2.2 Connection Number

In previous sections, connection number is introduced briefly. Binary Connection Number (BCN) is a special case of connection number, in this paper, more attentions are paid on BCN.

Definition 2 Given $A, B \in R^{+}, i \in [-1, 1]$, BCN has the form

$$\mu = A + Bi \tag{1}$$

where A is certain degree between the sets in SP, B is uncertain degree, the coefficient i is uncertain positive number whose value is between -1 and 1, and takes different values in different circumstance. BCN represents the relation of the two sets in SP, A_0 and B_0 are more similar as the value of A is bigger, and the relation is more uncertain as value of B is bigger. Given sample $x = (x_1, x_2, ..., x_n)$ whose real value is v. x and v make up a set pair H_0, Keqin Zhao proposed that BCN of H_0 can be calculated as formula (2).

$$\mu = \bar{x} + si \tag{2}$$

where $\bar{x} = average(x)$ is average value of sample x, s is the variance of x, and $i \in [-1, 1]$.

3 Application in Pattern Recognition

Nearest neighbor (NN) is one of the most popular classification rules, although it is an old technique. Given n classes, $w_j (j = 1, 2, ..., n)$ and a point $x \in R^l$, and N training points, x_j, $j = 1, 2, ..., N$, in the l-dimensional space, with the corresponding class labels. Given a point x, whose class label is unknown, the task is to classify x in one of the c classes [12]. NN is easy to apply, however, it is time-consuming, low-accuracy and sensitive to noise.

We apply BCN to pattern recognition. First, class $c_j (j = 1, 2, ..., n)$ is divided into $k_j (j = 1, 2, ..., n)$ subsets by k-means algorithm, then representing subset $c_{jh} (j = 1, 2, ..., n \; and \; h = 1, 2, ..., k_j)$ $\mu_{jh} = \bar{x}_{jh} + s_{jh} i_{jh}$ as where $\bar{x}_{jh}$ and s_{jh} is the average value and variance of points in subclass $c_{jh} (j = 1, 2, ..., n \; and \; h = 1, 2, ..., k_j)$ respectively. In statistics, the average represents certainty of samples and variance reflects uncertainty. Thus it is reasonable to represent c_{jh} as μ_{jh}.

However, the value of coefficient i_{jh} is uncertain [10]. In this study, we adopt GA as the search method to optimize i_{jh}. The detailed explanation for each step is presented as follows.

4 The Research Design and Experiments

4.1 Application Data

The data set used for this research was taken from the Wisconsin breast cancer (original) data set, which is one of the public medical database provided by UCI repository [13]. It is a database for classification algorithm testing, made publicly available by Dr. William H. Wolberg of University of Wisconsin Hospitals. The database consists of 699 instances, each instance has 10 attributes, and attributes 2 through 10 have been used to represent instances. Each instance has one of 2 possible classes: benign or malignant. There are 16 instances that contain a single missing attribute value, we have removed these instances before carrying out experiments. Thus we only use 683 instances in our experiments.

4.2 Research Design

We select 300 cases as training data to find $\bar{x}_{jh}$ and s_{jh}, as shown on phase 1 of Fig. 1, then the other 200 cases are selected for optimizing coefficient i. The rest 183 cases mixed with Gaussian noise ($\mu = 0$ and $\sigma = 0, 0.5\sigma', \sigma' or 1.5\sigma'$) is used for testing, σ' is standard deviation of training data.

In this study, the $k_j(j = 1, 2)$ parameter of k-means is set as $k_1 = 1$, $k_2 = 1$ namely class c_1 is divided into two subclasses. Thus, there are three subclasses in total, represented as $\mu_{11} = A_{11} + B_{11}i_{11}$, $\mu_{12} = A_{12} + B_{12}i_{12}$ and $\mu_{21} = A_{21} + B_{21}i_{21}$ respectively. To enable GA to find the optimal or near-optimal parameters, we should design the structure of a chromosome, a form of binary strings [14]. The structure of the chromosomes and population is represented in Fig. 2.

As shown in Fig. 2, the length of each chromosome is $72 \times 3 = 216$ bits. We encode $i_{jh}(j = 1, 2\, and\, h = 1, 2)$ with 72-bits, and the first bit is sign bit (0: minus, 1: plus). For the controlling parameter of the GA, we use 100 chromosomes in the population and set the crossover rate to 0.8 and mutation rate to 0.008. The number of generation is set to 1000.

4.3 Experimental Results

Table 1 shows the optimized value of i_{11}, i_{12} and i_{13}.

Table 2 describes the accuracy and program running time of NN and SPA based Pattern Recognition (SPR) which is produced when applying the parameters in Table 1. As we can see from Table 2, SPR has the higher level of accuracy than NN,

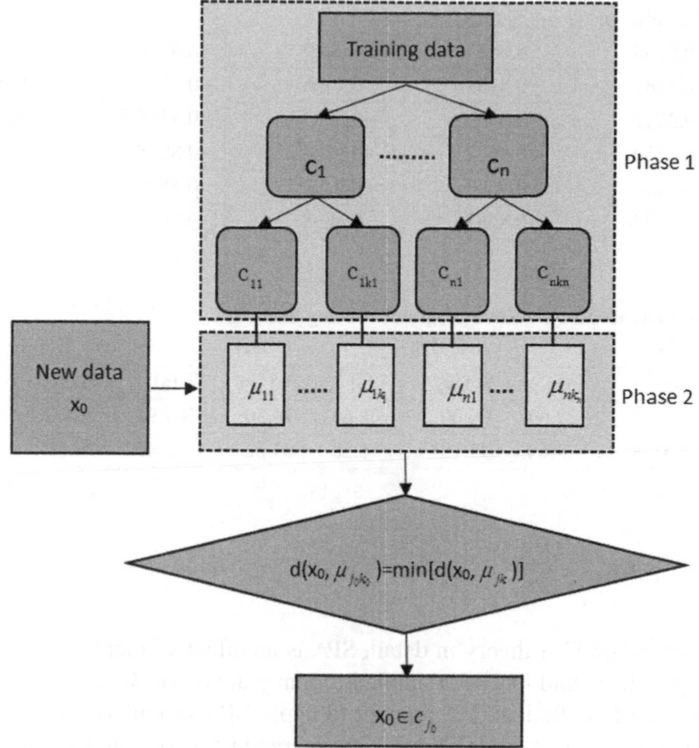

Fig. 1 Flowchart of this study

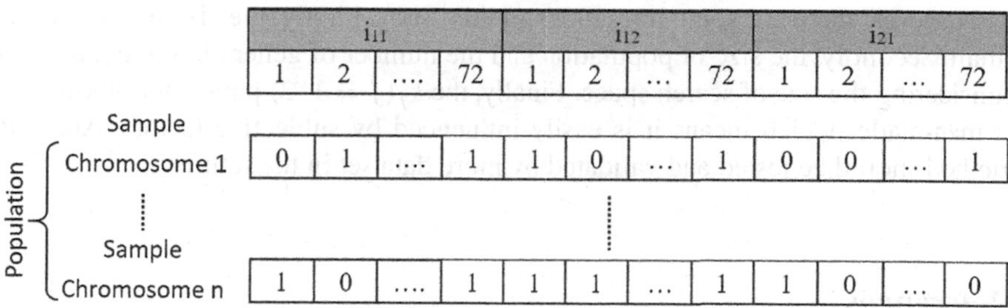

Fig. 2 Chromosome structure

and the superiority becoming more and more obvious as noise increasing. The results show that SPR improves the accuracy of NN by about 9.51 % while the variance of noise is equal to which means SPR has good stability against noise.

Table 1 The value of i_{11}, i_{12} and i_{13}

i11	−0.0709	0.6457	0.9449	−0.6614	−0.3386
i12	0.1181	0.8189	0.1496	−0.1181	−0.5512
i21	0.2047	0.5591	−0.9685	0.3465	−0.2126
i11	0.7323	−0.3622	−0.9764	−0.8898	
i12	−0.9449	−0.9606	−0.9291	0.5591	
i21	−0.4173	−0.8583	0.8898	−0.2677	

Table 2 Average accuracy (A) and programe running time (T)

	$\sigma = 0$		$\sigma = 0.5\sigma'$		$\sigma = \sigma'$		$\sigma = 1.5\sigma'$	
	A (%)	T (s)	A (%)	T (s)	A (%)	T (s)	A (%)	T (s)
NN (%)	99.00	127.64	90.00	108.64	74.50	102.91	70.00	110.35
SPR (%)	100.00	3.64	93.28	6.28	84.62	3.43	79.51	3.48

5 Conclusion

We have described SPA theory in detail. SPA is an effective method for processing uncertain problems and has been applied to many areas. However, it is difficult to find the value of coefficient. We attempt to apply SPA to pattern recognition, and represent class c as $\mu = \bar{x} + si$, where $\bar{x}$ is average value of c and s is the variance of c. The coefficient i is computed by GA. Experiment results show that our method enhance the accuracy and speed of NN algorithm.

However, there are some limitations in this study. Firstly, the size of database is small. Secondly, the size of population and the number of generation is small when considering the size of search space. Finally, the $k_j (j = 1, 2)$ parameter of k-means is man-made, which means it is easily influenced by subjective factors. And this method should be tested and validated by more data set in the future.

References

1. Zhao K (1994) Set pair analysis and its preliminary application. Explor Nat 13(47):67–72 (in Chinese)
2. Liu Y, Niu Y, Liu T (2005) Application of multi-radar data fusion with set pair analysis. J East China Shipbuilding Inst (Natural Science Edition) 19(2):64–67 (in Chinese)
3. Zhao K (2010) Decision making algorithm based on set pair analysis for use when facing multiple uncertain attributes. CAAI Trans Intell Syst 5(1):41–50 (in Chinese)
4. Li Z, Xia S, Cha J (2003) CBR-based same-indefinite-contrary product design and application. J Comput Aided Des Comput Graph 15(11):1397–1403 (in Chinese)
5. Huang D, Zhao K, Lu Y (2001) Classifying method of critical paths in network planning including unexpected incident and its applications. J Syst Eng 16(3):161–166 (in Chinese)

6. Huang D, Zhao K (1999) Using the connection number of the SPA to express and process the uncertainties in network planning. J Syst Eng 14(2):112–117 (in Chinese)
7. Wu T (2008) A value-fetching formula of nondeterministic coefficient in connection number and its applications. Bull Sci Technol 24(5):595–597 (in Chinese)
8. Wang X, Peng X (2002) The methods to fetch i in difference degree coefficient of set pair analysis and its applications. J Tianjin Univ Light Ind (4):56–58 (in Chinese)
9. Zhao K (1995) Disposal and description of uncertainties based on the set pair analysis. Inf Control 24(3):162–166 (in Chinese)
10. Yu G (2002) On indeterminate coefficient i in number of connexion. J Liaoning Normal Univ (Natural Science Edition) 25(4):349–356 (in Chinese)
11. Zhao K (2008) Application and basic principles of connection mathematics. J Anyang Inst Technol (2):107–110 (in Chinese)
12. Theodoridis S, Koutroumbas K (2010) Introduction to pattern recognition: a matlab approach. Academic Press, Burlington
13. The UCI Machine Learning Repository. http://archive.ics.uci.edu/ml/datasets/Breast+Cancer+Wisconsin+(Original)
14. Ahn H, Kim K (2009) Global optimization of case-based reasoning for breast cytology diagnosis. Expert Syst Appl 36(1):724–734

Application of a Comprehensive One Dimensional Wavelet Threshold Denoising Algorithm in Blasting Signal

Xia Liangmeng, Zheng Wu and Liang Mengdi

Abstract When collecting blasting signal, the blasting signal is inevitably mixed with a certain noise signal, and the noise signal will affect the analysis of the blasting seismic wave, and it is not conducive to formulate corresponding measures. The blasting signals are periodically and changeable, the previous denoising methods are not totally fit to blasting signals, sometimes the original signals are regarded as noise signal and are being removed. In this paper, a new algorithm combine the soft and hard threshold function is introduced and the Simulation experiments and results analysis show that it can not only effectively remove the noise signals, but also save the details of the blasting signals.

Keywords Blasting signal denoising · Wavelet transform · Soft threshold function · Hard Threshold function

1 Introduction

With the progress of the society and the construction of the city, more and more engineering blasting is used in the construction of urban houses, roads and mines. So the related safety assessment, vibration records and the requirements of the test requirements will be more stringent, no matter it is a blasting vibration meter or blasting instrument and other instruments the collected signals are inevitable mixed with noisy signals due to the instrument itself or the physical reason during the signal

X. Liangmeng · Z. Wu (✉) · L. Mengdi
School of Computer Science and Technology,
Wuhan University of Science and Technology, Wuhan, Hubei, China
e-mail: chinablast@qq.com; 1182789409@qq.com

X. Liangmeng
e-mail: 848002425@qq.com

L. Mengdi
e-mail: 467869227@qq.com

V.E. Balas et al. (eds.), *Information Technology and Intelligent Transportation Systems*, Advances in Intelligent Systems and Computing 455,
DOI 10.1007/978-3-319-38771-0_51

transmission process, so signal denoising is also very necessary. The model of the blasting signal is defined as [1]:

$$f(t) = g(t) + n(t) \tag{1}$$

where $f(t)$ is the collected data signal, $g(t)$ is the original signal, $n(t)$ is the noise signal. The essence of denoising is to separate $g(t)$ from $f(t)$.

At present, the threshold function method also known as wavelet threshold denoising method is most widely studied and applied, Donoho et al. has proved that the wavelet threshold denoising method is superior to other classical denoising method. The theoretical basis of the method is that: the signal with noise after wavelet transform, with the increase of the scale, the wavelet coefficients of the actual signal increase, but the corresponding wavelet coefficients of the noise decrease [1]. Therefore, selecting a threshold function and setting a reasonable threshold, we can filter out the noise, and finally the estimated coefficients of the wavelet transform can be obtained after the blasting signal. The basic step of the threshold function method is [2]:

(1) Calculate the orthogonal wavelet transform of the noisy signal.
(2) According to the wavelet coefficient calculate threshold and handle the threshold.
(3) Do inverse transform of the wavelet coefficients after processing.

The key step is to select the threshold and the threshold function. At present, the most common threshold function is hard threshold function and soft threshold function. For the hard threshold, the special solar blind wavelength, large laser power or narrow-band filter are usually used [3–8]. The hard threshold function is to compare the absolute value of the wavelet coefficients with the threshold. If the absolute value of the wavelet coefficients are smaller than the threshold, the wavelet coefficients are set to zero. If the absolute value of the wavelet coefficients are bigger than the threshold, the wavelet coefficients keep pristine. The hard thresholding may lead to discontinuous and abrupt artefacts in the de-noised signals, especially when the noise energy is significant. The brief description of the hard thresholding is shown as follows [9]:

$$\mathrm{hard}(w, \lambda) = \begin{cases} w(|w| \geq \lambda) \\ 0(|w| < \lambda) \end{cases} \tag{2}$$

For the Software algorithm methods, there are several denoising or filtering methods: Empirical mode decomposition, Kalman filtering method, wavelet transform and Fourier filtering method, etc. [10–12] the soft threshold function is to compare the absolute value of the wavelet coefficients with the threshold. If the absolute value of the wavelet coefficients are less than the threshold, the wavelet coefficients are set to zero. If the absolute value of the wavelet coefficients are bigger than the threshold, the coefficients are handled like what the formula (3) shows:

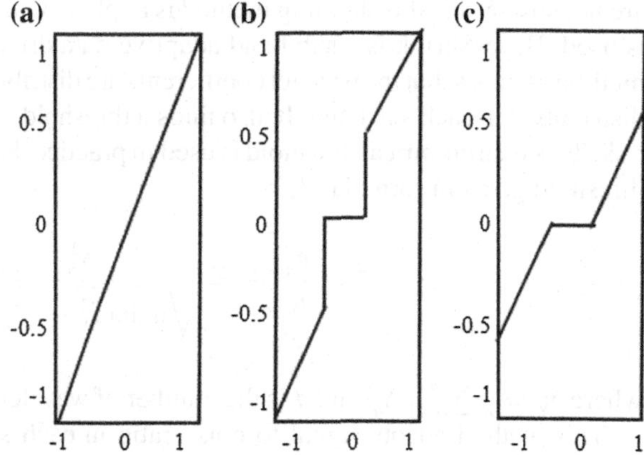

Fig. 1 **a** Original image. **b** Hard threshold. **c** Soft threshold method

$$\text{soft}(w, \lambda) = \begin{cases} \text{sgn}(w)(|w| - \lambda)(|w| \geq \lambda) \\ 0(|w| < \lambda) \end{cases} \tag{3}$$

Soft and hard threshold processing is shown in Fig. 1 [10].

The methods of calculating the thresholds are important for the signal denoising. If the threshold is too small, the effect of denoising is not obvious, but if the threshold is too big, the original signal will be modified. There exist three methods of estimating the thresholds, BayesShrink [11, 13], VisuShrink [14], and SureShrink [15].

VisuShrink is a universal thresholding method where a single threshold is applied on level of the wavelet coefficients entirely which is defined in formula (4).

$$\lambda = \delta_n \times \sqrt{2InN} \tag{4}$$

where N is the number of pixels in the image and δ_n is standard deviation of noise in image. SUREShrink is a sub band adaptive thresholding scheme where a different threshold is estimated and applied for each sub band based on Steins unbiased risk estimator (SURE). The function is given in formula (5).

$$\lambda = \text{arg minSURE(m, X)(m} \geq 0) \tag{5}$$

where the steins unbiased risk is minimized in formula (6).

$$\text{SURE(m, X)} = \text{d} - 2\{\text{i} : |\text{Xi}| \leq \text{m}\} + \text{d} \sum_{\text{i}=1}^{\text{d}} \min(|\text{Xi}|, \text{m})^2 \tag{6}$$

where X is the coefficients of the sub band X and d is the number of coefficients in the sub band. This optimization is straightforward, because the method is to order the wavelet coefficients in terms of magnitude and to select the threshold as the wavelet coefficients that minimizes the risk. As pointed out by Donoho, when the coefficients are not sparse, this thresholding method is applied. Otherwise the universal threshold is used. BayesShrink is a sub band adaptive data driven thresholding method. This method assumes that the wavelet coefficients are distributed as a generalized Gaussian distribution in each sub band. It also finds a threshold which minimizes the Bayesian risk. This is an empirical threshold is used in practice that is very close to the optimum threshold given in formula (7).

$$\breve{} = \frac{\delta^2_{noise}}{\delta_{signal}} = \frac{\delta^2_{noise}}{\sqrt{\max(\delta^2_Y - \delta^2_{noise}, 0)}} \tag{7}$$

where $\delta^2_{\mathrm{Y}} = \frac{1}{d}\sum_{i=1}^{d} X_i^2$ and d is the number of wavelet coefficients of sub band $\mathrm{Y}_{i,j}$.

This method adapts signal to noise ratio in each sub band and it uses a robust estimator of noise variance as median absolute value of the wavelet coefficients [16]. The noise valiance is estimated in formula (8).

$$\delta_{\text{noise}} = \frac{\text{median}(|Y_{i,j}|)}{0.6745} \tag{8}$$

where $Y_{i,j} \in$ sub band HH where $Y_{i,j}$ holds the coefficients in sub band HH which is the finest decomposition level.

2 Proposed Method

Although hard threshold and soft threshold function are two common methods of threshold processing, but in the application, there are still come shortcomings and limitations. For example, after the wavelet transform of the one-dimensional discrete blasting signal, detail coefficients is stored in the high frequency signal, and main information of the signal is stored in the low frequency part, and most of the noise signal is present in the high frequency part. Therefore, in the process of wavelet coefficients for different levels, different frequency bands of wavelet coefficients of the threshold selection of different threshold values will be more targeted. On the above analysis, in the processing of high frequency signal, the hard threshold algorithm can be used to remove the noise signal more effectively. For the low frequency coefficients, if the soft threshold function is used, it can also be used to keep the signal smoothness. Expressions as shown in formula (9):

$$new(w, \lambda) = \begin{cases} soft(w, \lambda) \text{ (w is low frequency coefficient)} \\ hard(w, \lambda) \text{ (w is high frequency coefficient)} \end{cases} \tag{9}$$

λ is wavelet threshold. In different wavelet level, λ is different, and the method of how to calculate wavelet threshold is designed by its basic data features.

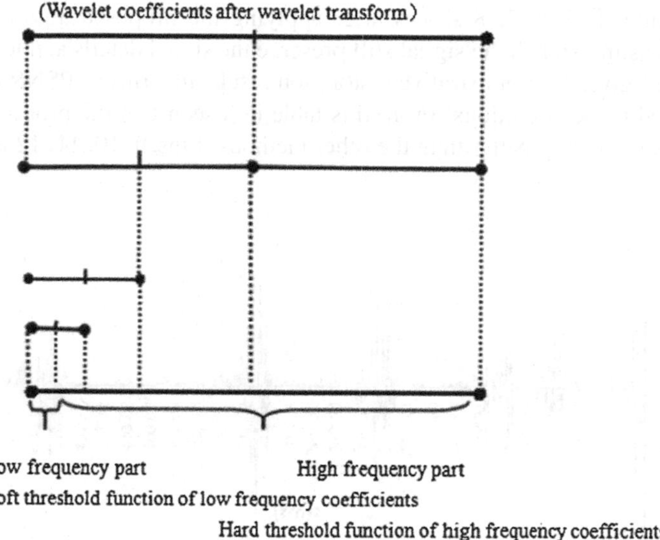

Fig. 2 Wavelet transform processing

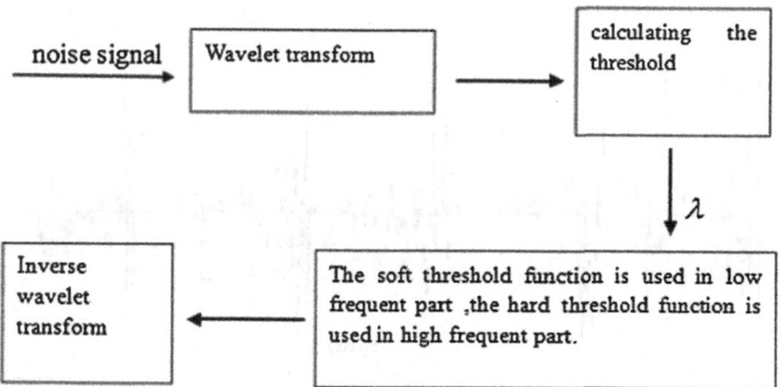

Fig. 3 Denoising flow chart

After 3 level wavelet transform, the high and low frequency coefficients are processed such as what shows in the Fig. 2.

The proposed denoising flow chart is shown below (Fig. 3).

3 Simulation Experiments and Results Analysis

In the simulation experiment, the selected wavelet is 97 wavelet, the signal is blasting signal, in order to make the experimental effect is obvious, the first is to add Gauss noise, and then the denoising method is used to compare the signal with the original signal. Figures 4, 5, 6, 7, 8 shows after applying the proposed algorithm on the collected blasting signal, the signal still preserve the signal details and textures very well. Table 1 gives the comparative restoration results in terms of PSNR (dB) under the specified noise conditions. From this table it is seen that the proposed method performs significantly better than the other methods (Figs. 9, 10, 11, 12 and 13).

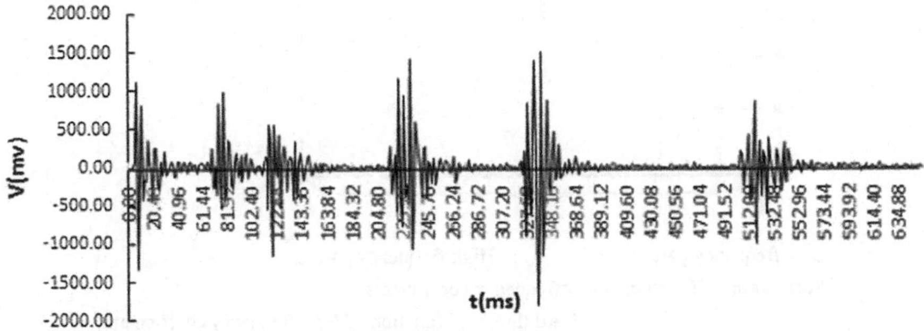

Fig. 4 Original blasting signal (a)

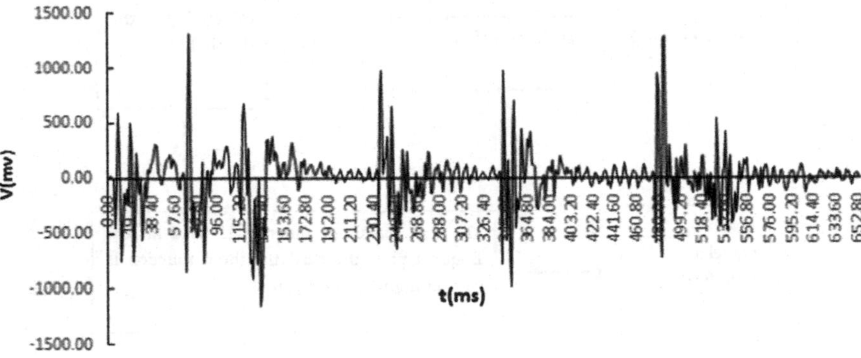

Fig. 5 Original blasting signal (a)

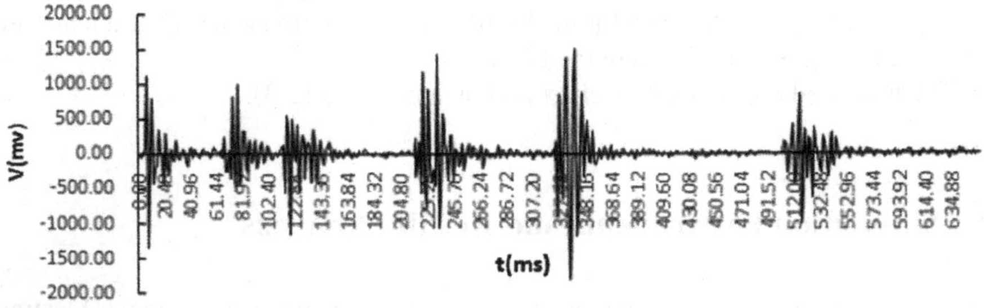

Fig. 6 Blasting signal after adding noise (a) (noise variance $\delta_n = 25$)

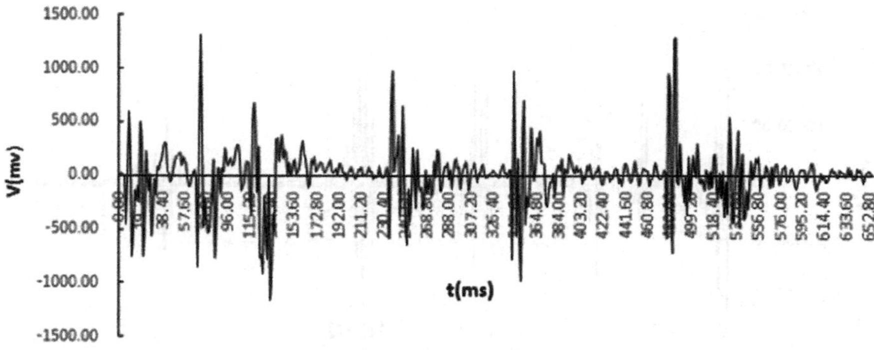

Fig. 7 Blasting signal after adding noise (b) (noise variance $\delta_n = 25$)

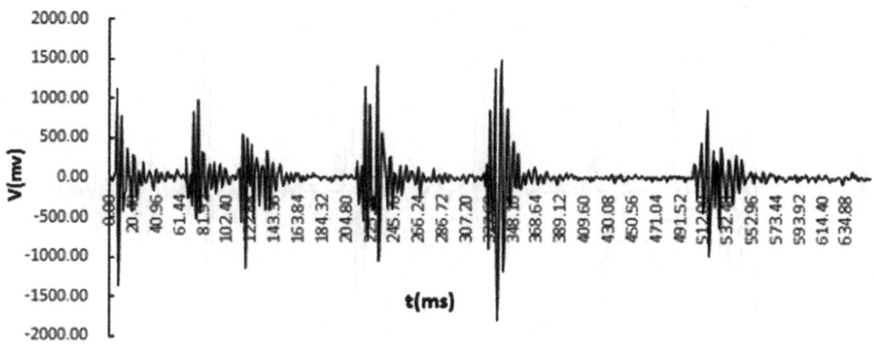

Fig. 8 Soft threshold to noise after the blasting signal (a)

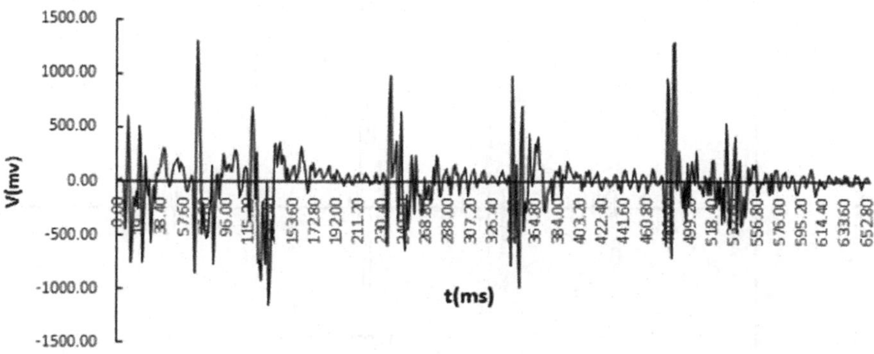

Fig. 9 Soft threshold to noise after the blasting signal (b)

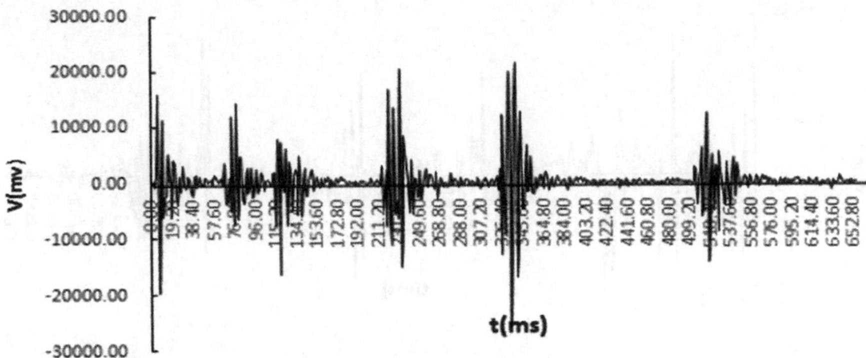

Fig. 10 Hard threshold to noise after the blasting signal (a)

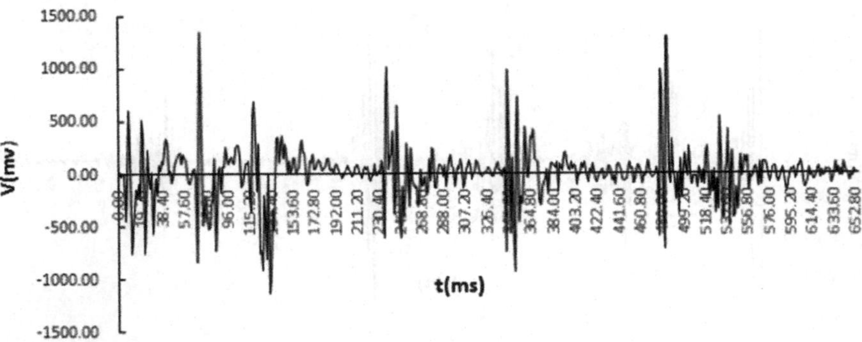

Fig. 11 Hard threshold to noise after the blasting signal (b)

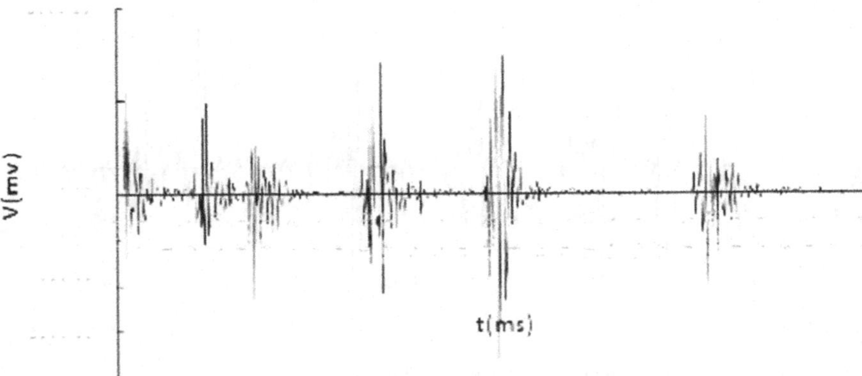

Fig. 12 Comprehensive threshold to noise after the blasting signal (a)

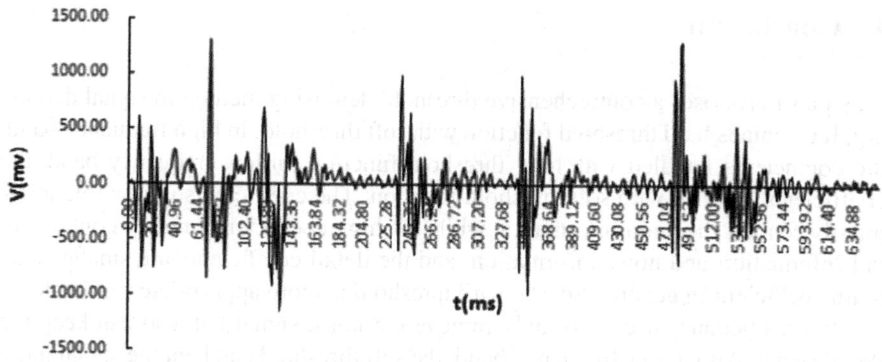

Fig. 13 Comprehensive threshold to noise after the blasting signal (b)

Table 1 Comparisons of qualitative results in PSNR for blasting signal collected in different time (131007 is October 7, 2013)

Date	Hard threshold function	Soft threshold function	Proposed threshold function
131007	23.09	23.71	24.12
131011	23.07	24.48	24.50
131015	21.60	23.34	23.38
131019	23.89	25.14	25.93
131023	23.31	24.25	24.77
131104	22.89	23.69	24.12
131108	22.16	23.45	24.22
131112	22.31	23.10	23.98
131116	21.52	22.79	24.14
131120	23.36	24.23	24.23
131124	22.16	23.59	23.94
131202	22.03	23.00	23.54
131206	22.00	22.03	23.93
131210	22.62	23.35	24.31

Table 2 Comparisons of qualitative results in PSNR for blasting signal corrupted by Gaussian noise (is noise variance)

	$\delta_n = 30$	$\delta_n = 60$	$\delta_n = 90$
Hard threshold	23.64	23.02	22.71
Soft threshold	24.10	23.63	22.83
Proposed threshold	24.63	24.12	22.90

4 Conclusion

This paper proposes a comprehensive threshold denoising method in signal denoising, It combines hard threshold function with soft threshold. In high frequency band, the coefficients handled with hard threshold function, in low frequency band, the coefficients handled with soft threshold function. The coefficients in low frequency mainly contain the main information while the high coefficients mainly contain detail information and noise information, and the detail coefficients are smaller than main coefficient in general, So The hard threshold is more appropriate for the detail coefficients because it can not only remove the noise signal but also can keep the detail signal. And in low frequency band, the soft threshold can kept the signal more smooth than hard threshold. The results of the simulation are presented in the array given in Table 1. Though the data of the table shows that the denoising effect is satisfactory but also to maintain the smoothness of the original signal. Thus it proves the validity of this method. However, there is a close relationship between the noise and the level of wavelet, the less the level is, the more obvious the effect is, which is to be improved in the future (Table 2).

References

1. Vittoria B, Domenico V (2006) Wavelet-based signal de-noising via simple singularities approximation 86:859–876
2. Tigaret CM et al (2013) Wavelet transform-based de-noising for two-photon imaging of synaptic Ca^2+. Transients Biophys J 5(104):1006–1017
3. Hua D et al (2007) Daytime temperature profiling of planetary boundary layer with ultraviolet rotational Raman Lidar. Jpn J Appl Phys Part 12 46(9):49–52
4. Hua D, Kobayashi T (2005) Ultraviolet Rayleigh-Mie Lider by use of a multicavity Fabry-Perot filter for accurate temperature profiling of the troposphere 30:64–74
5. Di Girolamo P, Summa D et al (2009) Measurement of relative humidity for the characterization of cirrus cloud microphysical properties. Atmos Chem Phys 9:8799–8811
6. Kwon KH, Gardner CS et al (1987) Daytime Lidar measurements of tidal winds in the mesospheric sodium layer at Urbana Illinois. J Geophys 92(A8):8781–8786
7. Rees D et al (2000) Daytime Lidar measurement of the stratophere and mesosphere at the ALOMAR obsevatory. Adv Space Res 26:893–902
8. Yonghua W (2003) Datetime Raman-Lidar measurements of water vapor over Hefei. Opt Lett 1(7):373–376
9. Nadernejad E et al (2012) Improving performance of wavelet-based image denoising algorithm using complex diffusion process. Image Sci J 60:208–218
10. Mukhopadhyay S, Mandal JK (2013) Wavelet based denoising of medical images using subband adaptive thresholding through genetic algorithm. Procedia Technol 10:680–689
11. Chipman H, Kolaczyk E, McCulloch R (1997) Adaptive Bayesian wavelet shrinkage. J Am Stat Assoc 92(440):1413–1421
12. Malfati M, Roose D (1997) Wavelet-based image denoising using a markov random field a priori model. IEEE Trans Image Process 6(4):549–565
13. Chang S, Yu B, Vetterli M (2009) Adaptive wavelet thresholding for image denoising and compression. IEEE Trans Image Process 9(9):1532–1546
14. Donoho D, Johnstone I (1994) Ideal spatial adaptation via wavelet shrinkage. Biometrikac 81(3):425–455

15. Donoho D, Johnstone I (1995) Adapting to unknown smoothness via wavelet shrinkage. J Am Stat Assoc 90(432):1200–1224
16. Alexandera ME et al (2000) Wavelet domain de-noising of time-courses in MR image sequences. Magn Reson Imaging 18:1129–1134
17. Chang S, Yu B, Vetterli M (2000) Spatially adaptive wavelet thresholding based on context modeling for image denoising. IEEE Trans Image Process 9(9):1522–1531

Histogram Thresholding in Image Segmentation: A Joint Level Set Method and Lattice Boltzmann Method Based Approach

Ram Kumar, F.A. Talukdar, Nilanjan Dey, Amira S. Ashour, V. Santhi, Valentina Emilia Balas and Fuqian Shi

Abstract The level set method (LSM) has been widely utilized in image segmentation due to its intrinsic nature which sanctions to handle intricate shapes and topological changes facilely. The current work proposed an incipient level set algorithm, which uses histogram analysis in order to efficiently segmenting images. The computational intricacy of the proposed LSM is greatly reduced by utilizing the highly parallelizable lattice Boltzmann method (LBM). The incipient algorithm is efficacious and highly parallelizable. Recently, with the development of high dimensional astronomically an immense-scale images contrivance, the desideratum of expeditious and precise segmentation methods is incrementing. The present work suggested a histogram analysis based level set approach for image segmentation. Experimental results on real images demonstrated the performance of the proposed

R. Kumar (✉) · F.A. Talukdar
Department of ECE, National Institute of Technology, Silchar 788010, India
e-mail: ramkumar.purnea@gmail.com

F.A. Talukdar
e-mail: fatalukdar@gmail.com

N. Dey
Department of IT, Techno India College of Technology, Kolkata, West Bengal, India
e-mail: nilanjan.dey@tict.edu.in

A.S. Ashour
Department of Electronics and Electrical Communications Engineering,
Faculty of Engineering, Tanta University, Tanta, Egypt
e-mail: amirasashour@yahoo.com

V. Santhi
School of Computing Science and Engineering, VIT University, Vellore, India
e-mail: vsanthi@vit.ac.in

V.E. Balas
Faculty of Engineering, Aurel Vlaicu University of Arad, Arad, Romania
e-mail: balas@drbalas.ro

F. Shi
College of Information and Engineering, Wenzhou Medical University,
Wenzhou 325035, People's Republic of China
e-mail: sfq@wmu.edu.cn

V.E. Balas et al. (eds.), *Information Technology and Intelligent Transportation Systems*, Advances in Intelligent Systems and Computing 455,
DOI 10.1007/978-3-319-38771-0_52

method. It is established that the proposed segmentation methods using Level set methods for image segmentation achieved 0.92 average similarity value and average 1.35 s to run the algorithm, which outperformed Li method for segmentation.

Keywords Histogram · Image segmentation · Lattice Boltzmann method (LBM) · Level set method (LSM)

1 Introduction

Imagehistogram is defined as a graphical representation of tonal/intensity distribution in digital images. It is used to plot a graph between the number of pixels for each tonal/intensity value. Image histogram helps the viewer to estimate the entire tonal/intensity distribution through careful examination of it. In computer vision, image histograms could be used as a tool for thresholding. It could be very much useful in analyzing for peaks and/or valleys in intensity values. The identified threshold value can then be used for edge detection, image segmentation, and co-occurrence matrices.

Image segmentation is defined as a process of partitioning a digital image into multiple smaller segments called regions. Image segmentation is a process of dividing an image into a set of homogeneous and consequential regions so that the pixels in each partitioned region possess an identical set of properties or attributes Regions are defined as collection of picture elements or pixels. The goal of segmentation is to simplify and/or transmute the representation of an image into smaller regions which further makes processing or analysis much easier.

Image segmentation is very essential to image processing and pattern apperception. It leads to final result of analysis with high quality. Image segmentation is a process of dividing an image into a set of homogeneous and consequential regions so that the pixels in each partitioned region possess an identical set of properties or attributes. One of the special kinds of segmentation is thresholding, which endeavors to relegate image pixels into one of two categories (e.g. foreground and background). At the terminus of such thresholding, each object of the image, represented by a set of pixels, is isolated from the rest of the scene. In this case, the aim is to find a critical value or threshold. Segmentation algorithms are predicated on different parameters of an image like gray-level, color, texture, depth or kineticism. In the domain of biomedical image processing, correct image segmentation would avail medicos greatly in providing visual designates for inspection of anatomic structures, identification of disease and tracking of its progress, and even for surgical orchestrating and simulation. Recently, medical image processing is utilized to identify the brain tumor. Various extensive works was conducted for medical image segmentation and thresholding [1–7].

The most straightforward approach is to pick up a fine-tuned gray scale value as the threshold and relegate each gray scale by checking whether it lies above or below this value. In general, the threshold should be located at the conspicuous and deep valley of the histogram. Especially for a well-defined image, its histogram has a deep valley between two peaks. Therefore, the optimum threshold value can be found in the valley region. One prodigiously simple way to find an opportune threshold is to find each of the modes (local maxima) and then find the valley (minimum) between them [8].

Theoretically, the optimal threshold value can be set according to the Bayes rule by practicing the pixel distribution of both classes. However, is not to separate distributions, but to a mixture both distributions as indicated in the histogram. Hence, it requires some assumptions about the contours of both distributions to simplify the problem.

Evaluation techniques are suggested in order to judge it. One of them approximates in the least square sense by a summation of Gaussian distribution, which is estimated from the histogram. A set of parameters to fit the image histogram to the probability models by minimizing the Mean Square Error (MSE) between the authentic probability density function and the mannequin can be obtained. An iterative cull method is utilized and predicated on the one of nonlinear optimizations [9]. As Such a method, nevertheless, uses as iterative computation, the final solution heavily depends on the initial value.

Many thresholding techniques utilized the criterion-predicated concept to cull the most opportune gray scale as the threshold value. One of the previous methods is Otsu's thresholding method that utilizes discriminant analysis to find the maximum separability of classes [10]. For every possibility of threshold value, Otsu (1979) evaluated the integrity of this value if utilized as the threshold. This evaluation includes the heterogeneity of both classes and the homogeneity of every class.

Kittler and Illingworth (1986) have used the criterion-based concept to obtain the minimum error threshold between Gaussian distribution used in the background and foreground [11]. Measure-based methods are efficient and efficient for determining a threshold value. The computation complexity increases exponential by increasing the number of threshold values. In addition, the methods work very well for bimodal or nearly bimodal histogram [12]. For uni-modal and multimodal histogram, however, the separation between both classes is not well-defined.

In earlier days, segmentations are carried out using two rudimental approaches namely edge-based and region-based. Both of them are complementary to each other based on feature used. Therefore, most recent methods, such as the Deformable Contour Methods (DCM) include both edge-predicated and region-predicated approaches. The DCMs additionally categorized into snakes and level set methods for carrying out the contour deformation process. These different methods are utilized for different dedicated quandaries and provide the desired features to efficaciously segment target contours from the image data. The snake method used curve evolution that based on minimization of the contour energy, which includes the internal energy from the contour and the external energy from the image. Snake is explicit in nature, thus it is unable to work on topological changes and time complexity. To evade this problem, Level set method, which considered a new concept is introduced. The level set method is a general framework for tracking dynamic interfaces and shapes. In computer vision and pattern recognition the level set method (LSM) had been widely used for image segmentation [13]. The attractive advantage of the LSM is its ability to extract complex contours and to automatically handle topological changes [14], such as splitting and merging.

Balla-Arabe et al. [13] presented the edge, region, and 2D gray histogram based level set model. A level set framework for segmentation and bias field correction of images with intensity in-homogeneity was discussed in [14]. Chan and Vese [15] proposed active contours to detect the region of interest (ROI) in a given image. The authors suggested method was based on techniques of curve evolution, Mumford-Shah functional for segmentation and LSM. Zhang et al. (2010) [16] designed an Unsigned Pressure Force (UPF) using the image characteristics inside and outside the contour. This procedure can effectively stop the evolving curve when segmenting object with weak edges or even without edges. It is suitable for parallel programming due to the local and explicit nature of the LBM solver. Balla-Arabé et al. (2011) [17] employed the LBM solver to solve the LSE by using a region based approach to stop the evolving curve. A multiphase LSM with LBM was discussed in [18]. This parallel programming will help to implementation in FPGA [19, 20] which results high speed automated segmentation [21].

The main objective is to stop the evolving contour when the entropy is minimum. Application of level set methods in image processing is discussed in [22, 23]. Balla-Arabe and Gao [24] conducted an adaptive and fast level set based multi-thresholding method for image segmentation. The authors used the advantages of thresholding methods, localized level set method and the lattice Boltzmann model, which are speed, effectiveness and high parallelizability; respectively. Balla-Arabé1 et al. (2014) [25] derived a method that combined the advantages of the LSM and the image thresholding technique in terms of the global segmentation, to be easily handle complex convergence towards the global minimum.

From the preceding related work, it is clear that LSM with LBM has been used in the field of fast image segmentation application. But no work has been taken place in image segmentation considering histogram thresholding based level set method with LBM.

The remaining sections are organized as follows. Section 2 introduced the level set method concept followed by Sect. 3 that described the proposed model formulation. The experimental results are presented in Sect. 4. Finally, the work is concluded in Sect. 5.

2 The Level Set Method

The LSM belongs to the active contours model (ACMs) which based on the geometric representation of the active contour instead of the parametric representation which is based on the Lagrangian framework.

The basic idea of the LSM is to evolve the zero level of the level set function (LSF) Ø in the image domain until it reaches the boundaries of the regions of interest

[15]. The active curve evolution is governed by the level set equation (LSE). It is a partial differential equation expressed as

$$\frac{\partial Ø}{\partial t} = |\nabla Ø| \left(\propto V + \beta \nabla \cdot \frac{\nabla Ø}{|\nabla Ø|} \right) \tag{1}$$

where, Ø is the level set function, V is the speed function, which drives and attracts the active contour towards the region boundaries. The second term of the right hand represents the curvature, which is used to smooth the contour. It is non-linear and computational expansive α and β user-controlling parameters.

For solving the LSE, most classical methods such as the upwind scheme are based on some finite difference, finite volume or finite element approximations and an explicit computation of the curvature [13]. These methods requires a lot of CPU time. Recently, the lattice Boltzmann method (LBM) has been used as solver for accelerating the PDE (partial differential equation) [12].

The LBM at first designed to simulate Navier–Stokes equations for an incompressible fluid [26]. Figure 1 illustrated the D2Q5 (two dimensions and five lattice speeds) LBM lattice structure.

The evolution equation of LBM is

$$f_i\left(\vec{r} + \vec{e_1}, t + 1\right) - f_i\left(\vec{r}, t\right) = \Omega \tag{2}$$

$$\Omega = \frac{1}{\tau}\left[f_i^{eq}\left(\vec{r}, t\right) - f_i\left(\vec{r}, t\right)\right] + \frac{2\tau - 1}{2\tau} \cdot \vec{F} \cdot \vec{e}_1 \tag{3}$$

To model the typical diffusion phenomenon, the following local equilibrium particle distribution is to be used [27, 28],

$$f_i^{eq}(\rho) = \rho A_i \text{ with } \rho = \sum f_i \tag{4}$$

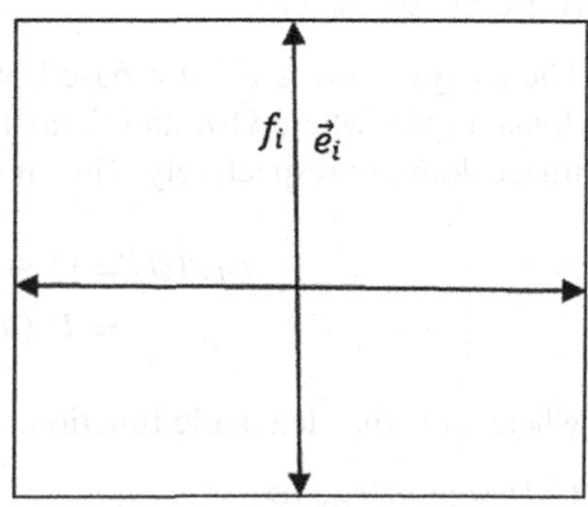

Fig. 1 Spatial structure of D2Q5 LBM lattice

3 Designed Level Set Method

In general, Level Set Methods could be categorized into two major classes' namely region-based methods and edge-based methods. In region-based models each region of interest is identified using a region descriptor to guide the motion of the active contour [29, 30]. In this section, the conception of the proposed fast convex level set algorithm is described in detail.

In the current work, energy functional has been designed for minimization is as

$$\varepsilon(\phi) = \varepsilon_{hist}(\phi) + \beta\varepsilon_{thr}(\phi) + \vartheta\varepsilon_{reg}(\phi) \tag{5}$$

where ϕ the set is function level, while β and ϑ are the positive controlling parameters.

A. Design of $\varepsilon_{hist}(\phi)$

The energy term $\varepsilon_{hist}(\phi)$ is the histogram based link with the image data, which defined as follows:

$$\varepsilon_{hist}(\phi) = \textstyle\int_{\Omega in}(\alpha m_1 + \beta m_2 - I)\exp\{\delta(I - I_m)\} \tag{6}$$

Here, H is the Heaviside function; I_m is the mean value and ϕ is level set function which is defined as the signed distance function [13]. The mean values m_1 and m_2 are defined as:

$$m_1 = \frac{\int_\Omega I(x, y)H(\phi)dxdy}{\int_\Omega H(\phi)dxdy} \tag{7}$$

$$m_2 = \frac{\int_\Omega I(x, y)(1 - H(\phi))dxdy}{\int_\Omega (1 - H(\phi))dxdy} \tag{8}$$

with $\alpha > 0$ $\beta > 0$

$$\alpha + \beta = 1$$

B. Design of $\varepsilon_{thr}(\phi)$

The energy term is texture based, which is defined as follows with Ω the image domain [25], where Ωin and Ωout are the inside and outside evolving curve of the image domain; respectively. The energy term is given by:

$$\begin{aligned}\varepsilon_{thr}(\phi) &= [1 - sign(\Phi_t)sign(\Phi)] \\ &= H(\Phi_t) + H(\Phi)(1 - H(\Phi_t))\end{aligned} \tag{9}$$

where, H is the Heaviside function.

C. Design of $\varepsilon_{reg}(\phi)$

The regularization term used as a constraint on the evolving contour, can be expressed as:

$$\varepsilon_{reg}(\phi) = \textstyle\int_\Omega |\nabla H(\phi)|dxdy \tag{10}$$

Therefore, the proposed energy functional in Eq. (5) can be rewritten as:

$$\begin{aligned}\varepsilon(\phi) = {} & \alpha \int_{\Omega} (m_1 - I)\, H(\phi)\, dxdy - \beta \int_{\Omega} (m_2 - I)\,(1 - H(\phi))\, dxdy \\ & + \lambda \int_{\Omega} (1 - J(\psi, m_{in}))\, H(\phi) + (1 - J(\psi, m_{out}))\,(1 - H(\phi))\, dx \\ & + \vartheta \int_{\Omega} \nabla H(\phi)\, |dx\end{aligned} \quad (11)$$

Using the gradient descent method, the level set equation can be recovered from the above defined energy functional

$$\partial\phi/\partial t = -\partial\varepsilon/\partial\phi \quad (12)$$

where $\partial\varepsilon/\partial\phi$ is the Gateaux derivative of ε.

$$\begin{aligned}\partial\phi/\partial t = {} & \delta(\phi)[(I - (\alpha m_1 + \beta m_2)) \exp\{\delta\,(I - I_m) \\ & + \cdots \beta\,(1 - H(\Phi_t))\}] + \vartheta div(\nabla\phi/|\nabla\phi|)\end{aligned} \quad (13)$$

The gradient projection method allowed replacing $\partial\phi$ by $|\nabla\,\varphi\,|$. In addition, since φ is a signed distance function (SDF), thus $|\nabla\,\varphi\,| = 1$.

The body force F can be expressed as:

$$F = \lambda\,(I - \alpha m_1 + \beta m_2) \exp\{\delta\,(I - I_m) + \beta\,(1 - H(\Phi_t))\} \quad (14)$$

where, λ and β are the controlling parameters used to adjust the impact of F on the active contour motion. The principal implementation steps of the proposed method are provided in the following algorithm.

Algorithm: Level set methods for image segmentation with LBM

Start
Initialize the distance function signed distance function
Compute the body force F in Eq. (14)
Resolve the LSE using LBM
Accumulate $f_i\,(\vec{r}, t)$ the values at each grid point which generates updated values and find the contour
Check convergence
If the evolving curve has not converged go back to step
Else End
End

---

4 Experimental Results and Analysis

Several experiments were carried out using various kinds of images in order to demonstrate the performance of the proposed method. Figure 2 demonstrated sample of the tested images, where the original images in (a), while (b) the segmented images and (c) the gray representation for the segmented images.

To evaluate the performance of the proposed method, the Dice's coefficient and Jaccards' coefficient measures set agreement was calculated. The Dice's coefficient is one of the measurements to measure the extent of spatial overlap between the two binary images. This coefficient is commonly used in reporting the performance of segmentation. A value of 0 indicates no overlap, while a value of 1 indicates perfect agreement (similarity). Higher numbers indicate better agreement.

Table 1 demonstrated comparing the designed method to Li method [14] in terms of the Dice's coefficient values.

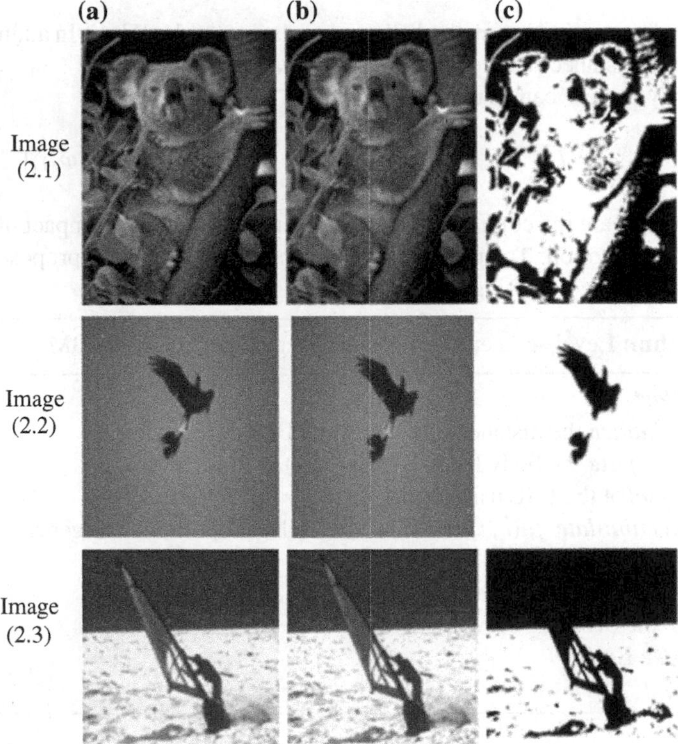

Fig. 2 Segmentation of different images **a** represent original images, **b** represent the evolution of the contour and **c** represents result of proposed method

It is established from Table 1 that the proposed method is outperformed Li method in terms of the Dice's coefficient values for all compared images. The average value of agreement achieved by the designed method is 0.92.

The special overlap between the pair of segmentations can be measured by using Jaccards coefficient.

It is established from Table 2 that the designed method is outperformed Li method in terms of the Jaccards 'coefficient values for all compared images. The average value of agreement achieved by the designed method is 0.78.

Comparing both algorithms with respect to the computational time was conducted in Table 3.

Table 3 proved that the proposed method is superior to Li method in terms of the computational time taken to run the segmentation process. The proposed method has been taken 1.35 s, while running Li method has been taken 83.97 s.

Researchers are interested with different image and signal processing applications [31–35]. In the current work, the preceding results established that the proposed segmentation methods using Level set methods for image segmentation achieved 0.92 average similarity value and average 1.35 s to run the algorithm. The results established that the proposed method is outperformed the segmentation method using Li method for segmentation.

Table 1 Dice's coefficient (D)

Image index	Proposed method	Li method [14]
Image 2.1	**0.94**	0.53
Image 2.2	**0.92**	0.22
Image 2.3	**0.89**	0.42
Average	**0.92**	0.39

Table 2 Jaccards' coefficient (D)

Image index	Proposed method	Li method [14]
Image 2.1	**0.84**	0.28
Image 2.2	**0.73**	0.39
Image 2.3	**0.79**	0.26
Average	**0.78**	0.31

Table 3 Total time taken to run the algorithm in seconds

Image index	Proposed method	Li method [14]
Image 2.1	**1.54**	92.65
Image 2.2	**1.28**	82.527
Image 2.3	**1.23**	76.72
Average	**1.35**	83.97

5 Conclusion

The current work proposed a histogram and thresholding based level set model. The use of the lattice Boltzmann method to solve the level set equation enables the algorithm to be highly parallelizable. The designed method is effective for segmenting objects with and without edges. Experimental results of different kinds of real images have been demonstrated the effectiveness of the designed model. The proposed method achieved 0.92 average similarity value with average 1.35 s running time, which outperform Li method in [14] that used the level set method for image segmentation.

References

1. Dey N, Roy AB, Pal M, Das A (2012) FCM based blood vessel segmentation method for retinal images. Int J Comput Sci Netw 1(3):49–55
2. Samanta S, Dey N, Das P, Acharjee S, Chaudhuri SS (2012) Multilevel threshold based gray scale image segmentation using cuckoo search. In: International conference on emerging trends in electrical, communication and information technologies-ICECIT, 12–13 Dec 2012
3. Samanta S, Acharjee S, Mukherjee A, Das D, Dey N (2013) Ant weight lifting algorithm for image segmentation. In: IEEE international conference on computational intelligence and computing research, Madurai, 26–28 Dec 2013
4. Bose S, Mukherjee A, Madhulika, Chakraborty S, Samanta S, Dey N (2013) Parallel image segmentation using multi-threading and k-means algorithm. In: IEEE international conference on computational intelligence and computing research, Madurai, 26–28 Dec 2013
5. Roy P, Chakraborty S, Dey N, Dey G, Ray R, Dutta S (2014) Adaptive thresholding: a comparative study. In: International conference on control, instrumentation, communication and computational technologies, 10–11 July 2014
6. Roy P, Goswami S, Chakraborty S, Azar AT, Dey N (2014) Image segmentation using rough set theory: a review. Int J Rough Sets Data Anal 1(2):62–74
7. Pal G, Acharjee S, Rudrapaul D, Ashour AS, Dey N (2015) Video segmentation using minimum ratio similarity measurement. Int J Image Min 1(1):87
8. Gonzalez RC, Woods RE (1993) Digital image processing. Addison-Wesley Publishing Company, Boston
9. Chi Z, Yan H, Pham T (1996) Fuzzy algorithms: with applications to images processing and pattern recognition. Word Scientific, Singapore
10. Otsu N (1979) A threshold selection method from gray-level histograms. In: IEEE Transactions on Systems, Man, and Cybernetics, vol SMC-9 , no 1, pp 62–66
11. Kittler J, Illingworth J (1986) Minimum error thresholding. Pattern Recognit 19:41–47
12. Tobias OJ, Seara R (2002) Image segmentation by histogram thresholding using fuzzy sets. IEEE Trans Image Process 11:1457–1465
13. Balla-Arabe S, Gao X, Wang B (2013) GPU accelerated edge-region based level set evolution constrained by 2D gray-scale histogram. IEEE Trans Image Process 10(1):1–11
14. Li C, Huang R, Ding Z, Chris J, Metaxas DN, Gore JC (2011) A level set method for image segmentation in the presence of intensity inhomogeneities with application to MRI. IEEE Trans Image Process 20(7):2007–2016
15. Chan T, Vese L (2001) Active contours without edges. IEEE Trans Image Process 10(2):266–277
16. Zhang K, Zhang L, Song H, Zhou W (2010) Active contours with selective local or global segmentation: a new formulation and level set method. Image Vis Comput 28(4):668–676

17. Balla-Arabé S, Wang B, Gao X-B (2011) Level set region based image segmentation using lattice Boltzmann method. In: Proceedings of the 7th International Conference on Computational Intelligence and Security, pp 1159–1163, Dec 2011
18. Balla-Arabé S, Gao X (2013) A multiphase entropy-based level set algorithm for MR breast image segmentation using lattice boltzmann model. In: Proceedings of the sino, foreign, interchange workshop intelligent science and intelligent data engineering, pp 8–16, Oct 2013
19. Lie I, Beschiu C, Nanu S (2011) FPGA based signal processing structures. In: 6th IEEE international symposium on applied computtional intelligence and informatics, Saci 2011, 5873043, pp 439–444
20. Nanu Sorin, Lie Ioan, Belgiu George, (2010) Musuroi Sorin,High speed digital controller implemented with FPGA, Buletinul Stiintific al UPT, seria Electronica si Telecomunicatii 55(69):17–22
21. Belgiu G, Nanu S, Silea I (2010) Arificial intelligence in machine tools design based on genetic algorithms application. In: 4th International workshop on soft computing applications, Sofa 2010, 5565623, pp 57–60
22. Balla-Arabé S, Gao X, Wang B (2012) A fast and robust level set method for image segmentation using fuzzy clustering and lattice Boltzmann method. IEEE Trans Syst Man Cybern Part B Cybern 99:1–11
23. TsaiI R, Osher S (2003) Level set methods and their applications in image science. Commun Math Sci 1(4):1–20
24. Balla-Arabe S, Gao X (2012) Image multi thresholding by combining the lattice Boltzman model and localized level set algorithm. Neurocomputing 93:106–114
25. Balla-Arabé1 S, Gao X, Xu L (2014) Texture-aware fast global level set evolution
26. Balla-Arabe S, Gao Xinbo (2014) Geometric active curve for selective entropy optimization. Neurocomputing 139:65–76
27. Zhao MY (2007) Lattice Boltzmann based PDE solver on the GPU. Vis Comput 24(5):323–333
28. Bhatnager P, Gross E, Krook M (1954) A model for collision processes in gases. I: small amplitude processes in charged and neutral one-component systems. Phys Rev 94:511
29. Li C, Xu C, Gui C, Fox M (2010) Distance regularized level set evolution and its application to image segmentation. IEEE Trans Image Process 19(12):3243–3254
30. Osher S, Fedkiw R (2003) Level set methods and dynamic implicit surfaces. Springer, New York
31. Sudha VK, Sudhakar R, Balas VE (2012) Fuzzy rule-based segmentation of CT brain images of hemorrhage for compression. Int J Adv Intell Paradig 4:256–267
32. Senthilkumar S, Piah ARM (2012) An improved fuzzy cellular neural network (IFCNN) for an edge detection based on parallel RK (5, 6) approach. Int J Comput Syst Eng 1(1):70–78
33. Chakraborty S, Acharjee S, Maji P, Mukherjee A, Dey N (2014) A semi-automated system for optic nerve head segmentation in digital retinal images. In: 2014 International conference on information technology, pp 112–117, Bhubaneswar, 22–24 Dec 2014
34. Bharathi S, Sudhakar R, Balas VE (2014) Biometric recognition using fuzzy score level fusion. Int J Adv Intell Paradig 6(2):81–94
35. Ghosh A, Sarkar A, Ashour AS, Balas-Timar D, Dey N, Balas VE (2015) Grid color moment features in glaucoma classification. Int J Adv Comput Sci Appl 6(9):1–14

Self-Calibration of Dead Reckoning Sensor for Skid-Steer Mobile Robot Localization Using Neuro-Fuzzy Systems

S. Rakesh kumar, K. Ramkumar, Seshadhri Srinivasan, Valentina Emilia Balas and Fuqian Shi

Abstract Wheel slip affects the accuracy of dead-reckoning based localization techniques as they introduce measurement errors in odometers. This investigation presents a new slip compensation scheme that uses neuro-fuzzy technique for self-calibration of odometer. The proposed self calibration procedure can be executed in robot navigating environment rather than having a separate test platform and neuro-fuzzy system employed can able to learn the dynamics of wheel slip from the data autonomously. The wheel slip data are generated using the standard data generated by the laser range finder with known landmarks and measurement data from the odometer. The proposed technique is implemented on a four wheel mobile robot navigating in a concrete terrain and the localization performance was evaluated with mean square error (MSE) of 0.0382 m for a 6.12 m run during training phase and 0.0442 m for a 4.20 m test run of the robot.

Keywords Dead reckoning · Slip compensation · Robot localization · Self calibration · Neuro-fuzzy

S. Rakesh kumar (✉) · K. Ramkumar
Electric Vehicle Engineering and Robotics (EVER) Lab,
School of Electrical and Electronics Engineering, SASTRA University,
Thanjavur 613401, Tamil Nadu, India
e-mail: srakesh@eie.sastra.edu

K. Ramkumar
e-mail: ramkumar@eie.sastra.edu

S. Srinivasan
Department of Engineering, University of Sannio, 82100 Benevento, Italy
e-mail: seshadhri@ieee.org

V. Emilia Balas
Aurel Vlaicu University of Arad, Arad, Romania
e-mail: balas@drbalas.ro

F. Shi
College of Information and Engineering, Wenzhou Medical University,
Wenzhou 325035, People's Republic of China
e-mail: sfq@wmu.edu.cn

V.E. Balas et al. (eds.), *Information Technology and Intelligent Transportation Systems*, Advances in Intelligent Systems and Computing 455,
DOI 10.1007/978-3-319-38771-0_53

1 Introduction

Environment uncertainties and sensor noise affect the accuracy of robot localization algorithms that are widely employed in applications such as material handling, docking, pick-and-place, and defense applications [1–6]. Although many approaches such as model based [7], extro-receptive sensor based [8], sensor fusion techniques [9–11] have been studied for localization, and dead-reckoning [12] based localization is widely employed due to its accuracy and simplicity. In spite of these advantages, accuracy of this technique is limited by the wheel-slip forces that are generated due to wheel-terrain interactions. Being a relative measurement technique, even a small measurement error will deteriorate the localization accuracy, as robot location is estimated by integrating relative positions of the robot. In particular, wheel-slip forces limit the deployment of autonomous robots in unknown and challenging environments. Therefore, techniques that improve the robustness of the dead-reckoning based localization method to wheel-slip [12, 13] are required for deploying autonomous mobile robots in unknown environments. This can be achieved by designing slip compensation technique which can compensate the wheel slip to improve the localization accuracy. The slip compensation technique [14] creates a model to learn the slip dynamics inherently and uses this model to generate wheel slip compensation. Therefore, wheel-slip compensation is preferred over slip control strategy.

Seeing the potential of the slip compensation in robot localization, researchers have investigated the problem and many approaches have been proposed. The available approaches can be broadly classified into three broad categories [15], they are: (i) model based filter design, (ii) sensor fusion systems and (iii) calibration algorithms. In model based filter approach, probabilistic filters such as Kalman filter and its variations (see, [8–10] and references therein) are widely employed to handle wheel slip uncertainties caused due to environmental variations. Although, model based compensation schemes are reliable, their performance is limited by the model fidelity and sensor accuracy. This is mainly due to the difficulty in obtaining accurate models and sensor measurements due to the uncertainties faced in real-world scenarios. The challenges with model inaccuracies and measurement uncertainties can be overcome by using sensor fusion systems equipped with more than one extroreceptive or proprioceptive sensors or both. Though, this approach provides good wheel slip compensation the computations involved are cumbersome.

The difficulties with the existing approaches can be restricted to a reasonable extend by calibrating the slip compensation model using accurate measurements as reference. The robot can be accurately localized using dead reckoning sensor with the tuned slip compensation model alone, thereby eliminating the need for additional filters and additional sensors as in other slip compensation technique. In spite of such advantages, the tuned slip compensation model may fail to capture the environmental variations. One way to make the approach robust is to recalibrate the compensation models [12, 13] considering the changes in environment. However, such an approach has not been investigated extensively in literature to our best knowledge [15]. Motivated by this research gap, this investigation aims to develop a wheel-slip compensation technique that uses self-calibration during robot moves in unknown environment. As a result, the proposed slip compensation technique can be used in unknown and challenging environments.

The proposed self-calibration technique uses the neuro fuzzy systems (NFS) for calibration of dead-reckoning sensors that are the inputs for building accurate wheel-slip compensation model. The proposed technique is implemented using experiments on a robot moving in an unknown environment and acquiring knowledge about wheel-slip. It has been pointed out in [7] that the wheel-slip depends on environmental parameters such as terrain characteristics, and wheel-terrain interaction. Additionally, the robot parameters such as wheel pressure, center of mass, robot velocity, type of motion etc., also influence the slip compensation technique. These dependencies on parameters make the wheel-slip compensation design challenging using analytical methods. On the other hand, data collected from experiments are useful inputs that can be exploited to build compensation models. Furthermore, the compensation models needs to be adapted for environmental conditions. Neuro-fuzzy systems that combine the learning features of neural networks with decision making capabilities of fuzzy systems in uncertainties emerge as a good candidate for the problem.

Main contributions of this investigation are (*i*) a NFS based slip compensation mechanism for dead-reckoning based localization algorithm, *(ii)* an automated system to generate wheel slip data, and *(iii)* illustration of the proposed methodology on a wheeled mobile robot intended to be deployed in defense application.

The paper is organized into three sections as follows. Section 1 describes the self calibration technique and methodology for designing a slip compensation model using the generated standard and measurement data. Section 2 illustrates the implementation of proposed technique in wheeled mobile robot (WMR) and their experimental outcomes are presented. Finally, the conclusion and future directions are discussed in Sect. 3.

2 Proposed Self Calibration Technique

Self-calibration [16, 17] is an automated procedure in which the condition unit of the sensor system is configured to minimize the error between the actual physical quantity and its measurement. To perform calibration, the sensor system is subjected to measure the priori known or accurately measured physical quantity called standard data. The measurements made by the sensor system are called as measured data. The error between these data is used by the calibration system to configure the conditioning unit. The configured conditioning unit will compensate for the error to reduce deviation between the standard and measured data.

The proposed self-calibration system consists of dead reckoning sensor and slip compensation model considered as sensing and conditioning unit respectively as shown in Fig. 1. A laser range finder (LRF) based localization [18] with the known landmarks can precisely estimate the robot location. It is used to generate standard data in 4 steps as follows. In the first step, the LRF scans the environment to acquire

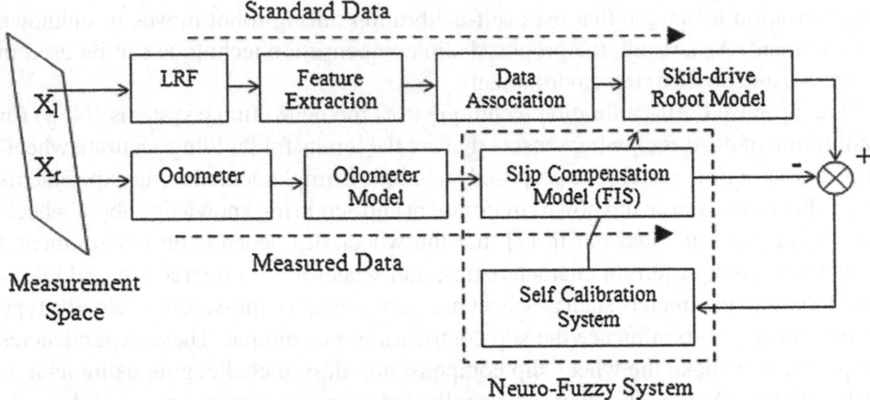

Fig. 1 Proposed self calibration of dead reckoning sensor

the relative position of the landmarks with reference to the robot (X_l). The coordinates of the landmarks are extracted in feature extraction step. In data association step, the extracted landmarks across each scans are associated to estimate the scene change. Using the estimated scene change, the robot location can be calculated using a mathematical model of a robot in the final step.

The odometer and its mathematical model are used to generate the measured data which are pronoun to error due to wheel slip. The localization error due to wheel slip is calculated by deviation of measured from the standard data. This error can be compensated by designing a fuzzy based slip compensation model using NFS. The learning algorithm of NFS acts as a calibration system to configure a slip compensation model which can reduce slip error in the dead reckoning sensor.

2.1 Generation of Standard Data

The standard data for localization are generated using the LRF with the known landmarks by measuring the relative distance of the landmarks in its vicinity. To localize the robot precisely, the experiments are conducted on a rectangular arena with four cylindrical pillars acting as landmarks placed at each corner. To enable the robot to distinguish landmarks, they are set to have variable cross section area with radius as in Fig. 2. The robot is set on a random motion in the arena ensuring at least one landmark will be in robot's vicinity over the run time. This enables the robot to be localized with respect to the visible landmark to generate standard data.

The LRF fitted on top of the robot scans the arena to create local map with respect to the robot. The coordinates of the landmarks are extracted from these local maps using feature extraction procedure. These landmarks between consecutive local maps

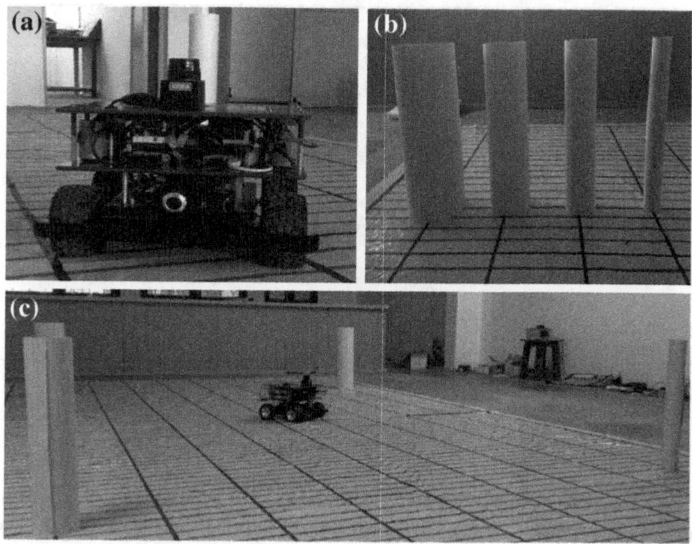

Fig. 2 Images of robot and landmarks used

are associated to estimate the scene change in data association. Finally, the robot can be localized using the scene change and a robot model.

Feature Extraction

In this step, the coordinates of the landmarks are extracted from LRF local maps. The LRF is intended to scan only the landmarks in the arena. But due to its long vision range, it scans the other obstacles that were beyond the arena. As a first step in feature extraction, the vicinity of the LRF is made adaptive to scan only the landmarks of interest and other obstacles can be filtered out. Next step in feature extraction is to determine the relative coordinates of landmark center as follows, the LRF can able to scan only a part of cross section area of landmarks using time of flight principle. This subjects the scanned surface to fall in anyone of these two category; On first category, when the relative position of LRF and landmark is large, the LRF will have better resolution and the landmark cross section appears like an arc (Refer Case: 1 in Fig. 3). In second category, the LRF is near the landmark, the resolution of LRF degrades due to lesser time of flight which causes the landmarks appears as straight line (Refer Case: 2 in Fig. 3). However on both categories, the center of the landmark is estimated by orthogonally projecting the robot bearing along the center of the scanned data as illustrated in Fig. 3.

Data Association

The extracted landmark coordinates between the consecutive local maps are associated to determine scene change in terms of rotation ($\Delta\theta_l$) and translation (Δr_l) as in '1' and '2'. The landmarks are associated by using nearest neighborhood search and Quickhull algorithm [19] without tessellation. This ensures only the common

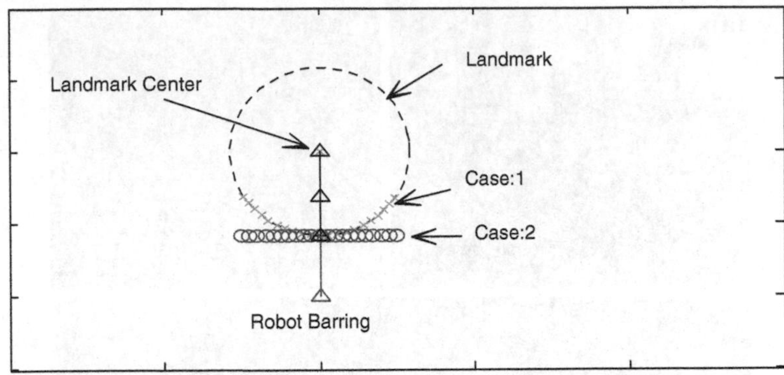

Fig. 3 Estimation of landmark center

landmarks between the local maps are associated for scene change estimation. The LRF scan rate is extremely large compared to the robot navigation speed which can ensure existence of common landmarks between two successive local maps.

$$\Delta r_l = \frac{1}{N_a} \sum_{a=1}^{N_a} (r_a(k) - r_a(k-1)) \tag{1}$$

$$\Delta \theta_l = \frac{1}{N_a} \sum_{a=1}^{N_a} (\theta_a(k) - \theta_a(k-1)) \tag{2}$$

where, r_a and θ_a are the set of radial distance and orientation of the associated landmarks respectively and N_a is the number associated landmarks.

Skid-Steer Robot Model

Using the scene change among the landmarks, the robot location can be calculated using a mathematical model of a skid-steer mobile robot. A skid steer robot [4] operates on any one of these two motions: linear motion and rotation motion. The mode of the robot can be identified by using the estimated scene change as follows, In linear motion, the robot moves with fixed bearing ($\Delta r_l \neq 0$; $\Delta \theta_l = 0$) whereas in rotation motion, the robot bearing changes with no translation (($\Delta r_l = 0$; $\Delta \theta_l \neq 0$). Thus, the overall skid steer robot model is given by '3' where the current location of the robot is characterized by its position (x_{rs}, y_{rs}) and its bearing (θ_{rs}).

$$\begin{bmatrix} x_{rs}(k) \\ y_{rs}(k) \\ \theta_{rs}(k) \end{bmatrix} = \begin{bmatrix} x_{rs}(k-1) + \Delta r_l \cos \theta_{rs}(k-1) \\ y_{rs}(k-1) + \Delta r_l \sin \theta_{rs}(k-1) \\ \theta_{rs}(k-1) + \Delta \theta_l \end{bmatrix} \tag{3}$$

2.2 Generation of Measurement Data

Dead reckoning technique uses odometer fitted to the robot wheel to generate measured data in which the error due to wheel slip has to be compensated. The odometer consists of a shaft encoder to measures the incremental wheel displacement called ticks which has been integrated to measure robot location. So any error in this tick measurement can get accumulated and can lead to a high degree of localization error. A mathematical model of odometer is formulated to generate the measured data from the measured wheel ticks.

Odometer Model

The odometer model as in '4' is employed to translate the left (d_{lw}) and right (d_{rw}) measured wheel ticks into measured robot location ($X_{rm} = \begin{bmatrix} x_{rm} & y_{rm} & \theta_{rm} \end{bmatrix}$). Similar to skid-steer robot model described earlier, the odometer model also operates on two motions (linear and rotational motion). The robot motion can be identified by the wheel direction as follows, if both wheels moves in same direction then robot is in linear motion and a rotation motion occurs when the wheels are in opposite direction. Unlike robot model, the odometer model uses a mode switch as described by '5' to switch between the linear motion (lm_{sw}) and rotational motion (rm_{sw}). This enables the odometer model to update robot translation and bearing during linear and rotational motion respectively. The magnitude of robot translation can be calculated by arithmetic mean of change in odometer displacement of both wheels (Δd_{odo}). In rotation motion, the robot rotates in its own axis by setting the wheels in opposite direction. This makes the robot to have a rotation with center of the robot and half the robot length (R_l) as rotation radius. The robot bearing is determined by measuring the arc distance traveled by the robot using the wheel ticks and rotation radius.

$$\begin{bmatrix} x_{rm}(k) \\ y_{rm}(k) \\ \theta_{rm}(k) \end{bmatrix} = \begin{bmatrix} x_{rm}(k-1) + lm_{sw}\Delta d_{odo}\cos\theta_{rm}(k-1) \\ y_{rm}(k-1) + lm_{sw}\Delta d_{odo}\sin\theta_{rm}(k-1) \\ \theta_{rm}(k-1) + rm_{sw}\Delta\theta_{odo} \end{bmatrix} \tag{4}$$

$$\begin{bmatrix} lm_{sw} \\ rm_{sw} \end{bmatrix} = \left\{ \begin{matrix} \begin{bmatrix} 1 \\ 0 \end{bmatrix} & if\ sign\,(\Delta d_{lw}) = sign\,(\Delta d_{rw}) \\ \begin{bmatrix} 0 \\ 1 \end{bmatrix} & if\ sign\,(\Delta d_{lw}) < sign\,(\Delta d_{rw}) \\ \begin{bmatrix} 0 \\ -1 \end{bmatrix} & if\ sign\,(\Delta d_{lw}) > sign\,(\Delta d_{rw}) \end{matrix} \right\} \tag{5}$$

where,

$$\Delta d_{odo} = \frac{\Delta d_{lw} + \Delta d_{rw}}{2}\ and\ \Delta\theta_{odo} = \frac{|\Delta d_{lw}| + |\Delta d_{rw}|}{R_l}$$

$$\Delta d_{lw} = d_{lw}(k) - d_{lw}(k-1)\ and\ \Delta d_{rw} = d_{rw}(k) - d_{rw}(k-1)$$

Table 1 Design specification of Neuro-fuzzy system

Parameters	Values/type
Type of fuzzy inference system	First-order Sugeno model
No of inputs	3
Types of input membership function	Gaussian bell MF
Type of output function	Linear type
No of rules	27 rules
Defuzzification method	Weighted average method
Learning method	Back propagation method

2.3 Self Calibration Using Neuro-Fuzzy System

The wheel slip data are calculated as a difference between the generated standard and measured data. It can be used to design a slip compensation model using NFS as in the proposed self calibration mechanism. NFS is a hybrid technique which combines the learning capability of artificial neural network (ANN) with the knowledge representation capability of FIS [13, 20]. This makes NFS an ideal tool to learn the complex slip dynamics from the wheel slip data and represent the knowledge for slip compensation. The learning algorithm of ANN acts as a calibration system to embed the knowledge of slip dynamics into a FIS based slip compensation model.

Slip Compensation Model

The slip compensation model is implemented using FIS to compensate the slip occurred during robot navigation and there by improves the accuracy of the dead reckoning sensor. As the mobile robot navigates in a two dimensional surface, two slip compensation models (*FISx, FISy*) have been designed to compensate localization error due to wheel slip on each dimension with specifications as shown in Table 1. The designed FIS uses parameters [21] such as velocity along each dimension (V_x V_y), its acceleration component (A_x A_y) and absolute difference in wheels velocities as input to determine the compensation velocity (ΔV_{cx} ΔV_{cy}). These compensation velocities are subtracted from measured robot velocity to eliminate wheel slip and the robot location is estimated as in '6'.

$$\begin{bmatrix} x_{sc}(k) \\ y_{sc}(k) \end{bmatrix} = \begin{bmatrix} x_{sc}(k-1) \\ y_{sc}(k-1) \end{bmatrix} + \begin{bmatrix} V_x - \Delta V_{cx} \\ V_y - \Delta V_{cy} \end{bmatrix} \tag{6}$$

3 Results and Discussions

The proposed self calibration technique is implemented on a four wheel mobile robot navigating in indoor environment with known landmarks for validating its performance. Experiments are carried out in two phases namely, training and testing phase. In training phase, the slip compensation model for the specified environment is calibrated using the generated wheel slip data. The performance of the calibrated slip compensation model is evaluated in testing phase with unknown wheel slip. In both these phases, data are generated by setting the robot to perform closed loop navigation with different linear and rotation velocities. Figure 4 illustrates the performance of the NFS based self calibrated dead reckoning sensor in training and testing phase. It is observed that an uncalibrated sensor encounters the slip error which gets accumulated as the robot navigates where as the calibrated sensor compensate for the slip in terms of velocities. So that slip error is compensated at every instant of time and thereby the drift in localization error has been minimized as seen in Fig. 4b, e. Further the performance of the proposed system is evaluated in terms of localization error which is a measure of the deviation of estimated from its actual robot location as in Fig. 4c, f.

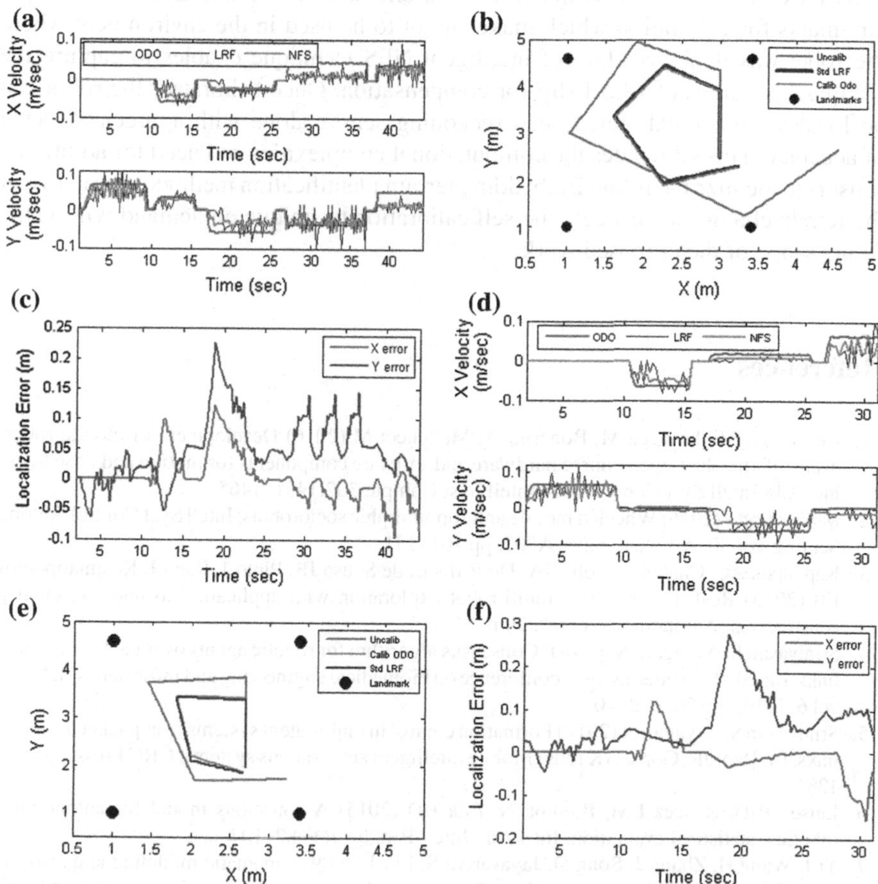

Fig. 4 Performance evaluation for training data (**a**)–(**c**) and for testing data (**d**)–(**f**). (*blue*—indicates the uncalibrated odometer, *green*—indicates the standard data from LRF and *red*—indicates calibrated odometer using NFS)

It has been observed that the maximum localization error occurred in both phases is around 0.28 m which is much lesser that the robot length ($R_l = 0.48$ m). This ensures that the calibrated dead reckoning sensor is able to keep in track of the robot position at every point of time without any need for additional sensors. A mean square localization error (MSE) of 0.0382 m for a 6.12 m run during training phase and 0.0442 m for a 4.20 m testing phase run has been observed which illustrates the performance of proposed self calibration technique.

4 Conclusion

The proposed work reports a self-calibration procedure using NFS for calibrating dead reckoning sensor. It employs robot's onboard sensor like LRF with simple landmarks for calibration which makes it apt to be used in the environment where the robot was deployed. Use of intelligent NFS technique enables to capture the complex dynamics of wheel slip for compensation. Once calibrated, the robot can be localized using calibrated dead reckoning sensor alone with appreciable level of accuracy. This eliminates the computational complexities and need for additional sensors to localize the robot. Embedding terrain identification methodology to detect the terrain change and execute the self calibration technique on demand will be the future scope of the proposed work.

References

1. Cucci DA, Migliavacca M, Bonarini A, Matteucci M (2016) Development of mobile robots using off-the-shelf open-source hardware and software components for motion and pose tracking. Adv Intell Syst Comput Adv Intell Syst Comput 302:1451–1465
2. McDermott J (2016) Wheeled mobile anthropomorphic sociorobots. Intell Syst Control Autom: Sci Eng vol 80, Article number A007, pp 101–131
3. Kapoutsis AC, Chatzichristofis SA, Doitsidis L, de Sousa JB, Pinto J, Raga J, Kosmatopoulos EB (2015) Real-time adaptive multi-robot exploration with application to underwater map construction. Autonomous robots, 29 p
4. Srinivasan S, Ayyagari, R. (2010) Consensus algorithm for robotic agents over packet dropping links. In: 2010 3rd international conference on biomedical engineering and informatics (BMEI), vol 6. IEEE, pp 2636–2640
5. Srinivasan S, Ayyagari R (2012) Formation control in multi-agent systems over packet dropping links. In: Raol JR, Gopal AK (eds) Mobile intelligent autonomous systems. CRC Press, pp 113–128
6. Luise DLD, Rancez LM, Biedma N, Elía BD (2015) A conscious model for autonomous robotics: statistical evaluation. Int J Adv Intell Paradig 7(1):82–107
7. Yi J, Wang H, Zhang J, Song D, Jayasuriya S, Liu J (2009) Kinematic modeling and analysis of skid-steered mobile robots with applications to low-cost inertial-measurement-unit-based motion estimation. IEEE Trans Robot 25(5):1087–1097
8. Cho B, Seo W, Moon W, Baek K (2013) Positioning of a mobile robot based on odometry and a new ultrasonic LPS. Int J Control Autom Syst 11(2):333–345

9. Canedo-Rodríguez A, Álvarez-Santos V, Regueiro CV, Iglesias R, Barro S, Presedo J (2016) Particle filter robot localisation through robust fusion of laser, WiFi, compass, and a network of external cameras. Inf Fus 27:170–188
10. Ramkumar K, Manigandan NS (2012) Stochastic filters for mobile robot SLAM problems - a review. Sens Transducers 138(3):141–149
11. Park J-T, Song J-B (2015) Sensor fusion-based exploration in home environments using information, driving and localization gains, Appl Soft Comput J 36:70–86
12. Lee H-K, Choi K, Park J, Kim Y-H, Band S (2008) Improvement of dead reckoning accuracy of a mobile robot by slip detection and compensation using multiple model approach. In: IEEE/RSJ International conference on intelligent robots and systems, 2008. IROS 2008. pp 1140, 1147, 22–26
13. Lee H, Jung J, Choi K, Park J, Myung H (2012) Fuzzy-logic-assisted interacting multiple model (FLAIMM) for mobile robot localization. Robot Auton Syst 60(12):1592–1606
14. Jung J, Lee H, Myung H (2013) Slip compensation of mobile robots using SVM and IMM. In: Advances in intelligent systems and computing - robot intelligence technology and applications 2012, vol 208, pp 5–12
15. Filliata D, Meyer J-A (2003) Map-based navigation in mobile robots: I. A review of localization strategies. Cogn Syst Res 4:243–282
16. Glueck M, Oshinubi D, Schopp P, Manoli Y (2014) Real-time Autocalibration Of Mems Accelerometers. IEEE Trans Instrum Meas 63(1):96–105
17. Hemerly EM, Coelho FAA (2014) Explicit solution for magnetometer calibration. IEEE Trans Instrum Meas 63(8):2093–2095
18. Yan R-J, Wu J, Yuan C, Han CS (2014) Extraction of different types of geometrical features from raw sensor data of two-dimensional LRF. J Inst Control Robot Syst 21(3):265–275
19. Barber CB, Dobkin DP, Huhdanpaa HT (1996) The quickhull algorithm for convex hulls. ACM Trans Math Softw 22(4):469–483
20. Jang JSR (1993) ANFIS: Adaptive-network-based fuzzy inference systems. IEEE Trans Syst Man Cybern 23:665–685
21. Gonzalez R, Rodriguez F, Guzman JL, Pradalier C, Siegwart R (2013) Control of off-road mobile robots using visual odometry and slip compensation. Adv Robot 27(11):893–906

Indian Sign Language Recognition Using Optimized Neural Networks

Sirshendu Hore, Sankhadeep Chatterjee, V. Santhi, Nilanjan Dey, Amira S. Ashour, Valentina Emilia Balas and Fuqian Shi

Abstract Recognition of sign languages has gained reasonable interest by the researchers in the last decade. An accurate sign language recognition system can facilitate more accurate communication of deaf and dumb people. The wide variety of Indian Sign Language (ISL) led to more challenging learning process. In the current work, three novel methods was reported to solve the problem of recogni-

S. Hore (✉)
Department of Computer Science and Engineering,
Hooghly Engineering and Technology College, Chinsurah, India
e-mail: sirshendu.hore@hetc.ac.in

S. Chatterjee
Department of Computer Science and Engineering, University of Calcutta,
Kolkata, India
e-mail: sankha3521@gmail.com

V. Santhi
VIT University, Vellore, India
e-mail: vsanthinathan@gmail.com

N. Dey
Department of Information Technology, Techno India College of Technology,
Kolkata, India
e-mail: neelanjandey@gmail.com

A.S. Ashour
Faculty of Engineering, Department of Electronics
and Electrical Communications Engineering, Tanta University, Tanta, Egypt
e-mail: amirasashour@yahoo.com

A.S. Ashour
College of CIT, Taif University, Taif, Kingdom of Saudi Arabia

V. Balas
Faculty of Engineering, Aurel Vlaicu University of Arad, Arad, Romania
e-mail: valentina.balas@uav.ro

E. Shi
College of Information and Engineering, Wenzhou Medical University,
Wenzhou 325035, People's Republic of China
e-mail: sfq@wmu.edu.cn

V.E. Balas et al. (eds.), *Information Technology and Intelligent Transportation Systems*, Advances in Intelligent Systems and Computing 455,
DOI 10.1007/978-3-319-38771-0_54

tion of ISL gestures effectively by combining Neural Network (NN) with Genetic Algorithm (GA), Evolutionary algorithm (EA) and Particle Swarm Optimization (PSO) separately to attain novel NN-GA, NN-EA and NN-PSO methods; respectively. The input weight vector to the NN has been optimized gradually to achieve minimum error. The proposed methods performance was compared to NN and the Multilayer Perceptron Feed-Forward Network (MLP-FFN) classifiers. Several performance metrics such as the accuracy, precision, recall, F-measure and kappa statistic were calculated. The experimental results established that the proposed algorithm achieved considerable improvement over the performance of existing works in order to recognize ISL gestures. The NN-PSO outperformed the other approaches with 99.96 accuracy, 99.98 precision, 98.29 recall, 99.63 F-Measure and 0.9956 Kappa Statistic.

Keywords Indian sign language · Neural network · Genetic algorithm · Evolutionary algorithm · Particle swarm optimization

1 Introduction

Sign language is a widely used medium of communication for deaf and dumb people. Developing systems to facilitate sign language handling is a significant subject of researchers' interest. Thus, recognition of sign languages is a challenging process due to the involvement of several technological aspects. Automated systems are urgent to serve the sign language interpretation purpose easily. Such systems facilitate the effective communication with deaf and dumb more reliable, accurate and robust processes. Sign languages have a wide variety all over the world, such as the American Sign Language, Australian Sign Language, German Sign Language, British Sign Language and Indian Sign Language (ISL). They differ in terms of morphology, grammar and syntax.

Large variation of ISL can be found in Indian sub-continent. However, the grammatical structure is same for all dialects [1]. In Indian sub-continent, there are approximately more than six million deaf and dumb people [2].

Extensive studies have been conducted using neural network in various applications [3–7]. Sign language recognition of ISL attracted researchers to improve recognition techniques in the last decade. A four step method to recognize alphabets of ISL has been proposed in [8]. In [9], a statistical approach to classify and recognize dynamic gestures of ISL in real time was proposed. Orientation of histogram has been used as a key feature for classification. Edge orientation of the sequence of dynamic ISL gestures has been calculated by employing a simple algorithm. K-nearest neighbor and Euclidean distance have been used for classification. The authors have reported accuracy range of 51.35 to 100 % and about 64.42 to 100 % while using Euclidean distance for 18 and 36 beans; respectively. Oszust and Wysocki (2010) [10] proposed an approach for automatic signed expressions recognition based on modeling gestures with subunits. The authors applied the evolutionary optimization technique to determine the cut points required for the proposed method. Krishnaveni and Radha (2011) [11] suggested a method for potential multilevel thresholds image segmentation method based on maximum entropy and PSO to estimate the human gestures. Results proved that computing using PSO can promptly convergence the results with high computational efficiency.

Recognition of South Indian Sign Language gestures have been proposed by Rajam et al. [12]. A set of 32 gestures, each depicting different postures of five fingers, has been used as the sign language. The authors have reported an accuracy of 98.125 % with 320 data instances for training and 160 for testing. Deora et al. [13] proposed a human computer interface capable of recognizing ISL gestures. The authors reported the recognition complexity due elaborating both hands and overlapping hands cases. A recent research has reported a Support Vector Machine based recognition system of ISL [14]. A novel method has been proposed in [15] to recognize ISL gestures for Humanoid Robot Interaction. The authors have proposed an efficient approach to communicate between a HOAP-2 robot and a human being. Feature extraction is carried out combining Genetic algorithm and Euclidean distance.

In the current work, three novel techniques have been proposed to recognize the ISL gestures. The NN has been trained using Genetic Algorithm, Evolutionary Algorithm and Particle Swarm Algorithm to enhance its performance. A set of 22 ISL gestures has been used to test the performance of the proposed work with ten images for each of the gestures. For the experimental results, 70 and 30 % of the data set has been used to train and test the NN, respectively.

The structure of the remaining sections is as follows. The background of the NN, GA and PSO techniques is introduced. The methodology followed by the results and discussion in Sects. 3 and 4; respectively. Finally, the conclusion is presented in Sect. 5.

2 Background

In this section the basics of the NN, GA and PSO are conducted as follows.

2.1 Artificial Neural Networks

Predominantly, artificial neural network (ANN) is used in numerous applications. It can be considered non-linear, highly parallel, robust, fault tolerant network. ANN has the ability to handle imprecise and fuzzy information easily [16].

In the ANN, neurons receive inputs (x_j) as stimuli from the environment. Then, these inputs are combined using their corresponding weights (w_j) to form 'net' input (net_j). The 'net' input is passed through a non-linear threshold filter to get the output (y) that can pass to another neuron in the network. The neuron can be activated if exceeds the threshold or bias 'θ_j' value of the concerned neuron. The net input is calculated using the Eq. (1), which computes the input for 'n' input signals by adding the dot products of weight (w) and strength (x) of each signal.

$$net_j = \sum_{i=1}^{n} w_{ij} x_i \tag{1}$$

In order to calculate the output signal, the calculated net input is compared against a threshold value as shown in Eq. 2.

$$y = \begin{cases} 1, & \text{if } net_j \geq \theta_j \\ 0, & \text{if } net_j < \theta_j \end{cases} \tag{2}$$

A perceptron learning rule is to be used in order to find the optimal weight vector. Several network architectures have been used to enhance the NN performance, such as the two-layer perceptron feed-forward network that can be used for the MLP-FFN experiments.

2.2 Genetic Algorithm

Genetic algorithm has been implemented to solve various optimization problems [17, 18]. The GA starts any problem solving using a set of initial solutions. It continuously applies crossover and mutations on the solutions to produce better offspring. The survival of any offspring depends on the fitness that based on the problem under concern. In each generation the best offspring survives and gradually produces better solutions in next generations until the required accuracy is achieved.

In the current work, determination of the NN initial weight vector is considered an optimization problem. Thus, the GA is applied to optimize the initial weight vector of the NN using the following algorithm, using the Root Means Square Error (RMSE) as a fitness function.

Algorithm: Genetic Algorithm

Start

Input: initial population represented by N numbers of chromosomes, which is randomly generated and takes value between '0' to '1'.

Calculating fitness values: a fitness function need to be defined for evaluating individual solution or chromosomes.

Selection

- The $RMSE_i$ value is calculated for each solution in population
- Calculate the average to find the $RMSE_{ave}$
- Select randomly the $RMSE_r$ from predefined closed interval $[0, RMSE_{ave}]$
- Calculate the difference $(RMSE_r - RMSE_i)$ for every solution.

If the obtained result is $\leq$ '0'
Then i^{th} individual is selected
End If

- Repeat the previous step until the number of solutions selected for next generation is equal to the number of solutions in the population.

Cross-over using the selected chromosomes
Mutation: Genes from randomly selected position are swapped to create new individual solution.
Termination: check the termination condition based on the selected condition
End

2.3 Particle Swarm Optimization

Particle Swarm Optimization (PSO) considers a population of particles (candidate solutions) in D-dimensional search space [19]. To estimate the particle's ability to realize the objective, each particle is associated with a fitness value. Primarily, the particles are located randomly in the search space. The swarm moves inside the search space to accomplish the optimal fitness. Particles are related to account for the best solution in the hyperspace, which denoted by '*pbest*'. The best value realized by any particle of the swarm is known as the Global best '*gbest*'. Each particle's position and velocity are initialized randomly. The fitness values of the particles are calculated after every iteration. The necessary adjustments are performed to the position/ velocity of each particle to move them to optimal fitness. The general algorithm of PSO to be used with the NN is as follows.

Algorithm: Particle Swarm Optimization

Start
The particles of the Swarm is placed at random positions with zero velocity
for n : 1 : Swarm size ***do***
Compute fitness
end for
for i : 1 : number of iterations ***do***
for j : 1 : Swarm size ***do***
Update *pbest*
Update *gbest*
Adjust position and velocity
Calculate fitness for the new population
end for
end for
End

Dixit et al. (2015) [20] introduced an exhaustive survey on nature inspired optimization algorithms including the evolutionary optimization strategy. Accordingly, the GA, EA and the PSO optimization algorithms were employed to support the NN for the Sign language recognition approach.

3 Experimental Methodology

In the current work, various models using Neural Network (NN), MLP-FFN, NN-GA, NN-Evolutionary, and NN-PSO have been utilized. The conjugate gradient algorithm has been used with the NN and MLP-FFN modules for learning [21]. The basic steps elaborated in the proposed methodology for sign language recognition are preprocessing, data cleaning and data normalization. Significant features/ attributes for classifying the dataset are extracted from the datasets accurately using effective features. If optimum feature set is combined with the Bayes [22] classifier it would result in minimum error for the given distributions. Therefore, theoretically the Bayes error would be considered as the optimum measure of the feature effectiveness. After the preprocessing step, data cleaning is required to resolve the missing values and noise issues. It is required to remove noise and fill up empty entries by suitable values through statistical analysis. Additionally, in order to reduce the distance between attribute values, data normalization is carried out by keeping the value of range between −1 to +1.

Afterward, the classification is conducted by dividing the datasets into two parts, namely (i) 70 % of the data as training dataset and (ii) 30 % as testing dataset. In training phase, the training dataset is supplied to different algorithms to build the required classification model. In testing phase, the classification models obtained from the training phase is employed to test the accuracy of the proposed models.

To measure the proposed performance, several metrics such as the correlation coefficient, Kappa statistic, Root Mean Square Error (RMSE), Mean Absolute Error (MAE), True Positive rate (TP rate), and F-measure were calculated [23]. These metrics are defined as follows. The RMSE is used to measure the difference between the values anticipated by a classifier and the values actually discovered from the surroundings of the system being modeled. The RMSE of a classifier prediction with respect to the computed variable v_{c_k} is evaluated by: $RMSE = \sqrt{\frac{\Sigma_{k=1}^{n}(v_{d_k} - v_{c_k})^2}{n}}$. Where v_{d_k}, denotes the originally observed value of kth object and v_{c_k} denotes the predicted value by the classifier.

The confusion matrix is a tabular representation of the classification performance. As illustrated in Table 1, each column of the matrix denotes the examples in a predicted class, while each row indicates the examples in an actual class, where (i) True positive (tp) is the number of positive instances categorized as positive, (ii) False positive (fp) is the number of negative instances categorized as positive, (iii) False negative (fn) is the number of positive instances categorized as negative and (iv) True negative (tn) is the number of negative instances categorized as negative.

Table 1 Typical example of confusion matrix of a binary classification problem

Actual class	Predicted class	
	Positive	Negative
Positive	tp	fp
Negative	fn	tn

Other metrics are derived from confusion matrix, namely: (i) the accuracy which defined as the ratio of the correctly classified instances sum to the total number of instances, which expressed as ($Accuracy = \frac{tp+tn}{tp+fp+fn+tn}$), (ii) the precision is defined as the ratio of correctly classified data in positive class to the total number of data classified to be in the positive class, which given by ($Precision = \frac{tp}{tp+fp}$), (iii) the recall (TP rate) is given by ($Recall = \frac{tp}{tp+fn}$), iv) F-measure which is a combined representation of the precision and the recall and represented by: ($F - measure = 2 * \frac{Precision * Recall}{Precision + Recall}$).

Another performance metric is the Kappa Statistic, which is a statistical measure denoted by k [24]. The value of k is estimated using: ($k = \frac{prob\,(0) - prob\,(E)}{1 - prob\,(E)}$). Where, $prob(0)$ is the probability of the observed agreements and $prob(E)$ is the same for agreements expected by chance.

4 Results and Discussion

Recently, various studies are concerned with neural network and optimization approaches in several applications [25, 26]. Meanwhile, the current study suggested a novel approach to optimize the input weight vector to the NN to achieve minimum error using optimization algorithms. The proposed work is tested with a dataset consists of 22 ISL gestures with 10 images for each gesture. Table 2 depicts the performance metrics values of the proposed algorithms along with NN and MLP-FFN classifiers.

Table 2 illustrated that with respect to the recall values, the NN is has attained 92.86 %, while using the NN-PSO is 98.29 %, which outperformed all other approaches. In addition, in terms of the F-measure values, the NN-PSO approach established its superiority compared to the other tested approaches. The obtained accuracy and precision for proposed the different compared approaches are demonstrated in Figs. 1 and 2; respectively.

Table 2 along with Figs. 1 and 2 illustrated that the accuracy of the NN approach is 93.64 %, which improved to be 96.7 % using the MLP-FFN approach. In terms of the accuracy, the results established that the NN-PSO based approach achieved 99.96 % accuracy, which outperformed both the NN-GA and the NN-EA. In terms of precision, the NN approach has attained the least precision value of 94.55 %, while the NN-PSO has achieved the superior precision value of 99.98 %. Finally, the kappa

Table 2 Calculated accuracy of proposed methodology

Algorithm / Performance measure	NN	MLP-FFN	NN-GA	NN-Evolutionary (NN-EA)	NN-PSO
Accuracy	93.64	96.7	97.83	98.86	**99.96**
Precision	94.55	97.25	98.43	99.83	**99.98**
Recall	92.86	94.67	97.71	97.75	**98.29**
F-measure	93.69	95.94	98.07	98.78	**99.63**
Kappa statistic	0.9127	0.9359	0.9772	0.9822	**0.9956**

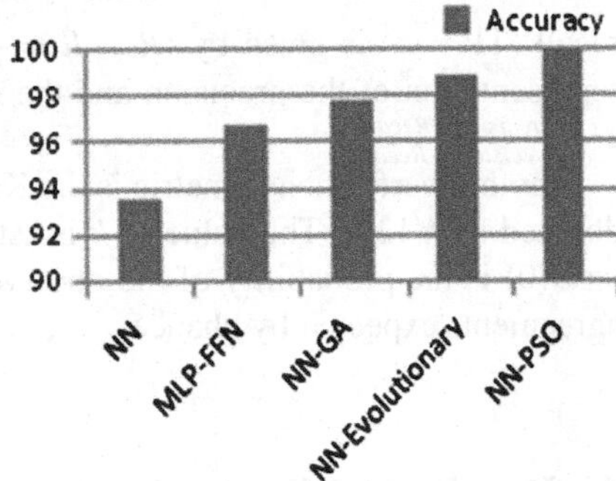

Fig. 1 Accuracy analysis for different proposed methodologies

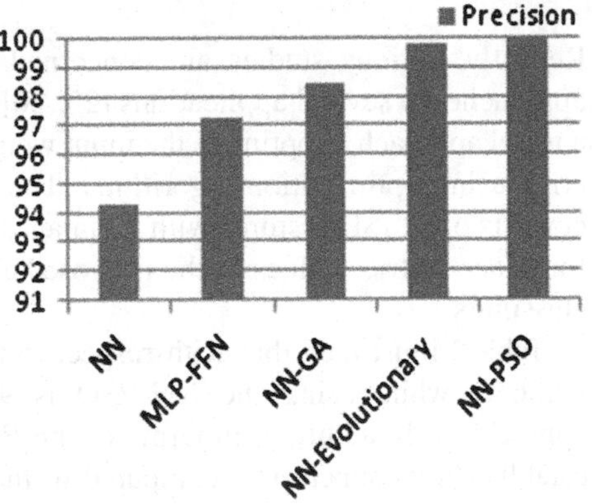

Fig. 2 Analysis of precision for different proposed methodologies

statistic profoundly establishes the improvement claims achieved by the NN-PSO as indicted in Fig. 3.

Figure 3 along with Table 2 proved that the NN-PSO gained the superior Kappa Statistic value of 0.9956 compared to the other methods. The confusion matrix of the testing phase for the NN (as an example) is illustrated in Table 3.

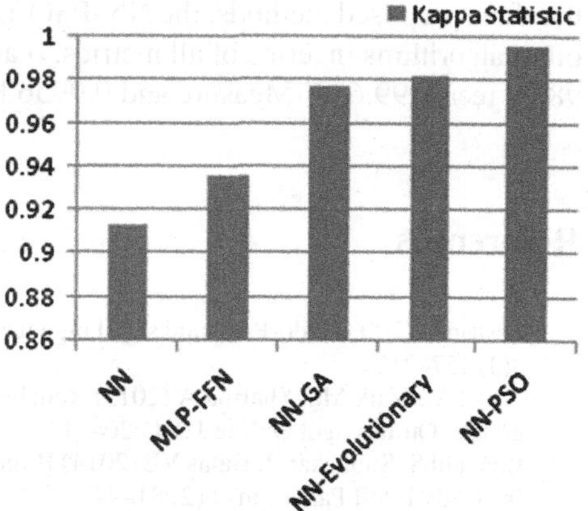

Fig. 3 Kappa statistic measure for different proposed methodologies

Table 3 Confusion matrix of testing phase for neural network

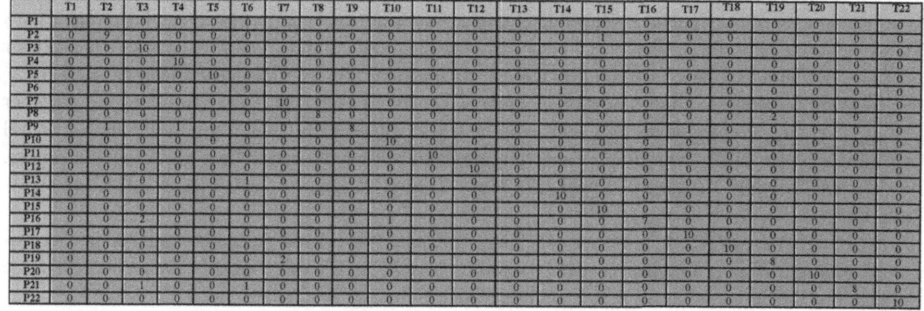

	T1	T2	T3	T4	T5	T6	T7	T8	T9	T10	T11	T12	T13	T14	T15	T16	T17	T18	T19	T20	T21	T22
P1	10	0	0	0	0	0	0	0	0	0	0	0	0	0	0	0	0	0	0	0	0	0
P2	0	9	0	0	0	0	0	0	0	0	0	0	0	0	1	0	0	0	0	0	0	0
P3	0	0	10	0	0	0	0	0	0	0	0	0	0	0	0	0	0	0	0	0	0	0
P4	0	0	0	10	0	0	0	0	0	0	0	0	0	0	0	0	0	0	0	0	0	0
P5	0	0	0	0	10	0	0	0	0	0	0	0	0	0	0	0	0	0	0	0	0	0
P6	0	0	0	0	0	9	0	0	0	0	0	0	0	1	0	0	0	0	0	0	0	0
P7	0	0	0	0	0	0	10	0	0	0	0	0	0	0	0	0	0	0	0	0	0	0
P8	0	0	0	0	0	0	0	8	0	0	0	0	0	0	0	0	0	0	2	0	0	0
P9	0	1	0	1	0	0	0	0	8	0	0	0	0	0	0	1	1	0	0	0	0	0
P10	0	0	0	0	0	0	0	0	0	10	0	0	0	0	0	0	0	0	0	0	0	0
P11	0	0	0	0	0	0	0	0	0	0	10	0	0	0	0	0	0	0	0	0	0	0
P12	0	0	0	0	0	0	0	0	0	0	0	10	0	0	0	0	0	0	0	0	0	0
P13	0	0	0	0	0	1	0	0	0	0	0	0	9	0	0	0	0	0	0	0	0	0
P14	0	0	0	0	0	0	0	0	0	0	0	0	0	10	0	0	0	0	0	0	0	0
P15	0	0	0	0	0	0	0	0	0	0	0	0	0	0	10	0	0	0	0	0	0	0
P16	0	0	2	0	0	0	0	0	0	1	0	0	0	0	0	7	0	0	0	0	0	0
P17	0	0	0	0	0	0	0	0	0	0	0	0	0	0	0	0	10	0	0	0	0	0
P18	0	0	0	0	0	0	0	0	0	0	0	0	0	0	0	0	0	10	0	0	0	0
P19	0	0	0	0	0	0	2	0	0	0	0	0	0	0	0	0	0	0	8	0	0	0
P20	0	0	0	0	0	0	0	0	0	0	0	0	0	0	0	0	0	0	0	10	0	0
P21	0	0	1	0	0	1	0	0	0	0	0	0	0	0	0	0	0	0	0	0	8	0
P22	0	0	0	0	0	0	0	0	0	0	0	0	0	0	0	0	0	0	0	0	0	10

From the preceding results, it is established that the NN-PSO outperformed all the other tested methods for the recognition of the ISL sign language.

5 Conclusion

The present work proposes three novel strategies to recognize Indian Sign Language using a dataset of 22 ISL gestures with 10 images for each. Based on the literature review, it is observed that the traditional learning algorithms significant problems as they have a chance of getting trapped into local optima while optimizing an objective. Thus the proposed optimization algorithms were elaborated to solve this problem. Moreover, the previous existing studies have used only the accuracy for the performance analysis, while it is observed that the accuracy of an algorithm varies greatly with the variation in the number of instances in different classes. Thus, precision, recall, F-Measure and kappa statistic have been considered as metrics to evaluate the proposed algorithms. Experimental results have revealed that among the three proposed methods, the NN-PSO approach has performed well compared to other algorithms in terms of all metrics. It achieved 99.96 accuracy, 99.98 precision, 98.29 recall, 99.63 F-Measure and 0.9956 Kappa Statistic.

References

1. Zeshan U (2003) Indo-Pakistani sign language grammar: a typologicaloutline. Sign Lang Stud 3(3):157–212
2. Naik SM, Naik MS, Sharma A (2013) Rehabilitation of hearing impaired children in India-an update. Otolaryngol Online J 3(1):20–31
3. Bharathi S, Sudhakar R, Balas VE (2014) Biometric recognition using fuzzy score level fusion. Int J Adv Intell Paradigms 6(2):81–94
4. Samanta S, Ahmed SKS, Salem MA, Nath SS, Dey N, Chowdhury SS (2014) "Haralick features based automated glaucoma classification using back propagation neural network." In: 2014 international conference on frontiers of intelligent computing: theory and applications (FICTA), special session: advanced research in computer vision, image and video processing, pp 14–15
5. Senthilkumar S, Piah ARM (2012) An improved fuzzy cellular neural network (IFCNN) for an edge detection based on parallel RK (5, 6) approach. Int J Comput Syst Eng 1(1):70–78
6. Maji P, Chatterjee S, Chakraborty S, Kausar N, Dey N, Samanta S (2015) Effect of Euler number as a feature in gender recognition system from offline handwritten signature using neural networks. In: 2nd international conference on computing for sustainable global development. INDIACom-2015, Bharati Vidyapeeth, New Delhi (INDIA), 11–13th March 2015, pp 1869–1873
7. Kotyk T, Ashour AS, Chakraborty S, Dey N, Balas VE (2015) Apoptosis analysis in classification paradigm: a neural network based approach. In: Healthy World conference 2015—a healthy World for a happy life, Kakinada (AP), India, pp 17–22
8. Ghotkar AS, Khatal R, Khupase S, Asati S, Hadap M (2012) Hand gesture recognition for indian sign language. In: 2012 international conference on computer communication and informatics (ICCCI), pp 1–4
9. Nandy A, Prasad JS, Chakraborty P, Nandi GC, Mondal S (2010) Classification of Indian sign language in real time, pp 52–57
10. Oszust M, Wysocki M (2010) "Determining subunits for sign language recognition by evolutionary cluster-based segmentation of time series", artificial intelligence and soft computing. Springer, Berlin
11. Krishnaveni M, Radha V (2011) Improved histogram based thresholding segmentation using PSO for sign language recognition. Int J Eng Sci Technol 3(2):1014–1020
12. Rajam PS, Balakrishnan G (2011) Real time Indian sign language recognition system to aid deaf-dumb people. In: 2011 IEEE 13th international conference on communication technology (ICCT), pp 737–742. IEEE
13. Deora D, Bajaj N (2012) Indian sign language recognition. In: 1st international conference on emerging technology trends in electronics, communication and networking (ET2ECN), pp. 1–5. IEEE
14. Agrawal SC, Jalal AS, Bhatnagar C (2014) Redundancy removal for isolated gesture in Indian sign language and recognition using multi-class support vector machine. Int J Comput Vis Robot 4(1):23–38
15. Nandy A, Mondal S, Prasad JS, Chakraborty P, Nandi GC (2010) Recognizing and interpreting Indian sign language gesture for human robot interaction. In: 2010 international conference on computer and communication technology (ICCCT), pp 712–717
16. Jain AK, Mao J, Mohiuddin KM (1996) Artificial neural networks: a tutorial. Comput, 1–44
17. Kumar BV, Karpagam M (2015) Differential evolution versus genetic algorithm in optimising the quantisation table for JPEG baseline algorithm. Int J Adv Intell Paradigms 7(2):111–135

18. Dey N, Ashour AS, Beagum S, Pistola DS, Gospodinov M, Gospodinova EP, Tavares JMRS (2015) Parameter optimization for local polynomial approximation based intersection confidence interval filter using genetic algorithm: an application for brain MRI image de-noising. J Imaging 1:60–84
19. Chakraborty S, Samanta S, Mukherjee A, Dey N, Chaudhuri SS (2013) Particle swarm optimization based parameter optimization technique in medical information hiding. In: 2013 IEEE international conference on computational intelligence and computing research (ICCIC), Madurai, 26–28 Dec 2013
20. Dixit M, Upadhyay N, Silakari S (2015) An exhaustive survey on nature inspired optimization algorithms. Int J Softw Eng Appl 9(4):91–104
21. Hornik K (1991) Approximation capabilities of multilayer feed forward networks. Neural Netw 4:251–257
22. Richard MD, Lippmann R (1991) Neural network classifiers estimate Bayesian a posteriori probabilities. Neural Comput 3:461–483
23. Stehman SV (1997) Selecting and interpreting measures of thematic classification accuracy. Remote Sens Environ 62(1):77–89
24. Carletta J (1996) Assessing agreement on classification tasks: the kappa statistic. Comput Linguist 22(2):249–254
25. Ashour AS, Samanta S, Dey N, Kausar N, karaa WBA, Hassanien AE (2015) Computed tomography image enhancement using Cuckoo search: a log transform based approach. J Signal Inf Process 6(3):244–257
26. Karaa WBA, Ashour AS, Sassi DB, Roy P, Kausar N, Dey N (2016) MEDLINE text mining: an enhancement genetic algorithm based approach for document clustering. Applications of intelligent optimization in biology and medicine. Springer International Publishing, Switzerland, pp 267–287

Study on Decision Fusion Identity Model of Natural Gas Pipeline Leak by DSmT

Yingchun Ye and Jinjiang Wang

Abstract The acoustic detection was novel method for natural gas pipeline leak. In order to improve accuracy and stability detection system, redundant structures of multiple sensor was necessary. The complex background noise and various working-condition adjective caused uncertainty, inadequacy and inconsistency of acoustic signal. In the process of multisource fusing identification, the high conflict among different sensor signal was inevitable. In this paper, the decision fusion model is built to identify natural gas pipeline leak. The decision fuse algorithm procedure includes signal preprocessing, feature extraction, basic relief assignment by BP neural network and decision fusion utilizing DSmT (Dezert–Smarandache Theory) and PCR5 rule. Experimental results show that the decision fusion model is effective and feasible. The information conflict of among different acoustic sensors is resolved perfectively. The fusion results for 150 group test samples indicate that the accuracy of leak detection reach 94.7 % under the given condition.

Keywords Multisource information fusion · DSmT · Natural gas pipeline · Leak detection · Conflict confidence · Acoustic

1 Introduction

The acoustic detection is analogous to human auditory sense from the view of bionics [1]. The relevant studies have shown that the low frequency of leak acoustic is difficult to be absorbed and spread further afield. Therefore time and frequency domain characteristics of low frequency acoustic signal can serve as characteristic parameters for pipeline leak identification. The diversity operation and complicated external

Y. Ye (✉) · J. Wang
China University of Petroleum, Beijing, 18 Fuxue Road, Changping,
Beijing 102249, China
e-mail: yeyingchun@cup.edu.cn

J. Wang
e-mail: jwang@cup.edu.cn

V.E. Balas et al. (eds.), *Information Technology and Intelligent Transportation Systems*, Advances in Intelligent Systems and Computing 455,
DOI 10.1007/978-3-319-38771-0_55

environment can produce noise which some noise are very similar to leak acoustic signal in the actual operation. The analysis that simply depend on data gathered from single acoustic sensor can result in reduced identification accuracy. The weakness of signal characteristic generated and imperfections of related pipeline parameters are the fundamental causes of affecting natural gas pipeline leak detection accuracy. The characteristic of weak leak signals usually appear in local domain of some special parameters. Because of pipeline structure and forces traveling, these parameters cant describe all pipeline status features steadily, accurately and comprehensively [2]. It is necessary that multichannel signals are fused to reduce conflict among multisource signal and improve identification accuracy. It is necessary that multichannel signals are fused to reduce conflict among multisource signal. DSmT (Dezert–Smarandache Theory) is novel algorithm to fusing multisource information [3]. In this paper, DSmT is used to fuse decision level data, which can improve identification accuracy.

2 DSmT Theory

DSmT is an extension and optimization to DST (Dempster–Shafer Theory) [4]. Based on DST framework, DSmT establish the generalized identification framework () and the concept of hyper-power set. The conflicting information is included in framework system, which can overcome limitations that DST cannot fuse highly conflict evidences [5].

2.1 Hyper-Power Set

If the number of elements in one set, the number of power set is $2n$. 2Θ is power set of authentication framework. Power set consist of all subsets of a set (including null set and corpora). If the number of elements in one set, the number of power set is $2n$. 2Θ is power set of authentication framework. In order to further extend fusion space, the concept of hyper-power set is proposed. If one finite set is $\Theta = \{\theta_1, \theta_2, \ldots, \theta_n\}$, the hyper-power set ($D\Theta$) of set ($\Theta$) must follow three principles: (1) $\phi, \theta_1, \theta_2, \ldots, \theta_n \in D^{\Theta}$; (2) If $\alpha_i, \alpha_j \in D^{\Theta}$, the intersection and union of α_i and α_j belong to $D\Theta$; (3) $D\Theta$ no longer includes other elements except elements which satisfies condition (1) and (2) [6]. If one set contains three elements ($\Theta = \{\theta_1, \theta_2, \theta_3\}$), the hyper-power set ($D\Theta$) of set ($\Theta$) is defined as Eq. 1.

$$
\begin{aligned}
D^{\Theta} = \{&\phi, \theta_1, \theta_2, \theta_3, \theta_1 \cap \theta_2, \theta_1 \cap \theta_3, \theta_2 \cap \theta_3, \theta_1 \cap \theta_2 \cap \theta_3, \\
&\theta_1 \cup \theta_2, \theta_1 \cup \theta_3, \theta_2 \cup \theta_3, \theta_1 \cup \theta_2 \cup \theta_3, (\theta_1 \cap \theta_2) \cup \theta_3, \\
&(\theta_1 \cap \theta_3) \cup \theta_2 (\theta_2 \cap \theta_3) \cup \theta_1, (\theta_1 \cup \theta_2) \cap \theta_3, (\theta_1 \cup \theta_3) \cap \theta_2, \\
&(\theta_2 \cup \theta_3) \cap \theta_1, (\theta_1 \cap \theta_2) \cup (\theta_1 \cap \theta_3) \cup (\theta_2 \cap \theta_3)\} \qquad (1)
\end{aligned}
$$

2.2 Generalized Confidence

Based on DSmT, every element in authentication framework of hyper-power set ($D\Theta$) is identification target, whose generalized confidence is presented as $m(.)$. Generalized confidence is a mapping from $D\Theta$ to the interval [0, 1] and is defined as Eq. 2.

$$\begin{cases} m(\phi) = 0 \\ \sum_{X \in D^{\Theta}} m(X) = 1 \end{cases} \tag{2}$$

2.3 Decision Rule

The decision rule of DSmT includes generalized belief function, generalized likelihood function and generalized pignistic probability function, those functions concept are all based on generalized confidence.

Assuming that there is $Y \in D^{\Theta}$, the generalized belief function can be calculate using to Eq. 3.

$$Bel(Y) = \sum_{X \in D^{\Theta}, X \subseteq Y} m(X) \tag{3}$$

The generalized likelihood function can be calculate using to Eq. 4.

$$\mathrm{Pl}(Y) = \sum_{X \in D^{\Theta}, X \cap Y \neq \phi} m(X) = 1 - \mathrm{Bel}(\overline{Y}) \tag{4}$$

The generalized pignistic probability function can be calculate using to Eq. 5.

$$\mathrm{GPT}(Y) = \sum_{X \in D^{\Theta}} \frac{C_M(X \cap Y)}{C_M(X)} m(X) \tag{5}$$

Where: $C_M(X)$ is the number of DSm, which is the number of complete separation unit based on X in Venn diagram. As shown in Fig. 1, Θ_1 contains four complete separation unit and $C_M(\Theta_1) = 4$.

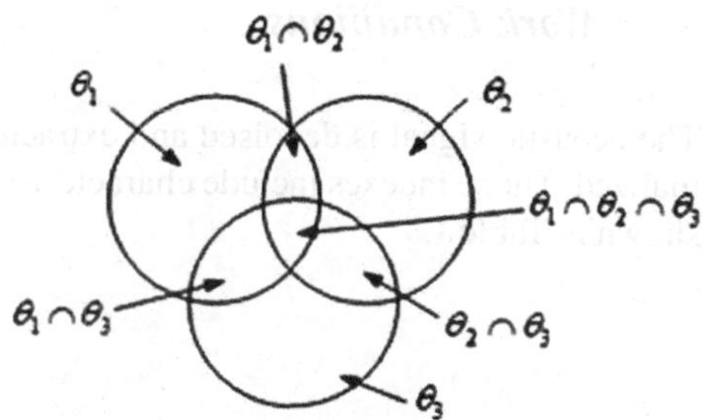

Fig. 1 Venn diagram of three element

The belief function only considers trusted part and is relatively optimistic. The likelihood function only considers untrusted part and is relatively pessimistic. The pignistic probability function considers trusted part, meanwhile uncertain part is prorated. The result is relatively neutral, and pignistic probability function is used to calculate generalized confidence in this paper.

2.4 Combination Rule

PCR (Proportional conflict Redistribution rule) redistributes conflict confidence to nonempty set by certain proportional relations. PCR is divided into three steps: calculating confidence of combination by Eq. 6; calculating confidence of global conflict and local conflict; redistributing conflict confidence to nonempty set by different proportional relations [7].

$$m_{\cap}(Y) = \sum_{A,B \in D^{\Theta}, A \cap B = Y} m_1(A) m_2(B) \forall Y \in D^{\Theta} \tag{6}$$

DSmT mainly includes five kind of PCR (PCR1-PCR5). PCR5 has the most complex calculation, but it is the most accurate and efficient conflict redistribution method on the mathematics. PCR5 distributes local conflict confidence between X and Y by their respective weightings in local conflict. The calculation method is shown as Eq. 7.

$$\begin{aligned} m_{PCR5}(Y) = & \sum_{\substack{X_1, X_2, \ldots, X_N \in D^{\Theta} \\ X_1 \cap X_2 \cap \ldots \cap X_N = Y}} m_1(X_1) m_2(X_2) \cdots m_N(X_N) + \\ & \sum_{\substack{X_1, X_2, \ldots, X_N \in D^{\Theta} \\ X_1 \cap X_2 \cap \ldots \cap X_N = \phi}} \begin{bmatrix} \frac{m_1(Y)^2 m_2(X_2) \cdots m_N(X_N)}{m_1(Y) + m_1(X_2) + \cdots + m_N(X_N)} + \cdots \\ + \frac{m_1(X_1) m_2(X_2) \cdots m_N(Y)^2}{m_1(X_1) + m_1(X_2) + \cdots + m_N(Y)} \end{bmatrix} \\ & \forall Y \in D^{\Theta} \backslash \{\phi\} \end{aligned} \tag{7}$$

3 Decision Model of Natural Gas Pipeline Leak Detection

3.1 Extracting Acoustic Signal Characteristic of Pipeline Work Conditions

The acoustic signal is denoised and extracted characteristic indexes which are normalized. Those indexes include characteristic of time and frequency domain and are shown in Table 1.

Table 1 The characteristic indexes of acoustic

Absolute mean value	Root mean square	Peak–peak value	Variance	Waveform index	Peak index
X_{am}	X_{rms}	X_{p-p}	X_{avr}	S	X_{cf}
Impulse factor	Margin factor	Kurtosis factor	Skewness	Energy ratio	
X_{imf}	L	X_{kf}	X_{sk}	η	

3.2 *Basic Belief Assignment Based on BP Neural Network*

The priori knowledge of conflict confidence is lack, so basic belief is assigned by BP neural network. The local BP neural network is established to each acoustic signal sensor, which is used to diagnose work conditions of gas pipeline as an independent evidence of DSmT. The output values of BP are transformed to basic belief assignment of corresponding work condition, which can avoid subjectivity [8].

If the jth output values in ith network is $o_i(j)$, the basic belief assignment of work condition (j) corresponding evidence (i) is calculated utilizing Eq. 8. Where: q is the number of elements in identification framework, which is the number of work conditions. The integral fusing process of multisource information for leak identification is achieved is shown as Fig. 2.

$$m_i(j) = \frac{o_i(j)}{\sum_{j=1}^{q} o_i(j)} \tag{8}$$

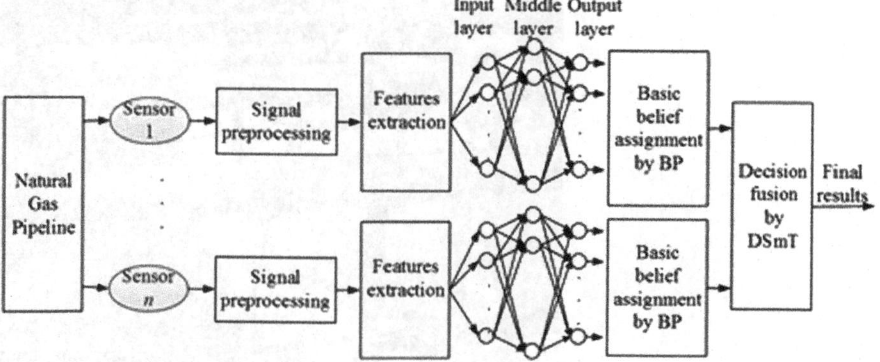

Fig. 2 Basic belief assignment by BP neural network

4 Experimental Verification

The experiment platform is comprised of two parts: test pipeline (its diameter is 168 mm and length is 2.7 km) and data acquisition. Test pipeline can provide various kinds work conditions simulation of natural gas pipeline (e.g. leak simulation, vibration of pipeline, pipeline distribution, normal processing et al.). The DAQ is used to exact gain leak acoustic wave. Experimental condition including: diameter of leak hole ranges from 1 to 5 mm, pressure ranges from 0.1 to 0.68 MPa. According to aeroacoustics analysis, numeric simulation and experiment, the jet noise has broadband frequency properties and nonlinear, the energy mainly concentrate low frequency range (<50 Hz). The sampling frequency is 200 Hz. The acoustic signal is removed noise using DTCWT-SVD method (Fig. 3).

4.1 *Test Verification*

Two same sensors are mounted on same position that can achieve redundant processing. Taking leak signal as an example, the distinguishing characteristic of different sensor at same work condition can be detected. Five continuous periods leak samples are selected that each period is one minute including 12,000 sample points. The radar

Fig. 3 Experimental device of Gas pipeline

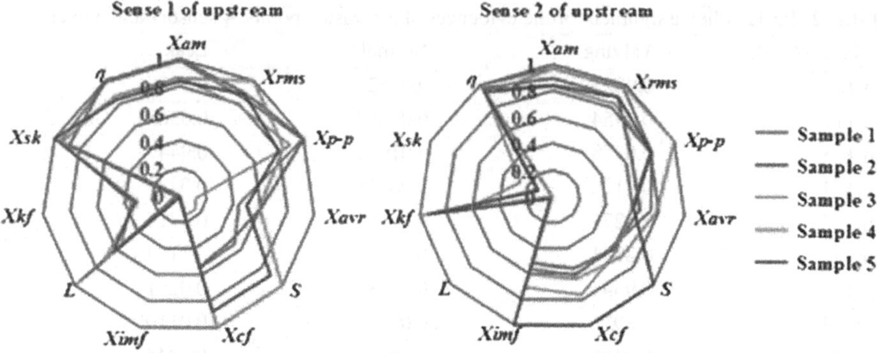

Fig. 4 Feature radar map generated from upstream leak acoustic samples

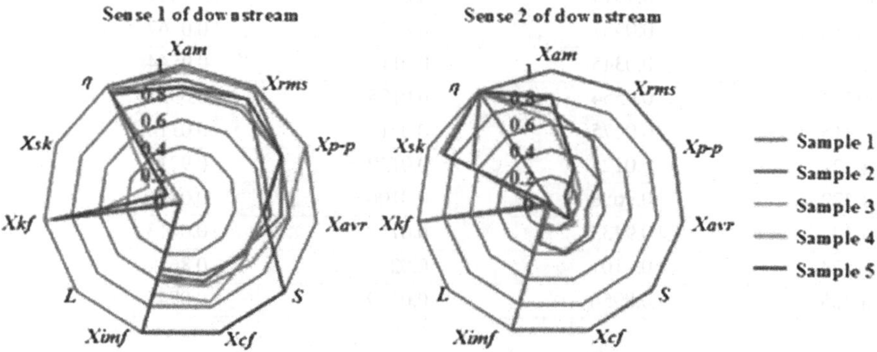

Fig. 5 Feature radar map generated from downstream leak acoustic samples

maps of upstream and downstream are drawn utilizing characteristic indexes. As seen in Figs. 4 and 5, characteristic indexes between two sensors exhibit different at the same time. The multisource acoustic signal of leak has conflict caused by complex background and nonlinear pipeline system.

Based in the simulation pipeline leak platform, the leak acoustic signals were collected from four independent sensors mounting at upstream and downstream. The experimental data consists five cycle that every cycle lasts a minute. The basic belief of each evidence are shown as Table 2. Where: i represents measuring point ($i = 1$ is output station, $i = 2$ is input station); j represents different sensor ($j = 1, 2$); k represents different cycle ($k = 1, 2, 3, 4, 5$). The results of the analysis indicated that basic belief of one sensor has high conflict for each work condition at different moments. For 1st sensor of 1st measuring point (m_{11x}), the confidences of valving is higher for m_{111}, $m_1 12$ and $m_1 14$, and the confidences of leak is higher for $m_1 13$ and $m_1 15$.

Table 2 Basic belief assignment of the evidences of 2 measure points, 4 sensors and 5 cycles

Evidence	Valving	Normal	Leak
m111	0.9021	0.0152	0.0827
m112	0.9354	0.0081	0.0565
m113	0.0358	0.0195	0.9447
m114	0.9126	0.0641	0.0233
m115	0.0271	0.0015	0.9714
m121	0.8871	0.0025	0.1104
m122	0.9541	0.0178	0.0281
m123	0.8976	0.0914	0.0110
m124	0.0107	0.1048	0.8845
m125	0.0014	0.0365	0.9621
m211	0.9314	0.0037	0.0649
m212	0.0987	0.8846	0.0167
m213	0.0345	0.0111	0.9544
m214	0.9554	0.0125	0.0321
m215	0.9675	0.0211	0.0114
m221	0.0126	0.0159	0.9715
m222	0.9688	0.0106	0.0206
m223	0.9433	0.0154	0.0413
m224	0.1101	0.0258	0.8641
m225	0.8956	0.0299	0.0745

Table 3 The final fusion result based on rule

D–S theory				DSmT (PCR5)			
Evidence	Valving	Normal	Leak	Evidence	Valving	Normal	Leak
m	1.0000	0	0	m	0.3669	0	0.6331

D–S (Dempster–Shafer Theory) and DSmT are used to fuse multisource information, the fusion result is shown as Table 3. The DS fusion result is valving and omits leak warning. There is high conflict for relief assignment of evidence between leak and valving, which is lead to misjudgment. DSmT and PCR5 can preferably fuse high conflict multisource information. Fusing result indicate that leak condition has maximum relief assignment (0.6331). The DSmT fusion result is right. 150 group samples at different work condition, including leak, valving and normal, are identified by DSmT and PCR5, the identification rate reaches 94.7 %.

5 Conclusion

The system framework of pipeline leak detection is constructed utilizing multiple acoustic sensor. In order to exactly describe work condition feature, the eigenvector consist of time and frequency domain indexes. Multisource acoustic signal is fused with DSmT theory and the initial basic relief is reasonably assigned by assignment based on BP neural network. In this paper, the decision model of pipeline leak is built and its algorithm procedure is present.

The conflict between different acoustic sensors is analyzed with experimental data. In order to eliminate the effect caused by conflict among evidence sources, the pignistic probability function is used to calculate generalized confidence and PCR5 rule is applied to redistribute conflict confidence.

Based on the leakage experimental platform, acoustic signal can be obtained at different work conditions. Compared to the fusing identification results applied respectively D–S and DSmT, DSmT model is accurately distinguish high conflict signal and significantly outperformed. Based on statistical analysis of 150 experimental data, the accuracy of leak detection reach 94.7.

Acknowledgments The paper supported by Science Foundation of China University of Petroleum, Beijing (No. KYJJ2012-04-25). Supported by Science Foundation of China University of Petroleum, Beijing (No. 2462015YQ0414). Supported by National Science foundation of China (No. 51504274).

References

1. Ellul IR (1989) Advances in pipeline leak detection techniques. Pipes Pipelines Int 34(3):7–12
2. Zhai XS, Hu JH, Xie SS, et al (2012) Diagnosis of aeroengine with early vibration fault symptom using DSmT. J Aerosp Power 27(2):301–306. In Chinese
3. Smarandache F, Dezert J (2004) Advances and applications of DSmT for information fusion. American Research Press, New Mexico
4. Talavera A, Aguasca R, Galvn B et al (2013) Application of Dempster–Shafer theory for the quantification and propagation of the uncertainty caused by the use of AIS data. Reliab Eng Syst Saf 11:95–105
5. Li XD, Dezert J, Smarandache F, et al (2011) Evidence supporting measure of similarity for reducing the complexity in information fusion. Inf Sci 181(10):1818–1835
6. Dezert J, Smarandache F (2011) On the generation of Hyper-power sets for the DSmT. In: The 6th international conference on information fusion, Cairns, pp 1118–1125
7. Li X, Huang X, Desert J et al (2007) A successful application of DSmT in sonar grid map building and comparison with DST-based approach. Int J Innovative Comput Inf Control 3(3):539–551
8. Xu CM, Zhang H, Peng DG, Zhang H et al (2012) Study of fault diagnosis of Integrate of D-S evidence theory based on neural network for turbine, Energy Procedia 16 Part C pp 2027–2032